Biological Risk Engineering Handbook

Infection Control and Decontamination

Biological Risk Engineering Handbook

Infection Control and Decontamination

Edited by
Martha J. Boss, CIH, CSP
Dennis W. Day, CIH, CSP

LEWIS PUBLISHERS

A CRC Press Company
Boca Raton   London   New York   Washington, D.C.

Cover micrograph courtesy of Janice Carr, Centers for Disease Control and Prevention.

Library of Congress Cataloging-in-Publication Data

Biological risk engineering handbook : control and decontamination /edited by
Martha J. Boss, Dennis W. Day.
     p. cm. -- (Industrial hygiene series)
     Includes bibliographical references and index.
     ISBN 1-56670-606-8 (alk. paper)
     1. Microbial contamination. 2. Sanitary microbiology. 3. Industrial hygiene. 4. Sanitary
engineering. 5. Industrial microbiology. I. Boss, Martha J. II. Day, Dennis W. III.
Industrial hygiene series (Boca Raton, Fla.)

QR48 .B487 2002
620.8′6--dc21
2002073064
CIP

Visit the CRC Press www.crcpress.com

No claim to original U.S. Government works
International Standard Book Number 1-56670-606-8
Library of Congress Card Number 2002073064
Printed in the United States of America 1 2 3 4 5 6 7 8 9 0
Printed on acid-free paper

Preface

Biological Risk Engineering — Infection Control and Decontamination provides a compendium of biological risk management information. Biological risk is of concern to us all. The biological risks we face vary and include biological contamination within our environment and, more personally, biological risk to ourselves through disease or the potential for disease.

This book deals with a subset of biological risk agents defined as bacteria, molds, yeasts, viruses, and prions. The term *biologicals* refers to these agents. Of these, the viruses and prions are not currently defined as independent life forms, and the extent to which these agents exhibit the characteristics of organic life are still being debated.

The intent of this compendium of information is to foster risk management decisions. In times of strength, we can manage many risks for ourselves and for those around us. As homeland security and other risk-management agendas are addressed politically, increasing emphasis will be placed on codifying biohazard management protocols. The biological risk regulatory process is expected to progress in a manner similar to the chemical risk regulations developed under Superfund. In fact, Superfund was always intended to include uncontrolled infectious substances. The authors of biological risk management regulations face a daunting challenge in that biohazardous agents, unlike chemicals, can reproduce.

As with most complex subjects, not all the authors included here or in the future will agree on everything. These differences were put aside to provide interdisciplinary discussions that hopefully will lead to sensible risk-management decisions. This text's authors are bacteriologists, biologists, industrial hygienists, environmental scientists, microbiologists, engineers, nurses, sanitarians, toxicologists, and safety professionals. All authors used their personal time and offered their professional opinions to shape the research and writing that resulted in this book. Whether they are in the public or private sector, one goal remained preeminent — to provide information to enhance the effectiveness of biological risk management and control.

Elizabeth Buckrucker, a reviewer, is a Project Manager for the Kansas City District Army Corps of Engineers, working in the Environmental Program branch. Elizabeth began her career in the U.S. Army and currently works as a civilian on environmental projects, including the U.S. Food and Drug Administration (FDA) Laboratory Decommissioning Program. In that capacity, Elizabeth met Dennis and Martha. Along with Donald Demers, current Chief of the FDA Safety Staff, and Renee Dufault, who is a Lieutenant Commander in the Public Health Service, they formulated the basic concepts for this book during their hours away from duty.

Biological Risk Engineering — Infection Control and Decontamination begins in Chapter 1 with a basic microbiological dictionary with emphasis on fungi and bacteria. Viruses and prions are also discussed. Illustrations of basic morphology and the appearance of mold cultures are provided. Chapters 2 and 3 provide sampling and laboratory procedural descriptions. For biological contaminant sampling (molds, bacteria, viruses), coordination between the sampling teams and the ultimate receiving laboratory is essential.

We then shift gears in Chapters 4 and 5 to interpretation issues associated with toxicological studies and ultimately risk assessment. Risk assessment quantitation had been more thoroughly developed for chemical risk, and the authors hope this volume will provide further impetus for synergistic studies related to risk assessment and management of biohazardous agents.

Because one of the exposure routes is inhalation, Chapter 6 deals with ventilation design. Should disruption occur in ventilation equipment or other building structures, maintenance will be required. Good design principles will ensure that maintenance can be safely and easily accomplished. Thus, it is emphasized in both Chapters 6 and 7 that correct design and ongoing maintenance using interdisciplinary expertise are essential.

Special requirements apply to laboratories, healthcare facilities, and other areas where immunocompromised patients may be exposed. Chapters 8 and 9, on infection control and medical settings, discuss these special requirements and current methods to reduce biological risk. The lead author for these chapters is Renee Dufault, who is a Lieutenant Commander in the Public Health Service. In response to concerns voiced by local hospitals about excessive fines for improperly regulated medical waste disposal in the District of Columbia (DC), Renee began working with the DC Hospital Association (DCHA) to improve disposal practices, as biohazardous waste was ending up at the DC waste transfer station. In coordination with a friend at the National Institutes of Health (NIH), Renee distributed regulated medical waste stream surveys at 22% of the district's hospitals. None of the hospitals surveyed provided the comprehensive training required by existing laws and regulations (OSHA, EPA, DOT) for waste handlers, including those who performed basic house-keeping services. Renee discovered that these housekeepers, who clean and disinfect all the patient care areas including isolation rooms, have a higher incidence of occupationally acquired tuberculosis than do nurses. Renee then researched nosocomial infections and was shocked by both the findings and lack of current and accurate data. With support from the EPA, the FDA, and especially the DCHA, Renee developed and presented the Environmental Services Professional Training Course, which will soon be available on the Internet. Renee's friend and colleague, Rita Smith, who is Georgetown Washington University Hospital's Infection Control Director, helped write a section on hospital infection prevention and control for Chapter 9. Ed Rau, another friend and colleague from the NIH, contributed the information on prions found in Chapter 8.

Decontamination and assessment are addressed in Chapter 10, which provides basic information and a sample of specifications, including statements or scopes of work that can be used as guidelines in developing specifications or purchase orders. Site-specific considerations will always take pre-cedence over any general guidance, and professionals must be consulted to provide site-specific interpretation and required design documents. Chapter 11, which discusses *Legionella* and cooling towers, is essentially a case study demonstrating how design, maintenance, and decontamination can be integrated into a seamless process.

Chapter 12 presents biocides given the various general chemical or physical alterations that constitute a biocidal (life-killing) effect. While biocide use is rarely the sole answer to mitigating biological risk, biocide usage remains an alternative. Chapter 13, on laws and regulations, discusses current regulations, patent utility requirements, and insurance processes. In particular, biocides and their approval are discussed. Chapter 14, on tuberculosis, is essentially a case study that compares OSHA and CDC guidelines. Both the CDC guidelines and current OSHA rule making will ulti-mately result in an OSHA regulation to control occupational exposure to tuberculosis. Finally, Chapter 15 presents security both from the standpoint of homeland security given current U.S. requirements, and from an individual laboratory perspective.

To put this book in perspective time-wise, Martha's father, Eugene Johnson, wrote his master's degree research paper in the 1950s on sanitation in the South Dakota schools. During that decade, the United States awakened to the prospect of controlling polio, even though the newly discovered virus was yet to be understood. Now, some 50 years later, we can identify some bacteria based on their viral phage loading and are just beginning to understand prions and the impact of bacteria and viruses on cancer initiation.

As time goes by we increasingly realize the vulnerability of our world, such as how quickly a viral or bacterial pandemic can envelop the Earth. Yet, despite this understanding, we continually forget the simplest of lessons. In the days since the 1950s, running water and indoor bathrooms have become commonplace in the continental United States. Yet, in Alaska in the heart of oil country, children still awaken each morning and carry honey pots to the local landfill. The raw sewage is no longer burned (to protect the air?) and the raw sewage is not treated (too expensive to build aboveground plumbing systems and waste treatment facilities?). The sewage flows to the nearby waters, marshes, and streams and the honey pot plastic bags float in the air like junkyard birds. An epidemic waiting to happen?

The answers to this and other questions are complicated, with politics, science, and the many facets of human existence commingled. The recent anthrax scare, the HIV pandemic, and the potential use of biological weapons are all high-profile issues that have their basis in a simple understanding: The world is alive, and the life forms compete. To guard humanity, we must protect and understand our world, and these efforts must be continual, rather than being initiated once a biohazardous agent is out of control. In the words of a wise man, the time to fix your roof is when the sun is shining.

Martha J. Boss

Dennis W. Day

About the Editors

Martha J. Boss is a practicing industrial hygienist and safety engineer living in Omaha, Nebraska and various airports throughout the United States. Many years ago, Martha won the Army Science award at the Des Moines, Iowa science fair. As fate would have it, Martha eventually worked for the Army and through the auspices of EPA grants was trained in industrial hygiene. All of this surprised Martha because she had intended to teach high school science and had prepared herself for that endeavor with a B.A. in biological education (University of Northern Iowa) and later a B.S. in biology (University of Nebraska).

During Desert Storm, Martha was tasked under the War Powers Act to assist in the preparation of a western Army base to be used to house and train special forces. Shortly thereafter, Martha was trained in what was then known as the U.S. Army Defense Ammunition Center and School, Technical Chemical Surety Materiel Course, AMMO-M-8. This course was offered to instruct personnel working at depots and arsenals on some of the issues associated with chemical warfare materiels. Martha then began an interdisciplinary set of assignments with her fellow Army industrial hygienists and engineers to assess chemical, biological, radioactive, and chemical warfare sites and to find solutions to the problems associated with them. The Army continued her training at such institutions as Johns Hopkins, Harvard, and other top centers through the nation.

After 5 years of traveling throughout the country to various very scary places, Martha decided to settle down in a regional engineering firm. After a couple of years, Martha realized she did not want to settle down and joined a national engineering firm where she is employed to this day. Martha is a Principal Toxicologist for URS Corporation and continues her practice as a Certified Industrial Hygienist and Certified Safety Professional (Safety Engineer). Martha is a member of the NEER (Nonlethal Environmental evaluation and Remediation Center), a Diplomat of the American Academy of Industrial Hygiene, serves as an Editorial Advisory Board Member for Stevens Publishing, and is a member of the American Industrial Hygiene Association.

Dennis W. Day is a practicing industrial hygienist and safety engineer living in Omaha, NE and various airports throughout the United States. Dennis began his career as a forester. For several years, he traveled through the forests of the East and South cruising timber. Then he decided to become a high school science teacher. Dennis used his B.S. in forestry (University of Missouri) to enable him to pursue additional studies in chemistry and biology (Creighton University) and become a professional teacher. After teaching for awhile Dennis was persuaded to join the Army Safety Office and ultimately the Omaha District engineering division.

Dennis continued for ten years to work with various Army, EPA, and Department of Defense missions. His work included sites throughout the nation and in Europe. Dennis concentrated his efforts on streamlining site assessment protocols, community outreach with protective action plans for chemical warfare sites, and training industrial hygienists entering the Army work force.

Dennis joined URS to develop an interdisciplinary industrial hygiene, safety, and engineering service to commercial and governmental clients. Dennis is the regional health and safety manager for URS Corporation and continues his practice as a certified industrial hygienist and certified safety professional (safety engineer). Dennis is a diplomate of the American Academy of Industrial Hygiene and a member of the American Conference of Governmental Industrial Hygienists, the American Industrial Hygiene Association, and the American Society of Safety Engineers.

Contributors

Harriet M. Amman, Ph.D., DABT
Washington State Department of Health
Olympia, Washington

Marwan Bader, MD, CIH
Oak Ridge National Laboratory
Knoxville, Tennessee

Martha J. Boss
URS Corporation
Omaha, Nebraska

Dennis W. Day
URS Corporation
Omaha, Nebraska

Renee Dufault, CHSP, CDR
Food and Drug Administration
Rockville, Maryland

James D. Hollingshead, MHS, CIH
URS Corporation
Santa Ana, California

Melanie Karst, Ph.D.
URS Corporation
Omaha, Nebraska

Jerry King, D.A.
Midwest Laboratories, Inc.
Omaha, Nebraska

R. Vincent Miller, Ph.D.
Aerotech Labs
Phoenix, Arizona

Richard C. Pleus, Ph.D.
Intertox
Seattle, Washington

Captain Edward Rau, MS, RS, CHSP, REM
National Institutes of Health
Rockville, Maryland

Heriberto Robles, Ph.D., DABT
URS Corporation
Santa Ana, California

Rita Smith, MSN, CIC
George Washington University Hospital
Takoma Park, Maryland

Brian Wight, PE
URS Corporation
Des Moines, Iowa

Chris Wrenn, BA
RAE Systems
Sunnyvale, California

Reviewers

Elizabeth Buckrucker
U.S. Army Corps of Engineers
Kansas City, Missouri

Donald Demers
Food and Drug Administration
Rockville, Maryland

Acknowledgments

We all extend a special thank you to Elizabeth Buckrucker, Donald Demers, Renee Dufault, and Richard Pleus. Elizabeth and Donald continually reviewed the developing work effort. Their efforts assisted the primary editors (Martha and Dennis) and each individual author. Similarly, Renee Dufault on the east coast and Richard Pleus on the west coast (United States) coordinated the work of various authors in their geographic regions.

A thank you is also extended to Dee Chambers for illustrations and Bridget Boss for graphic design. Picture and illustration contributors include Deniese Chambers (URS), Karin Galligan (SKC, Inc.), Aerotech Laboratories, Neil Carlson (University of Minnesota), Centers for Disease Control and Prevention (CDCP), RAE Systems, Daniel Behler, (Biotest), and Peter Pratt (Bioscience International). Additional thanks to Melanie Edwards (ASHRAE) and Dan Woodbury (Environmental Building News).

Contents

CHAPTER 1

Micro Dictionary

Dennis W. Day, Martha J. Boss, Jerry King, and Melanie Karst

CONTENTS

1-56670-606-8/03/$0.00+$1.50
© 2003 by CRC Press LLC

Biological contaminants are defined in terms of taxonomic names, staining characteristics, and pathological effect. This chapter presents the biologicals that are of concern to human effects and indoor air quality considerations.

1.1 BIOLOGICAL CONTAMINANTS

The term *biological* actually means "life-like," so these contaminants have the potential to grow and reproduce, as does all life. Biological contaminants discussed in this book include bacteria, fungi (yeast and molds), prions, and viruses. Bacteria, fungi, and protozoa all have a fluid-filled cell structure. Viruses have a protein coat over their genetic material and no fluid-filled layer. Prions are particulates that lack nucleic acids. This difference in structure can become less obvious when bacteria form cysts around themselves, thus creating a very dry, dormant bacterial life form. Bacteria encyst in order to survive periods of drought and other stresses. Fungi also have a dry form associated with their reproductive cycle. This form is called a *spore*. Spores are not seeds, but, like seeds, spores can lead to the formation of new mold colonies; thus, counting spores is the same as counting *colony-forming units*. (*Note:* Counting spores is similar to counting bacterial cells.)

1.2 BACTERIA

Bacteria are known to cause diseases either as pathogens or as opportunistic pathogens. The pathogenicity (ability to cause a disease) is determined by the bacterial and host defense responses. The bacterial genera or species are rarely identified in samples because of the cost of the analysis. Excessive bacterial counts may indicate that bacteria are successfully competing in lieu of other biologicals, including fungi. Competition with bacteria may cause fungal counts in these areas to be suspect.

Bacteria are essentially unicellular structures and are prokaryotic. They are not classified based on their ability to interbreed; instead, morphological characteristics are used to classify bacteria. Pure cultures of the same species may differ slightly. The term *strain* is used to define a group of cells in culture derived from a single cell.

Prokaryotic cells are similar chemically to eucaryotic cells (i.e., plant and animal cells). The following are the defining structural differences for prokaryotic cells compared to eucaryotic cells:

- DNA is not enclosed in a membrane (i.e., no cell nucleus).
- DNA is not associated with histone proteins (e.g., chromosomal proteins).
- Organelles are not enclosed in a membrane.
- Cell walls contain the complex polysaccharide peptiglycan.

Increasingly, analysis of nucleotide sequences in DNA and RNA, DNA hybridization, and cellular chemical component analysis are being used to classify bacteria. To date, not all bacteria have been classified, and current classification systems have not been verified using the improved scientific tools now available, such as:

- Staining (application of dyes prior to microscopic examination; includes differential staining)
- Biochemical tests
- Serology
- Phage typing
- Amino acid sequencing
- Protein analysis
- Nucleic acid base composition
- Nucleic acid hybridization
- Flow cytometry
- Genetic recombination

1.2.1 Morphology

The size range of naturally occurring bacteria is generally from 0.20 to 2 µm (microns) in diameter from 2 to 8 µm in length. Some bacteria are able to change shape and size in response to environmental conditions. Genetically, most bacteria are monomorphic, but some are pleomorphic even with unchanged environmental conditions. Basic shapes are assigned *singular/plural* names, as shown in the section headings that follow.

1.2.1.1 Coccus/Cocci

Coccus/cocci are the names for round, oval, elongated, or flattened spheres. After division, *diplococci* remain in pairs, *streptococci* remain in chain patterns, and *sarcinae* divide into three planes and remain attached in a cube pattern.

1.2.1.2 Bacillus/Bacilli

Bacillus/bacilli are staffs that divide only along their short axis. After division, *diplobacilli* appear in pairs, *streptobacilli* appear in chains, and *coccobacilli* are oval. (*Note:* When bacillus is capitalized and italicized, it refers to a genus, not a shape.)

1.2.1.3 Spiral

Spiral shapes include *vibrious*, which are curved like commas; *spirilla*, which are corkscrew shaped with rigid, outside flagella; and *spirochetes*, which are helical, flexible, axial filaments that move and are contained under an external flexible sheath.

1.2.1.4 Other

Other shapes include *stella* (star shaped), *halophilic archaeobacteria* (square), and *haloarcula* (triangular).

1.2.2 Endospore Formation

As a survival mechanism, some Gram-positive bacteria can form *endospores*, the cells of which dehydrate and form thick exterior walls with additional structural layers. These layers are formed interior to the outer cell membrane, and the endospore diameter may vary from the original vegetative cells. Endospores can survive extreme temperature ranges, lack of water, radiation, and the passage of time. They have germinated when rewarmed after a 7500-year resting period. One Gram-negative bacteria that forms endospores is *Coxiella burnetii*, which causes Q fever. The sporulation/sporogenesis process is as follows:

1. A triggering message is sent.
2. A newly replicated chromosome and a small section of cytoplasm are encapsulated by an ingrowth of plasma membrane (i.e., spore septum).
3. The spore septum matures to a double-walled membrane. When enclosure is complete, the entire structure is termed a *forespore*.
4. Thick peptiglycan layers are laid down between the two spore septum membrane layers.
5. A thick protein spore coat forms around the outside membrane of the spore septum.
6. When the spore is mature, the enclosing vegetative cell lyses and dies.
7. The endospore is freed.

Most of the water has been eliminated during sporogenesis, and metabolic activity has ceased within the spore. The spore essentially contains:

- DNA and RNA
- Ribosomes
- Enzymes
- Small molecules (few in number)
- Dipicolinic acid and calcium ions, which are essential for metabolic resumption

The process of returning the endospore to the vegetative state is termed *germination*, which is triggered by changes in the endospore coat that allow water to enter. Germination is a reanimation of the original cell, not a reproductive event.

1.2.3   Spirochetes

Spriochetes, which include some pathogenic genera, are coiled, with some resembling a spring (Figure 1.1).They are actively mobile due to their axial filaments, which are enclosed between the outer sheath and the cell body. They can be aerobic, facultatively anaerobic, or anaerobic. Spirochetes do not have flagella or endospores. *Treponema pallidum* causes syphilis, the genus *Borrelia* causes relapsing fever and Lyme disease, and the genus *Leptospira* (in animal urine) causes leptospirosis. Spirochetes are found in contaminated water, sewage, soil, and decaying organics.

1.2.4   Aerobic/Microaerophilic, Motile, Helical/Vibroid Gram-Negative Bacteria

This type of bacteria has a spiral, rigid, helical shape. The flagella are at one or both poles or are in tufts. Most of these bacteria are harmful and are found in an aquatic environment. Some are pathogenic. *Campylobacter fetus* causes abortion in domestic animals, foodborne *C. jejuni* causes intestinal disease, and *Helicobacter* causes ulcers in humans.

1.2.5   Gram-Negative Aerobic Rods and Cocci

Gram-negative aerobic rods and cocci are rod shaped with polar flagella. Some excrete extracellular, water-soluble pigments. Some are pathogenic.

1.2.5.1   Pseudomonas

Pseudomonas is problematic in hospital settings. It can grow on minute traces of carbon, including those found in soap residues or cap-liner adhesives. They are capable of growth in antiseptics such as quaternary ammonium compounds and are resistant to antibiotics. *Pseudomonas aeruginosa* causes urinary tract and skin infections, septicemia (blood poisoning), and meningitis (inflammation of the membranes that envelop the brain and spinal cord (Figure 1.2).

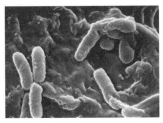

Figure 1.1   Spirochetes. (Courtesy of CDC Public Health Image Library.)

Figure 1.2   *Pseudomonas aeruginosa.* (Courtesy of CDC Public Health Image Library.)

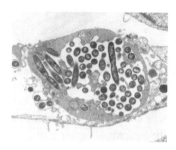

Figure 1.3 *Legionella pneumophila.* (Courtesy of CDC Public Health Image Library.)

1.2.5.2 Legionella

Six species comprise *Legionella*, which may inhabit water supply lines and water in cooling towers (Figure 1.3). Because *Legionella* does not grow in usual laboratory isolates, charcoal yeast agar is used instead. *Legionella* also does not stain with the usual histological staining techniques

1.2.5.3 Neisseria

Neisseria does not form endospores, is aerobic or facultatively anaerobic, and is parasitic on human mucous membrane. *Neisseria gonorrhea* causes gonorrhea, and *N. meningitis* causes meningococcal meningitis).

1.2.5.4 Moraxella

Moraxella, an aerobic coccobacillus, is egg-shaped (a structural intermediate between cocci and rods). *Moracella lacunata* causes conjunctivitis.

1.2.5.5 Brucella

Brucella causes brucellosis, characterized by fever, malaise, and headache and is also referred to as Gibraltar fever, Malta fever, Mediterranean fever, Rock fever, or undulant fever. *Brucella* is a small, nonmotile, obligate parasite that survives phagocytosis.

1.2.5.6 Bordetella

Bordetella is a nonmotile rod found only in humans. *Bordetella pertussis* causes whooping cough.

1.2.5.7 Francisella

Francisella is small and pleomorphic. It grows in complex media mixed with blood or tissue extracts. *Francisella tularensis* causes tularemia. Intermittent fever and swelling of the lymph nodes are characteristics of tularemia; also called rabbit fever.

1.2.6 Facultatively Anaerobic Gram-Negative Rods

These bacteria are often pathogenic and are medically important.

1.2.6.1 Enterobacteriaceae (Enterics)

Enterobacteriaceae inhabit intestinal tract of humans and other animals. They can be either motile or nonmotile; the motile forms have peritrichous flagella. Many have fimbriae (fringed border). Enterobacteriaceae have specialized sex pila that aid in transmittal of genetic information and may, thus, potentiate genetic susceptibility to antibiotics. They produce bacteriosins (proteins that lyse other bacteria cells).

1.2.6.2 Escherichia

The anaerobic *Escherichia* genus includes *E. coli* (Figure 1.4), which inhabits the intestinal tract of humans and other animals and can cause urinary tract infections and diarrhea.

1.2.6.3 Salmonella

Salmonella inhabits intestinal tract (poultry and cattle), contaminates food, and can cause salmonellosis. *Salmonella typhi* causes typhoid fever.

1.2.6.4 Shigella

Shigella causes bacillary dysentary, shigellosis, and traveler's diarrhea.

1.2.6.5 Klebsiella

Klebsiella pneumoniae causes septicemia in pediatric wards and pneumonia.

1.2.6.6 Serratia

Serratia marcescens produces a red pigment that can be used to trace the dispersal of biological warfare materials. *Serratia* causes nosocomial (hospital-acquired), urinary tract, and respiratory infections and has been found in catheters, saline irrigation solutions, and other solutions.

1.2.6.7 Proteus

Proteus is actively motile and causes urinary tract infections, infections of wounds, and diarrhea.

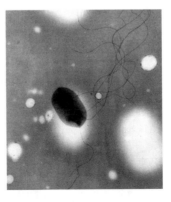

Figure 1.4 *Escherichia coli.* (Courtesy of CDC Public Health Image Library.)

1.2.6.8   Yersinia

Yersinia pestis causes bubonic plague.

1.2.6.9   Enterobacter

Enterobacter cloacae and *E. aerogenes* cause urinary tract and nosocomial infections.

1.2.6.10   Vibrio

Vibrio cholerae causes cholera, and *V. parahaemolyticus* causes gastroenteritis. *Vibrio* inhabits coastal waters and ultimately some shellfish. Transmittal to humans occurs when undercooked shellfish is eaten.

1.2.6.11   Pasteurellacea

Pasteurellacea cause septicemia and pneumonia. *Pasteurella multocida* can be transmitted to humans.

1.2.6.12   Haemophilus

Haemophilus commonly inhabits the mucous membranes of the upper respiratory tract, mouth, vagina, and intestinal tract. *Haemophilus influenzae* causes earaches, meningitis, epiglottiditis, septic arthritis in children, bronchitis, and pneumonia.

1.2.6.13   Gardinerella

Gardinerella vaginitis causes vaginal infections. (*Note:* The taxonomic classification is still in some dispute.)

1.2.7   Anaerobic, Gram-Negative, Straight, Curved, and Helical Rods

1.2.7.1   Bacterioides

Bacterioides do not form endospores and are nonmotile. They live in the oral cavity, upper respiratory tract, and genital tract and cause infections in puncture wounds, in surgical sites, and within the peritoneum (peritonitis).

1.2.7.2   Fusobacterium

Fusobacterium is long and slender and causes gum abscesses.

1.2.8   Dissimilatory Sulfate- or Sulfur-Reducing Bacteria

These bacteria are obligately anaerobic and use oxidized forms of sulfur (e.g., sulfates, elemental sulfur) to produce hydrogen sulfide (H_2S).

1.2.9   Anaerobic Gram-Negative Cocci

These cocci typically occur in pairs and are nonmotile. They do not form endospores. *Veillonella* is a component of dental plaque.

1.2.10 Rickettsias and Chlamydias

These are obligate intracellular parasites that are smaller than some viruses.

1.2.10.1 Rickettsia

Rickettsia are rod-shaped bacteria that are nonmotile and divide by binary fusion (Figure 1.5). They are transmitted to humans by insects and ticks. *Coxiella burnetii* causes Q fever and is transmitted by contaminated milk. A sporulated form may explain the resistance to pasteurization and antimicrobial chemicals. *Rickettsia prowazekii* causes endemic murine typhus and is transmitted by lice. *R. typhi* causes typhoid fever and is transmitted by fleas. *R. rickettsii* causes Rocky Mountain spotted fever and is transmitted by ticks.

1.2.10.2 Chlamydia

Chlamydia is coccoid and nonmotile and is transmitted by interpersonal contact or airborne respiratory routes. *Chlamydia trachomatis* causes trachoma, nongonococcal urethritis, and lymphogranuloma venereum. *C. pneumoniae* causes pneumonia.

1.2.11 Mycoplasmas

Mycoplasmas do not form cell walls. They are aerobes or facultative anaerobes and can produce filaments that resemble fungi. Cells are very small. *Mycoplasma pneumoniae* causes walking pneumonia. *Ureaplasma urealyticum* (occasionally) causes urinary tract infections; it splits the urea in urine.

1.2.12 Gram-Positive Cocci

1.2.12.1 Staphylococcus

Staphylococcus occurs in grapelike clusters and is an aerobe or facultative anaerobe. Staphylococci take many forms and grow under high osmotic pressure and low moisture.

1.2.12.1.1 Staphylococcus aureus

Staphylococcus aureus (Figure 1.6) produces many toxins. It can infect surgical wounds, can develop resistance to antibiotics, and is the agent of toxic shock syndrome. *S. aureus* produces enterotoxins that cause vomiting and nausea (food poisoning).

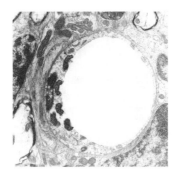

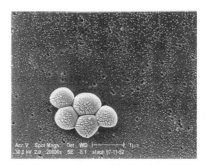

Figure 1.5 Rickettsia. (Courtesy of CDC Public Health Image Library.)

Figure 1.6 *Staphylococcus aureus.* (Courtesy of CDC Public Health Image Library.)

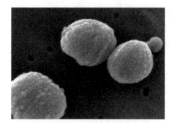

Figure 1.7   *Streptococcus pneumoniae.* (Courtesy of CDC Public Health Image Library.)

Figure 1.8   *Bacillus anthracis.* (Courtesy of CDC Public Health Image Library.)

1.2.12.2   Streptococcus

Streptococcus is spherical (Figure 1.7). It causes scarlet fever, pharyngitis, and pneumococcal pneumonia. Typically it appears in chains of 4 to 6 cocci, but 50 or more are possible. *Streptococcus* does not use oxygen but is aerotolerant. Some forms of *Streptococcus* are obligately anaerobic. They produce chemicals that destroy phagocytic cells and enzymes that digest connective tissue and spread infection. The enzymes lyse fibrin, thereby destroying the fibrous protein that is deposited in blood clots and normally would limit pathogen movement.

1.2.12.3   Endospore-Forming Gram-Positive Rods and Cocci

These endospores are resistant to heat and many chemicals.

1.2.12.3.1 Bacillus anthracis

Bacillus anthracis is nonmotile and a facultative anaerobe that can live in either aerobic or anaerobic conditions (Figure 1.8).

1.2.12.3.2 Clostridium

Clostridium is an obligate anaerobe. *Clostridium tetani* causes tetanus, *C. botulinum* causes botulism, and *C. perfringens* causes gas gangrene and food poisoning.

1.2.12.4   Regular Nonsporing Gram-Positive Rods

1.2.12.4.1 Listeria monocytogenes

Listeria monocytogenes survives within phagocytic cells and is capable of growth at refrigeration temperatures. *L. monocytogenes* can cause serious damage to a fetus resulting in stillbirth.

1.2.12.5   Irregular Nonsporulating Gram-Positive Rods

This type of bacteria is club shaped, pleomorphic, and sometimes age dependant. It is aerobic, anaerobic, or microaerophilic. *Corynebacterium diphtheria* causes diphtheria, *Propionibacterium acnes* is implicated in acne, and *Actinomyces isrealii* causes actinomycosis.

1.2.13   Mycobacteria

Rod-shaped mycobacteria are aerobic, do not produce endospores, and are nonmotile (Figure 1.9). Occasionally they exhibit filamentous growth. *Mycobacterium tuberculosis* causes tuberculosis, and *M. leprae* causes leprosy.

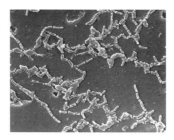

Figure 1.9 *Mycobacterium.* (Courtesy of CDC Public Health Image Library.)

1.2.14 Nocardioforms

1.2.14.1 Nocardia

Nocardia morphologically resembles *Actinomyces*. It is aerobic and forms rudimentary filaments to reproduce. *Nocardia asteroides* causes chronic pulmonary nocardiosis and mycetoma.

1.2.15 Gliding, Sheathed, and Budding and/or Appendaged Bacteria

These bacteria have prosthecae (protrusions such as stalks and buds) and include gliding, non-fruiting, gliding fruiting, budding, and sheathed types; chemoautotrophic bacteria; archaeobacteria; phototrophic; purple and green *Cyanobacteria*; *Actinomycetes*.

1.3 FUNGI

The term *fungi* refers to the taxonomic kingdom of Fungi. Fungi are nonmotile and eucaryotic, have cell walls, lack chlorophyll, and develop from spores. The spores can reproduce asexually or sexually. All fungi are chemoheterotrophs requiring organic food for energy. Fungi are either aerobic or facultatively aerobic and eucaryotic. As eucaryotic life forms, fungi have defined nuclear membranes and DNA within these nuclear boundaries. Fungi are carbon heterotrophs and absorb nutrients, including carbon-based preformed organics. Absorption occurs across the fungi cell walls, which are composed of chitin, chitosan, glucan, and mannan combinations. Given appropriate growing conditions, fungi are dimorphic (having two forms of growth) and can be found as either mold or yeast. This dimorphism may be temperature dependent. Mold germinates with branching hyphae and reproduces using spores. Yeast germinate as unicellular organisms and reproduce by budding.

1.3.1 Typical Mold Life Story

Molds develop from spores. When a spore settles on a hospitable surface, the spore swells and produces a germ tube (germination) that grows into a tiny, thread-like hypha (plural, hyphae). The hyphae form a tangled mass called a *mycelium* (Figure 1.10).The mycelium in turn produces aerial hyphae called *stolons* and root-like structures known as *rhizoids*. The rhizoids anchor the stolons in the substrate (living space and food source). As the mold matures, many upright fruiting bodies form above the rhizoids.

For asexual reproduction, the end of each fruiting body has a spore case, called a *sporangium*. A sporangium looks like a miniature pinhead and contains thousands of spores. When the spore case matures and breaks open, air currents carry the spores away (Figure 1.11). The asexual spores are genetic copies of the parent. For sexual reproduction a variety of methods are used to unite

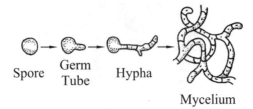

Figure 1.10 Fungal spore develops into hyphae and mycelia. (Courtesy of Deniese A. Chambers.)

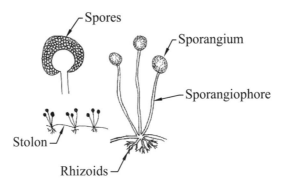

Figure 1.11 Structures of a *Rhizopus*. (Courtesy of Deniese A. Chambers.)

genetic material from two parent hyphae into a resultant spore. The sexual spores are genetically different from each parent. These spores may settle on damp food and grow, starting the reproductive cycle over again. Some molds, such as *Penicillium*, produce chains of spores at the tips of certain hyphae, called *conidiophores* (Figure 1.12).

1.3.2   Thallus and Hyphae

Vegetative structures are defined as those involved in catabolism and growth, rather than reproduction. The structures include thallus, which is a body consisting of long filaments of cells joined together; the filaments are hyphae (sing. hypha). Hyphae are actively growing and assimilative; new growth occurs as a linear elongation of the tip. Septate hyphae contain crosswalls known as septa (sing. septum). The septa divide the hyphae into uninucleate units. These units are structured like cells with openings to the next cell through the cell membrane. Due to these openings in the septa, these fungi are actually coenocytic (connected). Coenocytic hyphae contain no septa and are like long continuous cells with many nuclei. Hyphae grow by elongating at the tip; however, if the hyphae are damaged, any part of the hyphae may elongate to form new hyphae structures. Consequently, the presence of hyphae fragments may initiate the growth of molds even when spores are not present.

1.3.3   Mycelium

A mycelium is the mass of intertwined hyphae that forms when conditions for growth are suitable. Vegetative mycelium obtain nutrients. Reproductive aerial mycelium project above the surface of a growth medium and often bear reproductive spores.

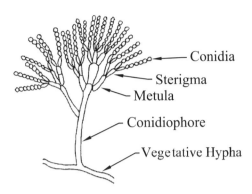

Figure 1.12 Structures of *Penicillium*. (Courtesy of Deniese A. Chambers.)

1.4 FUNGI REPRODUCTIVE STRUCTURES

Spores produce new mold through detachment and ultimate germination away from the parent structure. Spores are formed from the aerial mycelium. The anamorph structures denote asexual reproduction. The teleomorph structures denote sexual reproduction.

1.4.1 Asexual Spores

Asexual spores form from the aerial mycelium of one organism. *Arthrospores* are formed by fragmentation of septate hyphae; the resultant spore is actually a slightly thickened cell. *Blastopores* form as buds coming off the parent hyphae cell. *Chlamydospores* are formed by enlargement and rounding of a hyphal segment. The *conidiospore* is a unicellular or multicellular spore that is not enclosed in a sac and is produced in a chain at the end of a conidiophore. The term *conidia* means "dust;" these spores can move like dust through the air. *Sporangiospores* are formed within a sporangium (sac) at the end of aerial hyphae. The sporangium can contain hundreds of spores. Sporangia are the globular envelopes that encase the spores. The hyphae tips bearing the sporangia are sporangiophores.

1.4.2 Sexual Spores

Sexual spores form from the fusion of nuclei from two opposite mating parental strains from the same species. *Zygospores* result when nuclei from two morphologically similar cells fuse together; the spores have a thick wall. *Ascospores* result when nuclei from two morphologically similar or dissimilar cells fuse together. The initial spore divides to form a number of spores, which are produced in a spore sac or an ascus (sac-like structure). The structure holding the ascus is termed an *ascocarp*. *Basidiospores* are formed externally on a basidium (base pedestal).

1.5 FUNGI PHYLA

Fungi are divided into phyla based on morphology. The taxonomic classification may be modified with additional information (e.g., discovery of teleomorphs).

1.5.1   Deuteromycota/Fungi Imperfecta

The term *imperfecta* simply means that the taxonomic classification is imperfect in that no sexual spores have been identified. Reclassification occurs when sexual spores are identified.

1.5.2   Zygomycota/Conjugate Fungi

Zygomycota are saphrophytic and have coenocytic hyphae that are not divided by septa. *Zygo* is Greek for "yolk" or "joining." These fungi reproduce sexually by the physical blending of gametangia to form a zygosporangium. The initial contact of mycelia is followed by an exchange of chemical signals across hyphae to determine compatibility. If compatibility is established, the hyphae swell and fuse into gametangia. A septa forms to isolate the commingled multinucleate components, and fusion of the paired nuclei occurs in the newly formed zygosporangium. The zygosporangium outer walls thicken. After dispersion of the zygospore from the zygosporangium, a final meiotic division produces a mitosporangium. Asexual reproduction occurs with the formation of sporangiospores. The sporangiospores are produced in sporangiophores.

1.5.3   Dikaryomycota

Phylum Dikaryomycota contains molds that have hyphae. The dikaryomycotan hyphae have septa (crosswalls) at regular intervals. Depending on the type of mold, the hyphae become either ascomycetes or basidiomycetes. Most dikaryomycotan fungi have cell walls with chitin and septate hyphae. The hyphae have holes in their septa to permit the movement of cytoplasm. The hyphae can fuse with other similar hyphae in a process called *anastomosis*; during the fusion, nuclei are exchanged. Dikaryomycotan fungi have the ability in some cases to form a dikaryon. In the dikaryon, the nuclear material brought together during anastomosis does not immediately fuse to form a diploid zygote. The paired-off nuclei begin dividing, and the products of these divisions (sexually compatible haploid nuclei) are compartmentalized as dikaryotic nuclei.

1.5.3.1   Ascomycota/Sac Fungi

Sac fungi have sexual spores that are produced in an ascus (pl. asci), or sac-like structure. The paired nuclei of the sexual spores fuse, then undergo meiosis and mitosis to form eight haploid ascospores. Their asexual spores (called *conidiospores* from the word *conidia*, which means "dust") are produced in long chains from a conidiophore. The characteristic arrangement of the conidiospores is used to identify the different molds. Ascomycota include molds with septate hyphae and yeasts.

1.5.3.2   Basidiomycota/Club Fungi

Basidiomycota have sexual basidiospores that are produced on the basidia (sing. basidium) that often have sterigmate projections. The paired nuclei of the sexual spores fuse and then undergo meiosis to form four haploid basidiospores The sterigmata bear the haploid basidiospores, and their morphology varies, as does the number of spores per sterigmata. Asexual spores may also be produced by some genera.

1.6 YEASTS

Yeasts are monofilamentous, unicellular fungi and are spherical or oval. Yeast colonies grow as each individual yeast cell reproduces. The two types of reproduction are:

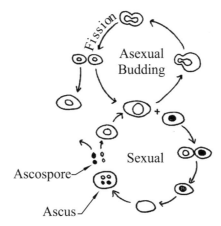

Figure 1.13 The life cycles of yeast, asexual and sexual. (Courtesy of Deniese A. Chambers.)

1. *Budding.* As the bud elongates, the nucleus of the parent cell divides. One nucleus migrates into the bud. Cell material provides a bridge between the two cells. If the buds do not detach, a pseudohydra is formed (Figure 1.13).
2. *Fission.* The parent cell elongates, the nucleus divides, and two daughter cells are produced.

When oxygen is present, yeasts respire aerobically. Without oxygen, yeasts are facultative anaerobes and can ferment carbohydrates for energy. Various yeasts are commonly identified in air samples. These yeasts are not known to be allergenic, but they may cause problems if a person has had previous exposure and has developed hypersensitivities. Yeasts present in sufficient concentrations may be allergenic for susceptible individuals. Yeasts grow when moisture and food are available at just the right temperature.

1.7 FUNGI OF CONCERN IN INDOOR ENVIRONMENTS

Molds can live for years in a dormant state. Molds are common, and their spores can be found everywhere. Fungi commonly found in ventilation systems and indoor environments include *Absidia, Acremonium, Alternaria, Aspergillus, Aureobasidium, Botrytis, Cephalosporium, Chrysosporium, Cladosporium, Epicoccum, Fusarium, Helminthosporium, Leptosphaerulina, Mucor, Nigrospora, Penicillium, Phoma, Pithomyces, Rhinocladiella, Rhizopus, Scopulariopsis, Stachybotrys, Streptomyces, Stysanus, Ulocladium, Yeast,* and *Zygosporium.* The 11 types of fungi typically found in homes are *Aspergillus, Cladosporium, Chrysosporium, Compacta, Epicoccum, Fonsecea, Penicillium, Stachybotrys, Trichoderma, Trichophyton,* and *Yeast.*

Fungi are ubiquitous in the environment, particularly in soil, and many are also part of the normal gastrointestinal and skin flora in humans and animals. In some areas of the United States, certain types of fungi are endemic and occur naturally in the soil. These soil fungi include *Histoplasma capsulatum,* found in some midwestern states, and *Coccidioides immitis,* which is found in the southwestern United States and parts of Central and South America. If the soil habitat of these fungi is disturbed by activities such as construction or natural disasters, the fungal spores become airborne; when they are inhaled, they can cause infection. Some can grow at subfreezing temperatures, but most become dormant. Snow cover lowers the outdoor mold count drastically but does not kill molds. After the spring thaw, molds thrive on the vegetation that has been killed by the winter cold. In the warmest areas of the United States, molds thrive all year and can cause year-round (perennial) allergic problems. In addition, molds growing indoors can cause perennial allergic rhinitis even in the coldest climates.

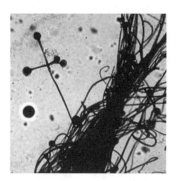

Figure 1.14 Photomicrograph of *Absidia*. (Courtesy of University of Minnesota Environmental Health and Safety, Minneapolis.)

1.7.1 *Absidia* (Zygomycete)

Absidia is a filamentous fungi that asexually produces sporangiospores and often causes food spoilage (Figure 1.14). The genus *Absidia* currently contains 21 species. The most commonly isolated species is *Absidia corymbifera*, the only recognized pathogen; also known as *Absidia ramose* and *Moocher corymbifera*. Health effects include the zygomycosis caused by *A. corymbifera*. Zygomycosis is an opportunistic mycosis that manifests with pulmonary, rhinocerebral, cutaneous, gastrointestinal, renal, or meningeal involvement. Disseminated zygomycosis may originate from these infections. Zygomycosis is observed in immunocompetent hosts.

1.7.2 *Acremonium* (Ascomycete)

Acremonium (Figures 1.15 and 1.16) is filamentous and is classified either as a deuteromycetes, as its sexual state is not well defined, or within the Ascomycota phylum due to its structural properties. It produces conidiophores and slender phialides. The conidia are hyaline (transparent) and one celled and are collected in a slime drop. Species of *Acremonium* are occasionally confused with isolates of *Fusarium*, *Verticillium*, and *Cylindrocarpon*. Synonyms include *Cephalosporium* spp. and *Gliomastix*. Health effects include onychomycosis (nail infections), keratitis, mycetomas (infections of cornea and nails), endophthalmitis, endocarditis, meningitis, peritonitis, and osteomyelitis. Occupants of residences may experience nausea, vomiting, or diarrhea. *Acremonium* produces trichothecene, a toxin that causes toxic health effects if ingested.

1.7.3 *Alternaria* (Ascomycete)

The word *Alternaria* is derived from the Greek *alteres,* meaning "dumbbell," or the Latin *alternare*, meaning "alternate." A number of very similar, related species are grouped together as *Alternaria* (Figures 1.17 and 1.18). Some species of *Alternaria* are the imperfect, asexual, anamorph spores of the Ascomycete pleospora. (See Color Figures 1 and 2.) The spores are multicelled and develop in chains, head to toe. They have multiple septa, both transverse and longitudinally. They vary in width and length according to species, generally from 8 to 75 µm in length. The length of some species such as *A. longissima* can be up to 500 µm (0.5 mm).

The conidiophores are dark, and the colonies are black, with variations from green black to greyish. They are ellipsoidal, with a beak that approaches one third of their total length. They are large (18–83 × 7–18 µm) and multicellular, with transverse and longitudinal septa.

Health effects include allergic reactions, due in part to its prevalence in the general environs; sinus infections or sinusitis (*Alternaria* colonizes the paranasal sinuses, leading to chronic hypertrophic sinusitis and may result in systemic and invasive disease); hypersensitivity pneumonitis;

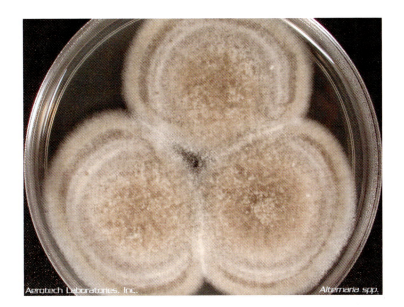

Figure 1
Culture of *Alternaria* spp.

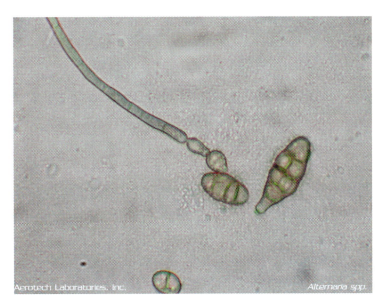

Figure 2
Photomicrograph of *Alternaria* spp.

Figure 3
Culture of *Aspergillus flavus*

Figure 4
Photomicrograph of *Aspergillus flavus*

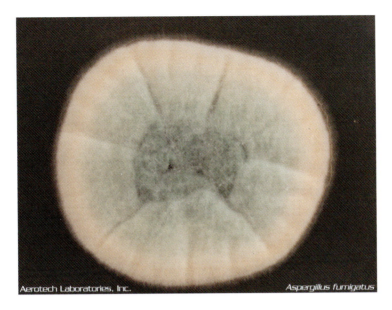

Figure 5
Culture of *Aspergillus fumigatus*

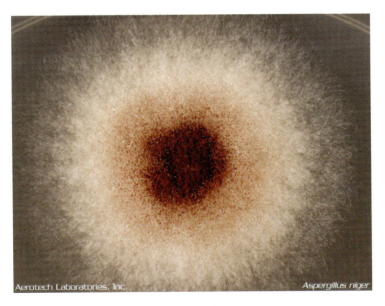

Figure 6
Culture of *Aspergillus niger*

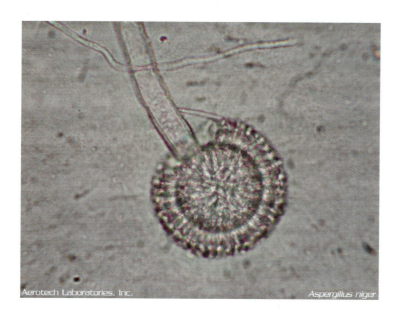

Figure 7

Photomicrograph of *Aspergillus niger*

Figure 8

Culture of *Aspergillus terreus*

Figure 9
Culture of *Aureobasidium*

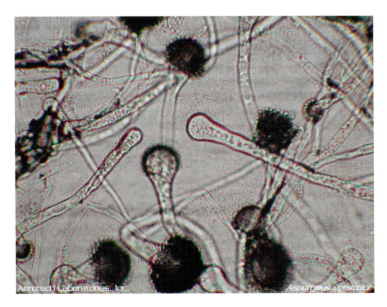

Figure 10
Culture of *Aspergillus versicolor*

Figure 11
Photomicrograph of *Aspergillus versicolor*

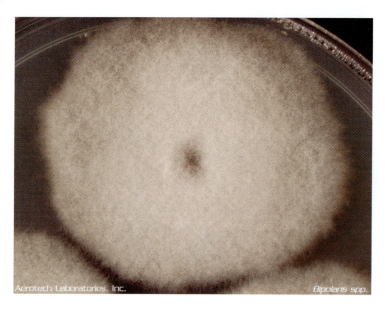

Figure 12
Culture of *Bipolaris*

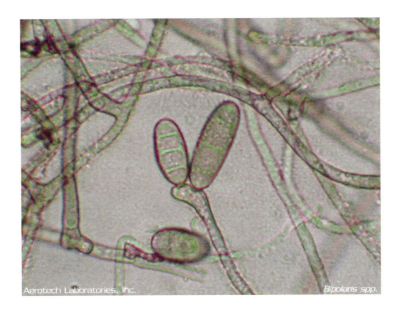

Figure 13
Photomicrograph of *Bipolaris*

Figure 14
Culture of *Botrytis*

Figure 15
Culture of *Cladosporium*

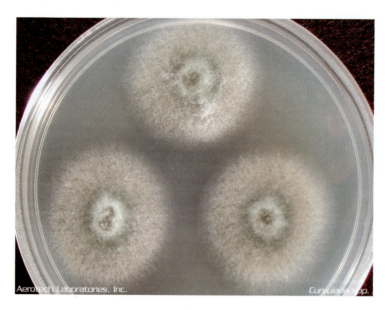

Figure 16
Culture of *Curvularia*

Figure 17
Photomicrograph of *Curvularia*

Figure 18
Culture of *Epicoccum*

Figure 19
Culture of *Fusarium*

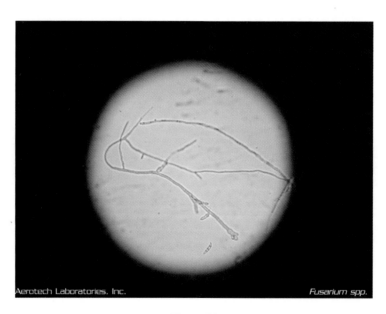

Figure 20
Photomicrograph of *Fusarium*

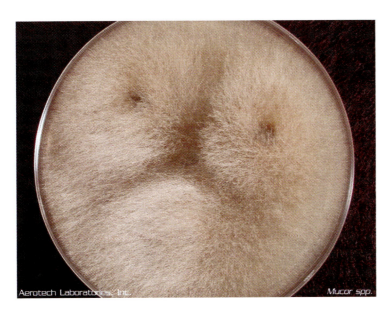

Figure 21
Culture of *Mucor*

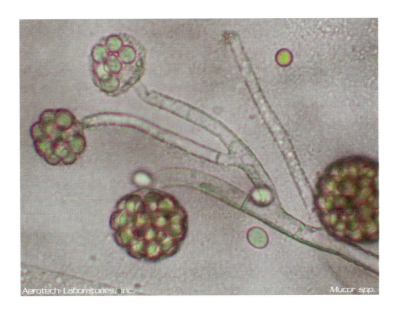

Figure 22
Photomicrograph of *Mucor*

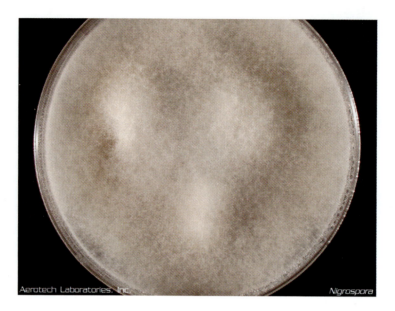

Figure 23
Culture of *Nigrospora*

Figure 24
Culture of *Paecilomyces*

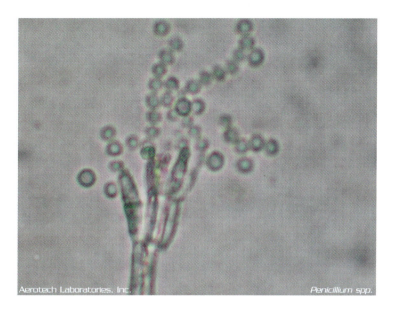

Figure 25
Photomicrograph of *Penicillium*

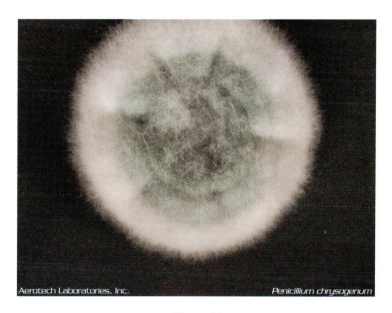

Figure 26
Culture of *Penicillium chrysogenum*

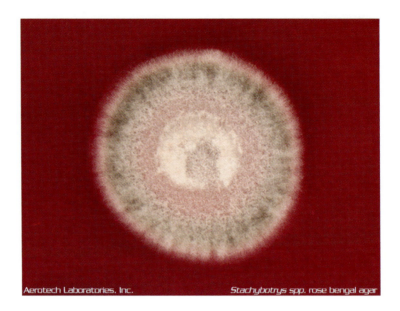

Figure 27
Culture of *Stachybotrys*

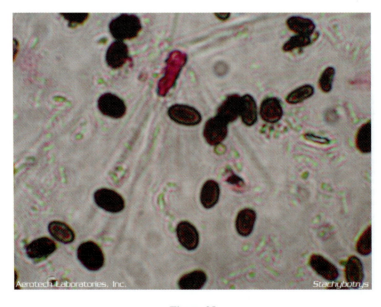

Figure 28
Photomicrograph of *Stachybotrys*

Figure 29
Photomicrograph of *Stachybotrys*

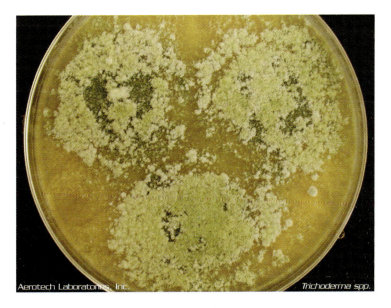

Figure 30
Culture of *Trichodema*

Figure 31
Culture of *Ulocladium*

Figure 32
Photomicrograph of *Ulocladium*

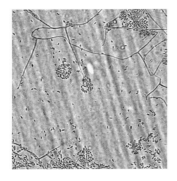

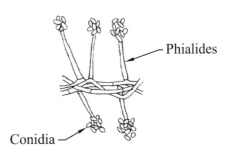

Figure 1.16 *Acremonium* structures. (Courtesy of Deniese A. Chambers.)

Figure 1.15 Photomicrograph of *Acremonium*. (Courtesy of University of Minnesota Environmental Health and Safety.)

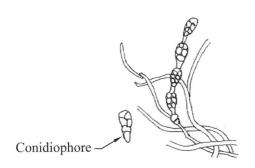

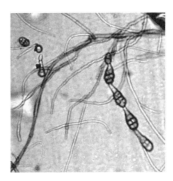

Figure 1.17 Structures of *Alternaria*. (Courtesy of Deniese A. Chambers.)

Figure 1.18 Photomicrograph of *Alternaria*. (Courtesy of University of Minnesota Environmental Health and Safety.)

extrinsic asthma; pulmonary edema; bronchiospasms; pulmonary emphysema; phaeohyphomycosis; onychomycosis; mycotic keratitis; ulcerated cutaneous infections; keratitis; visceral infections; osteomyelitis; middle ear infections (otitis media); farmer's lung (extrinsic allergenic alveolitis) in agricultural field workers; and baker's asthma due to its presence in flour.

Mycotoxins produced include alternariol, an antifungal; alternariol monomethylether (AME); mutagenics altertoxin I and II; alteneune; altenusin; and tenazonic acid.

1.7.4 *Aspergillus* (Ascomycete)

The word *Aspergillus* is derived from the Latin *aspergillium*, meaning "holy water stoop." The genus *Aspergillus* includes over 185 species. Around 20 species have so far been reported as causative agents of opportunistic infections in humans. *Aspergillus* is a filamentous ascomycota, or sac, fungus. The perfect names for *Aspergillus* are *Sartorya*, *Eurotium*, and *Emericella*. (*Note:* While teleomorphic states have been described only for some of the *Aspergillus* species, others are accepted to be mitosporic, without any known sexual spore production.)

Members of the *Aspergillus* genus are known as biodeteriogens (organisms that cause deterioration of materials). *A. niger* causes damage, discoloration, and softening of the surfaces of woods — even in the presence of wood preservatives. *A. niger* causes damage to cellulose materials, hides, cotton fibers, plastics, and polymers (i.e., cellulose nitrate, polyvinyl acetate, polyester-type polyurethane). In cases of extensive growth, colonies will grow into wood, plaster, and/or drywall, causing a soft bulging area.

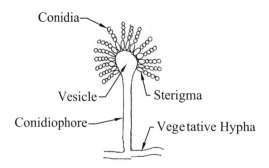

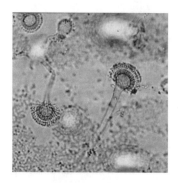

Figure 1.19 Structures of *Aspergillus fumigatus.* (Courtesy of Deniese A. Chambers.)

Figure 1.20 Photomicrograph of *A. fumigatus.* (Courtesy of the University of Minnesota, Minneapolis.)

This area lacks structural integrity and is subject to early deterioration. Growth areas may attract other airborne debris, particulates, and spores, some of which germinate in the newly hospitable substrate.

Aspergillus fumigatus is the most commonly isolated species, followed by *A. flavus* and *A. niger.* Among the other species less commonly isolated as opportunistic pathogens are *A. clavatus,* the *A. glaucus* group, *A. nidulans,* *A. oryzae,* *A. terreus,* *A. ustus,* and *A. versicolor.*

The conidia of *A. flavus* (see Color Figures 3 and 4) are lime green and fuzzy looking. Sclerota, if produced, are white and then brown. The conidia are found as unicellular, globose, dry chains with a size range of 3.5 to 6 μm. The upright conidiophores are simple, terminating in globose/clavate swellings. The conidia bear phialides at the apex.

The conidia of *A. fumigatus* (Figure 1.19; see also, Color Figure 5) are dark green. The aerial hyphae are colorless. They are globose to subglobose and rough walled to echinulate (set with small spines or prickles). Size range is from 2.5 to 3.0 μm. The conidiophores are short and green and bear columnar conidial heads.

The conidia of *A. niger* (Figure 1.21; see also, Color Figures 6 and 7) are brown, and bear warts, spines, or ridges. Sclerota if produced are large and cream colored. The conidia are unicellular, globose to subglobose. Size range is from 3.5 to 5.0 μm. The upright conidiophores are simple, terminating in globose/clavate swellings. They bear phialides at the apex.

The conidia of *A. ocraceus* are ochre-yellow and fuzzy (like felt). Sclerota, if produced, are finely rough walled and white in color, then lavender to purple. They are globose to subglobose. Size range is from 2.5 to 3.0 μm.

The conidia of *A. oryzae* are very light lime green and vary from smooth to rough. They are ellipsoidal, then globose to subglobose. Size range is from 4.5 to 8.0 μm

The conidia of *A. terreus* (Figure 1.22; see also, Color Figure 8) are light yellow and smooth. They are ellipsoidal, then globose and smooth. Size range is from 1.5 to 2.5 μm

The conidia of *A. versicolor* (see Color Figures 9 and 10) are emerald green and echinulate. Colony colors are white, then yellow, then yellow-green mixed with fleshy tone to pink. They are globose. Size range is from 2.5 to 3.5 μm

Among all filamentous fungi, *Aspergillus* is the most commonly isolated in invasive and opportunistic infections. Aspergillosis is now the second most common fungal infection requiring hospitalization in the United States. *A. niger* produces mycotoxins that can induce asthma-like symptoms. In situations where *A. niger* is found growing with *Penicillium* spp., massive inhalation of spores has been documented as causing an acute, diffuse, self-limiting pneumonitis (lung irritation). Healthy individuals can exhibit otitis externa (inflammation of the outer ear canal) as a result of *Aspergillus* growth. *Aspergillus* infections are opportunistic; immunosuppression is the major predisposing factor, secondary to bronchiectasis, carcinoma, other mycoses, sarcoid, and tuberculosis. Aspergillosis can vary from being local to systemic. Almost any organ or system in

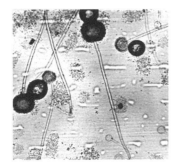

Figure 1.21 Photomicrograph of *Aspergillus niger.* (Courtesy of University of Minnesota Environmental Health and Safety, Minneapolis.)

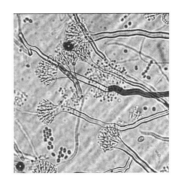

Figure 1.22 Photomicrograph of *Aspergillus terreus.* (Courtesy of University of Minnesota Environmental Health and Safety, Minneapolis.)

the human body can be involved, and any of the following can be observed: onychomycosis, sinusitis, cerebral aspergillosis, meningitis, endocarditis, myocarditis, pulmonary aspergillosis, osteomyelitis, otomycosis, endophthalmitis, cutaneous aspergillosis, hepatosplenic aspergillosis, and disseminated aspergillosis. Nosocomial aspergillosis occurs due to catheters and other devices.

Local colonizers can be found in previously developed lung cavities due to tuberculosis, sarcoidosis, bronchiectasis, pneumoconiosis, ankylosing spondylitis, or neoplasms, presenting as a distinct clinical entity called aspergilloma. *A. niger* is commonly associated with fungus ball growth in the lung, affects the kidneys, and produces allergic states. Some *Aspergillus* antigens are fungal allergens that may initiate allergic bronchopulmonary aspergillosis, particularly in an atopic host. Extrinsic asthma with edema and bronchiospasms is observed. Chronic cases may develop pulmonary emphysema.

Aspergillus flavus produces aflatoxin B_1, which is the most carcinogenic toxin caused by biologicals. Also produced are aflatoxin B_2, cyclopiazonic acid, kojic acid, aspergillic acid, 3-methyl butanol, 3-octanone, 1-octen-3-ol, and ethylene. *A. fumigatus* produces chanoclavine, ferricrocin,* festuclavine, fumagillin (anti-amoebae), fumifungin (antifungal drug), fumigacin** (helvolic acid), which acts against Gram-positive and Gram-negative bacteria), fumigaclavine, fumigatin,** fumi-toxins, fumitremogins, fusigen,** gliotoxin,** phyllostine,** sphingofungins, spinulosin,** trypa-cidin,** tryptoquivalins, and verrucologen.

Aspergillus niger produces asnipyrone A and B, aspergillin, asperrubrol, asperenones, auraspe-nones, 4-hydroxymandelic acid, maiformin C, some of the naphtho-γ-quinones, orobols, pyrophen, and tubigensin A and B. *A. niger* is used in bread and beer making to produce amylase and amyloglucosidase. *A. niger* is also used in the production of oxalic acid, fumaric acid, and citric acid. *A. ocraceus* produces flavacol, 4-hydroxymellein, neo-aspergillic acid, 1-octen-3-ol, 2-octen-1-ol, penicillic acid, viomellein (liver and kidney toxin), and xanomegnin acid (liver and kidney toxin). *A. oryzae* produces kojic acid, oryzacidin, 1-octen-3-ol, 3-octanone, 3-methylbutanol, cyclo-piazonic acid (toxin), and B-nitropropionic acid (toxin). *A. oryzae* is used in the production of enzymes, including amylase and amyloglucosidase. *A. terreus* produces aranotins, aspterric acid, aspulvinorones, astepyrone, asteriquinones, asterric acid, citreoviridin (neurotoxin), cytochalasin E., emodin, erdin, geodin, 6-hydroxymellein, mevinolins, patulin, quadrone, questin, sukichrin, terramides, terreic acid, terrein, terretrems (tremorgenic mycotoxin, which does not contain nitrogen and is produced by one strain of *A. terreus*), and terrotonin. *A. terreus* is used in the production of itaconic acid and mevinolin (lowers cholesterol levels). *A. versicolor* produces averufin, cyclopenin, cyclopenol, geosmin (mucosal irritant), sterigmatocystin (toxic and carcinogenic), and versicolorin.

---

* Exhibit known antibiotic activity.
** Metabolites with strong antibacterial action.

Some *Aspergillus* spp. produce various mycotoxins, including alfatoxins and ocratoxins, which are carcinogenic. Chronic ingestion has been proven to lead to carcinogenic potential, particularly in animals. Among these mycotoxins, aflatoxin may induce hepatocellular carcinoma and is mostly produced by *A. flavus* in contaminated foodstuffs.

1.7.5 *Aureobasidium*

The term *Aureobasidium* is derived from the Latin *aureus* ("golden"). *Aureobasidium* includes 14 species and one variety, with *A. pullulans* being the only well-known species. The common name for these molds is black yeast (see Color Figure 11). *Aureobasidium* grows under paint, resists some paint fungicides, and may colonize hair, skin, and nails in humans. The conidia are black and shiny upon maturation and are unicellular ovoids from 5.0 to 7.0 μm in diameter.

Health effects include phaehyphomycosis from exposure to *Aureobasidium pullulans*. This species may also contribute to keratomycosis; pulmonary mycosis with sepsis and other opportunistic infections; cutaneous mycoses, such as eumycotic dermatitis; peritonitis; nosocomial meningitis from exposure to *A. mansoni*; chromoblastomycosis (chronic cutaneous skin infection) from exposure to *A. pullulans*; and hypersensitivity pneumonitis. *Aureobasidium* is used in the production of the polysaccharide pullulan.

1.7.6 *Bipolaris*

(See Figures 1.23 and 1.24 and Color Figures 12 and 13.) The conidia of *Bipolaris* are brown; elliptical, straight, or curved; and several celled. The hyphae are dematiaceous septate. Health effects include phaeohyphomycosis (mycotic infections) and sinusitis.

1.7.7 *Botrytis*

(See Figures 1.25 and 1.26 and Color Figure 14) *Botrytis* (from *botrys*, the Greek word for "bunch of grapes") is a filamentous fungus isolated from decaying plants. It is a cosmopolitan feature, commonly reported from tropical and temperate areas. The conidia are colorless to pale brown; massed conidia will appear to be gray-brown in color. The color is often determined by the black irregular sclerotia produced. *Botrytis* is unicellular, spherical to ovoid. Size range is from 8 to 11 μm. Allergic reactions to *Botrytis* have been reported. *Botrytis* metabolites include botrydial and botryllin. Botrytis is used in the production of pectolytic enzymes and in wine making.

1.7.8 *Cephalosporium*

(See *Acremonium*.) This genus name is no longer used; the genus *Acremonium* was formerly known as *Cephalosporium*.

1.7.9 *Chaetomium* (Ascomycete)

Chaetomium asci are dense and contain four to eight brown ascospores. The ascospores collect outside the perithecium (fruiting bodies). Most species are strong decomposers of cellulose and occur wherever this substrate is abundant, such as in soil, dung, or rotting plants.

1.7.10 *Chrysosporium*

Chrysosporium conidia are unicellular, globose to pyriform, and occur in single or in short chains with a broad basal scar. Several species exist, with the most common ones being *C. merdarium* and *C. pannorum*. Chrysosporium causes rare skin infections in humans.

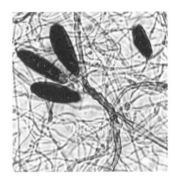

Figure 1.23 Photomicrograph of *Bipolaris*. (Courtesy of University of Minnesota Environmental Health and Safety, Minneapolis.)

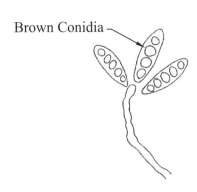

Figure 1.24 Structures of *Bipolaris*. (Courtesy of Deniese A. Chambers.)

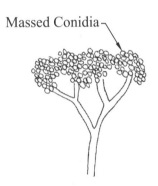

Figure 1.25 *Botrytis* structures; massed conidia gray-brown ("bunch of grapes"). (Courtesy of Deniese A. Chambers.)

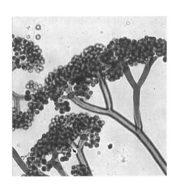

Figure 1.26 Photomicrograph of *Botrytis*. (Courtesy of University of Minnesota Environmental Health and Safety, Minneapolis.)

1.7.11 *Cladosporium*

(See Figures 1.27 and 1.28 and Color Figure 15.) *Cladosporium* (from *klados*, the Greek word for "branch") includes over 50 species. The most common ones are *C. elatum, C. herbarum, C. sphaerospermum*, and *C. cladosporioides*. *Cladosporium* is the most commonly identified outdoor fungus. Spores of *Cladosporium* are often found in higher concentrations in the air than any other fungal spore type. Spore transmittal may be on dry conidia chains. *Cladosporium* discolors interior porous surfaces and lives in refrigerator interiors, including water drainage reservoirs. It is a common indoor air allergen. Indoor *Cladosporium* spp. may be different from the species identified outdoors. *Cladosporium* spp. are commonly found on the surface of fiberglass duct liners in the interior of supply ducts. Spores are copious in number on branched conidiophores. They usually have distinctive scars at both ends: at the junctures to the spore at one end and to the conidiophore at the other. Although *Cladosporium* spores are often single celled, spores with a single transverse septum or several transverse septa are frequently seen. Size ranges from 4 to 20 μm in length. The conidia are dark. Colonies appear powdery with a velvety olive-green/brown color. They are two celled and typically lemon shaped, although the shape may vary, being ovoid, cylindrical, or irregular. They often occur as long, branched chains with variations in size and shape.

Cladosporium spp. can cause mycosis and are a common cause of extrinsic asthma (immediate-type hypersensitivity, type I). Acute symptoms include edema and bronchiospasms, and

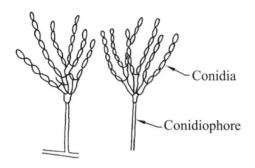

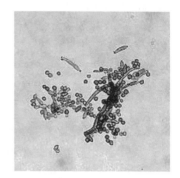

Figure 1.27 Structures of *Cladosporium*. (Courtesy of Deniese A. Chambers.)

Figure 1.28 Photomicrograph of *Cladosporium*. (Courtesy of University of Minnesota Environmental Health and Safety, Minneapolis.)

chronic cases may develop pulmonary emphysema. Health effects include allergies (asthma, hay fever); phaeohyphomycosis (mycotic infections); eye and skin infections; chromoblastomycosis lesions that are verrucoid, ulcerated, and crusted; onychomycosis (nails that generally split, flake, and grow too thick; abscesses; granulomata (cutaneous and subcutaneous tissue infections of skin, fascia, and bone); keratitis; sinusitis; and pulmonary infections, including pulmonary fungus ball.

1.7.12 *Coccidioides*

Coccidioides includes only one species: *Coccidioides immitis*. Clinical disease may occur in 90% of an exposed indoor population. Infections acquired outdoors in nature are asymptomatic in 50% of the cases. Spherules of the fungus may be present in clinical specimens and animal tissues, and infectious arthroconidia are found in mold cultures and soil samples. The size (2–5 nm) of the arthroconidia is conducive to dispersal in air and retention in the deep pulmonary spaces. The much larger size of the spherule (30–60 nm) considerably reduces the effectiveness of this form of the fungus as an airborne pathogen. Inhalation of arthroconidia from soil samples, mold cultures, or clinical materials after its transformation from the spherule form is the primary hazard. Accidental percutaneous inoculation of the spherule form may result in local granuloma formation. Coccidioidomycosis is the infection caused by the dimorphic fungus *C. immitis*. The disease is endemic only in regions of the Western Hemisphere. In the United States, the endemic areas include southern Arizona, central California, Southern New Mexico, and west Texas. The endemic region extends southward into Central and South America. An arid climate, alkaline soils, hot summers, few freezes, and yearly rainfalls ranging between 5 and 20 inches characterize this area. Outbreaks occur following dust storms, earthquakes, and earth excavation when dispersion of the arthroconidia is favored. Coccidioidomycosis is acquired from inhalation of the spores (arthroconidia). Once in the lungs, the arthroconidia transform into the spherical cells known as spherules. An acute respiratory infection occurs 7 to 21 days after exposure and typically resolves rapidly. The infection may alternatively result in a chronic pulmonary condition or disseminate to the meninges, bones, joints, and subcutaneous and cutaneous tissues. About 25% of the patients with disseminated disease have meningitis.

1.7.13 *Cryptococcus neoformans*

Cryptococcus neoformans is a yeast and is considered an opportunistic pathogen. *C. neoformans* var. *neoformans* colonizes pigeon and other bird droppings, and *C. neoformans* var. *gatti* colonizes the bark of the red gum tree. Both forms can cause disease in humans, although the former is the more common agent. Primary infection with *C. neoformans* follows exposure to an environmental source; inhalation of the fungus leads to an infection of the lungs. A transient colonization of the bronchial tree may result, or more extensive pulmonary involvement may occur. Bronchial infection may be self-limiting or chronic and may lead to dissemination to other parts of the body. The disseminated disease manifestations are dependent on the time frame of onset. Headaches, nausea, dizziness, decreased comprehension, impaired memory, and gait ataxia follow an increase in severity as invasion of the cerebral cortex, brain stem, cerebellum, and meninges progresses. Meningoencephalitis is the most common manifestation and the most severe symptom associated with *Cryptococcus* spp. infections. Onset of coma may be sudden and may be accompanied by respiratory arrest. If untreated, disseminated infection with *C. neoformans*, whether indolent or fulminant, may be fatal. Weathered pigeon excreta are considered a significant source. Unidentified alternative exposure routes are known to exist. The sexual stage of *C. neoformans* var. *neoformans* is *Filobasidiella neoformans* var. *neoformans*, and its basidiospores are a possible exposure vector agent. These basidiospores are dry and readily airborne. They vary in size from 1.8 to 3 μm in diameter. They are more readily deposited in the alveoli and are proven to be pathogenic when injected into mice.

1.7.14 *Cunninghamella* (Zygomycete)

Cunninghamella is a zygomycete with zygospores of the mucor type. The conidia are globose, one celled, and spiny or smooth. The conidiophores are colorless, simple, or branched. Enlarged tips bear the heads of the conidia. *Cunninghamella* asexually produces white mycelium. Health effects include disseminated pulmonary infections in immunocompromised hosts.

1.7.15 *Curvularia*

(See Figures 1.29 and 1.30 and Color Figures 16 and 17.) *Curvularia* conidia are dark with lighter end cells, fusiform, bent with central cell enlarged, and three- to five-celled. The conidiophores are brown and either simple or branched and bear the conidia apically. Health effects include allergenic reactions, corneal infections, and mycetoma.

1.7.16 *Epicoccum*

(See Figures 1.31 and 1.32 and Color Figure 18.) *Epicoccum* (from the Latin words *epi* for "over" and *coccus* for "sphere") is widely distributed and is commonly isolated from air, soil, and foodstuff. It is found in some animals and textiles. The genus *Epicoccum* includes only the species *E. purpurascens*, which is a pigmented (dematiaceous), mitosporic mold. The conidia are dark, globose, and verrucose. *Epicoccum* is several celled (15) and varies in size from 15 to 25 μm. The sporodochium (fruiting body) is visible to the naked eye. Epicoccum can grow at 37°C and thus can grow on the skin surface.

1.7.17 *Eurotium*

Eurotium (from *eurotiao*, the Greek word for "decay") is found in foodstuffs, leather, and cotton.

1.7.18 *Fonsecea compacta*

Fonsecea compacta is a pigmented (dematiaceous), filamentous fungus found in rotten wood and soil. It has no known teleomorphic phase and is saprophytic in nature. It causes infections in humans,

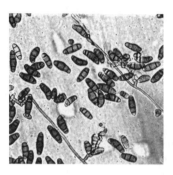

Figure 1.30 Scanning electron micrograph of *Curvularia geniculata*. (Courtesy of CDC Public Health Library and Janice Carr, Atlanta, Georgia.)

Figure 1.29 Photomicrograph of *Curvularias*. (Courtesy of University of Minnesota Environmental Health and Safety, Minneapolis.)

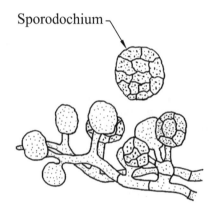

Figure 1.32 Photomicrograph of *Epicoccum*. (Courtesy of University of Minnesota Environmental Health and Safety, Minneapolis.)

Figure 1.31 Structures of *Epicoccum*; sporodochium is visible to the naked eye. (Courtesy of Deniese A. Chambers.)

and cold-blooded animals living in swamps may also be infected. *F. compacta* exhibits *in vivo* dimorphism; it produces a specific structure (sclerotic body) only in tissue and grows in mold form in laboratory conditions. The genus *Fonsecea* contains two species: *F. compacta* and *F. pedrosoi*.

Health effects include chromoblastomycosis, a posttraumatic, chronic infection of subcutaneous tissues. The etiologic agents of chromoblastomycosis are generally members of three genera of dematiaceous fungi that inhabit the soil: *Fonsecea*, *Phialophora*, and *Cladosporium*. Systemic invasion following chromoblastomycosis is very rare and presents with papules and verrucose, cauliflower-like lesions most commonly on lower extremities. Primary nasal chromoblastomycosis has also been reported. Paranasal sinusitis, keratitis, and fatal brain abscesses following hematogenous dissemination have been reported.

1.7.19 *Fusarium*

(See Figure 1.33 and Color Figures 19 and 20.) *Fusarium* (from *fussus*, the Latin word for "spindle") is a filamentous fungus found in normal mycoflora of commodities. Some *Fusarium* species have a teleomorphic state. The genus *Fusarium* currently contains over 20 species. The most common of these are *F. solani*, *F. oxysporum*, and *F. chlamydosporum*. The most virulent *Fusarium* species is *F. solani*. The two kinds of conidia are macroconidia (several-celled, canoe-

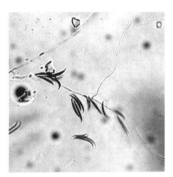

Figure 1.33 Photomicrograph of *Fusarium*. (Courtesy of University of Minnesota Environmental Health and Safety.)

shaped) and microconidia (unicellular ovoids). The mycelium is cotton-like in culture and is pink, purple, or yellow in color. The corn fungus *F. moniliforme* produces fusaric acid, which behaves like a weak animal toxin; however, when combined with other mold toxins, this fungus exaggerates the effects of the other toxins. All isolates of the *Fusarium*-type molds produce fusaric acid, suggesting that this compound is probably more prevalent in the environment than initially considered. These results indicate that analyses and toxicity studies should also include this toxin along with other suspect toxins under field conditions. Fumonisins might be teratogenic to humans.

Health effects include superficial and systemic infections in humans; infections due to *Fusarium* spp. are collectively referred to as fusariosis. Trauma is the major predisposing factor for development of cutaneous infections due to *Fusarium* strains. Disseminated opportunistic infections are found in immunosuppressed hosts, particularly in neutropenic and transplant patients. Other health effects include keratitis, endophthalmitis, otitis media, onychomycosis, cutaneous infections (particularly of burn wounds), mycetoma, osteomyelitis (particularly following trauma), sinusitis, pulmonary infections, endocarditis, peritonitis, central venous catheter infections, septic arthritis, disseminated infections, and fungemia. Nosocomial disseminated fusariosis in immunosuppressed patients may occur when *Fusarium* gets into hospital water distribution systems. *Fusarium* may also exist in the soil of potted plants in hospitals. These plants constitute a hazardous mycotic reservoir for nosocomial fusariosis. Ingestion of grains contaminated with the mycotoxins may give rise to allergic symptoms or may be carcinogenic in cases of long-term consumption. Fumonisins are the mycotoxins produced by *F. moniliforme* and *F. proliferatum* in maize; these mycotoxins may cause esophageal cancer. Another group of mycotoxins, zearalenones, may be produced by some *Fusarium* species growing in grains. Trichothecene toxins target the circulatory, alimentary, skin, and nervous systems. Ingestion of vomotoxin on grains imparts acute gastrointestinal illness. T-2 toxins are highly toxic and can severely damage the digestive tract with resultant hemorrhage. Zearalenone toxin targets the reproductive organs and is similar in structure to estrogen.

1.7.20 *Helminthosporium*

The only species is *Helminthosporium solani*, which is not a pathogen of humans or animal. The conidia develop laterally through pores beneath septa as the conidiophores are growing. Conidiophores are single, subhyaline to brown and ovate. The mycelium is dark. (*Note:* Isolates of *Bipolaris* and *Exserohilum* may be misidentified as *Helminthosporium*. *Helminthosporium* is rarely isolated in the laboratory.

1.7.21 *Histoplasma* (Ascomycete)

Histoplasma are Ascomycota; their perfect name is *Emmonsiella* or *Gymnoascus*. The genus *Histoplasma* includes one species, *H. capsulatum*, which has two varieties: *H. capsulatum* var.

capsulatum and *H. capsulatum* var. *duboisii*. A teleomorph is referred to as *Ajellomyces capsulatus*. *H. capsulatum* is the causative agent of a true systemic (endemic) mycosis known as histoplasmosis, which can be an acute benign pulmonary infection or chronic; disseminated disease is fatal. Following acquisition of the conidia by inhalation, the lungs are primarily involved. In disseminated infection, the reticuloendothelial system (RES) is most frequently involved. The fungus resides intracellularly in RES cells and may also rarely involve the thyroid gland; it may be isolated in fungemia. *H. duboisii* rarely involves the lungs but commonly involves the bones and skin; it is the causative agent of African histoplasmosis. Pulmonary infections have resulted from handling rhizomucor cultures. Collecting and processing soil samples from endemic areas have caused pulmonary infections in laboratory workers.

Encapsulated spores are resistant to drying and may remain viable for long periods of time. The small size of the infective conidia (< 5 μm) is conducive to airborne dispersal and intrapulmonary retention. Ten spores are almost as effective a lethal inoculum in mice as 10,000 to 100,000 spores. The infective stage of this dimorphic fungus (conidia) is present in sporulating mold form cultures and in soil from endemic areas. The yeast form is found in tissues or fluids from infected animals and may produce local infection following parenteral inoculation. Given the true systemic nature of histoplasmosis, otherwise healthy individuals are affected; however, dissemination and fatal course are more common in the immunocompromised and elderly. Chronic cavitary histoplasmosis is most commonly observed in individuals with underlying pulmonary disease.

1.7.22   *Leptosphaeria* (Ascomycete)

Leptosphaeria is an Ascomycete, and the asci contain eight spores. Among the many species, the most common ones are *L. senegalensis*, *L. tompkinsii*, and *L. coniothyrium*. The ascospores are colorless and vary in shape, being either oblong and ellipsoid or short and cylindrical. Transverse and longitudinal septa are lacking in some spores. A thin, gelatinous sheath browns with age. *Leptosphaeria* causes mycetoma, a clinical syndrome characterized by tumefaction, draining sinuses, and sclerotia (granules, grains). Localized infections involve cutaneous and subcutaneous tissue, fascia, and bone. Lesions consist of abscesses, granulomata, and draining sinuses. Following implantation of the etiologic agent, the primary lesion becomes locally invasive, indolent, and tumorlike. Patients may present with small, painless subcutaneous swellings. The lesions rupture, resulting in sinus tracts, swelling, and distortion of the infected body part. Sclerotia are present in pus and in tissue around the draining sinus tracts. Phaeohyphomycosis includes a group of mycotic infections characterized by the presence of dematiaceous (dark-walled) septate hyphae and sometimes yeast or a combination of both in tissue. Hyphae may be short to elongated, distorted, or swollen (toruloid hyphae) or regularly shaped, or any combination of the above. Yeast, when present, will be variable in size and most of the time will show budding. Infections of the eyes and skin are caused by the black fungi.

1.7.23   *Memnoniella*

Memnoniella is very closely related to *Stachybotrys*. The conidia are dark, globose, unicellular. *Memnoniella* occurs in long persistent chains. Conidiophores are dark and simple and bear phialide clusters at their apex. Mycotoxins produced include phenylspirodrimanes (griseofulvins).

1.7.24   *Mucor*

(See Figures 1.34 and 1.35 and Color Figures 21 and 22.) *Mucor* (from *mucor*, the Latin word for "fungus") is a zygomycete. This filamentous fungus may cause infections in humans, frogs, amphibians, cattle, and swine. Most *Mucor* are unable to grow at 37°C; the strains isolated from human infections are usually one of the few thermotolerant *Mucor* species. The genus *Mucor* contains

Sporangiospores

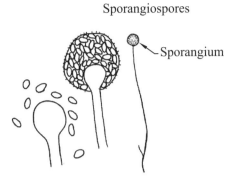

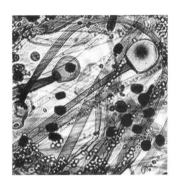

Figure 1.34 Structures of *Mucor.* (Courtesy of Deniese A. Chambers.)

Figure 1.35 Photomicrograph of *Mucor.* (Courtesy of University of Minnesota Environmental Health and Safety, Minneapolis.)

several species: *M. amphibiorum*, *M. circinelloides*, *M. hiemalis*, *M. indicus*, *M. plimbeus* (the cause of wood chip disease and furrier's lung), *M. racemosus*, and *M. ramosissimus*. Health effects include zygomycosis. Although the term *mucormycosis* has often been used for this syndrome, zygomycosis is now the preferred term for this angio-invasive disease. Patients with keto-acidosis resulting from diabetes mellitus, leukemia, and treatment with immunosuppressive drugs are susceptible to zygomycosis. Other health effects include mucocutaneous and rhinocerebral infections, septic arthritis, dialysis-associated peritonitis, renal infections, gastritis, and pulmonary infections. Predisposing infection factors include diabetic ketoacidosis, immunosuppression, des-feroxamine treatment, renal failure, extensive burns, and intravenous drug use. Infections produce vascular invasion that causes necrosis of the infected tissue and perineural invasion. (*Note:* Itraconazole prophylaxis in immunosuppressed patients may select the fungi in phylum Zygomycota as the cause of infections.)

1.7.25 *Nigrospora*

(See Color Figure 23.) *Nigrospora* is a filamentous dematiaceous fungus. Conidiophores forcibly discharge the conidia. *N. sphaerica* is the best-known species of the genus *Nigrospora*. The conidia are shiny black and vary from being unicellular egg shaped to flattened spheres. Produced singly, they have an equatorial colorless line or germ slit. *Nigrospora* has been isolated from cutaneous lesions of a leukemic patient and from a case with keratitis.

1.7.26 *Paecilomyces*

(See Color Figure 24.) *Paecilomyces* conidia are ovoid to fusoid, hyaline, composed of basipetal chains, unicellular, and 3–5 × 2–4 μm in size. The conidiophores look like *Penicillium*, except phialides taper to a long cylindrical neck.

Health effects include paecilomycosis, wood-trimmer's disease, pneumonia, allergic alveolitis, and humidifier-associated illnesses. *Paecilomyces* may produce arsine gas when growing on Paris green or other arsenic substrate.

1.7.27 *Penicillium* (Ascomycete)

(See Figure 1.36 and Color Figure 25.) *Penicillium* is an ascomycota. The perfect names are *Talaromcyes* and *Carpenteles*. Members of *Penicillium* subgenus *Penicillium* (with a few exceptions) may be indicative of indoor air quality problems. Teleomorphs are *Eupenicillium* and *Talaromyces*. *Penicillium* is a very large group of fungi valued as producers of antibiotics. *Penicillium* is

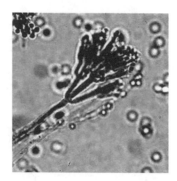

Figure 1.36 Photomicrograph of *Penicillium*. (Courtesy of University of Minnesota Environmental Health and Safety.)

commonly found in the soil, in the air, and on living vegetation, seeds, grains, and animals, as well as on wet insulation.

Penicillium is a fungus that grows when moisture, food, and just the right temperatures are available. Conidiophores arise from the mycelium as single, pencilate structures that end in a group of phialides. The spherical spores of *Penicillium* are produced in long, unbranched chains on each conidiophore. These usually fragment into individual spores, although chains of spores are seen periodically on slides. Although some species of *Penicillium* appear to reproduce solely by asexual means, some species of *Penicillium* are the anamorph (asexual) stage of the ascomycete genus *Talaromyces*.

Penicillium aurantiogriseum conidia are broadly ellipsoidal and gray-green with yellow bottoms. Conidia size range is 3.5 to 4.2 μm. *P. camemberti* conidia are hyaline or gray-green with light yellow bottoms and range in size from 3.2 to 5.0 μm. The mycelium starts out as white and may turn yellow or pink as aging occurs. *P. chrysogenum* (see Color Figure 26) conidia are blue to dark green; the conidia range in size from 2.3 to 4.0 μm. Colonies have an aromatic odor. *P. expansum* conidia are blue to dark green. Conidia range in size from 2.3 to 3.5 μm; colonies have an aromatic odor. *P. glabrum* conidia are globose, dull to dark green, and smooth to rough; the conidia range in size from 2.8 to 3.8 μm. *P. polonicum* conidia are broadly ellipsoidal and blue-green with yellow bottoms. Conidia range in size from 2.8 to 4.2 μm. *P. roqueforti* conidia are broadly globose and green; the conidia range in size from 3.5 to 4.5 μm.

Penicillium (from *penicillus*, the Latin word for "artist's touch") is the source of antibiotic lines that have aided humanity. *Penicillium*, as produced by Alexander Fleming in 1929, was a product of *P. notatum*. Since that time, other species of *Penicillium* have been used to create other antibiotics. As an example, griseofulvin is an antifungal antibiotic formed from a species of *Penicillium*. Not all species of *Penicillium*, however, are helpful. The route of entry into the body for invasive *Penicillium* infections is unknown; however, the respiratory route is used by many other fungi with abundant conidia. *Penicillium* may have abundant conidia, thus the respiratory route of entry is expected. Skin trauma has been associated with local infection but not with systemic disease. Infection via the digestive route is unusual for filamentous fungi.

Some *Penicillium* cause allergic reactions and other adverse health effects when dispersed through indoor air. *Penicillium* has been associated with hypersensitivity pneumonitis in some individuals when present in high concentrations. It is a rare agent of infection in cases of keratitis, peritonitis, and systemic disease. *P. marneffei* is the major pathogen in the genus and is the etiologic agent of penicilliosis marneffeii, a systemic disease in immunocompromised hosts. *P. marneffei* first proliferates in the reticuloendothelial system and then is disseminated. Lung and liver are usually the most severely involved organs. Other commonly involved areas include skin, bone marrow, intestine, spleen, kidney, lymph node, and tonsils. The reticuloendothelial system is made up of special cells called "phagocytes" located throughout the body; they can be found in the liver, spleen, bone marrow, brain, spinal cord, and lungs. When functioning correctly, phagocytes destroy disease-causing organisms by ingesting the organisms. An example of these cells are histiocytes, which try to ingest and kill *P. marneffei*. Unfortunately, when *P. marneffei* is not killed, the histiocytes carry the infection throughout the body. *P. marneffei* has two life formations and is the only *Penicillium* species that is dimorphic. The prevalence of one form over another is dependent on temperature. At 37°C, the fungus grows as yeasts forming white-to-tan, soft, or convoluted colonies. Microscopically, the yeasts are spherical or oval and divide by fission rather than budding. At 25°C, the fungus produces a fast-growing, grayish floccose colony. Microscopic examination reveals septate branching hyphae with lateral and terminal conidiophores that produce unbranched, broomlike chains of oval conidia. *Penicillium aurantio-griseum* metabolites include nephrotoxic glycopeptides (cause kidney karyomegaly), penicillic acid, terpenoid volatiles, and verrucosidin.

Pencillium camemberti metabolites include cyclopaldic acid, palitantin, fumigaclavine A and B, and rugulovasine A and B (which are produced by the wild-type *P. commune*). Cyclopaldic acid is produced by both types, thus the recommendation for storing white cheeses containing *P. camemberti* in a refrigerator at below 8°C.

Penicillium chrysogenum metabolites include chrysogine, meleagrin, penicillin, and roquefortine. The original name for *P. chrysogenum* was *P. notatum* — the original producer of penicillin. Since then, various mutations have been used in the production of penicillin F and G. *P. chrysogenum* can be found in buildings.

Penicillium expansum metabolites include chaetoglobsin C, citrinin (nephrotoxic), expansolide, patulin (affects phagocytes and is not degraded by pasteurization), and roquefortine C.

Penicillium glabrum metabolites include citromycetin and hepatotoxin. This species used to be called *Citromyces glaber*, in reference to its use in the production of citric acid.

Penicillium polonicum metabolites include anacine, nephrotoxic glycopeptides (which cause kidney karyomegaly), 3-methoxyviridicatin, penicillic acid, terpenoid volatiles, and verrucosidin.

Penicillium roqueforti metabolites include isofumigaclavine A and B, mycophenolic acid, and roquefortine C.

Mycotoxins produced include ochratoxin (which damages kidneys and liver and is a suspected carcinogen), citrinin (which causes renal damage, vasodilation, and bronchial constriction), glotoxin (an immunosuppressive toxin), and patulin (which causes hemorrhage in the brain and lung and extrinsic asthma).

1.7.28 *Phoma*

(See Figure 1.37.) *Phoma* grows on and discolors paint (pink, purple, red spots). Some have teleomorphs within *Leptosphaeria*. Dark colonies have pycindia (fruiting bodies) that contain one-celled spores. Health effects include mycetoma, phaeohyphomycosis (subcutaneous diseases), and keratitis.

1.7.29 *Pithomyces (Fungi imperfecti)*

The sexual state for this mold is not known; consequently, this genus is classified as *Fungi imperfecti*. The obsolete genus name *Scheleobrachea* was a synonym. *Pithomyces* is a dematiaceous (dark-walled) fungus. Spores are produced at the vegetative filament short side branches. Spores contain two to several cells. *Pithomyces* produces a mycotoxin that is pathogenic and causes facial eczema and liver damage. *P. chartarum* is the most common species; its spores are longitudinal and septa transverse.

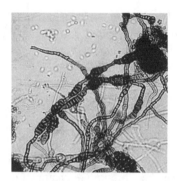

Figure 1.37 Photomicrograph of *Phoma*. (Courtesy of University of Minnesota Environmental Health and Safety, Minneapolis.)

1.7.30 *Rhinocladiella* (*Fungi imperfecti*)

The sexual state for this mold is not known; consequently, this genus is classified as *Fungi imperfecti*. *Rhinocladiella* is dematiaceous (dark-walled). The conidia are apical on new conidiophore growing points. It is subhyaline to dark, ovoid to oblong-ellipsoid, and mostly one celled. Conidiophores are simple or branched. Health effects include chromoblastomycosis (lesions, keloid formation), verrucous dermatitis, and brain abcess syndrome.

1.7.31 *Rhizomucor* (Zygomycete)

Rhizomucor has diminutive sporangiospores, rounded and thick walled, produced in globular sporangia at the hyphae tip. Health effects include allergenic and occupational allergy; mucorosis; lung, nasal sinus, brain, eye, and skin infection; zygomycosis (rhino-facial-cranial, lungs, gastrointestinal tract, skin); and vessel invasion that causes embolization and necrosis.

1.7.32 *Rhizopus*

(See Figure 1.38.) *Rhizopus* (from *rhiza*, the Greek word for "root," and *pous*, meaning "foot") *scopulariopsis* is a common species. Health effects include zygomycosis, blood vessel invasion leading to embolism and necrotic tissue, and sawmill lung (allergic aveolitis). Spores (sporangiospores) are rounded to oval, pale-brown, and ridged. *Rhizopus* ranges in size from 6.0 to 12.0 µm. Sporangia are shiny black.

1.7.33 *Rhodotorula*

Rhodotorula is a commonly isolated yeast, frequently from humidifiers and soil. Three common species are *R. glutinis*, *R. minuta*, and *R. rubra*. Health effects include a very rare pathogen reported in an endocarditis case and in a meningitis case. *Rhodotorula* may be allergenic to susceptible individuals when present in sufficient concentrations.

1.7.34 *Scendosporium*

Scendosporium hyphae are short or elongate, distorted or swollen. *Scendosporium* has many shapes. Health effects include phaeohyphomycosis and mycotic infections.

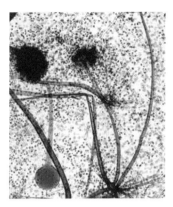

Figure 1.38 Photomicrograph of *Rhizopus*. (Courtesy of University of Minnesota Environmental Health and Safety, Minneapolis.

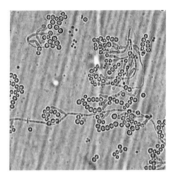

Figure 1.39 Photomicrograph of *Scopulariopsis*. (Courtesy of University of Minnesota Environmental Health and Safety, Minneapolis.)

1.7.35 *Scopulariopsis*

(See Figure 1.39.) The conidia of *Scopulariopsis* (from *scopula*, the Latin word for "broom") are globose to ovoid to flat-based, rose-brown to brown, and verrucose at maturity. Size range is from 5.0 to 8.0 µm. Health effects include infection of the skin and nails, occupational illness in the tobacco industry, and extrinsic allergic alveolitis. *Scopulariopsis* metabolites include ethylene and deacetoxycephalosporin C. *Scendosporium* may produce arsine gas when growing on Paris green paint or other arsenic substrate.

1.7.36 *Sporothrix*

Sporothrix conidia are globose to ovoid, unicellular, and hyaline. Apex of the condiophore bears conidia in a loose cluster, and it assumes various shapes. *S. schenckii* has caused a substantial number of local skin or eye infections in laboratory personnel. Most cases have been associated with accidents and have involved splashing culture material into the eye, scratching or injecting infected material into the skin, or being bitten by an experimentally infected animal. Skin infections have also resulted from handling cultures or necropsy of animals. No pulmonary infections as a result of laboratory exposure have been reported, although naturally occurring lung disease is thought to result from inhalation. Health effects include sporotrichosis.

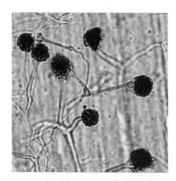

Figure 1.40 Photomicrograph of *Stachybotrys*. (Courtesy of University of Minnesota Environmental Health and Safety, Minneapolis.)

1.7.37 *Stachybotrys*

(See Figure 1.40 and Color Figures 27 through 29.) *Stachybotrys* (from *stachy*, the Greek word for "progeny," and *botrys*, meaning "bunch of grapes") *chartarum* is a greenish black fungus that grows on material with a high cellulose and low nitrogen content (such as fiberboard, gypsum board, lumber, ceiling tile, paper, dust, and lint) that becomes chronically moist or water damaged due to excessive humidity, water leaks, condensation, water infiltration, or flooding. Filamentous *Stachybotrys* is a type of slow-growing saprophytic fungus that is common in the northwestern United States. It grows in outdoor areas where the relative humidity is above 55%. The spores are brownish. Mycelial mats are pigmented dark olive-gray and appear to be a slimy mass with smooth margins; they may have either smooth or ridged surfaces. When the growth sporulates, the colony may appear to have a powdery surface.

About 15 species of *Stachybotrys* are known, and these have a worldwide distribution. Similar to various genera of filamentous fungi, *Stachybotrys* produces trichothecene mycotoxins (macrocyclic and trichoverroid trichothecenes) and satratoxins. Trichothecenes are potent inhibitors of DNA, RNA, and protein synthesis. They modulate inflammatory reactions, alter alveolar surfactant phospholipid concentrations, and may be acquired by ingestion of food products contaminated with the fungus or by direct inhalation of the spores.

Stachybotrys produces a hemolysin, stachylysin, which lyses sheep erythrocytes. The pathogenicity of *Stachybotrys* was first observed in cattle and horses in Russia in 1920. Stomatitis, rhinitis, conjunctivitis, pancytopenia, and neurological disorders developed in animals following ingestion of hay contaminated with *Stachybotrys*, and this syndrome was named stachybotrytoxicosis. This outbreak was the first to draw attention to *Stachybotrys* and its toxins.

The health effects of *Stachybotrys chartarum* were first noted in Russian and Eastern European farm animals that ate moldy hay. Much less is known about *S. chartarum* in indoor environments, such as homes or office buildings. If *S. chartarum* spores are released into the air, humans may develop symptoms such as coughing, wheezing, runny nose, irritated eyes or throat, skin rash, or diarrhea. Persons with chronic exposure to the toxin report cold- or flu-like symptoms with sore throat, diarrhea, headache, fatigue, dermatitis, intermittent local hair loss, and general malaise. The toxins may also suppress the immune system.

1.7.38 *Stemphylium*

Stemphylium is a diurnal sporulator. An alternating light and dark cycle is necessary for spore development. *Stemphylium* requires ultraviolet light for the production of conidiophores. The second developmental phase, when the conidia are produced, requires a dark period.

Stemphylium also requires wet conditions for growth. Spores vary in shape and size and range from 23 to 75 μm in length. The conidia are dark and variable in shape (globose, broadly ellipsoid, ovoid). Cross and longitudinal septa are observed and the major septum may be constricted.

1.7.39 Sterile Fungi

Sterile fungi are common to both outdoor and indoor air. These fungi produce vegetative growth but yield no spores for identification. Because these fungi are derived from ascospores or basidiospores, the spores of which are likely to be allergenic, these fungi should be considered allergenic.

1.7.40 *Streptomyces*

Streptomyces colonies are slow growing, aerobic, Gram-positive, non-acid-fast, glabrous or chalky, heaped and folded, and white, tan, gray, brown, or black in color. They have an earthy odor and cause white grain mycetoma.

1.7.41 *Stysanus*

The sexual state for this mold is not known; consequently, this genus is classified as *Fungi imperfecti*. The obsolete genus name *Cephalotrichum* was a synonym. *Stysanus* is dematiaceous (dark-walled). *Cephalotrichum stemonitis* (obsolete species name) is a synonym of *Doratomyces stemonitis*.

1.7.42 *Torula*

Torula conidia can be either one or several celled. They are dark and round and occur in chains. Health effects include allergenic responses.

1.7.43 *Trichoderma*

(See Figures 1.41 and 1.42 and Color Figure 30.) *Trichoderma* (from *trix*, the Greek word for "hair," and *derma*, meaning "skin") includes about 20 species, the most common being *T. viride*. *Trichoderma* causes lung infections. Skin, hair, and nail infections by these dermatophytid molds are among the most prevalent of human infections. Agents are present in the skin, hair, and nails of human and animal hosts. Contact with infected animals with inapparent or apparent infections is the primary hazard. Cultures and clinical materials are not an important source of human infection.

1.7.44 *Trichophyton*

Three genera of molds that contain this dermatophytosis-causing species are *Epidermophyton*, *Trichophyton*, and *Microsporum*. These dermatophyte molds are keratinophilic (grow on keratin) on a living host. Feathers, animal hooves, hair, and animal skin all can be used as substrates for the keratinophilic fungi. These agents are associated with the tineas, a series of named diseases that use Latin binomials for their naming structure. Tinea capitis is a fungal infection of the scalp due to dermatophytes such as *T. rubrum*. Hair of the head, eyebrows, and eyelashes may be involved. Endothrix infection begins by penetration of the hair; the organism then grows up the interior main axis of the hair where it fragments into arthroconidia. The typical causative agent is *T. tonsurans*.

Flavus (tinea flavus, a form of infection, as well as the clinical presentation of scutula) is the exception to the rule about arthroconidia formation. The typical causative agent is *T. schoenleinii*, which causes an endothrix-style growth but without the arthrocondia. Channels are formed within

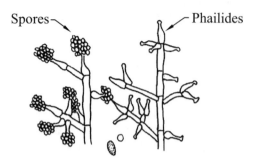

Spores ⌐   ⌐ Phailides

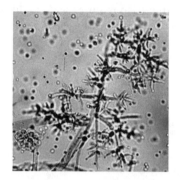

Figure 1.41 Structures of *Trichoderma*. (Courtesy of Deniese A. Chambers.)

Figure 1.42 Photomicrograph of *Trichoderma*. (Courtesy of University of Minnesota Environmental Health and Safety, Minneapolis.)

the hair shaft. This is useful diagnostically, as air bubbles move along these channels when an infected hair is immersed in a liquid. Ectothrix infection begins as endothrix and then extends back out through the hair cuticle (the outer wall of the hair) and forms a mass of arthroconidia both within and around the hair shift. The typical causative agent is *Microsporum canis*. Cases caused by *M. canis* occur sporadically and are acquired from puppies and kittens.

Tinea favosa is usually considered a variety of tinea capitis involving the scalp; however, this mycotic infection may also involve glabrous skin and nails. It is characterized by dense masses of mycelium. Epithelial debris forming yellowish cup-shaped crusts is called scutula, which develop at the surface of a hair follicle with the shaft in the center of the raised lesion. Removal of these crusts reveals an oozing, moist, red base. After a period of years, atrophy of the skin occurs, leaving a cicatricial alopecia and scarring. Scutula may be formed on the scalp or the glabrous skin.

Tinea corporis involves the glabrous (relatively hairless) skin. Infection is limited to the stratum corneum of the epidermis. Vellus hair (the fine hair present on glabrous skin) may be invaded, and the hair follicle may serve as a reservoir for the fungus

Tinea pedis, tinea manuum, and tinea cruris refer to tinea corporis that is limited to the foot, hand, and groin, respectively.

Tinea imbricata is an unusual form of tinea corporis caused by *Trichophyton concentricum*. It is characterized by ring-like growth in overlapping circles that may have an autosomal dominant genetic predisposition.

Tinea barbae is colonization of the bearded areas of the face and neck, hence is restricted to adult males. Most common causes are *Trichophyton mentagrophytes* and *T. verurucosum*. Zoophilic organisms are acquired from contact with cattle, dogs and other animals, which is the reason why tinea barbae is classically seen among dairy farmers and cattle ranchers. However, where antiseptic techniques are not used, person-to-person transmission can occur by being shaved in a barbershop. The superficial variety causes diffuse erythema and perifollicular papules and pustules that look exactly like bacterial folliculitis. Hair of the area may be affected with endothrix invasion that causes brittle, lusterless hair. Cases of the inflammatory type cause unilateral involvement of the chin, neck, or maxillary area with upper lip sparing. Nodular lesions cover with crusts, and the seropurulent material ends up having an abscess-like appearance. Hairs become loose and brittle. Permanent alopecia and important scaring may be the final consequences.

Tinea nigra is a superficial fungal infection of the stratum corneum and is a rare condition. Direct inoculation onto the skin from contact with decaying vegetation, wood, or soil seems to be the method of acquisition. Incubation periods may be as long as 20 years. The disease is asymptomatic in most cases and may be associated with pruritus. Brown to black nonscaly macules have well-defined borders that resemble silver-nitrate stains. They are unique or multiple, rounded or irregularly shaped. Size varies between 1 mm and 1.5 cm. Palmar surfaces are most often affected with tinea nigra palmaris, but lesions may occur on the soles with tinea nigra plantaris and other surfaces of the skin. Macules showing an uneven rate of spread and/or coalescence raise the suspicion of melanocytic nevi, junctional nevi, or melanoma. Other differential diagnoses include Addison's disease, pinta, stains from chemicals or dyes, and syphilis.

In the case of tinea unguium, when the infection is due to a dermatophyte, both ringworm of the nail and tinea unguium are sometimes used as synonyms for onychomycosis. The term *tinea unguium* is used specifically to describe invasive dermatophytic onychomycosis, such as invasion of the nail plate by a fungus. Infection may be due to a dermatophyte, yeast, or nondermatophyte mold.

1.7.45 *Ulocladium*

(See Figure 1.43 and Color Figures 31 and 32.) Among the nine species of *Ulocladium*, the most common one is *U. consortiale*, which contributes to the allergic burden of *Alternaria*-sensitive people. The conidia are solitary or in chains and are short and ellipsoidal with transverse and longitudinal septa.

1.7.46 *Zygosporium*

Zygosporium species comprise a small proportion of the fungal biota.

1.8 PRIONS

Prions are proteinaceous infectious particles that lack nucleic acids. Prions are composed largely, if not entirely, of an abnormal isoform of a normal cellular protein. In mammals, prions are composed of an abnormal, pathogenic isoform of the prion protein (PrP), designated PrP^{Sc}. The "Sc" superscript was initially derived from the word *scrapie* because scrapie is the prototypic prion disease. Because all of the known prion diseases of mammals involve aberrant metabolism of PrP similar to that observed in scrapie, use of the "Sc" superscript has been suggested by the Centers for Disease Control and Prevention (CDC) for all abnormal, pathogenic PrP isoforms. In this context, the "Sc" superscript is used to designate the scrapie-like isoform of PrP.

A chromosomal gene encodes PrP. No PrP genes are found in purified preparations of prions. PrP^{Sc} is derived from PrPC (the cellular isoform of PrP) by a posttranslational process whereby PrP^{Sc} acquires a high-sheet content. Neither prion-specific nucleic acids nor virus-like particles have been detected in purified, infectious preparations. In fungi, evidence for three different prions has

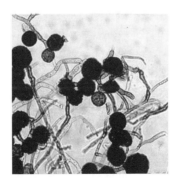

Figure 1.43 Photomicrograph of *Ulocladium*. (Courtesy of University of Minnesota Environmental Health and Safety, Minneapolis.)

been accumulated. The mammalian prions cause scrapie and other related neurodegenerative diseases of humans and animals. The prion diseases are also referred to as transmissible spongiform encephalopathies (TSEs).

1.9 VIRUSES

Viruses are obligatory intracellular parasites that contain either DNA or RNA (one or the other; never both). They have a protein coat that is sometimes enclosed by a lipid envelope. They multiply inside living cells. The viral DNA or RNA directs the cell to produce the chemicals that the viruses need. Essentially, the virus takes over the metabolic activity of the cell. The viruses can also direct the cells to produce structures to be used to transfer the viral DNA or RNA to other cells. In essence, the virus is alive only when it infects. Outside of living cells, the virus is inert. The only characteristics of life that the virus is capable of manifesting are the ability to multiply, direct processes in the cells that are infected, and mutate.

The essential viability of the virus in the hostile environment outside the cell is time marked. The smallest numbers of viruses detectable in cell cultures, the most sensitive hosts for many viruses, may be sufficient to infect susceptible individuals who consume them. Enteroviruses (polioviruses, coxsackieviruses [groups A and B], echoviruses, and hepatitis A virus), rotaviruses and other reoviruses (Reoviridae), adenoviruses, and Norwalk-type agents — a total of more than 100 different serological types — constitute the major enteric virus complement of human origin. Most of these viruses have been detected in sewage and in receiving waters over the years. Members of other virus groups have been recovered from human feces and urine, but none has been reported with great frequency or in large numbers in sewage or in receiving waters. Viruses of non-human sources abound in environmental waters. Some of these viruses, such as reoviruses, may infect humans; the significance of certain other viruses derived from non-human sources is as yet undetermined.

The numbers of viruses detected per liter of sewage range from less than 100 infective units to more than 100,000 infective units. In temperate climates, the numbers generally increase in the warmer months and decrease in the colder months, reflecting overall infection and excretion patterns in the community.

A number of techniques have been developed for recovering viruses from waters. The techniques include filter adsorption/elution, glass powder adsorption/elution, ultrafiltration, polyelectrolyte adsorption, aluminum hydroxide adsorption, protamine precipitation, hydroextraction, two-phase separation, organic flocculation, and alginate membrane filtration.

Because viruses do not multiply outside of susceptible living cells, dilution in hostile receiving waters and the toll of time eventually reduce the numbers of viruses to levels often barely detectable by the best techniques available, even when 1000-L quantities of water are tested. The numbers of viruses we now detect in environmental waters are probably an order of magnitude or more below the quantities actually present there. The numbers of viruses that reach recreational waters and intakes downstream of outfalls may thus be very large indeed. Detection varies given that:

1. Methods currently in use to concentrate viruses have low efficiencies.
2. Cell culture systems used for detecting viruses are usually sensitive to less than half of the virus types excreted by humans.
3. The plaque procedure usually used for detecting and quantifying viruses is relatively inefficient.
4. Viruses are present that have not yet been detected and identified.

In the publication *Human Viruses in Water, Wastewater and Soil* (Technical Report Series 639, World Health Organization Scientific Group, Geneva, Switzerland, 1979), the World Health Organization concluded that, while bacterial contamination of water and soils and the associated health risks have been thoroughly studied, attention is now increasingly being focused on the hazards associated with virus contamination of water. The contamination of water and soil by wastewater and human feces containing enteric viruses may pose public health problems. This problem occurs in areas of the world where the major waterborne bacterial diseases have been brought under control.

Over 100 different types of enteric viruses, all considered pathogenic to humans, have been identified. Concentrations in wastewater may reach 10,000 to 100,000/L. They have the ability to survive for months in water and in soil. The ingestion of a single infectious unit can lead to infection in a certain proportion of susceptible humans, and constant exposure of large population groups to even relatively small numbers of enteric viruses in large volumes of water can lead to an endemic state of virus dissemination in the community, which can and should be prevented. Bacteria are used as conventional indicators to evaluate the safety of potable water supplies. Enteric viruses may be present in water that manifests little or no sign of bacterial pollution. Viruses have been detected in the drinking-water supply systems of a number of cities, despite the fact that these supplies have received conventional water treatment, including filtration and disinfection, which are considered adequate for protection against bacterial pathogens. Viruses present in wastewater and sludge applied to land for irrigation, fertilization, or disposal purposes can survive in soil for periods of weeks or even months. Edible crops, contaminated by either contact with virus-laden soil or wastewater sprinkler irrigation, can harbor viruses for sufficient periods of time to survive harvesting and marketing; thus, their eventual consumption constitutes a potential health risk. Only limited data are available on the health risks resulting from the dispersion of viruses in aerosols created by sewage treatment and land disposal systems.

Ongoing virus control development has since resulted from these WHO recommendations. In light of the greater resistance of many enteric viruses to disinfection and other treatment processes compared to that of bacteria utilized as pollution indicators, drinking water derived from virus-contaminated sources should be treated by methods of proven high efficiency for removing or inactivating viruses, not only bacteria. Because of the ability of viruses to survive for long periods in seawater, it is recommended that coastal bathing and shellfish-growing areas be protected from contamination by wastewater and sludge. Virus monitoring of these areas is a desirable measure. Control procedures should be instituted in all situations in which wastewater or sludge is used for irrigation or fertilization to prevent the contamination of vegetables and fruits which are to be eaten raw. (Moreover, even though they may eventually be cooked, contaminated raw vegetables are likely to pollute other food in the kitchen.) Where it is planned to irrigate such crops or where sprinkler irrigation is to be used near populated areas, the effluent should be treated so that it reaches a high microbiological quality approaching that of drinking water. Because the factors that influence the movement of viruses in soil are still not fully understood, and because effluent and soil conditions vary so greatly, caution should be exercised if wastewater irrigation or land disposal takes place in the vicinity of wells supplying drinking water. Careful study of local conditions is required, and the cautious siting of such wells and routine virological monitoring of the water are advised as safety measures.

Further research into the health risks associated with viruses in water and soil is necessary. These studies should include the development and evaluation of methods of detecting viruses and alternative indicators of virus pollution (e.g., phages) and the improvement of treatment methods for the inactivation and removal of viruses from water and wastewater. The dissemination and survival of viruses in the natural environment should also be investigated. A standard method should be developed for the concentration and detection of viruses in large volumes of drinking water (e.g., 100 to 1000 L) based on a full evaluation in different laboratories of current techniques. Such an attempt would facilitate the development of virus-monitoring programs and would ensure a maximum degree of comparability of results. A laboratory quality-control system should be developed to enable participating laboratories to standardize their procedures.

These issues associated with virus transmittal and propagation in the environment continue to be a source of concern. The viruses that live in organisms and then are present in water streams, aerosolized droplet contaminant sources, and soils remain a concern to be addressed by future research efforts. (USEPA, 1984)

BIBLIOGRAPHY

Al-Doory, Y. and Domson, J.F., *Mould Allergy*, Lea and Febiger, Philadelphia, 1984.

Amdur, M.O., Doull, J., and Klaasen, C.D., Eds., *Cassarett and Doull's Toxicology — The Basic Science of Poisons*, 4th ed., Pergamon Press, New York, 1991.

Arora, D.K., Ajello, L., and Mukerji, K.G., Eds., *Humans, Animals, and Insects, Handbook of Applied Mycology*, Vol. 2, Marcel Dekker, New York, 1991.

Betina, V., *Mycotoxins: Chemical, Biological and Environmental Aspects*, Elsevier, Amsterdam, 1989.

Biological Contaminants in Indoor Environments, ASTM STP 1071, Morey, P.R., Feeley, J.C., Sr., and Otten, J.A., Eds., American Society for Testing and Materials, Philadelphia, 1990.

Burge, H.A. Fungus allergens, *Clin. Rev. Allergy*, 3, 319–329, 1985. Burge, H.A. *The Fungi, Biological Contaminants in Indoor Environments*, ASTM STP 1071, Morey, P.R., Feeley, J.C., Sr., and Otten, J.A., Eds., American Society for Testing and Materials, Philadelphia, 1990.

Crissy, J.T., Lang, H., and Parish, L.C., *Manual of Medical Mycology*, Blackwell Sciences, Cambridge, MA, 1995.

Fassatiova, O., *Moulds and Filamentous Fungi in Technical Microbiology: Progress in Industrial Microbiology*, Vol. 22, Elsevier, New York, 1986.

Flannigan, B., McCabe, E.M., and McGarry, F., Allergenic and toxigenic micro-organisms in houses, *J. Appl. Bacteriol. Symp.* 70(Suppl.), 61S–73S, 1991.

Funder, S. and Broggers, B., *Practical Mycology — Manual for Identification of Fungi*, Forlag, Oslo, Norway, 1953.

Godish, T., *Indoor Air Pollution Control*, Lewis Publishers, Chelsea, MI, 1989.

Grant, C., Hunter, C.A., Flannigan, B., and Bravery, A.F., The moisture requirements of moulds isolated from domestic dwellings, *Int. Biodeterioration*, 25, 259–284, 1988.

Ingold, C.T. and Hudson, H.J., *The Biology of Fungi*, 6th ed., Chapman & Hall, London, 1993.

Kindrick, B., *The Fifth Kingdom*, 2nd ed., Mycologue/Focus Texts, Newburyport, MA, 1992.

Kozak, P.P., Gallup, J., Cummins, L.H., and Gilman, S.A., Factors of importance in determining the prevalence of indoor molds, *Annu. Allergy*, 43, 88–94, 1979.

Larone, D.H., *Medically Important Fungi — A Guide to Identification*, 2nd ed., American Society for Microbiology, Washington, D.C., 1995.

Marasas, W.F.O. and Nelson, P. E., *Mycotoxicology — Introduction to the Mycology, Plant Pathology, Chemistry, Toxicology, and Pathology of Naturally Occurring Mycotoxicoses in Animals and Man*, Pennsylvania State University Press, University Park, 1987.

Morey, P., Foarde, K., Klees, J., and Streifel, A., Microbiological Contaminants, IAQ '93 Operating and Maintaining Buildings for Health, Comfort and Productivity, ASHRAE Conference, Philadelphia, Nov. 8–10, 1993.

National Research Council (NRC), Indoor Pollutants, National Academy Press, Washington, D.C., 1981.

Onion, A.H.S., Allsopp, D., and Eggins, H.O.W., *Smith's Introduction to Industrial Mycology*, 7th ed., Edward Arnold Ltd., London, 1981.

Rippon, J.W., *Medical Mycology — The Pathogenic Fungi and the Pathogenic Actinomycetes*, 3rd ed., W.B. Saunders, Philadelphia, 1988.

Sax, N.I. and Lewis, R.J., Sr., *Dangerous Properties of Industrial Materials*, 7th ed., Van Nostrand Reinhold, New York, 1989.

Sharma, R.P. and Salunkhe, D.K., Eds., *Mycotoxins and Phytoalexins*, CRC Press, Boca Raton, FL, 1991.

Smith, J. and Moss, M.O., *Mycotoxin Formation, Analyses and Significance*, John Wiley & Son, New York, 1985.

Volk, T., Fungi http://www.botany.wisc.edu/ Department of Biology, University of Wisconsin-La Crosse, 1995–2002.

Wilken-Jensen, K. and Gravesen, S., Eds., *Atlas of Moulds in Europe Causing Respiratory Allergy*, ASK Publishing, Denmark, 1984.

CHAPTER 2

Industrial Hygiene Sampling

Dennis W. Day, Martha J. Boss, R. Vincent Miller, and Chris Wrenn

CONTENTS

Sampling may not always be required in order to determine biological risk. Still, knowledge of sampling methods available will enable investigators and concerned parties to sample when required. Sampling includes both sampling for infective agents and sampling to determine other ambient air conditions.

2.1 SAMPLING: BIOLOGICALS AND GENERAL AIR QUALITY

Investigative sampling may be needed to determine biological quantitative levels caused by an amplifier that is problematic but difficult to identify visually. Examples include mold growth in heating, ventilation, and air conditioning (HVAC) systems and behind walls. Common indications for characterization and remediation of the discovered amplifiers are:

- Occurrence of symptoms consistent with adverse reaction to indoor molds
- Building management or administrative concerns that the amplifiers might cause symptoms in the future
- Indications of exacerbate materials degradation
- Noxious odors
- Cosmetic, esthetic, psychological, or political disadvantages of conspicuous decay

Sampling may be required before, during, and after decontamination efforts. Acceptable sampling methods and contamination levels must be determined by a competent person prior to the onset of either investigative or decontamination events. A sampling plan should be developed and reviewed by all parties.

Biocide application, if required, may also require sampling to determine the airborne, residual, and contact levels of the biocides. When pH-altering chemicals are used, sampling to determine the persistence of the pH-altering chemicals may be needed.

Only experienced samplers should be assigned this type of work. Experience may be through documented training with supervised on-site work (initially) or by virtue of education. All standard operating procedures (SOPs) should be reviewed with the project team. If field conditions warrant, the project team leader should make the decisions regarding any alteration in SOPs.

2.1.1   Regulatory and Industry Guidance Reviews

Because some states, including New York and California, now have or are developing regulatory requirements or guidelines for certain biological agents, a regulatory review is appropriate. Other state and local government agencies are rapidly developing acceptable criteria statements for indoor air investigations, including those that involve biological contaminants.

2.1.2   Sampling Scope

Sampling for aerosolized biologicals usually should be done in coordination with surface contact sampling. Photographs of investigative locations, sampling events, and incubated cultures may be included in the report. The sampling report may include only interpretation of the sampling results, with no conclusions as to required follow-up activities. Other reports include both the sampling information and such conclusions as the need for decontamination, engineering analysis, or rehabilitation of building areas. The scope of work must clearly define the ultimate report expectations and the distinction between these report types.

2.2 HEALTH AND SAFETY PRECAUTIONS

In some cases a site safety and health plan similar to that required by 29 CFR 1910.120 will be required. Uncontrolled biological risk agents are considered in the same way as uncontrolled chemical risk agents. When dealing with biological risk agents for which limited information as to human health risk is available, a risk assessment should be done incorporating all known human health information needed to adequately communicate hazards to samplers and to the client.

2.3 DATA GATHERING

The following sample data gathering routines may be required prior to any sampling:

- *Walk-through investigation.* The objective of this investigation is to note the current building status.
- *Interviews with affected parties (nonemployees).* An example would be clients or visitors to facilities.
- *Employee interviews.* Representative employees are made available for interview. These interviews are conducted to gather information regarding the interior building conditions and subsequent changes noted in the building interior environs. The company human resources office should provide each employee with company program documents related to the interviews for information gathering and release forms so that information gathered can be used in subsequent reports.

The purpose of these interviews is to gather historical information to compare with the current building status.

2.3.1 Document Review

Examples of documents to be reviewed include those associated with water leakage events, spill containment measures, maintenance activities, and employee complaints. This review will be used to refine the study criteria and project expectations and to provide the discussion and conclusion elements that address the impact of these occurrences given the current observed building status.

All available drawings, including as-built drawings and drawings that illustrate any renovation activities, should be provided, preferably electronically as computer-aided design drawings (CADDs), which can be generated in custom sizes. If paper copies are provided, half and quarter sizes are usually preferred to full-size drawings, as the samplers can more easily carry these from place to place during sampling efforts.

Maintenance records for the HVAC equipment, including those associated with cooling tower usage, should be reviewed. Water treatment chemical usage and results of testing to determine the effectiveness of this treatment should also be reviewed.

2.3.2 Sample Initial On-Site Investigation

The following is a sample statement of work for an investigation that will include interior air spaces and the HVAC system:

An initial meeting at the XYZ plant will be held to familiarize proposed project staff with the plant and processes. Company XYZ will provide plant escorts, who are authorized to provide access to all building system locations that are to be assessed. The plant escorts will accompany the investigative personnel.

In order to complete the initial investigation of the air handling units (AHUs), Company XYZ will provide facility maintenance personnel, who will disassemble the access panels on each AHU. Fans will be shut down and locked or tagged out by Company XYZ personnel. In the event that fans cannot be shut down during first shift activities, Company XYZ's facility maintenance personnel and the investigative team will conduct the air handling investigations after shift two. All equipment will be reassembled by Company XYZ personnel at the conclusion of the investigative effort for each air handling unit. The current assumption is that fourteen (14) AHUs will be inspected.

Ductwork associated with the AHUs is accessible through panels at seven (7) locations. These panels will be removed by Company XYZ facility maintenance personnel. All panels are located on the vertical face of the ductwork; prior to the removal of these panels, a sheet of 6-mil polyethylene (poly) will be placed by the contractor on the floor to receive these panels and any debris generated during panel removal. This poly sheet will be bagged by the investigative team for subsequent sanitary disposal by Company XYZ at the conclusion of the investigation for each ductwork area.

Flexible ductwork will be opened by removing the air outflow ceiling grids. Because these grids are located on the horizontal lower face of the duct, removal may thus cause debris spillage in the nearby area. The investigative team will position a poly sheet beneath each grid location, and the flexible ductwork grid and faceplate will be lowered into a receiving 6-mil poly bag during disassembly. Obviously soiled ductwork grids and faceplates will be cleaned or bagged for disposal and replaced by Company XYZ personnel. The investigative team will collect all poly sheeting and bags for sanitary disposal by Company XYZ.

Cooling water systems will be observed at the cooling tower rooftop locations, sump pump vaults, and in the water treatment area. No purging of the system will occur during this investigative effort. No valves will be opened in interior building locations. The investigative team will observe all system components that contain waters; however, plumbing lines that run within wall or floor interiors will not be examined. Current mechanical integrity will be determined based on visual evidence such as the presence of leaks and status of plumbing line materials; however, mechanical integrity will not be physically tested. All rooftop units are located in the center of a flat roof and no personnel will be within six (6) feet of a roof edge at anytime.

Documentation of the current status of these systems will be shown on the inspection checklists. All spaces to be entered have been evaluated as not being confined spaces or as no-permit-required confined spaces. In the event that permit-required confined spaces are identified and entry is required, the investigative team will initiate the protocols required by the Confined Space Program. These protocols include informing Company XYZ of a changed condition and the initiation of contract option one.

2.3.3 Sampling Areas

Air monitoring is accomplished in coordination with contact sampling, which may involve any of the following:

- Collecting bulk samples
- Collecting water samples
- Swabbing surfaces
- Applying agar plates to surfaces
- Vacuuming of small areas
- Tape sampling

All contact sampling requires consistent sampling techniques and, ultimately, either microscopic analysis or incubation of samples followed by microscopic analysis. Sampling events should document the following areas:

- Outdoors
- Indoor control areas assumed to be uncontaminated and at safe levels
- Indoor areas where contamination is suspected and where decontamination is required
- Ventilation systems suspected of either exhausting or supplying contaminated air to the indoor area
- Downwind of negative, high-efficiency particulate air (HEPA) filtration units used to exhaust decontamination areas
- Clean rooms used during decontamination for personnel decontamination access

Other areas may need to be sampled depending on the zoning of work and the activity in the building. Samples may be collected for total viable and nonviable airborne components. Summa canisters or sorbent tubes are used when microbe-produced volatiles are to be documented. Real-time instrumentation for microbial volatiles is also available that measures volatiles in the parts-per-billion range. Wall cavity checking may be required by probing and drawing air from wall cavity spaces, infrared photography, and/or moisture metering.

2.4 RECORDKEEPING

The daily monitoring log should contain the following information for each sample:

- Sampling and analytical method used
- Date sample collected
- Sample number
- Sample type
- Location/activity/name where sample collected

For air samples:

- Sampling pump manufacturer, model, and serial number; beginning flow rate; end flow rate; average flow rate (L/min)
- Calibration date, time, method, location, name of calibrator, signature
- Sample period (start time, stop time, elapsed time in minutes)
- Total air volume sampled (liters)
- Sample results
- Laboratory name, location, analytical method, analyst, and confidence level
- Printed name and a signature and date block for the industrial hygienist who conducted the sampling and for the certified industrial hygienist who reviewed the daily air monitoring log verifying the accuracy of the information

Sample results are time dependent. Thus, background sampling performed during a hot, humid day will not be consistent with interior results collected later during a rainstorm. Seasonal and climatic changes must be considered when comparing samples. The relative temperature and humidity should be recorded for the days on which sampling occurred, particularly if successive sampling days are required.

2.5 INDUSTRIAL HYGIENE MEASUREMENTS

In addition to the laboratory microscopy or incubation data, industrial hygiene measurements may need to be considered, particularly in the interpretation of air samples. Many industrial hygiene measurements are only understood in terms of ratios. The most common ratios are:

- Weight to weight
- Weight to volume
- Weight to area

To simplify matters, one (1) gravity is assumed and weight is commonly thought of as a measure of mass. So, the weight-to-weight ratio is expressed as grams or pounds. For solids and liquids that are assumed to be noncompressible, the weight-to-weight ratio makes sense. Gases are a state of matter that is very compressible and expandable, so weight-to-volume measures must be used.

In order to standardize air measures, the temperature and pressure measurements may be required. For normal temperature and pressure (NTP), the temperature is defined as approximately room temperature, 25°C (77°F). Industrial hygienists prefer NTP, while chemists prefer a cooler version known as standard temperature pressure (STP) for which the temperature is 21°C (70°F). Both NTP and STP use a pressure of one atmosphere, which is equivalent to 14.45 pounds per square inch (psi). One atmosphere is also described as 760 mmHg, because one atmosphere will push mercury up a barometer column 760 mm.

When measuring for biological risk, the following states and their measurement conventions must be considered:

- Gases expressed as a relative percentage of each to the total gases or to each other
- Liquids suspended in gases, mists, and vapors
- Liquids in liquids, mixtures, and miscible solutions
- Solids suspended in gases, fibers, fumes, dusts, and particulates
- Solids in liquids, mixtures, and emulsions
- Solids in solids — adsorption or absorption to particulates and mixtures

Mists are smaller than vapors. Essentially, mists are tiny droplets of liquid riding on a cushion of air. For solids in gases, various sizes of solids may ride on a cushion of air. Solids range in size, from smallest to largest, as fumes, dusts, and particulates.

For both liquids and solids in air, weight-to-volume measurements are used, and the most common units are milligrams per cubic meter (mg/m³). If we know the molecular weight of the compounds of interest, we can convert mg/m³ to parts per million (ppm), which is a weight-to-weight ratio. In some cases, calculating the volume in a weight-to-volume measurement is not possible. A good example of this is in wipe sampling. For wipe sampling, a standard area (usually 100 cm²) is wiped down. The contaminant from that area is suspended during analysis in a liquid. So, for this liquid suspension, a weight-to-volume and a weight-to-weight ratio can be obtained; however, this weight-to-volume ratio cannot be converted back to the original wipe sample contaminant load as weight to volume, because the only known dimension for the wipe sample is weight to area.

2.6 AEROSOLS

Sampling that involves air transmission vectors for biologicals is termed *bioaerosol sampling*. Aerosol dynamics must be understood in order to plan the appropriate sampling types and locations. The general properties of aerosols are presented in this section.

Aerosols are suspensions of solid or liquid particles in a gas (usually air). The particulate portion of an aerosol is referred to as particulate matter (PM). Particulate matter is a generic term applied to chemically heterogeneous discrete liquid droplets or solid particles. The size and electrostatic properties of an aerosol may determine its residency time in an airstream and subsequent availability for inhalation.

The metric unit used for describing PM is the micron or micrometer (μm; $1e^{-6}$ meters). The PM in an aerosol can range in size from 0.001 to > 100 μm in diameter. The following general information about particulates and any variance from this information should be considered when planning sampling routines:

- Visible particles constitute only about 10% of indoor air.
- Particle visibility depends on the eye itself — in other words, on the light intensity and quality, as well as background and particle type.
- Particles on furniture and those in a shaft of light are approximately 50 μm or larger.
- Particles as small as 10 μm may be seen using normal vision under favorable conditions.
- Approximately 98 to 99% of all particles by count are in the size range of 5 μm or less. These particles tend to remain in suspension or settle out so slowly that only quality electronic air cleaners and HEPA air cleaners are effective in removing these particles.
- The majority of harmful particles are 3 μm or less in size.
- Particles of 1 μm or less adhere to surfaces by molecular adhesion. Scrubbing is generally the only way to remove them.
- Larger particles tend to settle out of the atmosphere due to weight.
- Smaller, respirable particles remain virtually suspended in the air until breathed in.
- The average person breathes in about 16,000 quarts of air per day. Each quart contains some 70,000 visible and invisible particles. Thus, our lungs filter over a billion particles per day.
- The average home collects about 2 pounds of dust per week.
- A 9 × 12-ft carpet or rug will collect an average of about 10 pounds of dust per year.

2.6.1 Particulates

Particulates are generally categorized based on size:

- Coarse particles are > 2 μm in diameter.
- Fine particles are between 0.1 and 2 μm in diameter.
- Ultrafine particles are < 0.1 μm.

Most aerosol particles are *polydisperse* — they have a wide range of particle sizes that must be characterized by statistical measures. In some cases, such as for an inkjet printer, it is desirable to have a *monodisperse* aerosol with particles of equal size.

2.6.2 Solid-Particle Aerosols

Dust is formed by mechanical disintegration of a parent material; dust sizes range from less than 1 μm to visible. A fume is produced by the condensation of vapors or gaseous combustion products and are < 1 μm in size.

2.6.3 Liquid-Droplet Aerosols

Mist is formed by condensation or atomization; sizes range from < 1 μm to 20 μm. Fog is a visible mist with a high particle concentration.

2.6.4 Solid/Liquid Particle Aerosols

Smoke is a visible aerosol resulting from incomplete combustion and is < 1 μm in size. Smog is a photochemical reaction product, usually combined with water vapor, and is < 2 μm in size.

2.6.5 Suspended Particulate Matter

Suspended particulate matter (SPM) is a complex mixture of small and large particles of varying origin and chemical composition. PM10 particles range in size up to 10 μm in diameter, and PM2.5 particles range in size up to 2.5 μm in diameter. SPM varies in chemical composition, as particles can be made up of many components, including sulfates, nitrates, elemental carbon, organic compounds, metals, and soil dust. This variation in composition reflects the many sources of SPM.

2.7 AIR SAMPLING: METHODS FOR GENERAL PARTICULATES

Sampling for particulates requires that the particulates be filtered out or removed from the air stream by impaction (Figures 2.1 and 2.2). Because particulates that are suspended in the air stream come in many sizes, the first question is whether exposure standards are based on the respirable fraction or the total particulate levels. Total particulates are often analyzed by gravimetric methods.

2.7.1 Gravimetric Filter Weighing Procedure

The step-by-step procedure for weighing filters depends on the make and model of the balance. Consult the manufacturer's instruction book for directions. In addition, follow these guidelines:

- Smoking and/or eating must not take place in the weighing area; both generate extraneous particulate matter in the airstream.
- All filters are handled with tongs or tweezers; do not handle the filters with hands.

Figure 2.1 Filter used for particulate collection. (Courtesy of SKC, Inc., Eighty Four, PA.)   **Figure 2.2** Filter used for particulate collection. (Courtesy of SKC, Inc., Eighty Four, PA.)

- Desiccate all filters at least 24 hours before weighing and sampling, and change desiccant before the desiccant completely changes color (i.e., before the blue desiccant turns pink). Evacuate the desiccator with a sampling or vacuum pump.
- Zero the balance prior to use.
- Calibrate the balance prior to use and after every 10 samples.
- Immediately prior to placement on the balance, pass all filters over an ionization unit to remove static charges. (After 12 months of use, return the unit to the distributor for disposal.) When weighing the filter after sampling, desiccate first and include any loose material from an overloaded filter and cassette.
- Weigh all filters at least twice.
- If a difference of > 0.005 mg is found between the two weighings, zero the balance again, recalibrate, and reweigh.
- If a difference of < 0.005 mg is found between the two weighings, average the weights for the final weight.

Note: At all times take care not to exert downward pressure on the weighing pans, as such action may damage the weighing mechanism. When reassembling the cassette assembly, remember to add the unweighed backup pad. Record all the appropriate weighing information (in ink) in the weighing log.

2.7.2 Total Dust

Various filtration options for collecting particulates are available. Sampling options are defined based on the regulatory requirements and sampling environment. For example, one option is to collect total dust on a preweighed, low-ash polyvinylchloride (PVC) filter at a flow rate of about 2 liters per minute (L/min), depending on the rate required to prevent overloading, as evidenced by loose particulate in the filter cassette housing. The PVC filters are weighed before and after taking the sample. Personal sampling pumps must be calibrated before and after each day of sampling using a bubble meter method (electronic or mechanical) or the precision rotameter method (calibrated against a bubble meter) (Figure 2.3).

Figure 2.3   Placement of dust monitoring equipment for personnel sampling.

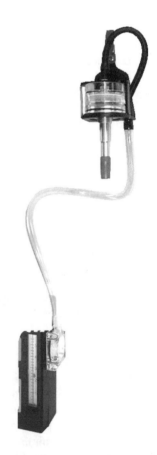

Figure 2.4   Cyclone adaptation for collection of respi-rable dust. (Courtesy of SKC, Inc., Eighty Four, PA.)

Figure 2.5   Cyclone adaptation for collection of respi-rable dust. (Courtesy of SKC, Inc., Eighty Four, PA.)

2.7.3 Respirable and Inhalable Dust

Respirable dust is the component of particulate in the airstream that may deposit within the gaseous exchange areas of the lung (see Figures 2.4 and 2.5). Respirable particles are just the right size to travel with inspired air down into the alveoli of the lung. Once in the alveoli, these particles may be a simple irritant or they may dissolve and provide chemicals in suspension with tissue fluids. These suspended chemicals are then available to exert toxic and carcinogenic effects (Figure 2.6).

Respirable dusts that do not go into solution pose another danger. These insoluble dusts and particulates and the fibers associated with respirable dusts are easy to breathe in; they move with ease deep into the lungs and once in the lungs may stay in the tissue bed forever.

For total particulate sampling results, usually the best guess is that 60% of the particles available in the airstream are ultimately respirable. The cut point for these particles is 50% at 4 μm, which means that 50% of the 4 μm particles are captured. When health effect and exposure limits are based on respirable dusts, either the 60% of total assumption must be made or special instrumentation must be used to segregate out just the respirable fraction of total dust. Inhalable dusts include all of those dusts from the general airstream that normal humans can bring into their respiratory tracts, which includes everything from the nose to the base of the lungs.

Inhalable dusts have a 50% cut point of 100 μm. Special inhalable-dust samplers are used to collect only inhalable dusts, and these samplers may vary as to the size of particulate collected. Cyclones of various types (aluminum, plastic) are used to collect respirable dust factions. Plastic cyclones are the only choice in acid gas-contaminated atmospheres. Respirable dust can be collected using a clean cyclone equipped with a preweighed low-ash PVC filter with a flow rate of 1.5 to 1.9 L/min.

2.7.4 Silica Respirable Dust: Cyclone Collection

When sporulated bacteria are purposely disbursed, chemicals such as silica may be added to facilitate the residence time of the bacteria spores in the air. Essentially the silica reduces the clumping sometimes associated with sporulated bacteria. Silica and other minerals also enhance the electrostatic neutrality necessary to potentiate airborne spore dispersal. Singular spores have the right size and electrostatic properties to float in the air and are thus available as respirable particulates. For these reasons, measurement of silica dust may be warranted as an indirect measure of disbursed sporulated bacteria.

Silica is collected only as a respirable dust. Aluminum cyclones are recommended to ensure that the cyclone material does not interact or become part of the sample; silica at sufficient velocity may etch a plastic cyclone. A bulk sample should also be submitted to provide a basis for comparing silica levels in stock to ultimate respirable levels of dust. All filters used must be weighed before and after sampling.

2.7.5 Direct-Reading Dust Monitors

Direct-reading dust monitors may be used to provide real-time data to predict room or area particulate loading. These instruments may also be necessary to quantitate respirator effectiveness and particulate loading in contained air spaces (such as those present within equipment housings), and for particulate shedding and component tests.

2.7.5.1 Condensation Nuclei Counters

Condensation nuclei counters are based upon a miniature, continuous-flow condensation nucleus counter that takes particles too small to be easily detected, enlarges them to a detectable size, and

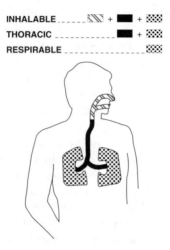

INHALABLE _____ ░ + ■ + ▨
THORACIC _____ ■ + ▨
RESPIRABLE _____ ▨

Figure 2.6   Schematic of lung areas. (Courtesy of SKC, Inc., Eight Four, PA.)

counts them. Submicrometer particles are grown to supermicrometer alcohol droplets by first saturating the particles with alcohol vapor. Particles in the sample pass through a saturator tube, where alcohol evaporates into the sample stream. The sample flow becomes saturated with alcohol vapor. The sample then passes into a cooled condenser tube, where the alcohol vapor supersaturates and condenses onto virtually all particles larger than 10 nm in diameter. The resulting droplets exit the condenser.

As the droplets pass through the sensing volume, the particles scatter the light. The light passes through a thin ribbon of laser light. It is then collected by an optical system and focused by the optical system. The scattered light is focused onto a photodetector or a photodiode, which generates an electrical pulse from each droplet. The concentration of particles is determined by counting the number of pulses generated. Individual airborne particles from sources such as smoke, dust, and exhaust fumes are counted. These instruments are insensitive to variations in size, shape, composition, and refractive index. The particle size and concentration range vary.

Isopropyl alcohol must be added to these units according to the manufacturer's instructions. Under normal conditions, a fully charged battery pack will last for about 5 hours of operation. Low battery packs should be charged for at least 6 hours, and battery packs should not be stored in a discharged condition.

2.7.5.2 Photodetection

Photodetectors operate on the principle of detecting scattered electromagnetic radiation in the near infrared and can be used to monitor total and respirable particulates. These devices measure the concentration of airborne particulates and aerosols, including dust, fumes, smoke, fog, and mist. Certain instruments have been designed to satisfy the requirements for intrinsically safe operation in methane–air mixtures. When the photodetector is not being operated, it is placed in the plastic storage bag, which should then be closed to minimize particle contamination of the inner surfaces of the sensing chamber.

After prolonged operation in or exposure to particulate-laden air, the interior walls and the two glass windows of the sensing chamber may become contaminated with particles. Repeated updating of the zero reference following the manufacturer's procedure will correct errors resulting from such particle accumulations; however, this contamination could affect the accuracy of the measurements as a result of excessive spurious scattering and significant attenuation to the radiation passing through the glass windows of the sensing chamber.

2.8 BIOLOGICALS: VIABLE VS. NONVIABLE

Mold spores are microscopic (2 to 10 μm) and are naturally present in both indoor and outdoor air. Some molds have spores that are easily disturbed and waft into the air and settle repeatedly with each disturbance. Other molds have sticky spores that will cling to surfaces and are dislodged by brushing against them or by other direct contact. Spores may remain capable of producing vegetative growth for years. In addition, whether or not the spores are alive, the allergens in and on them may remain allergenic for years. Analysis is based on recognizing the synecological assemblage of isolates consistent with the presence of indoor mold amplifiers. Locating and examining any mold amplifiers not detected in preliminary inspection is a logical follow-up step once sampling has revealed that these amplifiers are present.

2.9 MOLD SAMPLING: INDUSTRIAL HYGIENE PROTOCOLS

The following is a step-by-step procedure for mold sampling:

1. Assemble materials and equipment to be used. Segregate materials and equipment to be taken inside the building or area of concern. Use impermeable plastic bags whenever possible to containerize materials and equipment to be taken into the building. Do not use cardboard or other porous containers that cannot be readily decontaminated.
2. Mark each contact sample or strip agar blister pack with a unique sample number using a permanent marker (for example, a Sharpie® pen). Allow the ink to air dry before placing the blister pack in a plastic bag.
3. Use a large (quart or more), resealable, plastic freezer bag to package each contact sample or strip agar.
4. Assemble at least 10 of each type of sampling media in a large plastic bag. Package no more than 10 agar blister packs together for transfer to a contaminated area.
5. Assemble another bag to contain extra impermeable gloves (latex, 6-mil, or neoprene) and alcohol wipes. Alcohol wipes can be purchased in individual packets or made on site using paper towels and isopropyl alcohol. The alcohol-soaked paper towels are more effective for larger decontamination areas. Double bag all sources of alcohol and avoid direct alcohol contact with the agar blister packs.
6. Establish a staging area and set up a decontamination area in a biologically neutral location away from potential biological amplification sites.

Put on the personal protective equipment (PPE) in the following order:

1. First hooded Tyvek®
2. Boots
3. First and second layer of gloves (double gloving is optional in some situations), followed by duct taping boot and glove openings at the ankles and wrists (optional in some situations)
4. Respirator
5. Second hooded Tyvek® (optional in some situations)

Begin the sampling routine. Sample outside and in all assumed uncontaminated or amplified areas first. Then sample into progressively more contaminated areas. Use the same protocols for all sampling events, including the same pressure and motion when using contact plates and the same walking routines or static placement when using dynamic sampling devices carried with you.

2.9.1   Direct Detection

Procedures for the direct detection of mold amplifiers may be used either after an air sample has predicted the presence of amplifiers or as a preliminary survey. Common places where significant amplifiers can be visually identified are on, in, or under water-damaged walls; wallpaper or wallboard paper (whether painted over or not); backings of water-damaged carpets; HVAC coils, pans, and vanes; damp papers; walk-in refrigerators and incubators; windowsills; shower stalls; washroom fixtures; and moist organic materials, including any moist objects composed of cellulose. Exposed insulation may be visibly discolored with mold, as may the inner or outer surface of its covering paper. Paint, ceramic, grouting, or plastic may also support mold growth.

Tape, swab, contact, and grab samples may expose many normally settled elements that may not be significantly present in the air. Dust samples may reveal a long-term fungal deposition and negate air-current bioaerosol variability as seen in short-duration air samples.

2.9.2 Interior Wall Sampling

By making a small hole in wall planar surfaces or at baseboards, the interior wall area can be investigated. An air-sampling pump can be used to draw the air toward filter cassettes or mini-can Summa canisters.

2.9.3 Contact and Grab Sampling

The vial contact or specimen grab sampling routine (see Figure 2.7) typically is as follows:

1. Open the swab or contact plate blister pack over bag at first location to be sampled.
2. Sample mold by applying the contact plate to the area with gentle pressure or by swabbing. (*Note:* Obtaining a small sample of contaminated building (or other) material may also be required.)
3. Place the mold-contaminated contact plate into the blister pack or into the swab vial.
4. Place sample into another plastic container (Ziplock® bag). Seal the bag.
5. Dispose of gloves if contaminated by direct contact or if the presence of pathogenic fungi is suspected.
6. Place decontaminated gloves into a small waste bag.
7. Decontaminate any other tools used.
8. Decontaminate the outside of the sealed sample bag with alcohol wipes if it has been contaminated by direct contact or if the presence of pathogenic fungi is suspected.

Figure 2.7   Direct contact plates are momentarily applied to surfaces. (Courtesy of Biotest Diagnostic Corporation, Denville, NJ.)

2.9.4 Air Sampling: Bioaerosols

Air sampling is used to determine the bioaerosol type and concentration in the airstream of the sampled area. Air sampling for fungi and total particulates is used to identify types of culturable and nonculturable bioaerosols and bioaerosol concentrations. For direct air sampling, a designated flow of air is used. The air is either drawn toward a vacuum pump through a filter or past a staging assembly that may include impeller blades. When bioaerosols must be collected, extremely high flow rates may be required. The rule in general is that sonic flow requires a 0.5-atmosphere pressure drop. As with all pumps, the greater the pressure drop, the faster the intake of air toward that pressure void area.

Air samples for fungi and total particulates are taken by using a high-flow pump calibrated at a flow rate of 15 L/min and a collection time of 5 min for both room (ambient air) and outdoor reference samples. A collection time of up to 2 min may be used for samples collected inside wall cavities (e.g., WallChek™ samples). The air samples are collected in a sterile manner, sealed, labeled, and submitted to a microbiological laboratory for microbial identification. The laboratory uses direct microscopic examination of the cassettes and/or growth media to identify the type and concentration of culturable and nonculturable bioaerosols in the air. Any airborne fungi, as well as any other airborne particulates (e.g., pollen, fibers, skin cell fragments, or insect parts) will be collected during the sampling period.

Detecting general fungal materials such as chitin, glucan, and ergosterol may not allow discrimination of fungal elements from indoor and outdoor sources. Interpretable single-case results (as opposed to multi-case statistical trends) might be obtained only in cases where an extreme indoor buildup has occurred or the indoor accumulation of outdoor fungal material is otherwise known to be insignificant.

Air sampling is not appropriate for quantitative evaluation of *Stachybotrys* or certain other fungi that are poorly culturable from airborne spores because in heavily contaminated environments the sample may be overexposed, multiple spores may be counted as one after impaction, subsequent colony overgrowth may occur, or correction factors may be needed. Reducing the sampling time may result in correctly exposed media; however, results may be skewed if the sampling time effect is not factored into the ultimate sampling report conclusions. Contact and liquid dip agar plates are used to compare airborne levels to those present on surfaces or in liquid pools.

2.9.5 Example of Reuter Centrifugal Sampler (RCS) or SAS Sampling Sequence

RCS sampling routines are as follows (see Figures 2.8 and 2.9):

1. Open sample over a bag at first location to be sampled.
2. Insert agar strip or plate. Do not directly touch the agar media at any time. In the event that the agar is touched, discard that agar strip.
3. Sample for mold by running the instrument for the approved time duration.
4. Remove the agar. Do not directly touch the agar media at any time. In the event that the agar is touched, discard that agar strip.
5. Place the agar into the original sample bag.
6. Seal the bag.
7. Dispose of gloves if contaminated by direct contact or if the presence of pathogenic fungi is suspected.
8. Place decontaminated gloves into small waste bag.
9. Decontaminate the outside of sealed sample bag with alcohol wipes if contaminated by direct contact or if the presence of pathogenic fungi is suspected.
10. Place the alcohol wipes into a small waste bag.

Figure 2.8   SAS air sampler and viable impactor. (Courtesy of Bioscience International, Rockville, MD.)

Figure 2.9   Two versions of the RCS are available; the newer version has digital air flow program capability.

The sampling instruments, in some circumstances, may need to be decontaminated between sampling events. In the field the impeller assembly can be cleaned with isopropyl alcohol and thoroughly air-dried in a biologically neutral location. If further decontamination is required, it should be done at the issuing laboratory. In some circumstances, pathogenic sleeves must be used. Do not take the carrying case or battery charger into a contaminated environment. A remote control is provided to initiate the sampling event. At the conclusion of a sampling event, at a minimum, wipe down the sampling instrument's exterior with alcohol wipes. Use equipment decontamination pads to decontaminate temporary lighting and any other large equipment used. (*Note:* Lights are turned off prior to decontamination; the last set of lights may be decontaminated using handheld flashlights.)

2.9.6   Example of Exit Requirements

When exiting the area:

1. Place all used alcohol wipes into a small waste bag and seal the waste bag.
2. Exit area.
3. Decontaminate outer Tyvek® and respirator with alcohol wipes.
4. Remove outer Tyvek®.
5. Place used Tyvek® into a large waste bag.
6. Decontaminate inner Tyvek®, gloves, and boots.
7. Place decontaminated items into a waste bag.
8. Remove duct tape from wrists/ankles.
9. Remove boots, gloves, and Tyvek®.
10. Place used boots, gloves, and Tyvek® into a large waste bag.
11. Seal the large waste bag.
12. Decontaminate the respirator again prior to removal; place used alcohol wipes into a (new) small waste bag.
13. Remove respirator.
14. Use alcohol wipes to decontaminate hands.
15. Place all used decontamination pads into a small waste bag.
16. Bag all disposable equipment and dispose of it in an appropriate manner.
17. Bag all nondisposable equipment for further decontamination off-site.

2.10 VOLATILE SCREENING

Volatile screening may be needed to determine levels of microbial volatile organic compounds (mVOCs) or to monitor the biocides being used. For indoor air quality assessments, photoionization detectors (PIDs) or adsorbent media followed by gas chromatography/mass spectroscopy (GC/MS) lab analysis can be used to measure volatile organic compounds at low levels.

2.10.1 *Photoionization Detectors*

Time studies have repeatedly shown that in industrialized societies people spend very little time outdoors. While it is dependent on a number of factors, the proportion of time spent outdoors is often no more than 5%. The drive for greater HVAC efficiency has led to buildings that are sealed more tightly than ever, so less fresh air is admitted to dilute any airborne contaminants. Because people are spending upwards of 95% of their time inside sealed, high-efficiency buildings, complaints of sick building syndrome (SBS) are on the rise. According to the American Lung Association, the top five indoor air pollutants are:

- Carbon monoxide
- Formaldehyde
- Microbial contaminants (mold spores)
- Second-hand tobacco smoke
- Volatile organic compounds

To limit occupant exposure to these compounds, sources must be identified and either eliminated or controlled. A PID measures VOCs and other toxic gases in low concentrations from parts per billion (ppb) up to 10,000 parts per million (ppm, or 1% by volume). A PID is a very sensitive broad-spectrum monitor.

Photoionization detectors measure 0 to 10,000 ppm with resolution down to 1 ppb and therefore are a very appropriate means of measuring VOCs (and other toxic gases and vapors) at extremely low levels. The advantage of the PID is that, while it is not selective, it is a small, continuous monitor that can provide instantaneous feedback (less than 3 seconds). This instant feedback is critical because it provides users with the ability to deal with an indoor air quality problem instantly rather than letting the problem build while waiting for laboratory results.

A PID uses an ultraviolet (UV) light source (*photo*, light) to break down chemicals to positive and negative ions (*ionization*) that can easily be counted with a detector. Ionization occurs when a molecule absorbs the high-energy UV light, which excites the molecule and results in the temporary loss of a negatively charged electron and the formation of a positively charged ion. The gas becomes electrically charged. In the detector, these charged particles produce a current that is then amplified and displayed on the meter in parts per million or even in parts per billion. The ions quickly recombine after exposure to the electrodes in the detector to reform their original molecule. PIDs are not destructive; they do not burn or permanently alter the sample gas.

All elements and chemicals can be ionized, but they differ in the amount of energy required. The energy required to displace an electron and ionize a compound is called the ionization potential (IP), which is measured in electron volts (eV). The light energy emitted by a UV lamp is also measured in electron volts. If the IP of the sample gas is less than the electron-volt output of the lamp, then the sample gas will be ionized (Figure 2.10).

Photoionization detectors provide a direct means of detecting mVOCs. This makes PIDs an excellent choice for both portable indoor air quality surveys and permanent IAQ subsystems of an HVAC system. Recent advances in PID technology provide parts-per-billion resolution, thus providing immediate insight and diagnostics that have never been possible before in indoor air quality surveys. Innovations such as self-cleaning optics reduce long-term PID drift to a manageable level and can provide the means for HVAC total volatile organic compound (tVOC) control used alone or in an array of indoor air quality sensors (Figures 2.11 and 2.12).

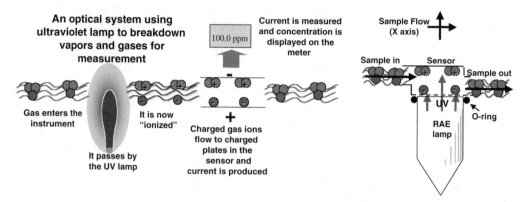

Figure 2.10   Photoionization detector lamp operation. (Courtesy of RAE Systems, Sunnyvale, CA.)

2.10.2   Photoionizaton Detectors and Indoor Air Quality

Photoionization detectors provide a direct means of quickly assessing indoor air quality. The urgency and complexity associated with sick building syndrome (SBS) have triggered a search for a practical (time- and cost-effective) assessment method using tVOC levels as practical standards (Godish, 1995). Such a total component concept has already gained acceptance in other health-related disciplines, such as using total suspended particles (TSPs) and total decibels (TdBs) as screening standards for particle and sound pollution, respectively. Pioneer work on using the tVOC level as a practical overall standard is not complete (Seifert, 1990; Molhave, 1990) and require further epidemiological research. Even so, measuring tVOCs is emerging as a more direct approach of surveying indoor environments for contamination.

2.10.2.1   Total Volatile Organic Compounds

Preliminary data indicate that tVOCs are good indicators not only for traditional contaminants (off-gassing products) but also for microbial actions (chemical releases from molds, fungi.). Recent studies on the chemistry of VOCs in indoor air (secondary emission and reactive species) and the effects of microbiological VOCs (Hess, 2001; Salthammer, 2000; Wolkhof, 2000) call for a need for further research. Until complete understanding is reached, researchers (Hara, 2000; Salthammer, 2000; Seifert, 1999) are refining the tVOC approach as practical screening method for exposure risk assessment to total VOCs in working and living environments.

2.10.2.2   General Guidelines

Isobutylene elicits a PID responsiveness about midpoint in the range of PID sensitivity; consequently, isobutylene is typically used to calibrate the PID. Because PIDs are calibrated to isobutylene, correction factors (CFs) are expressions of PID sensitivity to a chemical relative to isobutylene; $EL_{Chemical}$ is the exposure limit in chemical units (ppm). Unless otherwise indicated, the EL is typically an 8-hour time-weighted average (TWA):

$$CF = \frac{\text{PID Isobutylene Response} \times \text{Concentration of gas (ppmv)}}{\text{Conc. of isobutylene (ppmv)} \times \text{Response of gas on PID}}$$

$$EL_{Isobutylene} = \frac{\text{ELchemical (ppmv)}}{\text{CF Chemical}}$$

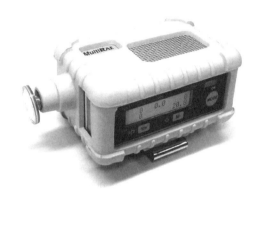

Figure 2.11 *ppb*RAE capable of continuous detection down to 1 ppb for volatile organic compounds (VOCs). Comparator circuitry in the sensor of the *ppb*RAE is used to zero-out background VOCs. (Courtesy of RAE Systems, Sunnyvale, CA.)

Figure 2.12 MultiRAE combines a PID with O_2, LEL, CO and other toxic sensors. (Courtesy of RAE Systems, Sunnyvale, CA.)

Thus, the exposure limit in units of isobutylene is the exposure limit in chemical units divided by the ratio of chemical units to isobutylene units.

Global consensus has resulted in the emergence of preliminary guidelines for tVOC standards for indoor air quality (Australian NHMRC, 1993; Finnish Society of Indoor Air Quality and Climate, 1995; Hong Kong EPA, 1999; Japan Ministry of Health, 2000; Seifert, 1999). Depending on location (home, school, workplace), recommended levels range from 200 to 1300 $\mu g/m^3$, or about 50 to 325 ppb toluene units, or approximately 100 to 650 ppb isobutylene units. The indoor air quality tVOC threshold for normal environments should not exceed 500 ppb (0.5 ppm) toluene units, which is equivalent to 1000 ppb (1 ppm) isobutylene units. Field experience suggests the following guidelines for the use of PIDs to assess indoor environments:

- <100 ppb isobutylene units for normal outdoor air
- 100 to 400 ppb isobutylene units for normal indoor air
- 500+ ppb isobutylene units for potential indoor air quality contaminants

2.10.2.3 Exposure vs. Time

Adsorptive tests may show that workers have been safe on average over an 8-hour day, yet elevated levels of chemical may be missed by the averaging affects of adsorptive testing. For example, consider an office building with a small print shop in the basement. During the winter, the building manager decides to save money by decreasing the amount of outside fresh air introduced into the HVAC system so that the air in the building is recirculated. Over the course of a work day, the mold odors build up and reach high levels. Workers in the building do not smell the vapors because these workers have grown accustomed to the odors through olfactory fatigue. Low solvent exposure in the morning coupled with the high exposure in the afternoon provides an average exposure that is acceptable; however, exposure vs. time data logging with a PID shows a trend of VOCs being absent in the morning with high levels in the afternoon. By adding the morning concentrations to the afternoon concentrations, adsorptive techniques missed this situation. If we were to plot the time vs. exposure for adsorptive tests, a straight line for the 8-hour work day would incorrectly indicate a static situation of little concern.

2.11 SUMMA CANISTER AND MINI-CAN

The Summa canister and mini-can can be preset to draw in a known volume of gas. The mini-can is worn by a worker or placed in a static location. Sample collection occurs without the use of an additional air-sampling pump. Another application is to use these devices during wall checking to determine mVOC levels (Figure 2.13).

2.12 ADSORBENT MEDIA FOLLOWED BY GC/MS LAB ANALYSIS

Various containers may be used to collect gases for later release into laboratory analytical chambers or sorbent beds. The remote collection devices include bags, canisters, and evacuation chambers. Remote collection refers to the practice of collecting the gas sample (intact chemically) at a site remote from the laboratory, where analysis will occur. This method of sample collection must always take into account the potential for the matter of the collecting vehicle to react with the gaseous component collected during the time between collection and analysis. For this reason, various plastic formulations and stainless steel compartments have been devised to assure that reactions with the collected gases are minimized. When bags are used, the fittings for the bags to the pumps must also be relatively inert and are usually stainless steel. Multiple bags may be collected and then applied to a gas chromatograph column using multiple bag injector systems.

Low-flow pumps are used to pull a sample through an adsorbent tube, or an evacuated stainless steel cylinder draws in the air to be sampled to provide continuous monitoring over an entire day or portion of day. These samples are sent to a laboratory. After analysis of these samples with gas chromatography/mass spectroscopy (GC/MS) one can tell exactly what the average concentration of chemical exposure was at the time the pump was used. To approximate the concentration vs. the time of exposure, multiple tubes or cylinders must be run through the pump during the working day. This leads to greater complication and cost. While these tubes and cylinders may be specific, adsorbent tubes and cylinders are reactive rather than proactive. Results can take days or weeks to return from the lab. Sampling followed by GC/MS testing is like a 35-mm camera. While they both produce excellent results, you must wait for the film to be developed. In addition, laboratory analysis is expensive (Figure 2.14). In his study, Hara (2000) found significant correlation between samples tested with a Tenax TA thermal desorption GC/MS and a RAE Systems *ppb*RAE.

Figure 2.13   Mini-can Summa canister. (Courtesy of Aerotech Laboratories, Phoenix, AZ.)

2.12.1 Solid Sorbent Tubes

Organic vapors and gases may be collected on activated charcoal, silica gel, or other adsorption tubes using low-flow pumps. Tubes may be furnished with either caps or flame-sealed glass ends. If using the capped version, simply uncap during the sampling period and recap at the end of the sampling period. Multiple tubes can be collected using one pump. Flow regulation for each tube is accomplished using critical orifices and valved regulation of airflow. Calibration of parallel or multiple Y-connected tube drawing stations must be done individually for each tube even in cases where the pump is drawing air through more than one tube in a parallel series. In instances where tubes are connected in series, only one calibration draw is done through the conjoined tubes that empty air directly into the other.

Sorbent tubes may be used simply to collect gases and vapors or to collect and react with the collected chemicals. Some of the reactions may produce chemicals that, when off-gassed, could harm the pumps being used to pull air through the sorbent media bed. In these cases, either filters or intermediate traps must be used to protect the pumps.

The following protocols should be followed:

- Immediately before sampling, break off the ends of the flame-sealed tube to provide an opening approximately half the internal diameter of the tube. Take care when breaking these tubes, as shattering may occur. A tube-breaking device that shields the sampler should be used.
- Wear eye protection when breaking ends.
- Use tube holders, if available, to minimize the hazards of broken glass.
- Do not use the charging inlet or the exhaust outlet of the pump to break the ends of the tubes.
- Use the smaller section of the tube as a back up and position it near the sampling pump.
- The tube should be held or attached in an approximately vertical position with the inlet either up or down during sampling.
- Draw the air to be sampled directly into the inlet of the tube. This air is not to be passed through any hose or tubing before entering the tube. A short length of protective tape, a tube holder, or a short length of tubing should be placed on the cut tube end to protect the worker from the jagged glass edges.
- Cap the tube with the supplied plastic caps immediately after sampling and seal as soon as possible.
- Do not ship with bulk material.
- For organic vapors and gases, low-flow pumps are required. With sorbent tubes, flow rates may have to be lowered or smaller air volumes (half the maximum) used when humidity is high (above 90%) in the sampling area or relatively high concentrations of other organic vapors are present.

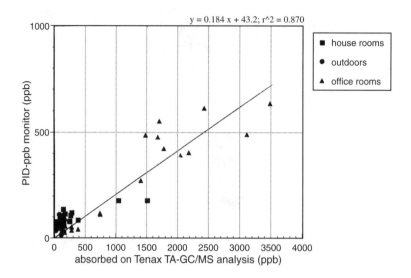

Figure 2.14 Graph of volatile monitoring to parts-per-billion level. (Courtesy of RAE Systems, Sunnyvale, CA.)

2.12.2 Detector Tubes

Detector tubes and pumps are screening instruments which may be used to measure more than 200 organic and inorganic gases and vapors or for leak detection (Figure 2.15). Some aerosols can also be measured. Detector tubes use the same media base (silica gel or activated charcoal) as do many sorbent tubes. The difference is that the detector tubes change color in accordance with the amount of chemical reaction occurring within the media base. The media base has been treated with a chemical that effects a given color change when certain chemicals are introduced into the tube and reside for a time in the media. The residence time for the reaction to occur and the volume of air that must be drawn through the detector tubes varies with the chemical and anticipated concentration. All detector tube manufacturers supply the directions for using their detector tubes as an insert sheet with the tubes.

Detector tube pumps are portable equipment that, when used with a variety of commercially available detector tubes, are capable of measuring the concentrations of a wide variety of compounds in industrial atmospheres. Each pump should be leak-tested before use and calibrated for proper volume at least quarterly or after 100 tubes.

Operation consists of using the pump to draw a known volume of air through a detector tube designed to measure the concentration of the substance of interest. The concentration is then determined by a colorimetric change of an indicator that is present in the tube contents. Detector tubes of a given brand are to be used only with a pump of the same brand; the tubes are calibrated specifically for the same brand of pump and may give erroneous results if used with a pump of another brand.

A limitation of many detector tubes is the lack of specificity. Many indicators are not highly selective and can cross-react with other compounds. Manufacturers' manuals describe the effects of interfering contaminants. Another important consideration is sampling time. Detector tubes give only an instantaneous interpretation of environmental hazards. This may be beneficial in potentially dangerous situations or when ceiling exposure determinations are sufficient. When long-term assessment of occupational environments is necessary, short-term detector-tube measurements may not reflect time-weighted average levels of the hazardous substances present.

Detector tubes normally have a shelf life of one to two years when stored at 25°C. Refrigeration during storage lengthens the shelf life. Outdated detector tubes (i.e., beyond the printed expiration

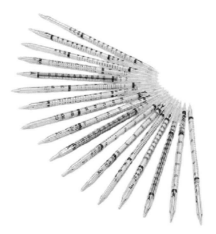

Figure 2.15   Detector tubes. (Courtesy of RAE Systems, Sunnyvale, CA.)

date) should never be used. The specific tubes are designed to cover a concentration range that is near the permissible exposure limit (PEL). Concentration ranges are tube dependent and can be anywhere from 1 one-hundredth to several thousand ppm. The limits of detection depend on the particular detector tube. Accuracy ranges vary with each detector tube. The pump may be handheld during operation (weight, 8 to 11 ounces) or may be an automatic type (weight, about 4 pounds) that collects a sample using a preset number of pump strokes. A full pump stroke for either type of short-term pump has a volume of about 100 mL. In most cases, where only one pump stroke is required, sampling time is about one minute. Determinations for which more pump strokes are required take proportionately longer.

Multiple tubes can be used with newer microcapillary detector tube instruments. Computer chips are programmed to draw preselected air volumes across these detector tubes. Levels are measured based on changes in light absorption across the microcapillary tubes.

2.12.3 Colorimetric Sorbent Packed Tubes

A variation on the detector tube technology is the use of sorbent packed tubes that change color in response to ambient airflow. The application of reactive adsorbing and/or absorbing chemicals onto test strips is also used to provide a general indication of airborne contaminant levels. An example is the ozone test strip used to monitor both outdoor and indoor ozone levels. Absorbent tubes respond slowly to changes in concentration and can miss or grossly underestimate some exposures. These transient exposures may affect people, but the slow response of adsorptive sampling techniques, coupled with their averaging, could underestimate or miss the volatile off-gassing. A PID can log these quick, high transient responses.

2.12.4 Vapor Badges

Passive-diffusion sorbent badges are useful for screening and monitoring certain chemical exposures, especially vapors and gases. Badges are available from the laboratory to detect mercury, nitrous oxide, ethylene oxide, and formaldehyde. Interfering substances should be noted. A variation on the vapor badge is available as a dermal patch. These dermal patches can also be used in the detection of semivolatiles such as pesticide residuals (Figure 2.16).

Figure 2.16 Dermal patches may be used on skin, personal protective equipment, or equipment surfaces. (Courtesy of SKC, Inc., Eighty Four, PA.)

2.12.5 Formaldehyde

Formaldehyde sampling can be accomplished using both passive and active (use of a pump) techniques (Figure 2.17). When long-duration sampling is required in indoor air investigations, passive sampling tubes may be the method of choice. Vapor badges can be used to monitor personnel exposures. Neither of these methods is recommended for acute exposure scenarios, as the sampling media will quickly become overloaded. In acute exposure scenarios, sampling with a sorbent tube attached to an air sampling pump or a detector tube attached to a pump/bellows is recommended. Attachment implies that the pump will be used to draw a known volume of air quickly into the media. This air will be at a concentration anticipated to provide information but below that which would overload the media.

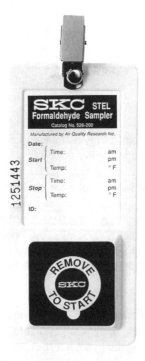

Figure 2.17 Vapor badge. (Courtesy of SKC, Inc., Eighty Four, PA.)

Figure 2.18 High-volume polyurethane foam sampler. (Courtesy of SKC, Inc., Eighty Four, PA.)

Figure 2.19 Impinger (liquid media) attached to a personal air sampling pump. (Courtesy of SKC, Inc., Eighty Four, PA.)

2.13 PESTICIDES, POLYAROMATIC HYDROCARBONS, AND POLYURETHANE FOAM

Both pesticides and polyaromatic hydrocarbons (PAHs) can be collected in polyurethane foam (PUF) tubes (Figure 2.18). PUF tubes are available for both high- and low-volumes sampling. The sampling volume requirement is determined by the regulatory requirements and the chemical properties of the anticipated sample.

2.14 ACID GASES OR CAUSTICS

Volatile acid gases may be an inappropriate designation. Acid gases are often generated during a reaction, and the latent volatility of the acid gas is really not the issue. In these cases, thermal volatilization based on boiling point predictions and mechanical dispersion may be of less importance than the rate of the reaction generating the acid gas or caustic. However, in addition to this reaction phenomenon, acid gases such as chlorine are given off when the liquid solution is distributed around an area. This is a classic case of the liquid–gas interface seeking an equilibrium. If air currents sweep the generated gas concentration away from this equilibrium site, the liquid will again yield molecules to the gas phase to achieve another equilibrium.

Acid gases and caustics, with their corrosive or caustic properties, can have health effects that include both toxicological and physical acute manifestations such as watering eyes and respiratory tract irritation. Because of these effects, sampling for acid gases and caustics must begin upon approaching the area of concern. Sampling for acid gases and caustics may use all of the techniques specified for any volatile. In addition, some acid gases and caustics are dispersed when precipitated out or adsorbed onto particulates; therefore, particulate sampling techniques are applicable. The reaction phenomena must always be considered during any sampling of acid gases or caustics. Any real-time instrumentation with unprotected metal sensors, lamp filaments, or sensor housings will often be rendered useless as the acid gases or caustics interact with the metals through reduction–oxidation (RedOx) reactions.

2.14.1 Impingers

Impingers may be used to bring acid/caustic-laced particulates into solutions that are retained within the impinger's vessel. Vapors, mists, and gases may also be introduced into the impinger solution. If the reaction within the impinger vessel could cause off-gassing, a filter or media barrier may be required between the air sampling pump and the impinger vessel tubing to the pump (Figure 2.19). Small impingers may be worn as personal sampling devices. The main concern with impingers as sampling devices, especially for personnel, is that the liquid could spill.

2.14.2 Sorbent Tubes

Sampling media must also be acid and caustic resistant. Sampling for acids and caustics is often discussed in terms of using silica gel sorbent tubes (Figure 2.20). The procedure for this sampling is the same as those for volatiles where charcoal tubes are often used. The essential problem with silica gel tubes is that these tubes tend to clog. Use of dual-flow tubes provides some assurance that if one tube plugs up the other might still remain effective so as to provide data from the sampling interval. In instances where silica gel tubes continue to clog, switching to larger bore silica gel tubes or altering the sampling interval (less time) may be needed. If this does not work, switching to charcoal tubes may be the only other solvent tube option. These sampling routines may be at odds with the recommended National Institute of Occupational Safety and Health (NIOSH) methods calling for the use of small-bore silica gel tubes at a low flow rate for extended periods of time. If so, decision logic must be documented and this documentation linked to the competency of the individual who devised the sampling plan.

2.14.3 Detectors

Various detector tubes are available for acid gases and caustics. Chemical-specific detectors are increasingly available as hard-wired permanent detectors based on electrochemical sensors. As with any other electrochemical sensor, recovery of the sensor after overdosing with a chemical may take time or may not be possible at all.

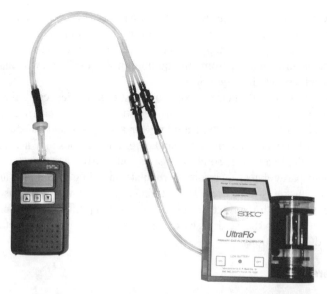

Figure 2.20 Orientation of dual sorbent tubes to sampling pump and calibration device. (Courtesy of SKC, Inc., Eighty Four, PA.)

Figure 2.21 Ozone indicator papers.

2.15 INDICATOR PAPERS OR METERS

Indicator papers or meters are particularly valuable where ozone, acid gases, and caustics may be of concern. Many sampling events require concurrent bulk sampling, and often these samples can be effectively characterized as to pH in the field (Figure 2.21).

2.16 VENTILATION ADEQUACY

In order to determine the potential for the spread of airborne biological components, ventilation adequacy should be measured. Duct diameters are measured to calculate duct areas. Inside duct diameter is the most important measurement, but an outside measurement is often sufficient for a sheet metal duct. To measure the duct, wrap the measuring tape around the duct to obtain the duct circumference, and divide that number by 3.142 to obtain the diameter of the duct. Hood and duct dimensions can also be estimated from plans, drawings, and specifications. If a duct is constructed of 2.5- or 4-foot sections, the sections can be counted (elbows and tees should be included in the length). Hood-face velocities outside the hood or at the hood face can be estimated with velometers, smoke tubes, and swinging-vane anemometers, all of which are portable and reliable.

2.16.1 Smoke Tubes

Nonirritant smoke can be used to visually determine air circulation paths. The general practice when using smoke is to squeeze off a quick burst of smoke, time the travel of the smoke plume over a 2-foot distance, and calculate the velocity in feet per minute. Smoke checking is useful to determine:

- Air circulation — Dispersal of smoke within several seconds from the center of the room suggests good air circulation, while smoke that stays essentially still for several seconds suggests poor circulation.
- Flow patterns in HVAC systems — Puffs of smoke released adjacent to HVAC vents, diffusers, and grilles give a general idea of airflow toward the supply and return system.
- Modulated variable air volume (VAV) systems — These systems must be checked during the *on* cycle.
- Air movement directly from supply diffusers to return grilles — Such air movement indicates an air short circuit; during a short circuit, both makeup air and dilutant air within the room may be compromised.
- Face velocity at the entry portal to confined areas (e.g., ductwork, hood windows, mini-enclosures)
- Integrity of glovebags and other bag isolation methods

2.16.2 Anemometers

The minimum velocity that can be read by an anemometer is 50 feet per minute (fpm). The meter should always be read in the upright position, and only the tubing supplied with the equipment should be used. Anemometers often cannot be used if the duct contains dust or mist because air must actually pass through the instrument for it to work. The instrument requires periodic cleaning and calibration at least once per year. Hot-wire anemometers should not be used in airstreams containing aerosols. To measure hood-face velocity, mark off imaginary areas (i.e., develop a sampling grid), measure the velocity at the center of each area, and average all measured velocities.

2.16.3 Static Pressure in the Hood

Static pressure in the hood (SPH) should be measured about four to six duct diameters downstream in a straight section of the hood take-off duct. The measurement can be made with a pitot tube or by a static pressure tap into the duct sheet metal.

2.16.4 Pressure Gauges

Pressure gauges come in a number of varieties, the simplest being the U-tube manometer. Inclined manometers offer greater accuracy and greater sensitivity at low pressures than U-tube manometers; however, manometers rarely can be used for velocities < 800 fpm (i.e., velocity pressures < 0.05 inch water gauge, or inH_2O). Aneroid-type manometers use a calibrated bellows to measure pressures. They are easy to read and portable but require regular calibration and maintenance.

2.16.5 Duct Velocity

Duct velocity measurements may be made directly (with velometers and anemometers) or indirectly (with manometers and pitot tubes) using duct velocity pressure. Airflow in industrial ventilation ducts is almost always turbulent, with a small, nonmoving boundary layer at the surface of the duct. Because velocity varies with distance from the edge of the duct, a single measurement may not be sufficient. However, if the measurement is taken in a straight length of round duct, four to six diameters downstream and two to three diameters upstream from obstructions or directional changes, then the average velocity can be estimated at 90% of the centerline velocity. (The average velocity pressure is about 81% of centerline velocity pressure.) A more accurate method is the transverse method, which involves taking six or ten measurements on each of two or three passes across the duct, 90° or 60° opposed. Measurements are made in the center of concentric circles of equal area. Air cleaner and fan condition measurements can be made with a pitot tube and manometer. Density corrections (e.g., temperature) for instrument use should be made in accordance with the manufacturer's instrument instruction manual and calculation/correction formulas.

2.16.6 Carbon Dioxide as an Indicator of Ventilation

Indoor carbon dioxide (CO_2) concentrations, whether from exhaled breath or other sources, can be used as an indicator of ventilation adequacy. Peak CO_2 readings should be taken between rooms, between air handler zones, and at varying heights above the floor. CO_2 can be measured with either a direct-reading meter (O_2/CGI) or a colorimetric detector tube. Test results should include relative occupancy, air damper settings, and weather. CO_2 should be collected away from any source that could directly influence the reading (e.g., hold the sampling device away from exhaled breath). Individual measurements should be short term.

Readings from outdoors and from areas with no apparent indoor air quality problems are frequently used as baseline or controls. Outdoor samples should be taken near the outdoor air intake. Measurements taken to evaluate the adequacy of ventilation should be made when concentrations are expected to peak.

If the occupant population is fairly stable during normal business hours, CO_2 levels will typically rise during the morning, fall during the lunch period, and then rise again, reaching a peak in mid-afternoon. In this case, sampling in the mid- to late afternoon is recommended. Sampling intervals should be chosen to detect CO_2 patterns throughout the work day. For residences, it may be necessary for the sampling to encompass a full 24 hours. Peak CO_2 concentrations above 1000 ppm in the breathing zone indicate ventilation problems. CO_2 concentrations below 1000 ppm generally indicate that ventilation is adequate to deal with the routine products of human occupancy.

2.16.7 Oxygen/Combustible Gas Indicators (O_2/CGI) and Toxin Sensors

Oxygen/combustible gas indicators (O_2/CGIs) and toxin sensors measure the lower explosive limits (LELs) of various gases and vapors. A platinum element or wire is the oxidizing catalyst. The platinum element is one leg of a Wheatstone bridge circuit. These meters measure gas concentration as a percentage of the LEL of the calibrated gas. The oxygen meter displays the concentration of oxygen in percent by volume measured with a galvanic cell. Some O_2/CGIs also contain sensors to monitor toxic gases and vapors. These sensors are also electrochemical (as is the oxygen sensor). Thus, whenever the sensors are exposed to the target toxins, the sensors are activated. Other electrochemical sensors are available to measure carbon monoxide, hydrogen sulfide, and other toxic gases. The addition of two toxin sensors, one for H_2S and one for CO, is often used to provide information about the two most likely contaminants of concern, especially within confined spaces. Because hydrogen sulfide and carbon monoxide are heavier than ambient air (i.e., the vapor pressure of hydrogen sulfide is greater than one), the monitor or the probe must be lowered toward the lower surface of the area being monitored. Other toxic sensors are available, and all are electrochemical; examples are sensors for ammonia, carbon dioxide, and hydrogen cyanide. These may be installed for special needs.

With the use of a remote probe, air sampling can be accomplished without lowering the entire instrument into the atmosphere. Thus, both the instrument and the person doing the sampling are protected. The remote probe has an air line (up to 50 feet) that draws sampled air toward the sensors with the assist of a powered piggyback pump. Without this arrangement, the O_2/CGI monitor relies on either a diffusion grid (passive sampling) or manual pumping of air toward the grid. All O_2/CGIs must be positioned so that the diffusion grids over the sensors and the inlet port for the pumps are not obstructed. For instance, do not place the O_2/CGI on your belt with the diffusion grids facing toward your body.

2.16.8 Toxic Gas Meters

These meters use an electrochemical voltametric sensor or polarographic cell to provide continuous analyses and electronic recording. Interference from other gases may be a problem. The sensor manufacturer's literature must be consulted in instances where mixtures of gases are to be tested. In operation, sample gas is drawn through the sensor and absorbed on an electrocatalytic-sensing electrode, after passing through a diffusion medium. An electrochemical reaction generates an electric current directly proportional to the gas concentration. The sample concentration is displayed directly in parts per million. Because the method of analysis is not absolute, prior calibration against a known standard is required. Exhaustive tests have shown the method to be linear; thus, calibration at a single concentration is sufficient. Sensors are available for sulfur dioxide, hydrogen cyanide, hydrogen chloride, hydrazine, carbon monoxide, hydrogen sulfide, nitrogen oxides, chlorine, ethylene oxide, and formaldehyde. These sensors can be combined with combustible gas and oxygen sensors in one instrument.

2.17 MOISTURE METERS AND PHOTOGRAPHY

Moisture monitoring and photography are used to determine levels of water or mold growth. Actively growing mold sites require water and have a higher water content than do dry building materials. Moisture meters rely on the increase in electrical conductivity caused by water intrusion. An electric signal is sent through a material, and this signal is then picked up by a receiving detector. Water and water-soaked materials have increased conductivity, thus the returning sampled signal is stronger than that of similar but dry materials. The comparison to dry materials of similar type is a requirement in order to evaluate the conductivity change. False-positive readings indicative of supposed water intrusion can occur due to metallic components in the layers of sandwiched water or electrical conduits. When these false-positive readings are next to actual positive readings, moisture-monitoring interpretation becomes more difficult. Infrared photography combined with thermography interpretation can further substantiate water intrusion and potential mold amplification areas. Thermography relies on the lower temperatures associated with water-soaked areas. Videography using fiberoptic probes behind building layers and in HVAC systems can also provide visual assistance when evaluating wet and/or mold growth sites.

REFERENCES AND RESOURCES

ASHRAE, Method of Testing Air-Cleaning Devices Used In General Ventilation for Removing Particulate Matter, ASHRAE Standard 55, American Society of Heating, Refrigerating, and Air Conditioning Engineers, Atlanta, GA, 1992.

ASHRAE, Ventilation for Acceptable Air Quality, ASHRAE Standard 62, American Society of Heating, Refrigerating, and Air Conditioning Engineers, Atlanta, GA, 1999.

ASHRAE, Method of Testing General Ventilation Air-Cleaning Devices for Removal Efficiency by Particle Size, ASHRAE Standard 52.2, American Society of Heating, Refrigerating, and Air Conditioning Engineers, Atlanta, GA, 2000.

Boss, M. and Day, D., *Air Sampling and Industrial Hygiene Engineering*, CRC Press, Boca Raton, FL, 2000.

Finnish Society of Indoor Air Quality and Climate, Classification of Indoor Climate, Construction, and Finishing Materials, Espoo, Finland, 2000.

Godish, T., *Sick Buildings*, Lewis Publishers, Boca Raton, FL, 1995, pp. 148–157.

Hara, K., Comparison among Three VOC Measuring Methods, (1) Absorbed on Tenax TA-Thermal Desorption-GC/MS Analysis, (2) Photoacoustics Gas Monitor, and (3) Photo-Ionization Detector, Internal Paper at Institute for Science of Labor, Japan.(2000)

Hess, K., *Environmental Sampling for Unknowns*, Lewis Publishers, Boca Raton, FL, 2001.

Hong Kong EPA, Guidance Notes for the Management of Indoor Air Quality in Offices and Public Places, www.info.gov.hk/epd, 1999, pp. 14–31.

Hurt, C., Ed., *Manual of Environmental Microbiology*, ASM Press, Washington, D.C., 1997.

Japan Ministry of Health, Indoor Air Quality Guideline: Interim Goal for tVOC Levels for Old and New Homes, Tokyo, 2000.

Molhave, L. et al., Volatile organic compounds indoor air quality and health, in *Proc. Indoor Air '90*, Vol. 5, 13–16, 1990,

National Health and Welfare, Canada, IAQ in Office Buildings: A Technical Guide, 1993.

NIOSH, *Pocket Guide to Chemical Hazards*, National Institute for Occupational Safety and Health Publications, Cincinnati, OH, 1997.

OSHA Technical Manual, OTM 1-0.15A, 1999.

Russian State Committee for Hygiene and Epidemiological Surveillance of the Russian Federation, Maximum Allowable Concentrations (MAC) of Harmful Substances, Russian Federation, Ministry of Science and Technology, Moscow, Russia, 1993.

Salthammer, T., Impact of secondary emissions and reactive species on indoor air quality, in *Proc. Seminar on VOCs in Indoor Air*, Tokyo, 133–165, 2000.

Salthammer, T. et al., Comparison of tVOC by GC/MS with Direct Reading Instruments, oral presentation at Healthy Buildings 2000 Conference, Espoo, Finland, 2000.

Seifert, B., Regulating indoor air, in Indoor Air '90, Toronto, Canada, July 29–August 3, 1990, *Proc. 5th Int. Conf. on Indoor Air Quality and Climate*, Inglewood Printing Plus, Aurora, Canada,1990.

Seifert, B., Richtwerte (tVOC) fur die Innenraumluft, *Bundesgesundheitsbl-Gesundheitsforsch-Gesundheitsschutz*, 42, 270–278, 1999.

USEPA, The Inside Story: A Guide to Indoor Air Quality, U.S. Environmental Protection Agency, Washington, D.C., 1992.

Wolkhof, P., VOCs in indoor air: their impact on the indoor air quality?, in *Proc. Semin. VOCs in Indoor Air*, Tokyo, 114–132, 2000.

CHAPTER 3

Biological Sampling and Lab Interpretation

R. Vincent Miller and Martha J. Boss

CONTENTS

1-56670-606-8/03/$0.00+$1.50
© 2003 by CRC Press LLC

A team consisting of microbiologists, industrial hygienists, toxicologists, and engineers is needed to evaluate indoor mold amplification. Integral to any risk assessment is the establishment of exposure, which, by definition, is dependent upon laboratory-based analyses and the accuracy of those analyses. In addition, for legally defensible data, it is imperative, before sampling methods and strategies are chosen, that the investigator have at least a rudimentary understanding of the available analytical techniques, their precision and accuracy, and their limitations and biases.

3.1 CHOOSING AN ANALYTICAL TECHNIQUE

During investigations, the analytical technique that will be utilized on the samples collected must be considered. Often, sampling is used to identify or verify that a problem exists and then estimate exposure or potential exposure. However, extrapolations of data from indoor environmental samples, especially in regard to exposure and potential health effects, may be very risky without some knowledge as to the particular technique being utilized to analyze the sample, limitations of that technique, and the competence of the laboratory and analysts. For instance, the presence of *Stachybotrys chartarum* in bulk or surface samples is often construed to be an exposure to the mold and (potentially) to trichothecene mycotoxins. However, such an interpretation may be erroneous in that surface and bulk samples do not give any indication as to airborne levels of the mold spores. Furthermore, many strains of *S. chartarum* do not even have the genetic capability to produce trichothecenes, so the presence of the organism cannot be used to predict mycotoxin exposure.

Figure 3.1 Air-O-Cell. (Courtesy of Zefon International, Inc., St. Petersburg, FL.)

Similarly, air-sampling data can give erroneous interpretations. For instance, air sampling via media-based impaction often misses *Stachybotrys*, unless a selective or semiselective medium is utilized. The use of a spore trap (Figure 3.1) such as the Air-O-Cell (Zefon International), Burkard (Burkard Manufacturing), or MK-3 (Allergenco) is more effective in detecting spores of *Stachybotrys* and assessing exposure to the mold itself, but again these data are not necessarily correlated to mycotoxin exposure. In order to properly interpret the data, the investigator must be acutely aware of the techniques used and the limitations of those techniques as utilized by the laboratory.

3.2 SAMPLING

3.2.1 Bulk and Surface Sampling

3.2.1.1 Dust Sampling

Because bacteria and mold are particulates and can also adsorb or absorb on dust particulates, quantification of particulate levels in the air may provide useful information. In the air or on surfaces, organic particulates share many of the same physical characteristics as inorganic particles from hazardous dusts. This characteristic has been demonstrated in military research on biological weapons and in civilian research to control the spread of infection in hospitals.

Dust sampling can:

- Provide information about the historical microbial populations within a building
- Provide sufficient sample volumes for mycotoxin and chemical (e.g., pesticides) analyses

When dust sampling, core samples taken from a room's spatial cavities must be taken precisely where mold occurs. Core samples are invasive, can cause structural damage, and have the potential for contaminating the building. Dust samples cannot be extrapolated to indicate potential airborne exposures.

3.2.1.2 Vacuum Sampling

For dust sampling, a small vacuum equipped with a HEPA (high-efficiency particulate air) filtration exhaust or a *dust-sock* attached to the front of the hose can be used to pick up debris. This vacuum sampling method pulls particulates from surfaces. Particulate can be impinged onto a filter, tape bed, agar, or liquid retention media. Transfer to the laboratory and subsequent analysis follow standard protocols for each media type. The process of vacuuming may cause either overestimation of particulate air entrainment or underestimation of microbe viability, due to desiccation or injury during the vacuuming. Surface vacuum samples cannot be extrapolated to indicate potential airborne exposures.

3.2.1.3 Bulk Sampling (Other than Dust)

Gross bulk sampling is a simple yet effective method for testing (and culturing) visible mold. Because samples are often taken only where mold is visibly present, mold that is not visibly obvious may be missed. Gross bulk samples cannot be extrapolated to indicate potential airborne exposures.

3.2.1.4 Tape Lifts

Tape sampling is used to directly pick up dust, fungal spores, and/or fungal structures on a sampled surface. The sticky surface of the tape often displays the mold materials, fungivorous mites, mite fecal pellets, and arthropods (e.g., book lice and small millipedes). Tape samples of an affected surface are taken using clear (not frosted) vinyl acetate adhesive tape. Then the following procedure is used:

1. Samples are taped flat in the interior and sealed in clean, plastic, recloseable sandwich bags (sterile bags are usually not needed) and appropriately labeled.
2. Samples are transported to a microscopic laboratory or other examination site.
3. At the examination site, the samples are peeled off the plastic and cut into convenient lengths.
4. The samples are then placed onto slides and stained for direct microscopic examination of the specimen attached to the underside of the tape.

This examination is used to determine the presence or absence of fungal spores as well as fungal structures such as hyphae (growth structures) or mycelia (filaments). Tape (adhesive tape) sampling is a simple and very rapid method for testing visible mold or surfaces that allows identification of many organisms to the genera level but not species level. It does not facilitate culturing for genera or species identification, and tape samples cannot be extrapolated to indicate potential airborne exposures.

3.2.1.5 Swabs

Swab sampling is used to determine the type and prevalence of fungi and bacteria that may be present on a sampled surface. Swabs are often used during clearance after remediation, following a blackwater (sewage) incursion to monitor coliforms or after visual detection of microbial growth. Swab samples of the affected surface area are taken using sterile culture collection swabs. Then:

1. Samples are sealed, labeled, and submitted to a microbiological laboratory.
2. Samples are cultured to encourage fungal growth on a specially prepared media.
3. Cultures are examined microscopically to determine the fungi type and prevalence.

Surface (swab) sampling:

- Is a simple method for testing (and culturing) visible and invisible molds and bacteria from surfaces; however, spore depositions vary due to their settling rates, according to their density and size
- Fails to detect nonviable spores, which can still carry mycotoxins and/or allergenic determinants
- Tends to select only spores and may leave intact fruiting structures such as conidiophores, pycnidia, or ascomata behind, which can render identification of species and even genera problematic; these fruiting structures can be recovered by cultivation, which requires additional time for growth of the organism(s)

Swab samples cannot be extrapolated to indicate potential airborne exposures.

3.2.1.6 *Contact-to-Agar Sampling*

Contact-to-agar sampling involves touching surfaces or visible mold with the surface of a microbial agar. Subsequent agar incubation may detect only the predominant or fastest growing viable mold, which may preclude assessing the entire population unless sample incubation is carefully observed throughout the incubation time frame. Contact plates are limited to an area equal to the size of the Petri dish or agar strip. Contact-to-agar samples have the advantage of directly transferring the microbials to agar. For the more delicate bacteria or mold, this direct transfer may retain viability to a greater degree than impaction or swabbing onto solid media. Contact-to-agar samples cannot be extrapolated to indicate potential airborne exposures.

3.3 AIR TESTING

3.3.1 Settle or Gravity Plates

Settle plates are not considered a valid method for airborne microbial sampling due to the fact that:

- Settle plates are subject to air movements and unpredictable particulate movements and depositions, again making it impossible to predict airborne exposures using this method.
- Spores have differential settling rates according to their weight and aerodynamic form. Settle plates are biased toward large conidia in indoor air, while the proportion of conidia belonging to important small-spored genera such as *Aspergillus* and *Penicillium* is underestimated.

Despite these limitations, settle plates are still used by some investigators, including infection control professionals. Settle plate sampling is not volumetric and therefore cannot be extrapolated to indicate potential quantitative airborne exposures.

3.3.2 Spore Traps

Spore traps are primarily used to determine total spore, pollen, mold vegetative material, and debris counts. The various spore trap air-sampling devices have different capture efficiencies given the same airstream and simultaneous sampling. Some of the more common spore trap impactors are:

- Slit samplers (Figure 3.2)
- New Brunswick slit-to-agar sampler
- Burkard suction slit impactor for direct particulate examination
- Sieve impactors

The main difference in the equipment used is in the ultimate capture media:

- *Membrane filters*: Spores may be trapped for later elution onto a growth medium; also used for polymerase chain reaction (PCR) analyses.
- *Adhesive covered glass slide*: Both the Burkard and Allergenco MK-3 utilize glass slides that require application of the grease-based adhesive by the user. Burkard Manufacturing also offers instruments that can obtain samples over various time frames.
- *Air-O-Cell filter cassette*: The Air-O-Cell cassette has the advantage of not requiring the user to apply the adhesive and comes as a ready-to-use unit.

The major disadvantages to spore trap techniques are the inability to:

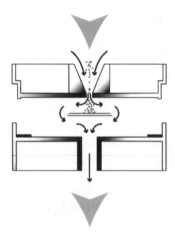

Figure 3.2   Cassettes schematic for slit samplers. (Courtesy of SKC Inc., Eighty Four, PA.)

- Distinguish certain spores from each other — an example is the single, small (2 to 5 μm), clear spores from *Aspergillus* or *Penicillium* or numerous another molds that, upon light microscopic examination, look the same
- Culture the spores for identification to the species or sometimes even genera level and/or lack of biochemical characterization (e.g., mycotoxin-producing capabilities), as the analysis is limited to what can be visually distinguished

3.3.2.1   Static Placement Impingers: May and Burkard

The impinger traps spores from an airstream in a viscous fluid for later plating onto growth media or for biochemical analyses such as PCR. Bacteriologists use the May impingers for the separation of particles according to their deposition sites in the respiratory system. The fractions are collected into a liquid where clumps can separate into single viable units. Sample overload is rarely a problem, and subsamples permit the use of a variety of culture methods. The original designs involved complex glass blowing and were difficult to clean and to reproduce accurately, whereas the Burkard version utilizes anodized aluminum alloy or stainless steel, which eliminates these drawbacks. The May and Burkard impingers separate particles into three fractions: <4 μm, 4–10 μm, and >10 μm.

Impingers have not been widely accepted in ordinary indoor mold and bacteria sampling work. Most potentially problematic airborne molds have highly water-repellent conidia. These conidia, upon contact with the aqueous media of the impingers, tend to exhibit a number of problems that affect efficient recovery including:

- Bouncing off the aqueous phase and passing through the vacuum pump
- Adhering to surface films and hydrophobic surfaces
- Clumping together in minute air pockets

3.3.3   Air-O-Cell Cassette System

The Zeflon Air-O-Cell cassette system is:

- Standardized for collecting single *grab* samples
- Very small, convenient, and easy to use (does not require user to greasing slides)
- Compatible with pumps commonly used in the field for indoor air quality (IAQ) investigations
- Totally disposable

The Air-O-Cell spore trap method:

- Allows detection of *Stachybotrys* and other mold genera that may not be recovered or are overgrown by more rapidly growing species using agar impaction techniques and culturing to determine viability
- Facilitates rapid turnarounds (2 to 3 days) for rapid profiling of buildings and issuing reports
- Yields a more complete representation of the microbial composition of total viable and nonviable spore levels (for spores that do not have to be alive to cause toxic or allergenic effects, total spore levels may be more indicative of exposure and thus the potential risk to human health)
- Does not provide information as to the viable spores present in the airstream, as both viable and nonviable spores are counted and subsequent culturing is not done

Sufficient studies have not been conducted to compare the Air-O-Cell cassette results with viable results; however, the two techniques give considerably different information and the use of both techniques together is the current recommendation.

3.3.3.1 Spatial Cavity Air Testing

Mold will grow only where sufficient moisture and an organic food source are available. A moisture meter is a valuable tool; however, cavities that have dried out or ones where the mold is growing on the opposite drywall surface may escape detection by moisture meters. The traditional method for identifying such cavities has been by core sampling, but core sampling can release aerosolized spores into living spaces and often yields false negative results, as the core must be obtained at exactly the location where active growth is occurring.

The Air-O-Cell cassette is used for testing spatial (wall, ceiling, and floor) cavities. A few feet or so of visible mold can often reveal 30 or more feet of hidden mold in a spatial cavity. The Air-O-Cell cassettes provide a powerful tool for sampling spatial (wall, ceiling, and floor) cavities via the WallChek™ technique:

1. Attach a simple adapter to a Zefon Air-O-Cell Cassette and Tygon hose assembly
2. Drill a 3/8-inch hole into the base of the wall. To prevent obscuring the spore trap with debris, the gypsum dust is removed from the hole and/or immobilized by vacuuming and moistening with water or alcohol
3. Insert the Tygon hose into the wall cavity and gently thump the surrounding wall to help detach spores for capture
4. Draw several cavity air volumes through the Zefon Air-O-Cell Cassette

WallChek™:

- Is rapid and easily done under field conditions with just an air pump
- Helps prevent further contamination of the structure
- Upon microscopic analysis, provides results within a few days of receipt at the analytical laboratory
- Easily detects most molds present within the wall cavities

Note that false negatives have been reported in wall cavity testing for *Stachybotrys*. This may be due to the fact that *Stachybotrys* spores are formed in a mucilaginous layer that can inhibit liberation of the free spores. Thumping on the wall to facilitate aerosolization of the spores prior to sampling may minimize the agglomeration effects that preclude spore liberation. Wall cavity testing does not indicate whether the mold spores collected are viable.

3.3.3.2 Viable Impaction Methods

The viable impactor was developed in the 1950s for biological warfare research. This method involved impaction of particles through precisely machined holes (400 in the case of the Aerotech A-6 or Andersen N-6) onto 100×15-mm Petri dishes filled with microbiological agar medium. The choice of agar medium is based on the target organisms. For instance, mold is often captured on a general medium such as malt extract agar or rose bengal agar, whereas many common bacteria are captured on blood agar (trypticase soy agar amended with 5% sheep's blood). The organisms are then allowed to grow, typically 3 to 5 days for bacteria and 7 to 11 days for mold species, and the organisms are then identified by microscopic and/or biochemical means.

Viable impaction samplers are essential to most investigations because they:

- Capture and quantify airborne bacterial populations
- Indicate the number of viable microorganisms present in air, which is particularly important with pathogens that must be viable to cause disease
- Identify certain mold genera that cannot be readily distinguished in direct microscopic techniques (e.g., spore traps) such as small clear conidia of *Penicillium*, *Aspergillus*, and a number of other genera
- Determine species that can give clues as to potential pathogenicity or toxigenicity (mycotoxins)

Limitations include:

- Turn-around times are long (3 to 11 days for cultures to grow).
- Viable impaction samplers may not give an accurate representation of exposure in the case of mycotoxins or allergens that do not require viability to adversely affect human health.
- Not all viable spores or cells physically captured will grow or these cells may be overgrown by more rapidly growing organisms.
- Recovery of certain mold species, including *Stachybotrys*, is poor due to their slower growth habit, reduced viability, and/or poor competitiveness as compared to other molds that occur in high numbers in indoor air, such as *Penicillium* or *Aspergillus*.
- Organisms sensitive to desiccation, such as *Legionella*, are not amenable to air sampling due to their high mortality.
- Some very slow-growing organisms, such as *Mycobacterium tuberculosis*, are difficult to capture with viable methods, as the medium almost certainly is overrun with competitors before colonies of the bacterium become visible.
- The impaction sampler may injure or kill some of the organism(s) or implant the organism, especially bacteria, too deep into the medium. This is especially true if the pump is not calibrated properly in order to achieve the correct volumetric collection rate or if the target organism is fragile and does not respond well to impaction techniques.
- The investigator may not guess correctly what microbes to target and may fail to provide a proper growth medium that allows the microbes or target microbes to grow.
- Microbes differ in their nutritional requirements and the environmental conditions, such as carbon source, water content, and temperature, required to grow.
- Microbes inherently have different growth rates, and fast growers can overgrow slow growers.
- Microbes also can inhibit growth of competitors by excreting molecules such as antibiotics and toxins into the growth medium, so even though a particular species has been captured that species will not produce a countable colony.

Sampling routines used with mobile samplers such as the SAS, Reuter centrifugal system (RCS), or the Anderson mobile samplers are the sampling instruments of choice.

3.4 CULTURING

The culture media must originally be sterile. Agar preparation chambers are used to ensure sterile initial media (Figure 3.3). All existing sampling media have recognized shortcomings. Thus, the aerobiological ideal of using a perfected, standardized sampling device with a perfected, standardized growth medium to evaluate potential fungal aerosol problems with reference to standard guidelines for acceptable numbers of colony-forming units (CFUs) may not be attainable. The investigator engaged in detecting potentially significant amplifiers must ensure that an adequate diversity of techniques is used to cover the diversity of possible amplifiers.

3.4.1 Media

Agar is added to the medium if a solidified growth platform is required. This agar is generally not in and of itself a nutrient; however, appropriate nutrients are added as needed for the particular target organism(s). For general culture of nonfastidious fungi and bacteria, a medium prepared from plant- or animal-derived material and approximately 2% agar will usually support growth.

3.4.2 Enrichment Culture and Specialized Growth Media

Enrichment is used to favor the growth of one organism over another. Although this qualifies as a selection process, the intent is not to select but to amplify small numbers of target microbes to the detection level. Successive transfers and enrichment may be needed to obtain pure cultures. For instance, fungi are more tolerant to acidic media, so acidification (below pH 6) can be used to aid in limiting bacterial contamination.

3.4.3 Selective and Differential Media

Selective and/or differential media are used in some clinical identification schemes for both fungi and bacteria. The best growth media is also the one where colony overgrowth and formation of spurious satellite colonies may ensue. Thus, shipping considerations and prompt incubation and evaluation in the laboratory are essential.

Fungi vary in their response to water activity in growth media; some prefer high water activity and some prefer conditions to be drier. Usually a combination of sampling methods and media is

Figure 3.3 Agar preparation chambers used to ensure sterile initial media. (Courtesy of Bioscience International, Rockville, MD.)

needed, especially if environments with wet/dry cycles are being considered. No one medium will optimize growth of both significant fungi adapted to high substrate water activity (e.g., *Stachybotrys*) and those requiring less water activity (e.g., *Eurotium, Wallemia*).

3.5 ISOLATION

3.5.1 Streaking

In order to obtain a pure culture, streaking may be needed. The streaking process proceeds as follows:

1. A sterile inoculation loop is dipped into a culture or sample that often contains more than one microbe.
2. The loop is streaked in a pattern over the nutrient media surface.
3. As the pattern is traced, microbes are rubbed off the loop.
4. Fewer and fewer cells are available to be rubbed off as the streaking pattern is concluded.
5. The last cells streaked grow into isolated colonies.
6. After incubation, until colonies are evident, isolated colonies are picked up with a new inoculating loop.
7. The isolated colonies are transferred to new growth media to form a pure culture.

An alternative to streaking is spiral plating.

3.5.2 Plate Counts

Visible cells are counted with the assumption that each viable microbe inoculated has grown into a visible colony without aggregation of cells. The original inoculum is assumed to be homogeneous. Serial dilution may be needed if overgrowth occurs from an inoculum source, as overgrowth prevents the proper counting of distinct colonies and may potentiate die back of sensitive organisms. Die back or occlusion of colonies can lead to false negative counts for the original sample. When the original sample bacterial loading from a media source is low, filtration may be needed. Filtration through a sieve concentrates the bacteria, which is then transferred to the nutrient and agar-filled plate.

3.5.3 Pour Plates

The plate counting procedure is as follows:

1. A sample suspension is prepared.
2. Dilutions of this suspension are poured into a Petri dish.
3. The nutrient media and agar are poured over the suspension, and the agar is kept liquid by placement in a water bath at 50°C.
4. Using gentle agitation, the sample is mixed with the nutrient and agar; cells will then grow within the agar as the agar solidifies.

Disadvantages of this method include:

- Damage to heat-sensitive materials
- Failure of colonies forming beneath the surface to exhibit the characteristics necessary for identification during differential-media-enhanced growth, specifically because the growth is not on the surface and unimpeded

3.5.4 Spread Plate

A prepoured solidified agar medium is used. Inoculum is poured on the surface and spread uniformly over the surface.

3.5.5 Dilution Testing and Spiral Plating

Spiral plating is a method to automatically prepare a sample dilution. This method does not rely on manual pipetting and may significantly increase repeatability and accuracy.

3.5.6 Staining: Putative Identification

Microscopes based on lens resolution are used to identify bacteria and fungi. This identification may occur from samples collected on-site or from samples cultured in the laboratory. With light microscopy the limit of resolution is the resolution of the light waves. Electron microscopy can be used to further define features, as the electron size now is the limit of resolution. Transmission and scanning electron microscopy are types of electron microscopy. For light microscopy, staining may be required. Various dyes may be used to identify the genus and species and specific structures.

3.5.7 Gram Stains

Differential staining is used to provide more information as to morphology. The most common differential stains for bacteria is the Gram stain. The procedure is as follows:

1. Bacteria suspension is applied as a smear and heat fixed.
2. Primary purple stain (crystal violet) is applied.
3. The smear is washed with water to remove excess dye.
4. The smear is covered with an iodine solution, which acts as a mordant. Mordants bind to the crystal violet, resulting in a larger complex that cannot pass through the small pores of the Gram-positive bacteria and thus becomes trapped, rendering the cells purple.
5. The smear is washed, and at this stage both the Gram-positive and Gram-negative bacteria are purple.
6. The smear is washed with an alcohol–acetone decolorizing solution, which removes the purple stain from Gram-negative bacteria.
7. The smear is washed with water; at this stage Gram-positive bacteria remain purple, and Gram-negative bacteria appear colorless.
8. Safrain counterstain is then added to render the Gram-negative bacteria pink.
9. The Gram stain relies on basic differences in the bacterial morphology — the bacterial cell walls.

Note that not all bacteria cells stain effectively, especially if the cells are old and not currently in an active growth phase.

3.5.8 Lactophenol Cotton Blue

Lactophenol cotton blue is a standard stain made by dissolving cotton blue in lactophenol. The stain is applied directly to specimens as the specimens are mounted on microscopic slides. This stain renders fungal cell walls red to blue in coloration.

3.5.9 Specialized Stains

Specialized stains and their uses include:

- Capsular stain, to visualize bacterial capsules
- Flagellar stains, to visualize bacterial flagella
- Fluorescent stains (usually bound to a species-specific antibody), to detect specific fungi, bacteria, or antigenic subgroups

3.6 SPECIAL GROWTH ATMOSPHERES

3.6.1   Anaerobic

Anaerobic organisms must be grown in atmospheres devoid of oxygen. In order to obtain these atmospheres, carbon dioxide or nitrogen is needed.

3.6.2   Carbon Dioxide Enrichment

Slightly elevated CO_2 levels are often provided for mammalian tissue cultures, required for culturing obligate organisms such as viruses and certain fastidious bacteria.

3.7 MEDIA COMMONLY USED IN INDOOR ENVIRONMENTAL STUDIES

3.7.1   Bacteria

- *Blood agar:* Trypticase soy agar (Difco) amended with 5% defibrinated sheep's blood; a good general medium for most Gram-positive and Gram-negative bacteria that is used to isolate and culture both airborne saprophytic and pathogenic organisms
- *Trypticase soy agar* (also known as *soybean-casein digest agar*): Another good general medium for saprophytic organisms
- *Buffered charcoal yeast extract agar* (BCYE): An agar with charcoal, yeast extract, and cysteine for the isolation and cultivation of *Legionella*

3.7.2   Fungi

- *Malt extract agar* (MEA): A good general medium for the isolation of most saprophytic fungi. Over ten published formulations of this medium are available, some containing peptone or yeast extract, others not. In unpublished laboratory studies conducted at Aerotech Laboratories, Inc., the differences between these different formulations was incremental, with formulations only slightly favoring one species over another; however, these differences were less than one order of magnitude. To standardize the industry, the American Conference of Governmental Industrial Hygienists (ACGIH) formulation (ACGIH, 1999) containing 20 g/L malt extract, 20 g/L dextrose, 1 g/L peptone, 15 g/L agar at a pH of 4.5 to 5.0 is recommended by Aerotech Laboratories.
- *Mycological agar:* A good general medium exhibiting incremental differences to MEA and potato dextrose agar (PDA), depending upon the species tested. Supports better growth and more profuse sporulation of *Stachybotrys* but does not give sufficient advantage to overcome overgrowth by other saprophytes.
- *Potato dextrose agar* (PDA): Another good general medium which, again in the studies at Aerotech Laboratories, Inc., was comparable to MEA, with incremental differences observed between different fungal species. Some species appear to sporulate better on PDA as compared to MEA, but again this appears to be an incremental improvement. The choice between MEA, PDA, and mycological agar may not be important and should be left to the discretion of the investigator.
- *DG:-18 agar:* A medium suggested for the isolation of xerotolerant fungi or those that can grow with reduced available water. This medium utilizes over 20% glycerol to reduce the available moisture, favoring aspergilli from the *Glaucus* group and some penicillia.

- *Cellulose agar*: Used to favor organisms that have potent cellulases such as *Stachybotrys*; however, the medium is difficult to use and evaluate due to the opacity of the medium and the poor growth exhibited by all fungi that grow on it. It does not give sufficient advantage to *Stachybotrys* to prevent overgrowth by other organisms during isolation from indoor environments.
- *Corn-meal agar*: Supports growth of many saprophytes and is reportedly better for the growth of *Stachybotrys*; however, as with cellulose agar, this agar does not support abundant growth, is slightly opaque, and does not give sufficient advantage to *Stachybotrys* to prevent overgrowth by other organisms.
- *Sabouraud* (2% dextrose agar): Named for the French dermatologist Raymond Sabouraud (1864–1938), this dextrose–peptone culture medium is used to grow certain fungi that sometimes contain antibiotics but are often pathogenic to humans and other animals. Onygenales, Herpotrichiellaceae, and Ophiostomatales grow on it. Sabouraud is also known as Sabouraud's medium, Sabouraud's dextrose agar, and Sabouraud's dextrose sugar.
- *Rose Bengal agar*: This rose-colored agar is restrictive of overgrowth for some molds and delays or represses sporulation to a lesser extent than Littman oxgall agar. May be relatively robust in shipping while permitting a relatively high level of *in situ* identification. Generates high-energy oxygen species on exposure to light, and illuminated medium may become lethal to some fungi.

3.8 GENERAL BACTERIOLOGY

Bacteria are identified based on colony morphology, nutritional requirements, and specialized tests (e.g., Colilert). For pathogenic organisms, Koch's postulates may be used to establish disease causation. Koch's tuberculosis causation requirements were codified by Loeffler, who produced these conditions for demonstrating the parasitic nature of a disease (Brock, 1961):

- The parasitic organism must be shown to be constantly present in characteristic form and arrangement in the diseased tissue.
- The organism that, from its behavior, appears to be responsible for the disease must be isolated and grown in pure culture.
- The pure culture must be shown to induce the disease experimentally.

Variants of these postulates have been used by generations of microbiologists to describe the necessary steps to show that a microorganism causes a disease.

3.8.1 *Legionella* Water and Biofilm Sampling

The sampling methods of choice for environmental samples include culturing, direct fluorescent screening, and immunochemical methods. All samples should be transported to the laboratory in insulated coolers to protect against temperature extremes. Samples that cannot be delivered to the laboratory within 24 hours of collection should be refrigerated.

For culture sampling, 100 mL of water is sufficient, unless a very low bacteria level is suspected and circumstances dictate concentration of the sample to obtain a detection limit of less than one organism per milliliter. If this detection limit is required, a 1-L sample should be collected. Swabs permit the sampling of biofilms, which frequently contain *Legionella*. The immunochemical methods are described later in this chapter as the RIA and ELISA methods. Sampling may also involve examining water for protozoa and ameba, as these organisms can harbor and therefore hide viable *Legionella* bacteria.

3.8.1.1 Culture

Culturing requires a 10-day incubation period. It is designed as a presence or absence test and is semiquantitative.

3.8.1.2 Direct Fluorescent Antibody Screen

Direct fluorescent antibody (DFA) detects nonviable *Legionella* bacteria and has been reported to cross-react with other Gram-negative bacteria.

3.8.1.3 Latex Agglutination

Latex agglutination is primarily for typing strains and is a clinical tool for the medical diagnosis of *Legionella* in a patient's body fluids. Using a protein group fraction derived from a *Legionella* surface antigen preparation, an antigenic latex diagnostic agent is produced. This agent is immunoglobin specific and the latex particles have one composition and size.

3.8.2 Tuberculosis, Other Mycobacteria

Bioaerosol sampling for mycobacteria may be attempted to verify airborne transmission during epidemiological investigations and studies to evaluate engineering controls; however, such sampling often meets with limited success due to the high probability of contamination that occurs during the extended (several weeks to months) culturing period required for these organisms to grow.

- Air sampling for mycobacteria requires specific, noncustomary air culture media such as Middle-brook 7H10 agar.
- The age of the sampling media is critical for good recovery of mycobacteria; ideally, the sampling media should be 2 to 3 days old at the time of sampling.
- The expected levels of bacteria and fungi should be estimated and several samples collected for different lengths of time at the same location in order to minimize the number of samples with fungal overgrowth.
- The investigator must clarify the purpose for the sampling, identify the specific organism of interest, and discuss the appropriate analytical technique with the laboratory before collecting samples.

3.9 FUNGI: GENERAL MYCOLOGY

Spores may float in the air and, upon contact with wet surfaces (especially wood- or fabric-based ones), the spores can germinate and produce hyphae. If the conditions are right and the proper nutrients are available, the hyphae can continue to grow, form a mass, and become visible to the naked eye. The vegetative mycelium gives rise to more spores, resulting in amplification whereby biological organisms continue to increase in number over time.

In general, fungi grow better with an acidic pH. The growth is usually not limited to the surface and can be embedded within a substrate (under the surface). Fungi are able to grow with a lower moisture content compared to the moisture required for bacterial growth; therefore, even a slight difference in temperature and surface moisture can facilitate the growth of fungi. Fungi are capable of using complex carbohydrates, such as lignin (wood); thus, with a little moisture, fungi can easily grow on wood or other complex organic materials such as painted walls and shoe leather.

3.10 MICROBIAL GROWTH MEASUREMENT

After growth has occurred, counting the microbial colonies (CFUs) is the technique used to provide quantitative information. The methods of counting vary with the type of initial sample obtained. Indirect methods such as turbidity, metabolic activity, and dry weight measurement may also be used.

3.11 FUNGI SPORE COUNTS

Yeasts are single cells that divide to form clusters. Molds consist of many cells that grow as branching threads, or hyphae. The reproductive parts of fungi are spores, which can be either sexual or asexual in origin. Spores differ in size, shape, and color among species and it is on this basis that fungi are traditionally taxonomically classified. Each spore can give rise to new mold growth, which in turn can produce millions of spores. Similar to pollen counts, mold counts may suggest the types and relative quantities of fungi present at a certain time and place. For several reasons, however, these counts probably cannot be used as a constant guide for daily activities.

The number and types of spores actually present in the mold count may change considerably in 24 hours. Weather and spore dispersal are directly related; for instance, rain can effectively remove spores from outside air. Over 60 species of fungi have been reported to cause chronic sinusitis and/or trigger asthma. Most species can potentially be allergenic to susceptible individuals. Identification of fungal species, not just fungal genera, may be required to assess risk and ultimately to develop standards as a basis for comparison. Fungal genus-only identification may result in inaccurate risk analysis as some species are more hazardous than others.

3.11.1 Colony-Forming Units

All contact, bulk, and swab samples that are cultured are usually reported as CFU per gram or CFU per unit area. Aerosolized mold spore counts are converted to colony-forming units per cubic meter (CFU/m^3), if done by viable impaction, or total spores per cubic meter for spore traps. Amplification of biologicals within a building, whether still producing CFUs per unit area or per gram that are less than background or control sample location may still be of concern if the species in the interior differ from those outdoors.

3.11.2 Comparisons

In general, background aerosolized levels of biologicals are defined as those obtained in outdoor locations away from any interior building venting areas. Control samples are defined as those taken in building areas where biological risk is not suspected due to the current status of building materials and systems in these areas. Empirical samples are defined as samples taken in areas where biological risk is potentially of concern.

Environmental (seasonal, climate, weather) variations must be taken into consideration when interpreting fungal concentration ratios. Outdoor fungal levels are influenced by climate and weather. Because of these variations, indoor fungal contamination may differ due to the intrusion of make-up air already influenced by these seasonal variations. Fungal aerosols also vary over time and from space to space. Space activity levels and the activity levels of the sampling humans must be considered. As much uniformity as possible should be achieved. Samplers should sample and move during sampling with defined consistency. Any variations from the defined norm must be documented with the other sampling information.

3.11.3 Normal Ranges

Some individuals and laboratories have developed ranges of certain biologicals that constitute normal conditions. Normal is defined as levels that are expected in certain environments; for example, the normal level in an operating room is less than the normal level in a carpeted hallway where school children are walking. However, normal ranges are dependent on geographic location and prevailing environmental conditions. For instance, normal maximum ranges for Phoenix, AZ, are a few hundred spores per cubic meter, whereas 30,000 spores per cubic meter are not uncommon in the Puget Sound area. Thus, recommended normal levels must be carefully interpreted.

3.11.4 Baseline, Background, and Control Ranges

Another approach is to measure baseline levels during an interval of time and place determined to represent normal conditions. All subsequent numbers as conditions change or become more abnormal are then compared to this baseline. The baseline may be the outdoor environmental levels and, if removed from ventilation pathways within a building, may be referred to as a background level. The term *background* implies that at the specified point in time and in the general geographic environs the levels of biologicals can be expected in certain numbers and ratios. Control ranges may also be defined as baseline. The term *control* implies that an area outside the boundaries of the investigated area is determined to be clean. All data are then compared to the control area. Of course, control areas may also be contaminated, which can lead to erroneous conclusions or, if discovered in time, require the selection of new control areas.

3.11.4.1 Amplification

In situations where one area is exhibiting growth or airborne suspension of biologicals in excess of another area, risk may be determined to exist even if levels are less than outdoor baseline levels and control levels. Amplification phenomena may be thought of as isoconcentrations that exist within an experimental area. The elevated isoconcentration areas may be defined either by identification of the biological types of concern or by identification of elevated levels, or both. These areas may be important indicators of developing problems due to their potentiation of biological growth.

3.11.4.2 Amerospores

Amerospores are small clear spores that range in size from 3 to 5 μm. When using direct microscopic methods, such as on spore traps, amerospores that commonly come from *Aspergillus* and *Penicillium*, but also from a myriad of other fungi, cannot be differentiated.

3.11.4.3 Quantitation Limits: Fungi

Existing quantitation limits are based on environmental (area) sampling; and/or short-term (grab) sampling measurements. Personal exposure assessment is difficult because few airborne fungal aerosol sampling devices can be comfortably attached to clothing in order to measure long-term exposure.

3.12 GOVERNMENT REGULATIONS — QUANTITATIVE

The Russian and Singapore governments have set quantitative regulatory limits on biologicals. Russia issued an official quantitative standard concerning fungi in air. In 1993, the State Committee for Hygiene and Epidemiological Surveillance of the Russian Federation revised their *Maximum Allowable Concentrations (MACs) of Harmful Substances*, which lists chemical and biological standards for industrial settings. Pharmaceutical industries appear to be the primary targets for these regulations. Fungal concentration limits are set for individual fungal and bacterial species based on allergenicity in animal models or on hazard class, or are otherwise not specified. Limits for some agents are based on metabolite or protein concentrations rather than culturable units.

3.13 GUIDELINES

3.13.1 The New York City Department of Health

The New York City Department of Health has developed Guidelines on Assessment and Remediation of Fungi in Indoor Environments as a guide for assessment and remediation of fungi (mold) in indoor environments. Fungal concentrations (high and widespread) in air that require action (e.g., remediation or immediate evacuation) are included by qualitative discussion.

3.13.2 Health Canada

Health Canada developed Indoor Air Quality in Office Buildings: A Technical Guide in 1993, which included these quantitation limits and resultant statements:

Toxigenic, Pathogenic Fungi Counts	Action
50 CFU/m^3 if one species	Investigate
≤150 CFU/m^3 if mixture of species	Allow
≤500 CFU/m^3 if common tree/leaf fungi	Allow in summer

3.13.3 OSHA

In the 1999 *OSHA Technical Manual (OTM)*, the Occupational Safety and Health Administration lists 1000 viable CFU/m^3 as being indicative of fungi contamination. OSHA also provides contamination ranges in the manual for *Legionella pneumophilia*, stating:

> The identification of predominant taxa, or at least fungi, is recommended in addition to determining the number of colony-forming units/m^3 of air (CFU/m^3). During growing seasons, outdoor fungus-spore levels can range from 1000 to 100,000 CFU/m^3 of air. …Levels in excess of the above do not necessarily imply that the conditions are unsafe or hazardous. The type and concentrations of the airborne microorganisms will determine the hazard to employees.

Since that time, lower levels have been suggested as being indicative of contamination. For instance, some investigators list 200 CFU/m^3 as an upper limit for *Aspergillus* in indoor air, with critical analysis being required for opportunistic or toxigenic genera at that level or lower. Given all the variables in sampling, all such limits must be justified as to decision logic, and the decision logic must be provided as needed during the reporting phase.

Most guidelines such as those of the American Industrial Hygienists Association (AIHA), American Conference of Governmental Industrial Hygienists (ACGIH), and the New York Department of Health do not recommend strict numerical values as indicators of contamination. Instead, the accepted standard for the industry is that the inside air should be as good as or better than outside air, both in numbers and in representative genera and/or species. The only caveat to this is when environmental conditions, such as rain or those encountered in the winter, preclude accurate representation of indigenous microbial populations.

3.13.4 Other Organizations

The American Conference of Governmental Industrial Hygienists (ACGIH) Committee on Bioaerosols previously established qualitative guidelines for microorganisms in indoor air; however, the committee currently does not recommend quantitative levels. Similarly, the World Health Organization (WHO) has published *Indoor Air Quality: Biological Contaminants*, which focuses on hazard assessment and preventative maintenance but not quantitation.

3.14 STANDARDIZATION: SAMPLING AND ANALYSIS

Standardization of sampling methodology, incubation routines, and microscopic analysis is necessary in order to ensure that sampling events yield comparable numbers. One problem encountered in the efforts to standardize risk is that no standard sampling and analytical protocols are in common use. Results may vary greatly due to:

- Sampling equipment
- Spatial and temporal location of samples
- Number and statistical significance of samples
- Equipment sensitivity and limits of detection
- Laboratory proficiency
- Culture media and incubation conditions
- Use of biosafety cabinets
- Laboratory airborne microbial levels
- Counting adequacy
- Data analysis and expertise of the investigative team

3.15 SPECIATION

Identification of an organism to the generic level for many species may be sufficient; however, on occasion identification to the species level may be necessary, especially during litigation. The potential health effects after exposure in a contaminated environment are dependent on the species and often even to the specific strain. Although the role of mycotoxins on human health in indoor environments certainly remains controversial, mycotoxins are often a major issue in litigation and potentially may have medical implications. As these toxins are dependent on the species, it may be imperative and in the best interests of the client for these organism to be correctly identified to species.

In the past, speciation has traditionally been an art reserved for highly trained mycologists, but evolving biochemical methods are helping to eliminate the subjectivity involved with classical techniques. Of the chemical methods currently available, the carbohydrate utilization test is the most advanced in regard to its available database and reproducibility. Speciation usually takes several weeks to complete if done correctly, by either the classical or any of the chemical techniques.

3.15.1 Classical Morphological Characterization

For some organisms such as penicillia and aspergilli, morphological characterization requires subculturing purified organisms on specialized media for 7 to 14 days. Following the culturing interval, microscopic examination by a trained mycologist is required. The examination is the official method for speciation. The major disadvantage is that this method is dependent on the subjective opinion of the mycologist.

3.15.2 FAME Analysis via Gas Chromatography

The fatty acid methyl ester (FAME) technique involves subculturing the organism on a specialized medium and then comparing the fatty acid profile to a library of known organisms. Being a chemically based test eliminates the subjectivity of the classical method; however, the database for fungi is limited, which reduces the accuracy of the test. In addition, as opposed to bacteria where fatty acid profiles are highly conserved, fungi are more variable in their fatty acid profiles.

3.15.3 Carbohydrate Utilization

Purified organisms are grown on specialized media and profiled based on their abilities to grow on individual carbon sources. A refinement to the carbon utilization test, utilized by the Biolog system, capitalizes on a color change if the carbohydrate is utilized, facilitating automated reading on a microplate reader. The chemical basis of the test eliminates the subjectivity and is thus more defensible. Currently, however, only a limited database is available for species identification.

3.15.4 Genetic Techniques

Most of the methods currently available or under development involve polymerase chain reaction (PCR). This method is a relatively simple technique by which genetic material is amplified many thousands of times, up to a million-fold, quickly and reliably, thus PCR is very sensitive. Species-specific systems are under development in a number of laboratories and should become available in the next few years. One extension of the PCR technique involves identifying microbes via genetic fingerprinting using a method known as random amplified polymorphic DNA (RAPD). Genetic material extracted from microbes is amplified nonspecifically (randomly), and microbes are identified by their specific banding patterns. This technique is the method of choice for identification of *Aspergillus* species by the Fungal Research Trust in England and many other investigators. As with the other chemically based techniques, this method eliminates subjectivity. Unfortunately, only the database for *Aspergillus* is available.

3.15.5 Polymerase Chain Reaction

Genetic sampling relies on a comparison of polymerase chain reaction components. Polymerase is the enzyme that provides a molecular assist for the assemblage of organic molecules into chains. The presence of certain genetic regions is indicative of a specific mold or bacteria. To date, PCR techniques have been commercialized for about 30 to 35 fungal species found in indoor air, including *Stachybotrys*. Commercial PCR systems are also available for *Legionella* and a number of human pathogens.

3.15.6 Random Amplified Polymorphic DNA (RAPD)

The random amplified polymorphic DNA (RAPD) method is used to locate random segments of the genomic DNA. The RAPD technique is the result of using PCR to amplify DNA synthesized from randomly derived primers. It uses a universal set of primers, and no preliminary work such as probe isolation, filter preparation, or nucleotide sequencing is necessary. The electrophoretic patterns generated from RAPD amplifications can be utilized to speciate organisms or for measuring mutational events.

3.16 CHEMOTAXONOMY

An emerging method for speciation involves chemotaxonomy, where enzyme profiles or secondary metabolites are separated by chromatography and analyzed by mass spectroscopy; pattern associations are made to species or even strains of organisms. One obstacle to the broader application of mass spectrometry in bacterial identification has been the high cost and expertise required for the analysis.

3.17 IMMUNOLOGICAL ASSAYS

Antigen concentrations can be quantified using immunologic methods. These methods measure the quantity of indicator molecules. In order for measurement to occur, the molecules must first be labeled using:

- Radioimmunoassays (RIAs)
- Radioallergosorbent test (RAST), which is used to determine exposures to an allergen such as a mold via detection of allergen-specific antibody production
- Enzyme-linked immunosorbent assay (ELISA), which uses an indicator molecule covalently coupled to an enzyme and is used extensively for allergen assays; utilizes a spectrophotometer to quantify the initial rate at which this enzyme converts a clear substrate to a correlated product
- Limulus amebocyte lysate (LAL), which is not an immunological technique in the strictest sense as it does not involve mammalian antibodies

LAL enlists the primitive immune system of the horseshoe crab, the amebocytes. The LAL assay involves activation of a serine protease by endotoxin, which results in a cascade activation of serine proteases. This protease activity is then detected via a chromogenic substrate. The most sensitive of these techniques is a chromogenic kinetic assay that compares samples to standard endotoxin concentrations.

3.18 ACCREDITATION

Laboratory accreditation:

- Provides a means of determining the competence of laboratories to perform specific types of testing, measurement and calibration
- Allows a laboratory to determine whether laboratory work is being performed correctly and to appropriate standards
- Provides formal recognition to competent laboratories, thus providing a ready means for customers to access reliable testing and calibration services

The general requirements for accreditation include:

- Organization and quality systems
- Document and record control
- Contract review, subcontracting
- Purchasing
- Client service
- Complaint resolution
- Nonconforming testing reporting
- Corrective and preventative action
- Internal audits
- Management reviews

The technical requirements include:

- Personnel qualifications and training
- Standard operating procedures (SOPs)
- Acceptable equipment and facilities
- Correct test methods and method validation
- Measurement traceability

- Sample receiving and traceability
- Quality assurance of test results
- Correct reporting results

Site inspection by a third party or representative of the accrediting body is often required. Most accreditation bodies have now adopted an international guide (ISO/IEC Guide 17025) as the accreditation basis for testing and calibration laboratories. International agreements, known as mutual recognition agreements, are crucial in enabling test data to be accepted among countries. The international mutual recognition agreements among accreditation bodies have enabled accredited laboratories to achieve a form of international recognition and allowed test data accompanying exported goods to be more readily accepted in overseas markets. This effectively reduces costs for both the manufacturer and the importers, as it reduces or eliminates the need for products to be retested in another country.

3.18.1 ISO Guide 58

ISO Guide 58, *Calibration and Testing Laboratory Accreditation Systems — General Requirements for Operation and Recognition*, is prepared by the International Standards Organization (ISO). For an ISO standard or guide to be approved, two thirds of the ISO members who have participated actively in the standards development process and 75% of all members who vote must grant their approval. Many of the industry organizations that evaluate laboratories use ISO Guide 58 as the criteria basis guide for laboratory accreditation.

3.18.2 ISO/IEC 17025

This standard contains the general requirements for the competence of testing and calibration laboratories. The standard applies to test laboratories, including research and development laboratories and calibration laboratories. ISO/IEC 17025 mentions a number of important management and organization aspects that must ensure the quality of the results of the tests and calibrations. This management includes the clear establishment of and accessibility for all employees of clear procedures. Examples of these procedures include:

- Approval and issuance of documents (among other things, test reports) by the laboratory
- Purchasing of goods and services
- Procedures for the control and correction of nonconformities in tests and/or calibrations
- Regular reviews to be carried out by the management of the quality management system
- Evaluations of the test and/or calibration activities
- A complaints-handling system

3.19 ACCREDITING ORGANIZATIONS

3.19.1 American Industrial Hygienists Association (AIHA)

The American Industrial Hygienists Association (AIHA) operates proficiency and accreditation programs specifically for indoor air quality microbiology laboratories. It has applied for international accreditation through ILAC:

- *Proficiency program*: The Environmental Microbiology Proficiency Analytical Testing (EMPAT) Program is for microbiology laboratories specializing in analysis for microorganisms detected in air, fluids, and bulk samples. Laboratories participating in the EMPAT program analyze pure cultures of fungi and bacteria. (*Note:* AIHA plans to eventually include mixed cultures of fungi and/or bacteria as well as samples of water, dust, and building material.)

• *Accreditation program:* Site visit, compliance with ISO 17025, and passage of the proficiency program for a minimum rating of 85% for three consecutive rounds are required to be rated as proficient for bacteria or fungi. Ratings are determined by averaging the performance of the three most recent consecutive rounds. In order to be judged as proficient, a laboratory must be rated proficient for all organism classes that the laboratory accepts for analysis.

3.19.2 International Laboratory Accreditation Cooperation (ILAC)

The International Laboratory Accreditation Cooperation (ILAC) is an international cooperation among the various laboratory accreditation schemes operating throughout the world. Founded 20 years ago, ILAC was formalized in 1996 when 44 national bodies signed a memorandum of understanding (MOU) in Amsterdam. This MOU provides the basis for further development of the cooperation and the eventual establishment of a multilateral recognition agreement among ILAC member bodies. The agreement was designed to further enhance and facilitate the international acceptance of test data and the elimination of technical barriers to trade. ILAC provides advice and assistance to countries that are in the process of developing their own laboratory accreditation systems. These developing systems are able to participate in ILAC as associate members and access the resources of ILAC's more established members. In conjunction with ILAC, specific regions have also established their own accreditation cooperations, notably in Europe (EAL) and the Asia–Pacific area (APLAC). These regional cooperations work in harmony with ILAC and are represented on ILAC's board of management. ILAC is encouraging the development of such regional cooperations in other parts of the globe.

3.19.2.1 European Cooperation for Accreditation (EA)

The European Cooperation for Accreditation (EA) is a multilateral agreement (MLA) for calibration.

3.19.2.2 Asia–Pacific Laboratory Accreditation Cooperation (APLAC)

The Asia–Pacific Laboratory Accreditation Cooperation (APLAC) multilateral mutual recognition arrangement (MRA) is an arrangement to guarantee that laboratories meet the accreditation policies accepted by the signatories. The American Association for Laboratory Accreditation (A2LA) and National Voluntary Laboratory Accreditation Program (NVLAP) are signatories to this MRA.

3.19.2.3 National Cooperation for Laboratory Accreditation (NACLA)

The National Cooperation for Laboratory Accreditation (NACLA) is a nonprofit corporation established to coordinate laboratory accreditation activities within the United States and to serve as the U.S. link to the worldwide lab accreditation system. NACLA evaluates accreditation bodies and grants recognition to accrediting bodies that conduct their assessments and accreditations in accordance with procedures given in the ISO/IEC Guide 58 and requirements of the ISO/IEC 17025 Standard.

3.19.2.4 North American Calibration Committee (NACC)

The goal of the North American Calibration Committee (NACC) is to develop mutual confidence in national calibration laboratory accreditation systems. The participants of the MOU are the

National Metrology Institutes (NMIs) of the United States (National Institute of Standards and Technology, NIST), Canada (National Research Council, NRC), and Mexico (Centro Nacional de Metrología, CENAM) and their recognized representatives. The representatives of the NMIs are the National Cooperation for Laboratory Accreditation (NACLA) in the United States, the Standards Council of Canada (SCC) in Canada, and the Entidad Mexicana de Acreditación (EMA) in Mexico. NACLA, SCC, and EMA intend to accept each other as having primary responsibility in their respective countries for the maintenance of accreditation systems for calibration. A memorandum of understanding establishing NACC was signed at the National Conference of Standards Laboratories International 2001 Conference in Washington, D.C.

3.19.2.5 American Association for Laboratory Accreditation (A2LA)

The American Association for Laboratory Accreditation (A2LA) accredits labs in various broad fields including biological, calibration, chemical, construction materials, electrical, and environmental fields. In addition to these broad fields, specifically tailored programs are available for animal drugs testing, asbestos, environmental lead (Pb), fertilizers, food chemistry, and food microbiology testing. Users of laboratory services are advised to seek the specific scope of accreditation from any accredited laboratory. The scope identifies the tests, types of tests, or calibrations for which the laboratory is accredited.

3.19.2.6 National Voluntary Laboratory Accreditation Program (NVLAP)

The National Voluntary Laboratory Accreditation Program (NVLAP) provides third-party accreditation to testing and calibration laboratories. The National Institute of Standards and Technology (NIST) administers the NVLAP.

3.19.2.7 International Conference of Building Officials (ICBO)

The International Conference of Building Officials (ICBO) is a not-for-profit service organization owned and controlled by its member cities, counties, and states. ICBO is dedicated to public safety in the built environment worldwide through development and promotion of uniform codes and standards, enhancement of professionalism in code administration, and facilitation of acceptance of innovative building products and systems. The founding purpose of ICBO in 1922 was the development of a code that all communities could accept and enforce. This goal was realized in 1927 with the publication of the first edition of the *Uniform Building*.

3.19.2.8 Clinical Laboratory Improvement Amendments (CLIA)

Congress passed the Clinical Laboratory Improvement Amendments (CLIA) in 1988 to establish quality standards for all laboratory testing to ensure the accuracy, reliability, and timeliness of patient test results regardless of where the test is performed. CLIA is user fee funded; therefore, all costs of administering the program must be covered by the regulated facilities. The Health Care Financing Administration (HCFA) assumes primary responsibility for financial management operations of the CLIA program. The Code of Federal Regulations (42 CFR, part 493) promulgated in 1992 (57 FR 7139, Feb. 28, 1992) implemented the CLIA as described herein. Part 493 (*Laboratory Requirements*) sets forth the conditions that all laboratories must meet to be certified to perform testing on human specimens under the CLIA. The term *laboratory*, as used in this standard, refers to a facility for the biological, microbiological, serological, chemical, immunohematological, hematological, biophysical, cytological, pathological, or other examination of materials derived from the human body for the purpose of providing information for the diagnosis, prevention, or treatment of any disease or impairment of, or assessment of the health of, human beings. These examinations also include procedures to determine, measure, or otherwise describe the presence or absence of various substances or organisms in the body. Facilities only collecting or preparing specimens (or both) or only serving as a mailing service and not performing testing are not considered laboratories.

3.19.2.9 National Laboratory System

The Centers for Disease Control and Prevention (CDC) Division of Laboratory Systems (DLS) firmly believes that development of a nationwide laboratory system that must be accredited by CLIA and that provides the communication, coordination, and testing capacity required to effectively detect and report outbreaks and exposures is crucial to the future health and safety of our communities. Recent federal initiatives addressing issues such as bioterrorism, food safety, and emerging infectious diseases have identified similar needs and may provide the funding required for the long-term success of a nationwide laboratory system.

3.19.2.10 Registration CDC Special Agent Transfer Program

The Department of Health and Human Services has published regulations regarding access, use, and transfer of select agents for research purposes. These regulations are designed to ensure that these infectious agents and toxins are shipped only to institutions or individuals equipped to handle them appropriately and only to those who have legitimate reasons to use them and to implement a system whereby scientists and researchers involved in legitimate research may continue transferring and receiving these agents without undue burdens.

RESOURCES

ACGIH, Source sampling, in *Bioaerosols: Assessment and Control*, American Conference of Governmental Industrial Hygienists, Cincinnati, OH, 1999.

AIHA, *Field Guide for the Determination of Biological Contaminants in Environmental Samples*, Dillon, H.K., Heinsohn, P.A., and Miller, J.D., Eds., American Industrial Hygiene Association, Fairfax, VA, 1996.

ALA, *Indoor Pollution in the Office*, American Lung Association, New York, NY, 2002.

American Academy of Pediatrics, Committee on Environmental Health, Toxic effects of indoor air molds, *Pediatrics*, 101, 712–714, 1996.

Atlas, R.M., *Handbook of Microbiological Media*, Vol. 2, CRC Press, Boca Raton, FL, 1996.

Australian National Health and Medical Research Council (NHMRC), Interim National Indoor Air Quality Goals Recommended by the National Health and Medical Research Council (NHMRC), Canberra, ACT, www.nhmrc.health.gov.au, 1996.

Brock, T.D., Ed., *Milestones in Microbiology*, Prentice-Hall, Englewood Cliffs, NJ, 1961.

Canadian Mortgage and Housing Corporation (CMHC), Clean-up Procedures for Mould in Houses, Canadian Government, Ottawa, Ontario, 1993.

Crook, B., Inertial samplers: biological perspectives, in *Bioaerosols Handbook*, Cox, C.S. and Wathes, C.M., Eds., CRC Press, Boca Raton, FL, 1995.

Health Canada, Indoor Air Quality in Office Buildings: A Technical Guide, A Report of the Federal Provincial Advisory Committee on Environmental and Occupational Health, Ottawa, Ontario, 1993.

Hurst, C.J.R., Crawford, L., Knudsen, G.R., McInerney, M.J., and Stetzenbach, L.D., *Manual of Environmental Microbiology*, 2nd ed., American Society of Microbiology Press, Washington, D.C., 2001.

Johanning, E., Ed., *Bioaerosols, Fungi, and Mycotoxins: Health Effects, Assessment, Prevention and Control*, Eastern New York Occupational and Environmental Health Center, Albany, NY., 1999.

Johanning, E. and Yang, C., Eds., *Fungi and Bacteria in Indoor Air Environments: Health Effects, Detection, and Remediation*, Eastern New York Occupational Health Center, Latham, NY.

Lacey, J. and Venette, J., Outdoor air sampling techniques, in *Bioaerosols Handbook,* Cox, C.S. and Wathes, C.M., Eds., CRC Press, Boca Raton, FL, 1995.

O'Andrea, C., Guidelines on Assessment and Remediation of Fungi in Indoor Environments, New York City Department of Health and Mental Hygiene, New York, NY, 2002.

OSHA, OSHA Technical Manual (OTM), Ted 1-0.15A, 1999.

Willeke, K. and Macher, J.M., Air sampling, in *Bioaerosols: Assessment and Control,* American Conference of Governmental Industrial Hygienists, Cincinnati, OH, 1999.

CHAPTER 4

Toxicology

Richard C. Pleus, Harriet M. Ammann, R. Vincent Miller, and Heriberto Robles

CONTENTS

1-56670-606-8/03/$0.00+$1.50
© 2003 by CRC Press LLC

Toxicology is the science that studies poisons. Usually the subjects of study are chemicals to which humans are exposed through contact with air, water, food, and soil. Chemicals can be studied for their effects from the points of view of determining either potency or exposure through inhalation, ingestion, or skin penetration. Biological contaminants also include chemicals such as irritants or naturally occurring poisons called *toxins*, which are produced by living organisms. Biological contaminants may include microorganisms that have the potential to do harm. A number of biological contaminants also have allergenic or infectious properties that are not evaluated the way toxic exposures of chemicals are; yet, the allergic or chemical properties may complicate the toxicity of chemical and other bio-contaminants.

4.1 DOSE–RESPONSE RELATIONSHIP: THE DOSE MAKES THE POISON

Toxicology is the scientific study of adverse effects of chemicals on living organisms. This science recognizes that chemical substances can be either beneficial or deleterious to a living organism. Paracelsus first articulated this relationship in the 15th-century: *All substances are poisons; there is none which is not a poison. The right dose differentiates a poison from a remedy.* Beneficial effects of chemicals include providing energy, nutrients, and protection to the organism. Adverse effects, however, can occur if the chemical concentration adversely influences how cells, tissues, and organisms function. The degree of harm or the influencing factors of toxicity are related to:

- Chemical and physical properties of the chemical (or its metabolites)
- Amount of the chemical absorbed by the organism
- Amount of chemical that reaches its target organ of toxicity
- Environmental factors and activity of the exposed subject (e.g., working habits, personal hygiene)
- Duration, frequency, and route of exposure
- Ability of the organism to protect itself from a chemical

One commonly hears of the concentration of a potentially hazardous agent in a medium (e.g., caffeine in coffee, benzene in air, dioxin in soil, lead in water, *Escherichia coli* in food). In addition to exposure concentration, characteristics of a chemical that affect absorption, metabolism, and excretion; its route of exposure; and duration of exposure are other elements that must be evaluated to determine risks of adverse effects. For a chemical to exert its effect, the chemical must be present in high enough concentrations at the target site to cause an adverse effect.

Most living organisms have defenses to protect them from the adverse effects of chemicals encountered daily. Mammals have a considerable number of defenses (e.g., liver detoxification, kidney excretion, skin barrier). Adverse effects occur when the dose received by the organism is high enough to overwhelm the organism's defense mechanisms.

The maximum dose that results in no adverse effects is called the *threshold dose*. Many chemical agents have a threshold dose. The concept of threshold implies that concentrations of exposure present are so low that adverse effect cannot be measured. Some notable exceptions occur, such as when a person develops an allergic reaction to a chemical (only specific chemicals are capable of causing allergic reactions).

Another exception, although controversial, is chemicals that cause cancer. Given our current lack of understanding of the mechanisms that lead to cancer initiation and development, regulatory agencies have adopted the position that any dose of a carcinogen has an associated risk of developing cancer. Scientifically, not all carcinogens are in fact capable of causing an effect at low doses; however, the problem is that no one knows what the dose must be in order to cause an effect, so to be safe the dose is set as low as practicable (usually at the limit of detection for instrumentation).

For biological exposures, the concept of a threshold dose applies to microbial organisms or their chemical metabolites. Toxicology applies to biological exposures by addressing:

- Chemicals released from living organisms (e.g., metabolic byproducts, secretion of toxins, volatile organic compounds)
- Aerosolized fragments of biological organisms (e.g., bacterial or fungal organisms, spores, hyphae, organismal structures)

The toxicity potential of various biological contaminants has been determined to differing extents. For example, volatile irritants that are part of everyday metabolism are no different from those produced by industrial or laboratory processes. For many of these solvents, potency is well characterized for various exposure routes. Other contaminants, such as bacterial or fungal toxins (e.g., mycotoxins), vary greatly in the extent of knowledge about their potency. Some, such as those commonly found in foodstuffs or those that may have pharmaceutical usefulness, have been well studied. For instance, aflatoxin, produced by *Aspergillus flavus* and some other molds, is among the most studied natural molecules known. Other toxins have had only crude comparative toxicity estimates made. Because of their potential economic importance, pharmaceutical companies test for toxins from molds and bacteria, and new toxins, as well as organisms not previously known to produce toxins, are actively investigated.

The concept of dose, then, encompasses two aspects:

1. Inherent potency (modulated by degree of absorption, defense, and removal of test animals or humans) to target organs
2. The amount and duration of exposure

4.2 POTENCY

4.2.1 Effective Dose

Effective dose is a term that is used to:

- Define the therapeutic levels for medications
- Denote the beginning of an adverse level in animal experiments
- Define the level at which a medication produces a desired effect
- Define the experimental dose at which a chemical causes a measurable effect

The therapeutic index for pharmaceuticals is obtained by dividing the median lethal dose by the median effective dose; the larger the ratio, the greater the relative safety of the drug.

4.2.2 LD_{50}

A dose concept that is used for crudely comparing the level of effect of various chemicals, the lethal dose 50% (LD_{50}), or median lethal dose, is the dose estimated to produce mortality in 50% of the exposed animals. LD_{50} only describes exposure levels that produce death and may differ with exposure routes and the animals being tested. For instance, guinea pigs tend to be more sensitive than rats or mice. Subtleties of target dose, metabolism, detoxification, or mechanism of action are not revealed by such experiments. Table 4.1 illustrates the variability in the LD_{50} of trichothecenes for mice vs. rats vs. guinea pigs.

Table 4.1   LD$_{50}$ Values (mg/kg) of Trichothecenes

Type	Trichothecenes	Mouse				Rat				Guinea Pig		
		i.v.	i.p.	s.c.	Oral	i.v.	i.p.	s.c.	Oral	i.p.	s.c.	Oral
A	T-2 toxin		5.2		10.5				5.2			3.06
	HT-2 toxin			9.0								
	DAS	12		23.0			1.3	0.75		7.3		
	Neosolaniol		14.5									
	Monoacetoxy Scirpenol								0.725			
B	Nivalenol		7.3	7.4	7.2	38.9						
	Diacetylnivalenol		9.6									
	DON		70.0	46.0								
	3-acetyl-DON		49.0	34.0								
	Trichothecin	300.0							250			
C	Roridin A	1.0										
	Verrucarin A	1.5	0.5						0.87			
	Verrucarin B	7.0										
	Verrucarin J		0.5									

Abbreviations: i.v., intravenous; i.p., intraperitoneal; s.c., subcutaneous.

Source: Adapted from Ammann, H.M., *Bioaerosols, Fungi and Mycotoxins: Health Effects Assessment, Prevention and Control*, Johanning, E., Ed., Eastern New York Occupational and Environmental Health Center, Albany, 1999. With permission.

4.2.3   Toxicological Interactions

The most current means of assessing toxicology of mixtures is to assume that the effects of mixtures are additive. This is not always the case, however. For example, some chemicals have effects that cancel out or reduce the toxicity of each other, and the toxicity of some individual chemicals is greater than the sum of each. Some mixtures, such as those resulting from various forms of combustion, have been approached with a concept of relative potency for carcinogenicity (Lewtas et al., 1987). Another way to assess effects of exposure to more than one substance is to design experiments where test subjects are exposed to more than one chemical substance. The purpose of these experiments is to see whether simultaneous exposure to two substances enhances or diminishes the effect of one chemical alone. Toxicologic interactions may be defined as additive, synergistic, or antagonistic. All may express:

- Response by the host to chemical/biological exposure
- Positive responses by the host to low doses of chemical/biological agents (e.g., enhanced resistance, enhanced biodegradation [i.e., enzyme activation], vaccination)
- Negative responses by the host to chemical/biological changes, such as cell death and tissue damage, altered organ function (e.g., olfactory paralysis caused by hydrogen sulfide or central nervous system intoxication by solvent inhalation), systemic toxicity, tissue irritation, abnormal immune responses (e.g., sensitivity, allergy, asthma), or cancer

Interactions may occur as the result of synergy, additivity, potentiation, or inhibition. The nature of the interaction may reflect the underlying mechanism so that two toxins acting on the same receptor are likely to have an additive rather than a synergistic effect or, alternatively, two toxins acting at related but different receptor sites may exhibit synergy.

The analysis for showing interactions must be based on dose–response relationships rather than concentrations. Because dose–response curves can have dramatically different slopes, combinatory analyses must be based on these curves. The most common analyses for interactions utilize isobolograms, which are based on dose–response curves of each toxin given separately and in combination. For example, consider the case where two cytotoxic compounds are being evaluated. The isobolograph plots compound A vs. compound B, and the combinations will give 100% of the endpoint cytotoxicity; a concentration of compound A that will give 25% cytotoxicity (from the

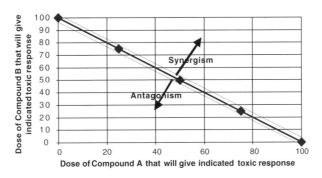

Figure 4.1   Isobologram of interaction of compounds A and B. (From Miller, R.V., Martinez-Miller, C., and Bolin, V., *Proc. Tenth Int. IUPAC Symp. on Mycotoxins and Phycotoxins*, Ponson and Looijen, Wageningen, 2000. With permission.)

dose–response curve of compound A) is added to a concentration of compound B that will give 75% of the endpoint cytotoxicity. If the combination yields 100% of the cytotoxicity, then the compounds are additive; if the combination gives more cytotoxicity, then the interaction is synergistic; and if the interaction is less than 100%, then the interaction is antagonistic.

More complex analyses are usually done using response surface analyses integrating isobolographic principals. Again, these combinations are made based on dose-response relationships, not on concentration. Such isobolographic analyses have been widely used for some time in the study of drug and pesticide interactions in the pharmaceutical and agrichemical industries, respectively. Figure 4.1 shows a simple isobologram.

An additive effect, in its most simple form, means a sum of the toxic effects produced by the chemicals. An antagonistic effect, in simple form, means a decrease in effect (e.g., a classic example of antagonism is the use of an antidote to a poison). A synergistic effect is a multiplication of effects. Because interactions are actually very complex, these terms are used as generalities when describing interactions. Measurement of interactions requires highly complex, three-dimensional characterizations such as isobolographic analysis.

4.2.4   Entry into the Body

Biological and chemical agents enter the body through several portals of entry, including:

• Oral ingestion
• Inhalation
• Dermal absorption
• Injection (subcutaneous, intramuscular, or intravenous)

In natural settings outside of the laboratory, exposure occurs from:

• Breathing air that contains the chemical or biological agent (inhalation exposure)
• Consuming food or water that contains the agent (oral or ingestion exposure)
• Contact and penetration of the skin (dermal exposure)

In the laboratory, chemicals may be deliberately introduced via all routes of exposure so that the effect of route of entry and subsequent dose to the target organ can be evaluated. Examples of laboratory methods include injection or instillation of a chemical or biological agent:

• Into the bloodstream (intravenous, i.v.)
• Into the membrane that lines the abdominal cavity (intraperitoneal, i.p.)

- Under the skin (subcutaneous, s.c.)
- Into the muscle (intramuscular, i.m.)

These routes vary in the time and extent of distribution of the introduced chemical, and this variability may affect the dose that gets to the target organ. Each of these portals of entry provides a route of exposure and has barriers to entry.

4.2.5   Barriers to Entry

Barrier to entry are defined by some type of defense mechanism (e.g., a physical barrier such as the keratin layer of the skin or the destruction of biological and chemical agents in the intestinal tract), and they influence the amount of chemical that actually gets to the organ or system (target organ of toxicity) where harm can occur. For many of the hazards in the environment, inhalation, ingestion, and dermal exposure are the only routes of exposure. The extent to which these routes allow chemicals to be adsorbed into the body depends on the degree of contact these exposure routes have with the vascular system, health of the system and the body, the amount of surface area available for contact, and the physical and chemical nature of the chemicals.

4.2.6   Metabolism, Activation, and Detoxification

A chemical enters the body by absorption (via one of the exposure routes), is distributed to tissues in the body, can be biotransformed (metabolized), and may be excreted (exits the body). In general, each of these processes can be considered as a protective mechanism, a barrier, a means of detoxifying, or a physical defense — all working to protect the body from harm, all with differing degrees of effectiveness. In some cases, metabolism will increase the potency of a toxin. The various defenses against harmful effects are related in some part to the biological port of entry through which exposure occurs.

4.2.7   Excretion

Excretion, along with metabolism, is one of the major tools used by organisms to protect themselves against potentially toxic compounds. Excretion is the elimination of absorbed foreign substances. The major function of the liver and kidneys is the excretion of nonvolatile, water-soluble substances. Volatile substances are eliminated mostly through the lungs. Non-water-soluble substances, if transformed into water-soluble substances in the liver, can be eliminated in the urine. Non-water-soluble substances that cannot be transformed are excreted very slowly through the bile and feces. To a lesser extent, chemicals can also be excreted through sweat and breast milk. For example, lactating mammals can excrete non-water-soluble substances (e.g., DDT or polychlorinated biphenyls) in mother's breast milk. The excretion rate of chemical substances is of toxicological importance. For many noncarcinogenic chemicals, the dose of a chemical that exceeds a threshold dose can be interpreted as the body's ability to transform and/or excrete the chemical. For example, consumption of alcohol at a rate faster than the liver can transform the alcohol and the kidneys can eliminate the metabolites of the alcohol results in alcohol intoxication.

4.3 EXPOSURE

For an adverse effect to take place, the following conditions have to be met:

1. The subject must be exposed to the potentially toxic agent.
2. The potentially toxic agent must be present in a form that is available for introduction into the body by any of the natural routes of exposure.
3. Exposure conditions must be favorable so that the potentially toxic compound is absorbed by the organism.
4. The exposure dose and duration are high enough to result in toxic doses at the target organ.

In this section, terms used to determine exposures to hazardous agents are defined. To accurately estimate a chemical exposure and reduce the uncertainty associated with this exposure estimation, some toxicologists endeavor to improve the scientific methods by which such exposure assessments are accomplished. Improvements have been made in determining exposure factors, exposure models, and exposure measurement technologies. For example, computer models predict future exposure scenarios from dose information and from experience in past human exposure studies.

The important information to consider when assessing the potential hazard posed by a chemical or biological organism includes the inherent potency (for biological agents, this would be the toxicity, pathogenicity, or potential for allergenicity of the organism or the metabolic products of the organism), dose received, and length of exposure. The concept of dose includes the amount of chemical absorbed into the body, time, and the target organ.

4.3.1   Acute, Subacute, Subchronic, and Chronic Exposure

Terms such as *acute*, *subacute*, *subchronic*, and *chronic* are used to indicate duration and frequency of exposure. Typical guidelines associated with these terms are:

- Acute exposure is short term, usually < 24 hours; for animal inhalation studies, acute exposure is 4 hours.
- Subacute exposure is repeated exposure to a chemical for 30 days or less.
- Subchronic exposure lasts for 30 to 90 days.
- Chronic exposure exceeds 3 months.

For human exposures in building interiors, acute exposure usually means a one-time exposure, while chronic exposure occurs over longer intervals, usually at least months to years.

4.3.2   Severity and Duration

While the terms *severity* and *duration* would seem to apply only to duration of exposure, some implication of degree of exposure (short-term, high dose; long-term, low dose) may also be implicit. These implications have some bearing on the severity and duration of effect. Severity and duration of effects are implied in other concepts related to dose. Another way of considering a threshold dose is to think in terms of a level at which the body's defenses are overcome, and damage begins to be observable or even measurable.

4.3.3   Single Pathway Exposure

Single pathway exposure refers to a subject being exposed to an agent by a single route of exposure. For example, a hazardous agent is introduced into a subject by only one of the portals of entry (i.e., inhalation).

4.3.4   Multimedia Exposures

Multimedia exposures occur when a subject is exposed to an agent by more than one medium. Most commonly, media include food, air, soil, and water. So, if a subject is exposed to more than one medium, the subject might be eating food and drinking water that contains a similar hazardous agent.

4.3.5 Multipathway Exposures

Multipathway exposure refers to a subject being exposed to an agent by more than one portal of entry. For example, a hazardous agent could be introduced into a subject through breathing, such as by inhaling emissions downwind of a combustion facility, and by eating meat containing the chemical as a result of emissions from the combustion facility depositing on plants used to feed livestock.

4.4 ROUTES OF EXPOSURE

4.4.1 Inhalation

When inhaled, microscopic fungal spores and sometimes fragments of fungi may cause health problems. Small mold spores (see Figure 4.2) may evade the protective mechanisms of the nose and upper respiratory tract and reach the lungs. Once in the alveolar region of the lungs, immune cells of the organisms can detect the microscopic spores. The immune cells attack the invading organisms. The attack by the immune cells causes collateral damage to alveolar cells. The repeated attack and damage may cause lung diseases, including emphysema and possibly asthma. Symptoms associated with asthma include the buildup of mucus, wheezing, and difficulty in breathing. Less frequently, exposure to spores or fragments may lead to a lung disease known as hypersensitivity pneumonitis.

4.4.2 Dermal Exposures

The skin is a target organ for many irritating and potentially toxic chemicals as well as for many pathogenic organisms. The skin is a complex organ with many and varied functions and abilities. Some of the most important functions of the skin include regulating body water, electrolyte, and temperature balances; acting as a shock absorber; providing a barrier against foreign objects, organisms, and chemicals; and providing protection against harmful effects of ultraviolet light. For these reasons, biological and chemical agents that affect the skin can also affect various organs and may, in fact, compromise the well-being of the organism.

Intact skin is not a perfect barrier, and some chemicals and organisms are able to cross the skin barrier without having an effect on the skin. The ability of some chemicals to cross the skin without directly affecting the skin itself is used today to administer medications through skin patches. The protective ability of the skin may be diminished by skin damage (e.g., cuts, abrasions, psoriasis, acne). In such cases, pathogenic organisms and potentially toxic chemicals may enter the body through the damaged area without having a direct effect on the surrounding skin. This effect is of toxicological importance as the dermal doses required to produce an adverse effect in an individual with damaged skin are lower than the doses needed to produce the same effect in an individual with healthy skin.

As with any toxicological phenomena, adverse effects produced in the skin are directly related to the amount of chemical applied to the skin as well as to the exposure duration. However, unlike other pathways of chemical exposure, dermal uptake can be enhanced by increasing the skin surface area in contact with the chemical; covering the area of application (occlusion); applying the chemical in abraded or damaged skin; co-applying certain organic solvents, oils, and lotions; or co-applying irritating or corrosive substances.

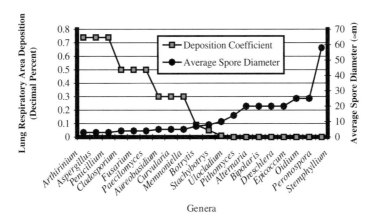

Figure 4.2   Spore deposition coefficients of fungal genera found in indoor environments.

4.4.3   Ingestion Exposures

For indoor biological exposure agents, inhalation and dermal routes are the primary pathways of exposure; however, because airway clearance of particulate pollutants involves swallowing mucous that the respiratory system cilia sweep toward the oropharynx, ingestion can be a minor pathway of exposure.

4.5 EFFECTS FROM EXPOSURE

The manifestation of adverse effects falls into four general categories: altered immune response (allergy), irritation, infection, and toxicity.

4.5.1   Altered Immune Response (Allergy)

The Institute of Medicine (part of the National Academy of Sciences) stated that allergy is the most common chronic disease of humans (Pope et al., 1993). Allergy can include such symptoms as those resembling hay fever, sneezing, runny nose, red eyes, watery eyes, skin rash (dermatitis), cough, sneezing, fatigue, digestive problems, dizziness, difficulty breathing, and headache (due to sinus congestion), as well as other skin reactions. Serious allergic illness such as asthma and less frequently hypersensitivity pneumonitis may occur.

Allergic reactions may occur only after repeated exposure to a specific biological allergen. The reaction may occur immediately upon reexposure or after multiple exposures over time. As a result, people who have noticed only mild allergic reactions or no reactions at all may suddenly find themselves very sensitive to particular allergens. Repeated exposure has the potential to increase sensitivity.

Bioaerosols contain many potentially allergenic substances. Generally, such substances are called antigens and are usually proteinacious, although some small molecules can join with adjuvants and elicit allergic reactions. Among allergenic agents in bioaerosols are:

- Pollens
- Bacteria
- Amebae
- Algae
- Insects and their body parts and effluvia (e.g., dust mite fecal allergens)
- Molds

Fungi or similar microorganisms may cause other health problems in which allergy may play a role. Fungi may lodge in the airways or in the deep compartments of the lung and grow into a compact sphere known as a *fungus ball*. In people with lung damage or serious underlying illnesses, *Aspergillus* may grasp the opportunity to invade and actually infect the lungs or the whole body. The occurrence of allergic aspergillosis suggests that other fungi might cause similar respiratory conditions. In some individuals, exposure to certain fungi can lead to asthma or to an illness known as allergic bronchopulmonary aspergillosis (ABPA). This condition, which occurs occasionally in people with asthma, is characterized by wheezing, low-grade fever, and coughing of brown-flecked masses and mucous plugs. Skin testing, blood tests, x-rays, and examination of the sputum for fungi can help establish the diagnosis.

Inhaling or touching mold or mold spores may cause allergic reactions in sensitized individuals. Allergic responses include hay-fever-type symptoms, such as sneezing, runny nose, red eyes, and skin rash (dermatitis). Allergic reactions may occur only after repeated exposure to a specific biological allergen. The reaction may occur immediately upon reexposure or after multiple exposures over time. As a result, people who have noticed only mild allergic reactions or no reactions at all may suddenly find themselves very sensitive to particular allergens. Repeated exposure has the potential to increase sensitivity. Fungus spores and fragments can produce allergic reactions in sensitive individuals regardless of whether the fungus is dead or alive.

4.5.2 Asthma

According to the Institute of Medicine, asthma prevalence and incidence are increasing for reasons not clearly known (Pope et al., 1993). Asthma is a serious respiratory disease characterized by inflammation of airways, with and without symptoms, obstruction of airways from airway constriction, and secretion of thick mucus that results in difficulty in breathing during an asthmatic attack. Asthma is a complex disease that varies in individuals. Allergic sensitization to environmental antigens appear to play a role both in the initiation of asthma as a disease and in the initiation of asthmatic attacks. Exposure to cold, to respiratory irritants, odors, and even exercise can initiate asthmatic attacks, depending on the characteristics of disease in the individual.

4.5.3 Hypersensitivity Pneumonitis

Inhalation of spores from fungus-like bacteria (e.g., actinomycetes) and from molds can cause the lung disease termed hypersensitivity pneumonitis, which may develop following either short-term (acute) or long-term (chronic) exposure to molds. The disease resembles bacterial pneumonia. Hypersensitivity pneumonitis is often associated with specific occupations and develops in people who live or work in environments with high concentrations of aerosolized fungus and bacteria. Symptomatically, hypersensitivity pneumonitis resembles bacterial or viral infections such as the flu or pneumonia and may lead to serious heart and lung problems.

4.5.4 Irritant Effects

Exposure to irritant substances can cause irritation of the mucous membrane in the eyes and respiratory system or irritation of the nerve endings, resulting in strange sensations and cognitive and other central nervous system changes (described more fully in Chapter 5). Microbial volatile organic compounds (mVOCs) are compounds produced by molds; they are vaporous and are released directly into the air. Because these compounds often have strong and/or unpleasant odors, they can be the source of odors and irritants associated with molds. Exposure to VOCs has been linked to symptoms such as headaches, nasal irritation, dizziness, fatigue, and nausea. Measurement of mVOCs is considered by some researchers to be a diagnostic tool for determining mold growth in a building.

4.6 TOXICITY

Both bacteria and mold can produce biological poisons known as toxins.

4.6.1 Bacterial Endotoxins

Endotoxin is the name given to a group of heat stabile lipopolysaccharide molecules present in the cell walls of Gram-negative bacteria that have a certain characteristic toxic effect. The lipid portion of each molecule is responsible for the molecule's toxicity and can vary among bacterial species and even from cell to cell. Endotoxin is common in the environment due to the ubiquitous nature of Gram-negative bacteria. Exposure to elevated levels of endotoxin primarily occurs through exposure to aerosols from specific reservoirs such as cotton mills, metalworking fluids, wastewater treatment facilities, indoor swimming pools, air washers, or humidifiers and in any other occupational settings where Gram-negative bacteria can flourish. When inhaled, endotoxins may create an inflammatory response in humans that can result in fever, malaise, alterations in white blood cell counts, headache, respiratory distress, and even death.

4.6.2 Bacterial Exotoxins

Bacteria can produce exotoxins that invade host cells and cause the adverse effects recognized as disease symptoms. For instance, *Bacillus anthracis* produces exotoxins that cause the disease anthrax. Other soil bacteria such as *Clostridium botulinum* and *Clostridium tetanii* can also produce toxins and secrete these toxins into the environment. Such bacteria are generally associated with exposure through contact with soil and ingestion or skin penetration exposures and not with inhalation exposures in indoor environments. Unusual exposures could result from bioterrorism exposures, as the spores of some of these bacteria have been developed for use as weapons. These spores can be genetically altered or chemically treated to concentrate the toxins or ground to a size effective for air dispersion. Effective air dispersion is defined as long float time in airstreams and/or resuspension potential. The intent of particulate size alteration is ultimately to make the spore more available over a given time frame for potential inhalation.

4.6.3 Fungal Toxins

Molds can produce potentially toxic substances called mycotoxins. Many common environmental fungi produce secondary metabolites that are potentially toxic to eukaryotic cells. The term *mycotoxin* is commonly used to refer to these compounds. Some mycotoxins cling to the surface of mold spores or are found in dust. More than 200 mycotoxins have been identified from common molds, and many more remain to be identified. Some of the molds that are known to produce mycotoxins are commonly found in moisture-damaged buildings. Exposure pathways for mycotoxins can include inhalation, ingestion, or skin contact. Although some mycotoxins are well known to affect humans and have been shown to be responsible for human health effects, little information is available for many mycotoxins.

Aflatoxin B_1 is perhaps the most well known and studied mycotoxin. Aflatoxin B_1 can be produced by the molds *Aspergillus flavus* and *A. parasiticus* and is one of the most potent carcinogens known. Ingestion of aflatoxin B_1 can cause liver cancer, and some evidence exists that inhalation of aflatoxin B_1 can cause lung cancer. Aflatoxin B_1 has been found on contaminated grains, peanuts, and other human and animal foodstuffs; however, *A. flavus* and *A. parasiticus* are not commonly found on building materials or in indoor environments.

Many symptoms and human health effects attributed to inhalation of mycotoxins have been reported, including mucous membrane irritation, skin rash, nausea, immune system suppression, acute or chronic liver damage, acute or chronic central nervous system damage, endocrine effects, and cancer. More studies are needed to obtain a clear picture of the health effects related to most mycotoxins.

The production of mycotoxins by fungi and the accumulation of mycotoxins in fungal spores are dependent upon environmental conditions (e.g., substrate, temperature, and humidity) and the species and strains of fungi and the presence of competitive organisms. Detection of a fungal species known to be toxigenic does not imply mycotoxin exposure.

Much of the information on the human health effects of inhalation exposure to mycotoxins comes from studies done in the workplace and some case studies. These studies have revealed such mycotoxin effects as immunosuppression, carcinogenesis, cytotoxicity, neurotoxicity (including acute or chronic central nervous system damage), mucous membrane irritation, skin rash, nausea, acute or chronic liver damage, and endocrine effects. These effects may be independent of infection or stimulation of antibodies (in contrast to the mycobacterial mycotoxins).

Some molds can produce several compounds of toxicological importance. Molds such as *Aspergillus versicolor* and *Stachybotrys chartarum* (formerly *S. atra*) are known to produce potent toxins under certain circumstances. In addition, preliminary reports from an investigation of an outbreak of pulmonary hemorrhage in infants suggest an association between pulmonary hemorrhage and exposure to *S. chartarum*.

Information on ingestion exposure, for both humans and animals, is more abundant than for inhalation. A wide range of health effects has been reported following ingestion of moldy foods, including liver damage, nervous system damage, and immunological effects.

4.7 MYCOTOXIN TYPES (INDOORS)

Over 20 mycotoxins have been detected in indoor environments, but some of the more common and relevant mycotoxins include trichothecenes, produced by certain species of *Stachybotrys*, *Trichoderma*, and *Fusarium*; aflatoxin and sterigmatocystin, produced by a number of species of *Aspergillus*; ochratoxin, produced by various species of *Aspergillus* and *Penicillium*; and griseo-fulvins, produced by certain species of *Memnoniella* and *Penicillium* (Macher et al., 1999; Jacobsen et al., 1993). Recent advances in technology have given laboratories the ability to test for specific mycotoxins without employing cost-prohibitive gas chromatography or high-performance liquid chromatography techniques. Currently, surface, bulk, food and feeds, and air samples can be analyzed for the mycotoxins given in Table 4.2. Other mycotoxins of clinical significance are also provided in the table.

More research is needed on other mycotoxins, including penicillic acid, roquefortine, cyclopiazoic acid, verrucosidin, rubratoxins A and B, PR toxin, luteoskyrin, cyclochlorotine, rugulosin, erythroskyrine, secalonic acid D, viridicatumtoxin, kojic acid, xanthomegnin, viomellein, chaetoglobosin C, echinulin, flavoglaucin, versicolorin A, austamide, maltoyzine, aspergillic acid, paspaline, aflatrem, fumagillin nigragillin chlamydosporol, and isotrichodermin, among others. More research is required in this field to better understand the relationships of fungal contamination, mycotoxin production on building substrates, and building-related disease.

4.8 RESEARCH NEEDS

Essential dose–response information is needed to correlate numbers of fungal spores with a particular chemical composition to health effects in humans. Acceptable dose information would doubtless be arduous to acquire even if laboratory tests could be devised. Surrogate tests such as tests for the responses *in vitro* of human cells (e.g., alveolar macrophages) are in their infancy, and animals lack the ability to corroborate or deny the persistent, subjective symptoms commonly reported in cases of indoor mold proliferation. The need for objective measures of adverse responses to mold inhalation is great, and devising such measures would be an important step in developing scientific correlates between spore counts and the need for remediation of buildings.

Table 4.2 Mycotoxins Produced by Some Fungi

Aflatoxin	Aflatoxin is a potent carcinogen and has been associated with a wide variety of human health problems. The FDA has established maximum allowable levels of total aflatoxin in food commodities at 20 parts per billion. The maximum level for milk products is even lower at 0.5 parts per billion. Primarily *Aspergillus* species produce aflatoxin.
Alternariol	Alternariol is a cytotoxic compound derived from *Alternaria alternata*.
Citrinin	Citrinin is a nephrotoxin produced by *Penicillium* and *Aspergillus* species. Renal damage, vasodilatation, and bronchial constriction are some of the health effects associated with this toxin.
Fumonisin	Fumonisin is a toxin associated with species of *Fusarium*. Fumonisin is commonly found in corn and corn-based products, with recent outbreaks of veterinary mycotoxicosis occurring in Arizona, Indiana, Kentucky, North Carolina, South Carolina, Texas, and Virginia. The animals most affected were horses and swine, resulting in dozens of deaths. Fumonisin toxin causes "crazy horse disease," or leukoencephalomalacia, a liquefaction of the brain. Chronic low-level exposure in humans has been linked to esophageal cancer. The American Association of Veterinary Laboratory Diagnosticians (AAVLD) advisory levels for fumonisin in horse feed is 5 ppm.
Gliotoxin	Gliotoxin is an immunosuppressive toxin produced by species of *Alternaria, Penicillium, Aspergillus*, and *Stachybotrys*.
Ochratoxin	Ochratoxin is primarily produced by species of *Penicillium* and *Aspergillus*. Ochratoxin damages the kidneys and liver and is also a suspected carcinogen. Ochratoxin may impair the immune system.
Patulin	Patulin is a mycotoxin produced by *Penicillium, Aspergillus*, and a number of other genera of fungi. Patulin is believed to cause hemorrhaging in the brain and lungs and is usually associated with apple and grape spoilage.
Satratoxin H	Satratoxin H is a macrocyclic trichothecene produced by *Stachybotrys chartarum, Trichoderma viridi,* and other fungi. High doses or chronic low doses are lethal. This toxin is abortogenic in animals and is believed to alter immune system function; it makes affected individuals more susceptible to opportunistic infection.
Sterigmatocystin	Sterigmatocystin is a nephrotoxin and a hepatotoxin produced by *Aspergillus versicolor.* This toxin is also considered to be carcinogenic, especially in the liver.
T-2 toxin	T-2 toxin is a trichothecene produced by species of *Fusarium* and is relatively potent. If ingested in sufficient quantity, T-2 toxin can severely damage the entire digestive tract and cause rapid death due to internal hemorrhage. T-2 has been implicated in the human diseases alimentary toxic aleukia and pulmonary hemosiderosis. Damage caused by T-2 toxin is often permanent.
Vomitoxin or deoxynivalenol (DON)	Vomitoxin, chemically known as deoxynivalenol, a trichothecene mycotoxin, is produced by several species of *Fusarium.* Vomitoxin has been associated with outbreaks of acute gastrointestinal illness in humans. The FDA advisory level for vomitoxin for human consumption is 1 ppm.
Zearalenono	Zearalenone is also a mycotoxin produced by *Fusarium* molds. Zearalenone toxin is similar in chemical structure to the female sex hormone estrogen and targets the reproductive organs.

Tolerance to molds appears to vary biologically among individuals and appears to relate at least partially to the vagaries of allergic sensitization. In the absence of any direct indicators of mold bioaerosol numbers exceeding human tolerance levels, a reasonable indicator of potentially significant problems would seem to be the coincidence of (1) symptoms attributed to building air quality and compatible with mold exposure (nonspecific upper respiratory or flu-like symptoms, mucous membrane irritation, exacerbation of asthma, wheezing, and shortness of breath, with remission within hours of leaving building and recurrence upon reentry into building), and (2) evaluation of levels of toxigenic or allergenic species measured indoors and outdoors in a suspect building where, after adequate study, significant indoor mold amplifiers are not thought to exist.

Ideally, standards for fungi in indoor air should be based on the health effects of such exposure. Information on human dose–response relationships for fungi in air, however, is currently not readily available.

REFERENCES AND RESOURCES

ACGIH (2002) *TLVs and BEIs: Threshold Limit Values for Chemical Substances and Physical Agents and Biological Exposure Indices*, American Conference of Governmental Industrial Hygienists, Cincinnati, OH.

AIHA (2001) *Emergency Response Planning Guidelines and Workplace Environmental Exposure Level Guides Book*, American Industrial Hygiene Association, Fairfax, VA.

Ammann, H.M. (1999) IAQ and human toxicosis: Empirical evidence and theory, in *Bioaerosols, Fungi and Mycotoxins: Health Effects, Assessment, Prevention and Control*, Johanning, E., Ed., Eastern New York Occupational and Environmental Health Center, Albany.

Betina, V. (1989) *Mycotoxins: Chemical, Biological, and Environmental Aspects*, Vol. 9, Elsevier, New York.

Buck, W.B. and Cote, L.-M. (1991) *Handbook of Natural Toxins*, Vol. 6, *Toxicology of Plant and Fungal Compounds: Trichothecene Mycotoxins*, Marcel Dekker, New York.

Burge, H. (1996) *Indoor Air and Human Health: Health Effects of Biological Contaminants*, CRC Press, Boca Raton, FL.

Etzel, R.A., Montaña, E., Sorenson, W.G., Kullman, G.J., Allan, T.M., and Dearborn, D.G. (1998) Acute pulmonary hemorrhage in infants associated with exposure to *Stachybotrys atra* and other fungi, *Arch. Pediatr. Adolesc. Med.*, 152, 757–761.

Gravesen, S., Frisvad, J.C., and Samson, R.A. (1994) *Microfungi*, Munksgaard, Copenhagen, Denmark.

Gravesen, S., Nielsen, P.A., Iversen, R., and Nielsen, K.F. (1999) Microfungal contamination of damp buildings: examples of risk construction and risk materials, *Environ. Health Persp.*, 107(suppl. 3), 505–508.

Hammond, P.B. and Coppock, R. (1990) *Valuing Health Risks, Costs, and Benefits for Environmental Decision-Making*, National Academy Press, Washington, D.C.

IOM (2000) *Clearing the Air: Asthma and Indoor Air Exposures*, National Academy Press, Washington, D.C.

Jacobsen, B.J., Bowen, K.L., Shelby, R.A., Diener, U.L., Kempppainen, B.W., and Floyd, J. (1993) Mycotoxins and Mycotoxicoses, Circ. ANR-767, Alabama Cooperative Extension System, Alabama A&M and Auburn Universities.

Johanning, E., Ed. (1999) *Bioaerosols, Fungi and Mycotoxins: Health Effects, Assessment, Prevention and Control*, Eastern New York Occupational and Environmental Health Center, Albany, NY.

Johanning, E. and Yang, C. (1995) Fungi and bacteria in indoor air environments; health effects, detection and remediation, in *Proc. Int. Conf.*, Saratoga Springs, NY, October 6–7, Eastern New York Occupational and Environmental Health Center, Albany, NY.

Lewtas, J., Complex mixtures of air pollutants: characterization of cancer risk of polycyclic organic molecules (POM), *EHP*, 100, 211–218, 1993.

Macher, J., Ammann, H.A., Burge, H.A., Milton, D.K., and Morey, P.R., Eds. (1999) *Bioaerosols: Assessment and Control*, American Conference of Governmental Industrial Hygienists, Cincinnati, OH.

Miller, R.V., Martinez-Miller, C., and Bolin, V. (2000) *A Novel Risk Assessment Model for the Evaluation of Fungal Exposure in Indoor Environments*, Proc. Tenth Int. IUPAC Symp. on Mycotoxins and Phycotoxins: Ponsen and Looijen, Wageningen.

Morey, P.R., Feeley, J.C., and Otten, J.A., Eds. (1990) *Biological Contaminants in Indoor Environments*, American Society for Testing and Materials, Philadelphia, PA.

Pope, A.M., Patterson, R., and Burge, H.A., Eds. (1993) *Indoor Allergens*, National Academy Press. Washington, D.C.

Samson, R.A., Flannigan, B., Flannigan, M.E., Verhoeff, A.P., Adan, A.C.G., and Hoekstra, E.S., Eds. (1994) *Health Implications of Fungi in Indoor Environments,* Air Quality Monographs, Vol. 2, Elsevier, New York.

Samson, R.A., Hoekstra, E.S., Frisvad, J.C., and Filtenborg, O. (2000) *Introduction to Food- and Airborne Fungi*, Sixth ed., Centraalbureau voor Schimmelcultures (CBS), Utrecht, The Netherlands.

WHO (1990) *Mycotoxins*, Environmental Criteria 105, World Health Organization, Geneva.

Wyllie, T.D. and Morehouse, L.G., Eds. (1978) *Mycotoxic Fungi, Mycotoxins, Mycotoxicoses: An Encyclopedic Handbook*, Vols. 1–3, Marcel Dekker, New York.

CHAPTER 5

Risk Assessment

Harriet M. Ammann, R. Vincent Miller, Heriberto Robles, and Richard C. Pleus

CONTENTS

1-56670-606-8/03/$0.00+$1.50
© 2003 by CRC Press LLC

This chapter discusses general concepts of exposure and risk assessment, their applications and shortcomings for indoor environments, and some evolving or alternative concepts that may aid in assessing the potential health consequences of exposure to contaminants in indoor environments.

5.1 EXPOSURE AND RISK ASSESSMENT

The assessment of biological and chemical exposure is a central component to any health evaluation involving an environmental contaminant. Industrial hygienists and toxicologists have extensively studied the health effects of acute or short-term exposure to a number of chemicals (and chronic effects for a few), primarily in industrial occupational situations. As a result, permissible exposure levels for workers in the industrial workplace have been established that are based on the statistical adverse response of the majority of individuals to the contaminant. However, in 1999, the American Conference of Governmental Industrial Hygienists (ACGIH) determined that threshold limit values (TLVs) for biological contaminants could not be recommended because:

- The mixture of biological contaminants is very complex and varies from setting to setting.
- The methods of measuring components (viable and nonviable) of biological contaminant mixtures do not translate to meaningful numbers that can be used for exposure assessment.
- The susceptibility of exposed persons varies too much to be able to set a safe level for most workers (the definition of a TLV).

The ACGIH determined therefore that assessment of exposure to biological contaminants depends on:

- The judgment of professionals, including industrial hygienists, building scientists, toxicologists, epidemiologists, medical personnel, and others with profound knowledge regarding buildings, exposures and effects after a careful analysis
- The use of common sense in investigating problem buildings

Hampering quantitative risk assessment are the difficulties of sampling bioaerosols (i.e., air suspensions of spores, bacteria, payments, and products), lack of knowledge about the specific health effects of individual toxic and irritative substances produced by microorganisms that grow in damp indoor spaces, and the effects of exposure to the organisms themselves.

The lack of knowledge about interactions among all the agents that comprise exposure within indoor spaces makes quantitative assessment of risk even more problematic. Such agents include not only toxic substances such as mycotoxins (produced by fungi) and bacterial endotoxins (that have at least limited dose–response information from animal experiments and occupational studies) but also infective and allergenic substances and chemical air pollutants that are often found in higher concentrations indoors than outside. This complex exposure to the indoor mixture complicates the analysis of effect from any one agent.

Indoor environments pose a particularly complex system, with chemicals and biological agents originating from both external and internal sources. Exposure and risk assessments in these environments are further complicated by these facts:

- Individual contaminants may not reach acute toxicological thresholds.
- Complex mixtures of contaminants with diverse endpoints are formed.
- Some long-term or chronic exposures have not been well studied.

5.2 RISK

Risk is the probability that harm, injury, or disease will occur as a consequence of exposure to a particular hazard. Risk, in human health terms, is comprised of the evaluation of:

- Information on the hazardous properties of substance(s)
- Quantification of hazard through dose–response assessment
- Evaluation of the extent and duration of human exposure
- Characterization of the possible consequences resulting from such exposure

To accomplish this, a systematic approach must be taken to organize and analyze scientific information to evaluate the hazard potential from specified exposures (National Academy of Science, 1994); however, the process requires that many assumptions be made due to lack of specific knowledge about either basic toxicological or pathogenic mechanisms or specificity of exposure. Quantitative assessments are attempted when some degree of knowledge is available about the toxicity, dose–response relationship, or pathogenicity of the specific agent and extent of exposure.

Assessments are limited to qualitative descriptions without such data; however, in both quantitative and qualitative risk assessments, default values that can introduce large uncertainties into the estimate are often necessary. Because the numbers that result from risk assessment, particularly quantitative risk assessment, give the appearance of certainty, assumptions and defaults must be clearly defined. Both quantitative uncertainty analysis, where possible, and qualitative uncertainty analysis, when numerical estimates are not possible, should be included in risk assessments so that the process is transparent to the reader. The limitations of the assessment and a description of the analysis must be provided.

5.3 QUANTITATIVE PARADIGMS

Standard methods are available for the measurement of many chemicals, and exposure paradigms have been developed. Some chemicals have good toxicological information, and dose–response relationships have been worked out for at least one of the three general pathways of exposure: inhalation, oral, and dermal, and some indirect pathways. The principles developed for risk assessment of chemical substances can, to a large degree, be applied to biological contaminants if sufficient toxicological and/or pathological and exposure information has been obtained. That is:

- Known hazards described in the scientific literature can be evaluated.
- Exposures can be estimated or modeled.
- Risk can be characterized.

Dose–response relationships form the quantitative portion of hazard assessment. Generally, observing measurable effects in any of the following has elucidated these relationships:

- Laboratory experiments involving controlled exposures of animals
- Controlled exposure of humans
- Occupational case studies
- Epidemiological studies of humans

As a result, a dose–response curve can be drawn that allows limited extrapolation or interpolation to exposures not included in the analysis and extrapolation to organisms (i.e., humans) that were not experimentally exposed.

Controlled human exposures are limited to low-level exposures that are not thought to do permanent harm and are limited by ethical considerations. Epidemiological investigations (i.e., animal exposures) are limited in their power of effect detection by the size of the population being exposed and analyzed. The smaller the population analyzed, the smaller the power of the analysis to detect effect.

Underlying such analyses are assumptions that what is true for the experimental animal is true for humans, and that what is true of the exposed human population being studied is true for other human populations. Paradigms for assessing chemical exposures have been developed for those chemicals that have been studied. Many of these have resulted in the establishment of threshold limits by ACGIH, NIOSH, OSHA, EPA, and AIHA (ACGIH, 2001; AIHA, 2001; Hammond and Coppock, 1990; NRC, 1983; USEPA, 1992). These paradigms are based on dose–response curves developed for animals and extrapolated to humans or on human occupational studies.

Another basis for standards could be the concentration required to induce a specific physiological dysfunction, such as reduced pulmonary function, into a certain percentage (often 10%) of a test population. In general, these paradigms follow the general dose equation (USEPA, 1992):

$$\text{Potential dosage} = \Sigma C_i \cdot E_i \cdot D_i$$

where C_i is the concentration of organism or chemical (e.g., toxin) at time i; E_i is the exposure concentration by ingestion, surface contact, or inhalation rate at time i; and D_i is the duration of exposure in hours at time i. An estimated dosage can then be derived by substituting in the:

- Average concentration (C_{ave})
- Exposure rate (ER_{ave})
- Total duration (ED)

Resulting in the following equation:

$$\text{Potential dosage} = C_{ave} \cdot ER_{ave} \cdot ED$$

The Environmental Protection Agency (EPA) developed a risk paradigm for inhalation that incorporates more information about variables that influence risk. In developing their reference concentrations (RfCs), the EPA has incorporated information that addresses some of the uncertainties that arise due to differences between experimental animal species and humans (USEPA, 1994).

No observed adverse effect levels (NOAELs) and lowest observed adverse effect levels (LOAELs) are extracted from the best chronic animal exposure study available and converted to human equivalent concentrations (HECs). For gases, concentration units must be converted from ppm to mg/m^3. Human equivalent concentrations are calculated by converting experimental exposure durations to 24-hour equivalents, taking into account the breathing rate and respiratory surface area impacted in the experimental animal relative to that of humans.

An RfC is then calculated by incorporating uncertainty and modifying factors into the NOAEL$_{[HEC]}$. A reference concentration is defined as an estimate (with uncertainty spanning perhaps an order of magnitude) of a daily exposure to the human population (including sensitive subgroups) that is likely to be without an appreciable risk of deleterious effect during a lifetime (USEPA, 1994):

$$\text{RfC} = \text{NOAEL}_{[HEC]}(\text{mg/m}^3)/(\text{UF} \times \text{MF})$$

where UF is the uncertainty factor and MF is the modifying factor. The uncertainty factor usually is a tenfold factor intended to account for the uncertainties due to variation in susceptibility within the human population, uncertainty in extrapolation from experimental animal data to human effect, uncertainty in converting extrapolation data from less than lifetime to lifetime exposures, and the inability of any single study to address all adverse outcomes in humans.

Reference concentrations can also be calculated for particulate contaminants, such as fine particles from combustion. Because of differential deposition throughout the lung, deposition depends on particle size and behavior, expressed as mean aerodynamic diameter. Particles deposit in the upper portion of the bronchial tree by impaction, farther down in the tree by sedimentation, and in the terminal bronchioles and alveoli by diffusion. A term for regional deposition must be included when calculating effects from particles.

Modifying factors are greater than 0 and less than or equal to 10; they have a default value of 1 (one). Modifying factors allow the incorporation of evaluations of scientific uncertainties, such as the number of animals tested or endpoints accounted for, but not incorporated, in the risk equation.

Other EPA risk paradigms are taken from the risk assessment guidelines for Superfund (USEPA, 1989). The inhalation exposure paradigm for airborne chemicals is:

$$\text{Intake} \left(\text{mg/kg - day}\right) = \frac{\text{CA} \times \text{IR} \times \text{ET} \times \text{EF}}{\text{BW} \times \text{AT}}$$

where:

CA = Contaminant concentration in air (mg/m^3); a site-specific or modeled value

IR = Inhalation rate (m^3/hour), with an adult average of 20 m^3/day; other values can be obtained from the *Exposure Factors Handbook* (USEPA, 1992)

ET = Exposure time (hours/day) specific to the individual (e.g., work day)

EF = Exposure frequency (days/year) specific to the individual

ED = Exposure duration; 70-year lifetime by convention

BW = Body weight (kg); 70-kg adult, average

AT = Averaging time (period over which exposure is averaged, in days) calculated for noncarcinogenic effects by $ED \times 365$ days per year; for carcinogenic effects, by 70-year lifetime $\times$ 365 days per year

Note that individual variations or susceptibilities are not very well addressed in the above paradigms.

Risk assessment is most frequently performed for assessing the effects from exposure to individual agents, with the realization that humans are not exposed to compounds one at a time or in isolation from other routes of exposures. The risk assessment of chemical mixtures is still problematic for many reasons, including the fact that the composition of mixtures changes in real-life exposures. Effects of mixtures are often addressed by assuming that effects of the mixture components are additive (at least across similar endpoints) or that synergism can occur among the components.

5.4 QUALITATIVE PARADIGMS

For the majority of chemicals and biological contaminants, thorough systematic analyses of toxicological and/or pathological effects simply have not been done. This fact leads to a major risk assessment limitation in that the analyses are reduced to qualitative analyses. A qualitative assessment is, by definition, more uncertain for agents that have inadequate information available or for mixture components that are not measured (this is also true for quantitative assessments).

5.5 HAZARD IDENTIFICATION

5.5.1 Medical Evaluation and Surveillance

The first step in attempting to characterize risk for indoor exposures is the evaluation of adverse effects in potentially exposed individuals. Such an evaluation includes an analysis of complaints, as documented through a differential diagnosis by a physician or other medical professional (Hodgson, 1995). Diagnoses result from reviewing the patient's history, the patient's symptoms, and a medical evaluation that records the signs and symptoms. This evaluation may be based on the actual signs and symptoms observed or on test results (e.g., physical or biochemical laboratory tests).

Many differing medical conditions and exposures can share both signs and symptoms. Symptoms are not so much nonspecific as common to different exposures or underlying pathologies. The history and differential diagnosis can assist in distinguishing symptom causation. For example, a headache can result from mechanical injury to the head, tension, sinus obstruction, or carbon monoxide or other toxic exposure, such as solvents or mycotoxins, among many others.

Without a careful patient history, the causality cannot be ascertained with medical certainty. As a result, many of the routine clinical screening tests such as blood chemistries alone are often of little value in the medical assessment (Rose et al., 1999). Surveillance of persons exposed under similar circumstances may be necessary to determine associations with environmental conditions. An environmental appraisal may be crucial to the medical evaluation. This appraisal must encompass the suspect building, any other buildings (e.g., residence or workplace) frequented by affected individuals, and other exposures from occupations, hobbies, or avocations.

The appraisal must be structured to identify exposures that may cause and/or exacerbate the health effects noted in the medical evaluation. Regrettably, most physicians do not make environment assessment house calls nor are they trained to do so; therefore, good medically based environmental appraisals are usually lacking, particularly for private residences. In addition, complete environmental investigations of all the indoor environments to which affected individuals are exposed (which includes a careful building walk-through to identify potential sources and judicious use of sampling and analyses for both chemical and biological contaminants) are costly and often neglected.

5.5.2 Hazards in Indoor Air

All air breathed under natural conditions is composed of mixtures of chemical compounds. In the ambient air, the nature of the mixture depends on proximity to sources of various contaminants, such as industrial or mobile sources. Some contaminants are thought to be ubiquitous throughout the country and are addressed by National Ambient Air Quality Standards (NAAQS) which, by law, are health-based standards. These standards regulate particulate matter, sulfur and nitrogen dioxides, carbon monoxide, ozone, and lead. At present, all other toxic ambient air pollutants are regulated by source control. Chemicals breathed by human beings indoors are not regulated except in the industrial workplace, where the acute exposure of some is limited through the Occupational Safety and Health Administration (OSHA). Other indoor exposures to chemicals are not regulated.

5.5.3 Chemicals (of Nonbiological Origin)

Chemicals breathed indoors can be divided into several large categories: combustion products, volatile organic compounds, and irritant compounds.

5.5.3.1 Combustion Products

Combustion produces thousands of compounds, the highest concentrations of which are fine particles (less than 1 μm in aerodynamic diameter), carbon monoxide, oxides of sulfur, and nitrogen oxides. All of these have been extensively studied and health criteria have been developed for them. Specific information regarding the components of other combustion mixture components has not been extensively developed. Many toxic compounds generated during combustion are known to adsorb to the surface of fine particles and are available to be carried deep into the lung. One hypothesis put forward to explain the toxicity of fine particles to the lung and heart, and for their role in lung cancer, is that such adsorbed toxins rather than the pesticides themselves play a large role in these disease processes.

Carbon monoxide (CO) prevents blood from carrying sufficient oxygen to cells to maintain adequate metabolism. High oxygen demand on organs such as the heart and lung is most quickly and severely affected by CO. The NAAQS for CO is based on this effect of CO on the most sensitive human population, cardiac patients.

Both nitrogen oxides and sulfur oxides are upper airway irritants. Sulfur dioxide (SO_2) adsorbs to particulate co-pollutants that carry the compound deep into the lung, where the SO_2 becomes a lower airway irritant that can initiate and exacerbate asthma. Nitrogen oxide effects decrease both the physical and immunological defenses of the lung, making some populations, especially children, more susceptible to infectious organisms.

Combustion sources indoors are room-vented appliances (gas stoves, ovens, and heaters); backdrafting vents for stoves, fireplaces, or gas water heaters; outside sources such as attached garages, indoor parking areas that vent to occupied spaces through elevator shafts, and other stack-effect pathways, and improperly placed air intakes.

5.5.3.2 Volatile Organic Compounds

Many volatile organic compounds (VOCs) are found in indoor spaces in higher concentrations than in the ambient air, even in that of industrial areas (USEPA, 1987). These higher concentrations are due to tightening of buildings for purposes of energy conservation without providing for adequate ventilation. Prominent indoor sources include emissions from paints, varnishes, plastics, cleaning solvents, office products, and construction materials.

Many VOCs are toxic to the nervous system and are respiratory and eye mucous membrane irritants. When considered as singular chemical constituents, the concentrations of most individual VOCs indoors may be higher than in the ambient air but usually not at levels that exceed individual industrial workplace standards for the VOCs that have such standards.

Work performed by the EPA and Danish colleagues (Otto et al., 1990) has shown that the aggregate VOC concentration of all mixture components may result in both neurotoxic and irritative effects, even when the individual components are not at sufficient levels to cause measurable toxic effects. In other words, the additive or synergistic toxic effect of chemical mixtures often exceeds the effect of any singular chemical component of the mixture.

Many of the VOCs used for cleaning (e.g., alcohols, ammonia, and complex solvents such as limonene and pinene) are also produced by certain molds. The presence of molds and bacteria can complicate the question of exposure to VOCs because primary and secondary metabolites from these organisms can contribute to the total VOC burden to the occupant.

The neurotoxic endpoints that seem to be most affected at low exposure levels are those that affect the olfactory sense and the common chemical sense (neurasthenic sense) that responds to pungency (Schiffman et al., 2000). The common chemical sense resides in the trigeminal, vagus, and glossopharyngeal spinal nerves. The sensory nerve endings respond to irritative stimuli, while the motor portion responds by smooth muscle contraction, secretion from excretory glands, and central nervous system effects that can include impairment of attention and memory and a variety of fight or flight responses.

5.5.3.3 Irritants

In addition to the mucous membrane and nerve irritation brought about by exposure to VOCs, other irritant compounds (including aldehydes, ketones, and other semivolatiles) can lead to mucous membrane irritation, resulting in inflammation, and can then involve sinus blockage and drainage, sore throats, irritated eyes, and respiratory symptoms. Such compounds (for example, formaldehyde) can originate with combustion; can off-gas from building materials such as particle board, oriented-strand board, plywood, glues, and adhesives; or can off-gas from finished fabrics in curtains and upholstery.

5.6 BIOLOGICAL CONTAMINANTS (BIOAEROSOLS)

Biological contaminants, depending on their amount and potency, can have effects on health singly or in concert. Biological contaminants can include:

- Organisms such as bacteria, algae, protozoa, fungi (as molds), which may be allergenic or infectious
- Nonorganismal infectious particles such as viruses
- Products from animals, such as cat dander, dust mite feces, cockroach effluvia, plants (pollens)
- Enzymes and metabolic products from microorganisms
- Bacterial endotoxins
- Fungal exotoxins (mycotoxins)
- Microbial VOCs (mVOCs)

Depending on their manner of dispersion, many biological contaminants may also be classified as bioaerosols. The effects of infection and allergy caused by biological contaminants may exacerbate the irritative and toxic effects of other agents because of additive or synergistic effects. Additive effects occur when the effects of two or more agents result in the numerical sum of the agents acting on a particular system alone. Synergistic effects occur when the sum of agents acting on a particular system or organ is greater than the numerical sum and may in fact result in multiples of the individual effects.

5.6.1   Bacteria

While bacteria are generally known for their infectious qualities, bacteria can produce toxins as a part of their infectious processes (e.g., the toxins produced by *Bacillus anthracis* that allow the bacteria to invade animal cells). Such bacterial exotoxins also can be the agents of detrimental effects that constitute the disease (e.g., other *Bacillus anthracis* toxins, and *Diphtheria* toxins). Gram-negative, rod-shaped bacteria also have toxins that are part of their cell wall that are released into the environment when the bacterial cell is disrupted. These toxins are known as bacterial endotoxins and have been implicated in respiratory diseases of workers, including hypersensitivity pneumonitis, which is a serious disease of the lung that causes progressive loss of lung function with continuing exposure to the etiologic agent.

5.6.2   Molds

Molds can have an impact on human health, depending on:

- Species involved
- Infectious or allergenic nature of the mold species
- Metabolic products being produced by these species
- Amount and duration of the individual's exposure to mold parts or products
- Specific susceptibility of the individuals exposed

Health effects generally fall into four categories:

- Allergy
- Infection
- Irritation (mucous membrane and sensory)
- Toxicity

Allergy is the most common effect from mold exposure. Infection is a hazard for some mold species found indoors, for certain sensitive populations. Irritation and toxicity are other potential effects. The potential of the agent to induce toxic effect depends on:

- Species involved
- Strain of the species (which can determine its metabolic products)
- Environmental conditions
- Presence of competitive organisms

5.6.3   Viruses

Airborne or droplet-borne viruses cause influenza, colds, measles, rubella, encephalomyelitis, parotitis (mumps), pneumonia, varicella (chickenpox), and Hanta virus syndrome (Otten and Burge, 1999). Most of these diseases are associated with specific buildings and spread within building systems by nonmechanical transmission. Properly operated HVAC systems and dry surfaces are not considered the primary method of transmission, although a study indicating HVAC system transmission of measles has been published (Riley et al., 1978). Most of these viruses are transmitted through short-distance droplet spread originating from the infected individual or direct contact with an infected human or animal. Poor ventilation, however, leads to increased aerosol concentrations, indirectly resulting in increased disease incidence.

5.7 ALLERGY

One of the most common responses to exposure to biological pollutants is allergy. Several indoor allergens, including other microorganisms, dust mites, cockroaches, and effluvia from domestic pets (such as birds, rodents, dogs, and cats) and rodent pests, have been implicated in allergic disease (Pope et al., 1993). Allergy symptoms can be exacerbated by exposure to multiple allergens.

People who are atopic, that is, who are genetically capable of producing an allergic response, can develop allergies to specific antigens (foreign proteins) with sufficient exposure. Clinical responses to very low antigen exposure levels in the future may result from an original sensitizing exposure that did not have an observable initial effect. This process is termed *sensitization*, and the individuals are said to be sensitized. Allergy reactions can include skin reactions such as rashes and hives; respiratory responses such as inflammation, with excess production of mucus from affected membranes; allergic sinusitis; and severe diseases (e.g., asthma, hypersensitivity pneumonitis). Allergic reactions can range from mild, transitory responses to severe, chronic illnesses.

The Institute of Medicine (Pope et al., 1993) estimates that one in five Americans suffers from allergic rhinitis (type I response), the single most common chronic disease experienced by humans. Additionally, about 14% of the population suffers from allergy-related sinusitis, while 10 to 12% of Americans have allergy-related asthma. About 9% experience allergic dermatitis (type IV response). A much smaller number, less than 1%, suffers serious chronic diseases such as allergic bronchopulmonary aspergillosis (ABPA) or hypersensitivity pneumonitis (type III response).

As an aside, along with allergies, allergic fungal sinusitis is not uncommon among individuals residing or working in moldy environments (Ponikau et al., 1999). Debate continues as to whether this fungal sinusitis is solely an allergic reaction or if it has an infectious component.

5.8 INFECTION

Some molds found growing indoors as the result of moisture problems; for instance, *Aspergillus parasiticus* and *A. fumigatus* can cause infections of the lung and other systems in susceptible people. Asthmatics can develop allergic bronchopulmonary aspergillosis, which has elements of both allergy and infection. Immunocompromised individuals or those with massive exposures can develop aspergillosis, an infection of the lung or other systems, as well as aspergilloma (fungus ball of the lung). Other infectious fungi include *Coccidioides*, *Geotrichum*, *Cryptococcus*, *Nocardia blastomyces*, and *Histoplasma* (Mandel et al., 1996). These organisms are generally found in soil or bird and bat droppings and can be a problem in buildings where soil or guano contamination occurs; they do not grow (amplify) indoors due to excess moisture.

Coccidioidomycosis is a dustborne fungal disease affecting many inhabitants of arid regions in the southwestern United States, especially the San Joaquin Valley of California, hence its common name of Valley Fever. The causative organism is *Coccidioides immitis*. Almost two thirds of infections are without symptoms, and one third manifest as severe respiratory infections including inflammation of the lung. The disseminated form of the disease can be fatal. A rash, thought to be a hypersensitivity reaction to the infecting organism, often accompanies respiratory infections.

Histoplasmosis is caused by *Histoplasma capsulatum*. Histoplasmosis infects up to 90% of persons in the midwestern United States in its benign form; the chronic pulmonary disease has about a 30% mortality rate if untreated. *Histoplasma* is carried by birds and bats and can be found indoors in buildings that have accumulated bird and bat guano.

North America blastomycosis is caused by *Blastomyces dermatitis*, which is found in soil. Blastomycosis can be localized in the skin or can be systemic. It has a high mortality rate without treatment.

Geotrichum, *Nocardia*, and *Cryptococcus* can all cause primary pulmonary and other systemic infections. *Cryptococcus* can infect any system, including the skin, and usually enters the body through the respiratory tract. *Cryptococcus* has particular affinity for the central nervous system, causing meningitis. *Nocardia* can enter through abrasions in the skin (usually of the feet) or through the respiratory system, and can also metastasize to the brain, causing abscesses. These are soil organisms.

5.9 IRRITATION

Volatile and semivolatile products produced by molds, either alone or together with VOCs produced by building materials, paints, solvents, and combustion can irritate the mucous membranes of the eyes and respiratory tract and the nerve endings of the common chemical or neurasthenic sense, as previously discussed. Some of these VOCs (e.g., alcohols and aldehydes and ketones) are products of primary metabolism and are produced throughout the life of the microbe. Others, which tend to be more complex molecules, have a characteristic moldy or musty odor and are produced through secondary metabolism.

5.10 TOXICITY

Molds produce some compounds that are toxic to other organisms; the compounds that are toxic to other microorganisms are generally called antibiotics. Toxic substances called mycotoxins are mold poisons that are toxic to plants and animals, including humans. Metabolism is either primary or secondary.

5.10.1 Primary Metabolism

Primary metabolism is the day-to-day metabolic activity that uses enzymes excreted into the environment to digest nutrients. These nutrients are reabsorbed and converted to energy building blocks to make proteins, nucleic acids, fats, and other cellular building blocks. In addition to these nutrients, water is essential to the growth of microbial organisms.

5.10.2 Secondary Metabolism, Antibiotics, and Mycotoxins

Secondary metabolites are complex molecules of various kinds that are produced only as needed by the organism to compete for ecological niches shared with other microbes or more complex animals and plants. Many fungi produce secondary metabolites that are directly toxic to eukaryotic cells such as those of plants and animals (and other fungi). The term *mycotoxin* is commonly used to refer to these compounds. The production of mycotoxins by fungi and accumulation of mycotoxins in fungal spores are dependent upon environmental conditions (e.g., substrate, temperature, and humidity) and the species and strains of fungi. Some molds can produce several toxins. Secondary metabolism costs the cell extra energy and is used only as needed. Mycotoxins are not produced throughout the life of the organism, but tend to be produced around the time of sporulation, at least in aspergilli and penicillia (Larson and Frisvad, 1994). Both antibiotics and mycotoxins are substances produced through secondary metabolism. Mycotoxins can be found in the substrate on which molds grow, and in dust.

5.11 SOURCE DISCOVERY AND RISK ASSESSMENT

Most of the test methods currently employed for capturing bioaerosols were developed from the early to middle 20th century and include source sampling (Martyny et al., 1999), air sampling (Willeke and Macher, 1999), and spatial cavity sampling. Whether or not any of these sampling techniques is applied depends entirely on the question being asked. Questions range from very simple such as "Is it mold or not?" to very complex such as "Is there exposure to toxic mold?" The first question is easily answered through a tape lift. The second question requires much more information and much more complicated and expensive testing. At best, quantifying exposure may not be possible, and an association between known symptoms and known toxic effects and the potential for exposure may be the only conclusion that can be drawn. Determination of exposure requires knowing the following about the toxic substance:

- Nature of the substance (effect and toxic impact)
- Extent of dispersal within an environment
- Duration and availability for exposure occurrence

Thus, the sampling methodology and other analyses must determine the nature, extent, and duration of exposures.

Focusing on a particular species such as *Stachybotrys* or *Aspergillus* alone does not allow for a sufficient analysis of bioaerosols that might be impacting occupants' health. Similarly, presence of such an organism in bulk samples does not necessarily indicate exposure. Finding particular species such as *Stachybotrys chartarum, Aspergillus versicolor*, or *A. sydowii* is an indicator of long-term or severe moisture problems. *Stachybotrys* has a low nitrogen requirement and can grow on cellulose materials such as hay, straw, and paper. *Stachybotrys* grows readily on wet straw or hay in agricultural environments, but is rarely found in nonagricultural outdoor air samples. *Stachybotrys* grows readily on damp paper products and sheetrock indoors, but because this growth is wet and slimy when growing *Stachybotrys* rarely aerosolizes unless it has dried and been disturbed.

Toxigenic, allergenic fungal genera (e.g., *Penicillium* and *Aspergillus*) are the agents associated with some indoor air problems. Their small, nondescript conidia are difficult to assess accurately with light microscopy. These devices tend to be biased toward the identification of larger, distinctively shaped and/or dark-pigmented structures.

Common fungi possessing such large or dark spores (e.g., *Cladosporium, Alternaria, Pithomyces*, and *Bipolaris*) and conspicuous allergens (e.g., basidiospores of the bracket fungus *Ganoderma applanatum* or *Ustilaginaceous* smut teliospores) may also be counted with accuracy. However, these smuts are from outdoor sources and have little relevance to the major indoor air questions.

5.12 AIR SAMPLING AND RISK ASSESSMENT

Air always contains a mixture of contaminants. In damp buildings, multiple species of molds, bacteria, other microbes and their metabolic products can be found in addition to sources of chemical contaminants such as volatile compounds and particles. Molds that disseminate their spores through the air do so in *blooms*. Blooms are episodic and their periodicity is not predictable. Commonly, samples are taken in a few locations in a building (and outdoors for comparison) at one or more times during one day or, in some cases, over multiple days. Such strategies may miss or hit blooms, which would give a skewed impression of the average exposure usually used in risk assessments.

Timing of the sampling is a critical factor. The sampling event can easily miss a period of bloom for one or more mold species. Other molds, such as *Stachybotrys*, are wet and slimy when growing; yet when dry are easily aerosolizable and inhalable when disturbed through activity within the building. Because moisture intrusion may be influenced by weather, many damp and moldy buildings have periods of time when drying occurs. This drying affects the molds contaminant levels and dispersion.

5.13 AGENTS AND AGENT MODE OF ACTION

5.13.1 Chemical Agents: Toxicity

Toxicity of chemical agents depends on the interruption of homeostatic mechanisms, which are defined as the biochemical and physiological interactions that maintain life. A basic tenet of the science of toxicology is that all things are toxic, depending on the degree of exposure. An underlying principle of this concept is that organisms have defense mechanisms that work against toxic exposures. Defense mechanisms include such physical barriers as the waterproof skin layer or the mucociliary escalator in the respiratory system, which traps particles that impact the branching walls of the bronchial tree. Mucus secreted in the bronchial tree traps these particles, and cilia, with constant upward movement, push the particle-containing mucus out to the oropharynx, where the mucus is swallowed. Other defense mechanisms include the blood–brain barrier or biochemical defenses, such as chelating substances (for toxic metals), antibodies that bind foreign proteins, and detoxifying metabolic pathways that make toxic molecules more soluble to facilitate their excretion.

Such physical and chemical defense mechanisms ensure that the physiological function and well-being of organisms can be maintained under environmental conditions of low exposures. When the *threshold* for one or more defense mechanisms is overwhelmed, the physiological balance is upset and systems become damaged. Damage to systems may not be detectable until an observable threshold is reached. With increasing exposure, the imbalance becomes greater, and permanent damage can ensue. The degree and permanence of the damage depend on which system is damaged, the ability of the organism to heal, the nature of the contaminant, and the kind of damage caused by the contaminant. Most noncarcinogenic chemicals have a threshold below which a normal adult can recover without chemical insult. Some chemicals have low thresholds that, for practical purposes, are virtually indistinguishable from chemicals that do not have threshold qualities. An example of such a low threshold chemical is lead.

5.13.2 Carcinogenicity

The Environmental Protection Agency and some other scientists think that chemicals that cause cancer do not have a threshold. Risk assessment for chemicals that cause mutation in the DNA molecule assumes that no threshold exists. Such an assumption is based on the *one-hit* hypothesis, which states that a single mutation can result in a cell transformation that leads to cancer. Measuring a single hit to a DNA molecule is not currently technically possible; consequently, risk measurement at low dose exposures requires interpolation of a dose–response curve. This curve is developed from measurable tumor incidence in animals responding to higher doses of exposure. The assumption of one-hit makes the zero value a data point that affects the slope of the dose–response curve. The probabilistic risk of cancer is determined by the slope of the dose–response curve and is reported as a *unit risk*, which is defined as the concentration at which a lifetime daily exposure will create a probability of cancer risk in an individual of one in a million, assuming a lifetime of 70 years.

Not all carcinogenic compounds cause mutations in DNA. Some have other mechanisms that interact with receptors and initiate a cascade of events that results in tumor promotion. Other substances, such as arsenic, cause cancers by mechanisms not yet understood. No animal model exists to explain how arsenic acts, yet arsenic is a potent lung, skin, bladder, kidney, and liver carcinogen.

Even though the human (and animal) body has defenses against many of the processes that lead to tumor formation that could (and probably do) constitute a threshold effect, risk assessment often uses the one-hit model. The purpose of such risk assessment is to prevent exposure, often through regulatory or clean-up action. Risk estimation is used to provide a margin of safety for a population that includes highly susceptible individuals.

5.13.3 Irritation

Irritation is one form of noncarcinogenic effect caused by a large number of compounds of nonbiologic and biologic origin. Irritation from environmental agents is generally separated into mucous membrane irritation and irritation of nerve endings of the neurasthenic or common chemical sense. Mucous membrane irritation of the eyes and respiratory and other membranes can lead to inflammation, which is characterized by redness, pain, heat, and/or swelling.

Mucous membrane inflammation makes membranes leaky, resulting in excess fluid secretion (and loss). Such leakiness also reduces the effectiveness of the membrane barrier. Contaminant molecules, antigens, and infectious agents may have easier entry into the body when membranes are inflamed. Chronic inflammation can lead to healing processes that involve the building of scar tissue and lead to loss of function. For instance, in the lung, chronic inflammation can lead to thickening or fibrosis that prevents effective gas exchange, leading to loss of lung function (Nielsen et al., 1995).

Irritation of the common chemical sense that responds to pungency, not odor of chemicals, occurs in the sensory nerve endings of the trigeminal, vagus, and glossopharyngeal nerves (Schiffman et al., 2000) when these nerves are exposed to a number of chemicals including VOCs and semivolatile compounds. Impairment of cognitive function, paresthesias (weird sensations of tingling, itching, formication), changes in reflexes and coordination, and alterations in mood (anxiety and irritability) have been reported. Changes in breathing rate and depth and smooth muscle contraction in the upper respiratory tract have also been measured. Such reactions are mediated through the limbic system of the central nervous system (CNS) and are thought to be part of the *fight or flight* protective reactions of the CNS.

5.14 BIOLOGICAL AGENTS

Mycotoxins, endotoxins, and immunosuppressive compounds that are produced by organisms in the contaminated indoor environment may work in concert to produce effects on health. Combinations of various classes of mycotoxins and other bioaerosols present in a contaminated environment may act in an antagonistic, an additive, or a synergistic manner to cause health effects (Burge and Ammann, 1999). Immunosuppressive and combined effects of mycotoxins and other bioaerosols may account for the unexplained health effects reported by the individuals exposed to moldy environments.

5.14.1 Bacterial Endotoxins

5.14.1.1 Endotoxins

Endotoxin is the name given to a group of heat-stable lipopolysaccharide molecules present in the cell walls of Gram-negative bacteria that have a certain characteristic toxic effect. The lipid portion of each molecule is responsible for the molecule's toxicity and can vary between bacterial species and even from cell to cell. When inhaled, endotoxin creates an inflammatory response in humans that may result in fever, malaise, alterations in white blood cell counts, headache, respiratory distress, and even death. Endotoxin is common to the environment due to the ubiquitous nature of Gram-negative bacteria. Exposure to elevated levels of endotoxin primarily occurs through exposure to aerosols from specific reservoirs such as cotton mills, metalworking fluids, wastewater treatment facilities, indoor swimming pools, air washers, humidifiers, and any other occupational settings where Gram-negative bacteria can flourish.

5.14.2 Mycotoxins

Mycotoxins are toxic substances produced by molds. Mycotoxins cling to the surface of mold spores and can be found within spores. They are produced as molds and fungi grow and are found on environmental substrates where molds are growing and in the dust from such substrates. Molds in indoor environments and their health effects have gained considerable attention in recent years. Due to some well-publicized cases of mold exposure in which mycotoxins were thought to play a role, much attention has centered on mycotoxins and the role that they may play in the symptoms that have been reported by exposed individuals. More than 300 species of molds have been identified as being able to produce mycotoxins. More than 200 mycotoxins have been identified from common molds and many more remain to be identified. Some of the molds that are known to produce mycotoxins are commonly found in moisture-damaged buildings.

Some molds can produce several toxins, and some molds produce mycotoxins only under certain environmental conditions. The reason why molds produce toxins is that these poisons are useful in inhibiting or killing off competitors that share the same ecological niche. Penicillin antibiotics were first discovered by observing the rings of bacterial growth inhibition in the media on which the bacteria were growing. Molds capable of producing mycotoxins do so when in a mixture of microorganisms, as is often the case when molds and bacteria grow in damp indoor environments. Molds isolated and grown in pure cultures (cultures containing only one species/strain of organisms) will stop making toxins after a few generations. The current explanation for this is that, without competition, the mold need not invest energy useful for survival in making poisons that have no target. *Aspergillus* and *Penicillium* species are known to produce potent toxins under certain circumstances associated with sporulation (Larson and Frisvad, 1994).

Toxins are composed of a variety of chemicals that vary depending on the species of the mold capable of producing them. A single species can produce one or more toxic molecules, with some producing as many as ten or more different chemicals. The molecules produced vary in toxic potency, mechanism, target species, and target organs.

Not all the toxins have been investigated to the same degree. For instance, some have only been tested for cytoxicity, which is a relatively crude measure of effect involving testing the toxin against isolated cells or tissues in culture (Gareis, 1995). Different strains of the same mold will make differing amounts of a given mycotoxin, with some even making none. Certain strains may not always produce mycotoxins in the field, making it impossible to predict mycotoxin levels based solely on spore concentrations in air (Jarvis and Hinkley, 1999; Jarvis et al., 1998). Mycotoxins comprise a diverse group of chemical compounds (Sorenson, 1993), including:

- Ergot alkaloids, from *Claviceps purpurea* and species of *Aspergillus*, *Rhizopus*, and *Penicillium*
- Substituted coumarins (aflatoxins), from *Aspergillus flavus* and *A. parasiticus*
- Ochratoxins, from several species of *Aspergillus* and *Penicillium*
- Quinones (citrinins), from several species of *Aspergillus* and *Penicillium*
- Anthoquinones (e.g., rugulosin), from *Penicillium islandicum*
- Trichothecenes (sesquiterpenes with a trichothecane skeleton, olefinic groups at C-9 and C-10, and epoxies at C-12 and C-13; macrocyclic trichothecenes have a carbon chain between C-4 and C-15 in an ester or ether linkage [e.g., T-2 toxin, DON, satratoxins G and H; verrucarins B and J, trichoverrins A and B]) from *Fusarium*, *Stachybotrys*, and *Trichoderma*, among others
- Substituted furans (e.g., citreoviridin), from *Penicillium citreo-viride*
- Epipolythiodioxoperazines (gliotoxin), from at least six species of *Aspergillus*, *Penicillium*, and *Stachybotrys* (Jarvis, 1995; Jarvis et al., 1998)
- Lactones, lactams (patulin), stachybotrylactones, stachybotrylactams (Jarvis, 1995; Jarvis et al., 1998)
- Estrogenic compounds (e.g., zearalenone)

This list of chemical structures of mycotoxins is not exhaustive. The mycotoxins differ in their absorption, toxicokinetics, toxicodynamics, target organs, metabolism, detoxification, and elimination due to differences in chemical structure. They also differ in potency, ranging from a lethal dose 50% (LD_{50}) in fractions of milligrams per kilogram to hundreds of milligrams per kilogram. Those mycotoxins that have particularly great economic, pharmaceutical, medical, or military importance have been studied to a greater degree than less potent ones. For these mycotoxins, potency has been established, mechanisms are known, and/or target organs or systems have been established for individual toxins.

Toxins have been tested against specific animals (e.g., rats or guinea pigs) to determine lethal potency. Such measures provide a crude way of comparing the relative potency and use the incidence of death in a population that brings about 50% mortality (LD_{50}). For some mycotoxins, potency investigations have compared the incidence of a particular effect over a range of doses and a dose–response curve is available. A dose–response curve is particularly useful for risk assessment, because the curve allows interpolation of doses to exposures encountered by humans indoors.

Because toxigenic molds do not produce mycotoxins all the time, the presence of mold in a building does not necessarily mean that mycotoxins are present in large quantities. While mold spores are routinely sampled from tape-lift, bulk, or even air samples, these samples are not routinely analyzed for mycotoxins.

Recent work in Denmark and Finland has demonstrated that mold and spores may not be the only exposure medium that needs to be taken into account to determine whether mycotoxin exposure is occurring. The substrate on which the mold grows can contain mycotoxins, as the molds secrete these exotoxins into their environment to inhibit others in their ecological niche. Gravesen et al. (1999) and Tuomi et al. (2000) have found mycotoxins in building materials with mold growth and in dust derived from such substrates as sheet rock.

Many of these toxins have pharmaceutical value, and fungi are being actively investigated for their ability to make such substances; consequently, new toxins and physiologically active molecules are frequently being isolated. The U.S. Army Medical Research Institute for Infectious Disease (USAMRIID) has investigated a number of toxins for their potential to be used as weapons. Other toxins have been investigated because of their large economic impact on agricultural animals and crops.

Specific limits in food have been set for aflatoxins (produced by *Aspergillus flavus*) by the U.S. Food and Drug Administration (FDA) and for zearalenone (from *Fusarium* and some other molds) by Health Canada (Kuiper-Goodman et al., 1987).

5.14.2.1 Microbial Volatile Organic Compounds

Molds and bacteria produce gaseous metabolic products collectively referred to as microbial volatile organic compounds (mVOCs). Some of the mVOCs are volatile intermediates of primary energy metabolism and are primary solvents. Many of these emitted chemicals are identical to those originating from solvent-based building materials and cleaning supplies, including alcohols, aldehydes, ketones, hexane, methylene chloride, benzene, and acetone. Some microorganisms can also produce ammonia and other nitrogen-containing compounds (amines) and organic acids such as butyric acid. Molds produce more complex products of secondary metabolism, including terpenes. These secondary metabolites are generally the molecules that give wet buildings their characteristic moldy, mildewy, or earthy odors. Their production has been studied in penicillia and aspergilli and is associated with active growth and sporulation (Larson and Frisvad, 1994). Health effects from mVOCs have not been specifically studied but are implicated in health effects associated with trigeminal nerve irritation and odor-related health complaints.

5.14.2.2 Exposure Pathways

Exposure pathways for mycotoxins can include inhalation, ingestion, or skin contact. Although some mycotoxins are well known to affect humans and have been shown to be responsible for human health effects, little health information is available for many mycotoxins. Studies have included the effects from various exposure routes, including intravenous (i.v.), intradermal (i.d.), intramuscular (i.m.), and intraperitoneal (i.p.) routes, as well as more natural dermal, ingestion, or inhalation routes.

Effect often varies, depending on the degree of access of the exposure route to blood or lymph pathways. These pathways provide a means of distribution to target tissues for the specific poisons. With the exception of mycotoxins examined for military use, the bulk of research with animals has focused on the ingestion route. Ingestion of mycotoxin-contaminated feed and fodder presents ongoing problems for livestock health in agriculture and has a huge economic impact on the agriculture industry. The U.S. Department of Agriculture and agriculture-related regulatory agencies in other countries fund and conduct studies on mycotoxins in food. The World Health Organization has also focused its efforts at investigating and helping to control mycotoxin exposure to humans, particularly in developing countries where grain and other food storage does not prevent mold contamination (WHO, 1990).

A few studies have investigated inhalation of mold and products and have found that inhalation produces more potent effects than ingestion (Cresia et al., 1985, 1986, 1987, 1990) and effects as potent as i.v. administration (Pang et al., 1988a,b). Such research is particularly important to the examination of indoor molds because inhalation and dermal exposures are most likely in such environments. Those mycotoxins that have particularly great economic, pharmaceutical, medical, or military importance have been studied to a greater degree than less potent ones. For these mycotoxins, potency has been established, mechanisms of action have been explained, and particular cellular or tissue targets have been identified.

5.14.2.3 Effects of Dose

The doses necessary to establish effect levels for various symptoms reported by individuals in damp and moldy buildings have not been established. Epidemiological studies capable of identifying a causal or even a strong relationship between exposure and effect have not yet been designed.

Epidemiological studies in buildings are also greatly limited by the small populations present in the buildings, by variable exposure for individuals in any given building, and by confounding exposure routes found when comparing damp with dry buildings. Risk assessment for exposure to mycotoxins in damp buildings is hindered by the inability to describe the effects of mixtures of toxins and other physiologically active molecules found in contaminated buildings and by the difficulty encountered in defining what people are exposed to in damp buildings and in accurately measuring exposure to occupants.

Animal experiments are unlikely to detect the incidence of such effects because of the low power of such studies. Low power is defined herein as investigations involving a small number of animals and animal homogeneity. This resultant small degree of test population variability requires that fairly high doses be administered to observe an effect (e.g., ten animals per dose group).

5.14.2.4 Assessment

Methodology is currently available only for detecting aflatoxins in human tissues or bodily fluids. The remainder of the mycotoxins can only be assessed in the body by indirect methods of detecting antibodies to the toxins, which only indicate recent exposure, not effect. Assessment depends on whether or not any of the toxins from the mold spore mixture are detected. Such assessment assumes that the investigator has properly characterized the mixture of molds growing, airborne, and producing toxins. Spore capture and isolation is not the only part of exposure that needs to be measured. Recent research has shown that building materials and dust resulting from such materials can contain toxins in the absence of spores. Spore sampling alone, therefore, may not be sufficient to characterize mycotoxin exposure (Gravesen et al., 1999; Tuomi et al., 2000).

5.14.2.5 Mycotoxin Types Indoors

Over 20 mycotoxins have been detected in indoor environments but some of the more common and relevant mycotoxins include trichothecenes, produced by certain species of *Stachybotrys*, *Trichoderma*, and *Fusarium*; aflatoxin and sterigmatocystin, by a number of species of *Aspergillus*; ochratoxin, by various species of *Aspergillus* and *Penicillium*; and tremorgens and griseofulvins, by certain species of *Memnoniella* and *Penicillium* (Burge and Ammann, 1999; Jacobsen et al., 1993). Recent advances in technology have given laboratories the ability to test for specific mycotoxins without employing cost-prohibitive gas chromatography or high-performance liquid chromatography (HPLC) techniques. Currently, surface, bulk, food and feeds, and air samples can be analyzed relatively inexpensively for the following mycotoxins: aflatoxin; ochratoxin; trichothecenes, including T-2 toxin, fumonisins, deoxynivalenol (DON, vomitoxin), satratoxins, and verrucarins; zearalenone; citrinin; alternariol; gliotoxin; patulin; and sterigmatocystin. Other mycotoxins include penicillic acid, roquefortine, and cyclopiazoic acid. Verrucosidin, rubratoxins A and B, PR toxin, luteoskyrin, erythroskyrine, secalonic acid D, viridicatumtoxin, kojic acid, xanthomegnin, viomellein, chaetoglobosin C, echinulin, flavoglaucin, versicolorin A, austamide, maltyzine, aspergillic acid, paspaline, aflatrem, fumagillin, fumitrems A and B, nigragillin, chlamydosporol, isotrichodermin, anguidine, and many more. More research is required to understand the relationship among fungal contamination, mycotoxin production, and exposure, and building-related disease.

5.14.2.6 Stachybotrys

The apparent association of pulmonary bleeding and deaths of infants with *Stachybotrys chartarum* (Etzel and Dearborn, 1999; Fung et al., 1998) stimulated much of the current public attention being paid to mycotoxins. Intense controversy exists as to whether sufficient mycotoxin exposure occurs in contaminated indoor environments to cause health effects (Auger et al., 1999; Gordon et al., 1999; Robbins et al., 2000; Rylander, 1999). Regarding the studies to date, only a very few acute-effect inhalation studies have been conducted. Several of the acute-effect inhalation studies indicate elevated toxicity compared to toxicity by ingestion of mycotoxins. Other studies show inhalation potency to be equivalent to that of intravenous injection. Results from acute-effect inhalation studies cannot be extrapolated to predict the health effects of the chronic long-term exposures experienced by individuals in contaminated indoor environments. Studies indoors have been hampered by having to reconstruct past exposures (as these studies have been performed after people are already sick) and not being able to completely characterize the extent of exposure. Lack of knowledge of the effect of exposures to mixtures has also played a role.

5.15 EXPOSURE ASSESSMENT

5.15.1 Models

Models are used to attempt to estimate exposures when measurement in media characterizes exposure incompletely and biomarkers for exposure are lacking.

5.15.2 Contact-Point Model

Exposure to mold in indoor environments has been linked to a number of adverse health effects. Some researchers have suggested that some effects may be due to mycotoxin exposure, but the role of mycotoxins in these situations remains highly controversial. Unfortunately, most knowledge on mycotoxins has been obtained following acute rather than chronic exposures.

Recently, a novel risk assessment model was proposed that is based on localized effects at the initial site of exposure, the lungs (Miller et al., 2000). The model factors in inhalation rates, lung capacity, exposure time, and deposition, based on the aerodynamics and particle size of the spore. Assumptions of the model involve:

- Daily inhalation rates determined empirically for the affected individual
- Rough inhalation rate estimates using inhalation rate models developed by the EPA (USEPA, 1997)
- Utilization of sedentary rates that are probably closer to the average inhalation rate (unless heavy physical labor is occurring)

Lung surface areas should be determined empirically for the affected individual. The assumption is that lung surface area averages from 28 m^2 at rest to 93 m^2 at deepest inspiration for an adult, and 6 m^2 at rest to 19 m^2 at deepest inspiration for an infant (Benjamin, 1996).

Mold spores, as with any particulate, are preferentially deposited (or excluded) in the thoracic or respirable (gas-exchange) regions based on the size of the particle (Nardell and Macher, 1999). Refer to Figure 5.2 for the predicted depositions of various mold spores into the lower recesses of the lung. Using the assumptions given above, the potential inhalation dose of spores or toxin can be estimated using equations provided in the *Guidelines for Exposure Assessment* (USEPA, 1997) amended to take into account the amount of expected deposition based on particle size.

5.16 MEDICAL ASPECTS

Medical evaluation of environmental exposures in general tends to be the purview of occupational physicians. Allergists and some other medical specialists, as well as a growing number of physicians, have begun to take an interest in evaluating environmental exposures and are moving to include environmental exposure questions in their patient histories and differential diagnoses. In general, however, few medical practitioners ask questions pertaining to occupation, hobbies, living conditions, or other sources of potential exposures to any chemical including metals, com-

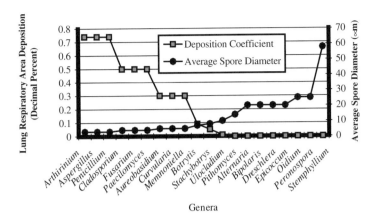

Figure 5.2   Spore deposition coefficients of fungal genera found in indoor environments.

bustion products such as carbon monoxide, and even more rarely to damp indoor environments that can grow molds or bacteria.

In 1993, the Institute of Medicine (IOM), part of the National Academies of Sciences, queried medical schools that had occupational health training programs as to how much instruction their medical students received regarding environmental exposures. On the average, in such schools across the country, the amount of time was 4 hours. Given this lack of emphasis in medical training, medical professionals may not know how to use environmental exposure questions in their diagnoses. The 1993 IOM publication *Indoor Allergens* examined the role of exposures to common allergens indoors but put little emphasis on molds as allergens. In 2000, the IOM released *Clearing the Air,* which deals with allergic disease (primarily asthma) and discusses mold exposure from this perspective. Aside from that, only a limited number of investigations have focused on mold exposure indoors. Difficulties with characterizing in any quantitative way the exposures of people having health effects have limited the ability of medical practitioners to gain specific diagnostic parameters (such as laboratory tests) to elucidate what causes illness in people with mold exposures in indoor environments.

5.16.1   Medical Assessments

5.16.1.1   Infection

The presence of fungi in lavage fluids from lungs or sinuses, together with characteristic symptoms of lung or sinus involvement, is to a large degree pathognomonic for aspergillosis and fungal sinusitis. Human systemic fungal infection (histoplasmosis, blastomycosis, or coccidio-mycosis) have diagnostic criteria and are more likely to be recognized as fungal infections in those areas where such infections are endemic. In areas where they are not frequently seen they may simply be acknowledged as pneumonia (with lung involvement). Diagnoses may be entirely missed if the attending practitioner has no experience with these diseases. Much depends on the practitioner having knowledge of the disease and asking the right questions when taking the patient's history.

5.16.1.2   Fungal Syndrome

Other diagnoses such as fungal syndrome take into account the spectrum of symptoms — respiratory, neurologic, immune and others — reported in published case and epidemiological studies and the clinical experience of those who have treated patients exposed to fungal mixtures indoors. Aside from drawing inferences about exposure from limited environmental data and characteristic symptoms and ascertaining that respiratory and fatigue symptoms are not due to allergy, the finding of this constellation of symptoms depends on clinical experience.

5.16.1.3   Allergy

Diagnostic tools are available to detect allergy. Being able to distinguish allergic from irritative or toxic reactions is important because allergy symptoms are treatable and, if specific allergens are known, triggers can be avoided. Tests for allergy are limited by the availability of test sera that reflect the spectrum of molds likely to be encountered indoors. However, allergy tests can often determine atopy, and some presumption can be made if a patient is allergic to any test substance. The patient may also be allergic to molds from previous exposures, even if mold allergy tests are negative. The mold allergy tests may show negative results due to the limitation of available mold sera.

5.16.1.4   Markers for Exposure

Immunoglobulin E (IgE) antibody serological tests have limited value in diagnosis but can determine whether the subject has recently been exposed to specific molds. Unfortunately, only exposure can be determined, not etiology of any health effect, with the current state of the art. While mycotoxins can be readily detected in environmental samples, they cannot as yet be detected in bodily fluids, except for aflatoxin adducts. Finding such DNA, RNA or protein adducts in blood or urine is a marker of exposure to aflatoxin and to some degree an indicator that damage is being done, but such a finding does not give a quantifiable estimate of damage due to this mycotoxin.

5.16.1.5   Mycotoxin Effects

Information regarding human health consequences from exposure to individual mycotoxins is not known except from extrapolation of controlled animal exposure experiments. Exposure to toxins in the field (whether in agriculture or indoor exposures) is usually to multiple toxins for which the combined exposures or concomitant exposures to other air contaminants have not been characterized. A number of sources report that exposure to combinations of mycotoxins (i.e., citrinin and penicillic acid) can act synergistically (Sorensen, 1993). Single toxins can mistakenly be emphasized, even though exposure to multiple toxins from a single mold species capable of making multiple toxins, or to a mixture of molds making multiple toxins, is far more likely in environments where multiple molds are growing. Other effects (allergy or irritation) can also complicate the disease picture.

Information from field exposures of animals indicates that immune suppression is a characteristic LOAE outcome from mycotoxin exposure (Smith and Moss, 1985). This immune suppression manifests as a decreased resistance to infectious disease and a failure to thrive. Observations in studies of office workers (Johanning et al., 1996; Hodgson et al., 1988) show that the incidence, severity, and time course of infectious disease are greater in offices with toxigenic mold contamination than in buildings without mold. Exposure to one or more trichothecene toxins, which are among the most potent natural product inhibitors of protein syntheses (antibodies are proteins) known, or to lactams and lactones, which specifically target the immune system, could produce such effects.

5.16.2   At-Risk Groups

People vary in their susceptibility to environmental insult due to:

- Age
- Genetic predisposition
- State of health
- Nutrition
- Gender
- Amount and kinds of exposure

Whether or not an individual is at risk from environmental exposure depends on individual susceptibility and the amount, kind, and duration of exposure to contaminants. For example, the young and old differ from healthy adults in potential risk. Fetuses and children are more susceptible to toxins and to many infectious agents because their systems are growing rapidly, and their defense mechanisms are incompletely developed. Children also breathe more air per body weight than adults do, so in effect they are exposed to more air contaminants. In general, children are more active than adults, and because lung ventilation increases with exercise they may also breathe more on that account. Children are also more likely to come in contact with sources of mold in carpets, from which spores and dust can be aerosolized and breathed, but they also tend to touch many surfaces and put dirty hands into their mouths. Because young children crawl and get dirty, they are also likely to have more skin exposure. Older adults may be more susceptible due to loss of defense mechanisms, the ravages of survived diseases, a lack of exercise, or poor nutrition.

REFERENCES AND RESOURCES

ACGIH, *ACGIH 2001 TLVs and BEIs: Threshold Limit Values for Chemical Substances and Physical Agents and Biological Exposure Indices*, American Conference of Governmental Industrial Hygienists, Cincinnati, OH, 2001.

AIHA, *The AIHA 2001 Emergency Response Planning Guidelines and Workplace Environmental Exposure Level Guides Book*, American Industrial Hygiene Association, Fairfax, VA, 2001.

Ammann, H.M., Microbial volatile organic compounds, in *Bioaerosols, Fungi and Mycotoxins: Health Effects, Assessment, Prevention and Control*, E. Johanning, Ed., Eastern New York Occupational and Environmental Health Center, Albany, NY, 1999, pp. 84–93.

Benjamin, G.S., The lungs, in *Fundamentals of Industrial Hygiene*, B.A. Plog, J. Niland, and P.J. Quinlan, Eds., National Safety Council, Itasca, IL, 1996, pp. 35–51.

Betina, V., *Mycotoxins: Chemical, Biological, and Environmental Aspects, Bioactive Molecules*, Vol. 9, Elsevier, New York, NY, 1989.

Burge, H. A. and Ammann, H. M., Fungal toxins and β $(1 \rightarrow 3)$-D-glucans, in *Bioaerosols Assessment and Control*, Macher, J., Ed., American Conference of Governmental Industrial Hygienists (ACGIH), Cincinnati, OH, 1999.

Cometto-Muòiz, J.E. and Cain, W.S., Efficacy of volatile organic compounds in evoking nasal pungency and odor, *Arch. Environ. Health*, 48(5), 309–314, 1993.

Cresia, D.A., Thurman, D., Wannemacher, R.W., and Bunner, D.L., Pulmonary toxicology of T-2 mycotoxin, *Toxicologist*, 5, 233, 1985.

Cresia, D.A., Thurman, J.D., Jones, L.J., III, Nealley, M.L., York, C.G., Wannemacher, R.W., Jr., and Bunner, D.L., Acute inhalation toxicity of T-mycotoxin in mice, *Fund. Appl. Toxicol.*, 8(2), 230–235, 1987.

Cresia, D.A., Thurman, J.D., Wannemacher, R.W., and Bunner, D.L., Acute inhalation toxicity of T-2 mycotoxin in the rat and guinea pig, *Fund. Appl. Toxicol.*, 14, 54, 1990.

Cresia, D.A., Wannemacher, R.W., and Bunner, D.L., Acute inhalation toxicology of aerosols of T-2 toxin in solution and as a suspension, *Toxicologist*, 6, 62, 1986.

Crook, B., Inertial samplers: biological perspectives, in *Bioaerosols Handbook*, Cox, C.S. and Wathes, C.M., Eds., CRC Press, Boca Raton, FL, 1995.

Etzel, R.A., Montaña, E., Sorenson, W.G., Kullman, G.J., Allan, T.M., and Dearborn, D.G., Acute pulmonary hemorrhage in infants associated with exposure to *Stachybotrys atra* and other fungi, *Arch. Pediatr. Adolesc. Med.*, 152, 757–761, 1998.

Etzel, R.A. and Dearborn, D.G., Pulmonary hemorrhage among infants with exposure to toxigenic molds: an update, in *Bioaerosols, Fungi and Mycotoxins: Health Effects, Assessment, Prevention and Control*, Johanning, E., Ed., Eastern New York Occupational and Environmental Health Center, Albany, NY, 1999, pp. 79–83.

Fatterpekar, G. et al., Fungal diseases of the paranasal sinuses, *Semin. Ultrasound Comp. Tomogr. Magn. Res.*, 20(6), 391–401, 1999.

Frisvad, J.C., and Gravesen, S., *Penicillium* and *Aspergillus* from Danish homes and working places with indoor air problems: identification and mycotoxin determination, in *Health Implications of Fungi in Indoor Environments*, Samson, R.A., Flannigan, B., Flannigan, M.E., Verhoeff, A.P., Adan, A.C.G., and Hoekstra, E.S., Eds., Air Quality Monographs, Vol. 2, Elsevier, New York, NY, 1994.

Fung, F., Clark, R., and Williams, S., Stachybotrys, a mycotoxins-producing fungus of increasing toxicological importance, *J. Toxicol. Clin. Toxicol.*, 36(1–2), 79–86, 1998.

Gareis, M., Cytotoxicity of samples originating from problem buildings, in *Proc. Int. Conf. Fungi and Bacteria in Indoor Environments: Health Effect, Detection and Remediation*, E. Johanning and C.S. Yang, Ed., Saratoga Springs, NY, October 6–7, 1994, 1995, pp. 139–145.

Gravesen, S., Frisvad, J.C., and Samson, R.A., *Microfungi*, Munksgaard, Copenhagen, Denmark, 1994.

Gravesen, S., Nielsen, P.A., Iversen, R., and Nielsen, K.F., Microfungal contamination of damp buildings: examples of risk construction and risk materials, *Environ. Health Perspect.*, 107(suppl. 3), 505–508, 1999.

Hammond, P.B. and Coppock, R., *Valuing Health Risks, Costs, and Benefits for Environmental Decision-Making*, National Academy Press, Washington, D.C., 1990.

Henson, P.M. and Murphy, R.C., *Mediator of the Inflammatory Process*, Elsevier, New York, NY, 1989.

Hinckley, S.F., Jiang, E.P., Mazzola, E.P., and Jarvis, B., Atratones: Novel diterpoids from the toxigenic mold *Stachybotrys atra*, *Tetrahedon Lett.*, 40, 2725–2728, 1999.

Hintikka, E.-L., Human stachybotrystoxicosis, in *Mycotoxic Fungi, Mycotoxins, Mycotoxicoses: An Encyclopedic Handbook*, Vol. 3., T.D. Wyllie and L.G. Morehouse, Eds., Marcel Dekker, New York, NY, 1978, pp. 87–89.

Hodgson, M.J., The medical evaluation, in *Effects of Indoor Environment on Health: State of the Art Reviews*, J. Seltzer, Ed., Hanley & Belfus, Philadelphia, 1995.

Hodgson, M.J., Morey, P., Leung, W.-Y., Morrow, L., Miller, D., Jarvis, B.B., Robbins, H., Halsey, J.F., and Storey, E., Building associated pulmonary disease from exposure to *Stachybotrys chartarum* and *Aspergillus versicolor*, *J. Occup. Environ. Med.*, 40, 241–249, 1998.

IOM, *Clearing the Air: Asthma and Indoor Air Exposures*, National Academy Press, Washington, D.C., 2000.

Jacobsen, B.J., Bowen, K.L., Shelby, R.A., Diener, U.L., Kemppainen, B.W., and Floyd, J., Mycotoxins and Mycotoxicoses, Circular ANR-767, Alabama Cooperative Extension System, Alabama A&M and Auburn Universities, 1993, pp. 1–17.

Jarvis, B.B., Mycotoxins in the air: Keep your building dry or the bogeyman will get you, in *Proceedings of the International Conference: Fungi and Bacteria in Indoor Environments, Health Effects, Detection and Remediation*, Johanning, E. and Yang, C.S., Eds., Saratoga Springs, NY, October 6–7, 1994, Eastern New York Occupational and Environmental Health Center, Albany, NY, 1995.

Jarvis, B.B. and Hinkley, S.F., Analysis for *Stachybotrys* toxins, in Johanning, E., Ed., *Bioaerosols, Fungi and Mycotoxins: Health Effects, Assessment, Prevention and Control*, Eastern New York Occupational and Environmental Health Center, Albany, NY, 1999, pp. 232–239.

Jarvis, B.B., Salemme, J., and Morais, A., *Stachybotrys* toxins. 1. *Natural Toxins*, 3, 10–16, 1995.

Jarvis, B.B., Sorenson, W.G., Hintikka, E.-L., Nikulin, M., Zhou, Y., Jiang, J., Wang, S., Hinkley, S., Etzel, R.A., and Dearborn, D., Study of toxin production by isolates of *Stachybotrys chartarum* and *Memnoniella echinata* isolated during a study of pulmonary hemosiderosis in infants, *Appl. Environ. Microbiol.*, 64(10), 3620–3625, 1998.

Johanning, E., Biagini, R., Hull, D.L., Morey, P., Jarvis, B., and Landbergis, P., Health and immunology study following exposure to toxigenic fungi (*Stachybotrys chartarum*) in a water-damaged office environment, *Int. Arch. Environ. Health*, 68, 207–218, 1996.

Kuiper-Goodman, T., Scott, P.M., and Watanabe, H., Risk assessment of the mycotoxin zearalenone, *Reg. Toxicol. Pharmacol.*, 7, 253–306, (1987.

Lacey, J. and Venette, J., Outdoor air sampling techniques, in *Bioaerosols Handbook*, Cox, C.S. and Wathes, C.M., Eds., CRC Press, Boca Raton, FL, 1995.

Land, C.J., Rask-Anderssen, A., Werner, S., and Bardage, S., Tremorgenic mycotoxins in conidia of *Aspergillus fumigatus*, in Samson, R.A. and Flannigan, B., Eds., *International Workshop, Health Implications of Fungi in Indoor Environments*, Elsevier, Amsterdam, 1994, pp. 317–324.

Larson, T.O. and Frisvad, J.C., Production of volatiles and presence of mycotoxins in conidia of common indoor *Penicillia* and *Aspergilli*, in *Health Implications of Fungi in Indoor Environments*, Samson, R.A., Flannigan, B., Flannigan, M.E., Verhoeff, A.P., Adan, A.C.G., and Hoekstra, E.S., Eds., Air Quality Monographs, Vol. 2, Elsevier, New York, NY, 1994, pp. 251–279.

Leopold, D.A., Nasal toxicity: end-points of concern in humans, *Inhalation Toxicol.*, 6, 23–40, 1994.

Macher, J.M., Ed., *Bioaerosols: Assessment and Control*, American Conference of Governmental Industrial Hygienists, Cincinnati, OH, 1999.

Mandell, G.L., Bennet, J.E., and Dolin, R., Eds., *Principles and Practice of Infectious Diseases: Mycoses*, 4th ed., Churchill-Livingstone, Philadelphia, 1996, pp. 2288–2393.

Martyny, J., Martinez, K.F., and Morey, P.R., Source sampling, in *Bioaerosols: Assessment and Control*, Macher, J., Ammann, H.M. Burge, H.A., Milton, D.K., and Morey, P.R., Eds., American Conference of Governmental Industrial Hygienists, Cincinnati, OH, 1999.

McLaughlin, C.S., Vaughan, M.H., Campbell, I.M., Wei, C.M., Stafford, M.E., and Hansen, B.S., Inhibition of protein synthesis by trichothecenes, in *Mycotoxins in Human and Animal Health*, Rodrick, J.S., Hesseltine, C.M., and Mehlman, M.A., Eds., Pathotox, Park Forest, IL, 1997, pp. 263–273.

Meggs, W.J., RADS and RUDS — the toxic induction of asthma and rhinitis, *Clin. Toxicol.*, 32(5), 487–501, 1994.

Miller, J.D., Mycotoxins, in *Organic Dusts: Exposure, Effects, and Prevention*, Rylander, R. and Jacobs, R.R., Eds., Lewis Publishers, Boca Raton, FL, 1994.

Miller, R.V., Martinez-Miller, C., and Bolin, V., *Proc. Tenth Int. IUPAC Symp. on Mycotoxins and Phycotoxins: A Novel Risk Assessment Model for the Evaluation of Fungal Exposure in Indoor Environments*, Ponsen and Looijen, Wageningen, 2000.

Miller, R.V., Martinez-Miller, C., and Bolin, V., Application of a novel risk assessment model to evaluate exposure to molds and mycotoxins in indoor environments, *Proc. Second NSF International Conf. on Indoor Air Health, Trends and Advances in Risk Assessment and Management: Application of a Novel Risk Assessment Model to Evaluate Exposure to Molds and Mycotoxins in Indoor Environments*, NSF International, Ann Arbor, MI, 2001.

Nardell, E. and Macher, J., Respiratory infections — transmission and environmental control, in *Bioaerosols Assessment and Control*, Macher, J., Ed., American Conference of Governmental Industrial Hygienists (ADGIH), Cincinnati, OH, 1999.

National Academy of Sciences, *Science and Judgement in Risk Assessment*, National Academy Press, Washington, D.C., 1994.

Nielsen, G.D., Alarie, Y., Poulsen, O.M., and Nexø, B.A., Possible mechanism for the respiratory tract effects of non-carcinogenic indoor-climate pollutants and based for their risk assessment, *Scand. J. Work Environ. Health*, 21, 165–178, 1995.

NIOSH, *NIOSH Pocket Guide to Chemical Hazards*, National Institute for Occupational Safety and Health, Washington, D.C. (updated periodically), 1987.

NRC, *Risk Assessment in the Federal Government: Managing the Process*, National Academy Press, Washington, D.C., 1983.

Otten, J.A. and Burge, H.A., Viruses, in *Bioaerosols Assessment and Control*, Macher, J., Ed., American Conference of Governmental Industrial Hygienists (ACGIH), Cincinnati, OH, 1999.

Otto, D., Molhave, L., Rose, G., Hudnell, H.K., and House, D., Neurobehavioral and sensory irritant effects of controlled exposure to a complex mixture of volatile organic compounds, *Neurotoxicol. Teratol.*, 12, 649–652, 1990.

Pang, V.F., Lambert, R.J., Felsburg, P.J., Beasley, V.R., Buck, W.B., and Haschek, W.M., Experimental T-2 toxicosis in swine, following inhalation exposure: effects on pulmonary and systemic immunity and morphological changes, *Toxicol. Pathol.*, 15, 308–319, 1988a.

Pang, V.F., Lambert, R.J., Felsburg, P.J., Beasley, V.R., Buck, W.B., and Haschek, W.M., Experimental T-2 toxicosis in swine, following inhalation exposure: clinical signs and effects on hematology, serum biochemistry, and immune response, *Fund. Appl. Toxicol.*, 11, 100–109, 1988b.

Ponikau, J.U. et al., The diagnosis and incidence of allergic fungal sinusitis, *Mayo Clin. Proc.*, 74, 877–884, 1999.

Pope, A.M., Patterson, R., and Burge, H.A., Eds., *Indoor Allergens*, National Academy Press, Washington, D.C., 1993.

Riley, E.C., Murphy, G., and Riley, R.L., Airborne spread of measles in a suburban elementary school, *Am. J. Epidemiology*, 107, 421–432, 1978.

Rose, C.S., Antigens, in *Bioaerosols Assessment and Control*, Macher, J.M., Ed., American Conference of Governmental Industrial Hygienists (ACGIH), Cincinnati, OH, 1999.

Schiffman, S.S., Walker, J.M., Dalton, P., Lorig, T.X., Raymer, J.H., Shusterman, D., and Williams, C.M., Potential health effects of odor from animal operations, wastewater treatment, and recycling by-products, *J. Agromed.*, 7(1), 7–81, 2000.

Schleibinger, H., Böck, R., and Rüden, H., Occurrence of microbiologically produced aldehydes and ketones (mVOCs) from filter materials of HVAC systems: field and laboratory experiments, ASHRAE '95, American Society of Heating, Refrigeration, and Air Conditioning Engineers, Atlanta, GA, 1995.

Shusterman, D., Critical review: the health significance of environmental odor pollution, *Arch. Environ. Health*, 47(1), 76–91, 1992.

Smith, J.E. and Moss, M.O., *Mycotoxins Formation, Analysis, and Significance*, John Wiley and Sons, New York, NY, 1985.

Sorenson, W.G., Mycotoxins: toxic metabolites of fungi, in *Fungal Infections and Immune Response*, J.W. Murphy, Ed., Plenum Press, NY, 1993, pp. 469–491.

Tuomi, R., Reijula, K., Hemminki, K., Hintikka, E.-L., Lindoos, O., Kalso, S., Koukila-Kähkölä, P., Mussalo-Rauhamaa, H., and Haahtela, T., Mycotoxins in crude building materials from water-damaged buildings, *Appl. Environ. Microbiol.*, 66(5), 1899–1904, 2000.

USEPA, The Total Exposure Assessment Methodology (TEAM) Study, EPA/600/6–87/002a, Office of Acid Deposition, Environmental Monitoring, and Quality Assurance, Office of Research and Development, U.S. Environmental Protection Agency, Washington, D.C., 1987.

USEPA, Risk Assessment Guidance for Superfund, Vol. I, Human Health Evaluation Manual (Part A), interim final, EPA/540/1–89/002, Office of Research and Development, U.S. Environmental Protection Agency, Washington, D.C., 1989.

USEPA, Exposure Factors Handbook, EPA/600/Z-92/001, Office of Research and Development, National Center for Environmental Assessment, U.S. Environmental Protection Agency, Washington, D.C., 1992.

USEPA, Methods for the Derivation of Inhalation Reference Concentrations and Application of Inhalation Dosimetry, EPA/600/8–90/066F, Office of Research and Development, U.S. Environmental Protection Agency, Washington, D.C., 1994.

USEPA, Guidelines for Exposure Assessment, EPA/600/P-95/002Fa, U.S. Environmental Protection Agency, Washington, D.C., 1997.

WHO, Environmental Criteria 105, World Health Organization, Geneva, 1990.

Willeke, K. and Macher, J.M., Air sampling, in *Bioaerosols: Assessment and Control. Source Sampling*, Macher, J., Ammann, H.A., Burge, H.A., Milton, D.K., and Morey, P.R., Eds., American Conference of Governmental Industrial Hygienists, Cincinnati, OH.(1999)

CHAPTER 6

Ventilation Systems

Martha J. Boss and Dennis W. Day

CONTENTS

1-56670-606-8/03/$0.00+$1.50
© 2003 by CRC Press LLC

In order to understand biological hazard mitigation, basic ventilation concepts and equipment usage must be understood. These concepts are presented in general terms here, and in more specific terms in Chapter 14.

6.1 INDOOR AIR QUALITY IMPROVEMENT METHODS

The three most common means for improving indoor air quality (IAQ), in order of effectiveness, are:

Source control: Eliminating or controlling the sources of pollution
Ventilation: Diluting and exhausting pollutants through outdoor air ventilation
Air cleaning: Removing pollutants through proven air cleaning methods

Of the three, the first approach, source control, is the most effective. This involves minimizing the use of products and materials that cause indoor pollution, employing good hygiene practices to minimize biological contaminants (including the control of humidity and moisture and occasional cleaning and disinfection of wet or moist surfaces), and using good housekeeping practices to control particulates.

The second approach, outdoor air ventilation, is also effective and is commonly employed. Ventilation methods include installing an exhaust fan close to the source of contaminants, increasing outdoor airflows in mechanical ventilation systems, and opening windows, especially when pollutant sources are in use.

The third approach, air cleaning, is not generally regarded as sufficient by itself but is sometimes used to supplement source control and ventilation. Air filters, electronic particle air cleaners, and ionizers are often used to remove airborne particles, and gas adsorbing material is sometimes used to remove gaseous contaminants when source control and ventilation are inadequate.

6.2 SOURCE CONTROL

Source control or reduction may involve adding additional ventilation systems and enclosing the areas where contaminant generation is occurring. One of the initial advantages of any closed-duct or closed-area ventilation system is that the heating and cooling mechanisms may be located separate from the living spaces. Given the limitations of the human sensory system, source reduction devices must be monitored by more than just sensory input (i.e., seeing or smelling the contaminant or experiencing skin irritation). Modern logic control systems and contaminant detection systems serve to monitor the day-to-day operation of more sophisticated systems. All too often, however, these systems are juxtaposed with the in-place older systems and adequate monitoring does not occur. In-place monitors are also subject to degradation, and not all chemicals can be monitored via in-place systems.

6.3 VENTILATION HOODS

If hoods are used as a means of source control, hood placement must be close to the emission source to be effective. The design elements discussed here are general design practices; site-specific ventilation design by a qualified professional is required to ensure ventilation system efficacy.

The maximum distance from the emission source should not exceed 1.5 duct diameters. The approximate relationship of capture velocity (V_c) to duct velocity (V_d) for a simple plain or narrow flanged hood should be calculated as follows.

- If an emission source is one duct diameter in front of the hood and the duct velocity (V_d) = 3000 feet per minute (fpm), then the expected capture velocity (V_c) is 300 fpm. At two duct diameters from the hood opening, V_c decreases by a factor of 10. Varying hood conformations and air entry designs will alter this calculation.
- For simple capture hoods, if the duct diameter (D) is 6 in., then the maximum emission source distance from the hood should not exceed 9 in. Similarly, the minimum capture velocity should not be less than 50 fpm.

System effect loss, which occurs at the fan, can be avoided if properly designed or sized ductwork is in place. Use of the six-and-three rule ensures better design by providing for a minimum loss at six diameters of straight duct at the fan inlet and a minimum loss at three diameters of straight duct at the fan outlet. System effect loss is significant if any elbows are connected to the fan at the inlet or the outlet. For each 2.5 diameters of straight duct between the fan inlet and any elbow, the loss (measured in cubic feet per minute, or cfm) will be 20%. Stack height should be 10 ft higher than any roof line or air intake located within 50 ft of the stack. For example, a stack placed 30 ft away from an air intake should be at least 10 ft higher than the center of the intake.

Ventilation system drawings and specifications generally use standard forms and symbols, such as those described in the Uniform Construction Index (UCI). Plan sections include electrical, plumbing, structural, or mechanical drawings (UCI, Section 15). The drawings come in several views: plan (top), elevation (side and front), isometric, and section. Elevations (side and front views) give the most detail. An isometric drawing is one that illustrates the system in three dimensions. A sectional drawing provides duct or component detail by showing a component cross-section. Drawings are usually drawn to scale (check dimensions and lengths with a ruler or a scale to be sure that this is the case); for example, 1/8 inch on the sheet may represent one foot on the ground.

6.4 DESIGN ALTERNATIVES

Professional engineers and equipment manufacturers offer many design alternatives to achieve ventilation goals. When reviewing the design scope of work and ultimately the design drawings and specifications, consider the project background and objectives and project scope (what is to be included and why). Look for conciseness and precision. Mark ambiguous phrases, legalese, and repetition. Ask these questions and document the answers:

- Do the specifications spell out exactly what is wanted and what is expected?
- Do plans and specifications adhere to appropriate codes, standards, requirements, and policies?
- Do plans and specifications recommend good practice as established by the industry?
- Will the designer be able to design, or the contractor build, the system from the initial plans and specifications?
- Will the project meet requirements of the Occupational Safety and Health Administration (OSHA) and guidelines of the American National Standards Institute (ANSI) if built as proposed?
- Will maintenance personnel be able to access equipment to ensure proper operation and to perform required cleaning and, if needed, decontamination?

Maintain a project file that includes the answers to these questions and the design documents. Require that designers and/or contractors mark up a set of design drawings to illustrate any changes that occur during construction. Require that the system be empirically tested to determine airflow rates, structural integrity, and humidity variations. Also ensure that as-built drawings are prepared and request a copy. This copy should be kept on file both at the building and in the engineering and/or environmental health and safety office.

6.5 POTENTIAL BIOLOGICAL CONTAMINANTS

Biological exposures that contaminate building interiors have a potential additional hazard in that the biological risk can amplify through reproduction in our homes, industries, and in our bodies. The same heating, ventilation, and air conditioning (HVAC) system that distributes conditioned air throughout a building can distribute dust and other pollutants, including biological contaminants. Dirt or dust accumulation on any air-handling system component (cooling coils, plenums, ducts, or equipment housing) may lead to air supply contamination.

Indoor air contaminants include but are not limited to particulates, pollen, microbial agents, and organic toxins. These contaminants can be transported by the ventilation system or originate in the following ventilation system parts: wet filters; wet insulation; wet undercoil pans; cooling towers; and evaporative humidifiers. People exposed to these agents may develop signs and symptoms related to humidifier fever, humidifier lung, or air conditioner lung. In some cases, indoor air quality contaminants cause clinically identifiable conditions such as occupational asthma, reversible airway disease, and hypersensitivity pneumonitis.

6.6 AIR INTAKE

During the past 25 years, interest in constructing energy-efficient buildings has increased. Some current construction practices can trap pollutants that normally form inside the building with those brought inside with everyday traffic. The combination of heating, cooling, and ventilation systems that recycle existing indoor air and windows that do not open can result in greater concentrations of indoor pollutants because fresh outside air, which serves to dilute the trapped pollutants, is not admitted.

To provide replacement or make-up air, a variety of systems are used to move air into and out of a facility. The basic systems rely on the creation of pressure differentials to move air. A suction fan system is often used to create a partial vacuum. Through various intakes, air rushes in toward the lower pressure area. The side where the partial vacuum was created in an air-handling system is the suction or return side; the side where the air is being forced into the facility is the supply side.

Various devices are used to provide equalization and appropriate airflow. The American Society of Heating, Refrigeration, and Air Conditioning Engineers (ASHRAE) requirements specify minimum fresh air exchanges per hour for normal office-type occupancy. When interior sources of industrial or commercial air pollutants are present, source reduction is usually the remedy of choice vs. general ventilation to dilute both the source and source receiving areas.

Designs are often complicated by the need to conserve energy and reuse interior air streams that have already been tempered (heated or cooled) and may have been humidified. Heat recovery may include systems to channel heat from HVAC systems and service water heating, use of economizer cycles, mixing of reusable air with fresh air, and various forms of insulation. Advanced designs of new homes are starting to feature mechanical systems that bring outdoor air into the home. Some of these designs include energy-efficient heat recovery ventilators (also known as air-to-air heat exchangers).

The rate at which outdoor air replaces indoor air is the *exchange rate*, which measures how many times the complete volume of air inside the house is replaced with fresh outside air. In typical U.S. homes, the average exchange rate is 0.7 to 1 complete air exchanges per hour. In tight homes, the exchange rate can be as low as 0.02 complete air exchanges per hour.

Unfortunately, in an effort to reduce energy costs during the 1970s and thereafter, nonstandard methods of energy conservation were used. The first step after identifying indoor air quality issues should be to conduct a joint air quality study and HVAC system evaluation. Indoor air quality studies should be conducted in parallel with an evaluation of the current mechanical system usage, operation, and maintenance.

6.7 TURNKEY ISSUES: BIOSAFE BUILDINGS

The following general principles will help ensure biosafe buildings:

- Install and use exhaust fans that are vented to the outdoors in kitchens and bathrooms. Vent clothes dryers outdoors. These actions can eliminate moisture that builds up from everyday activities. Another benefit to using kitchen and bathroom exhaust fans is that these fans can reduce organic pollutant levels that vaporize from hot water used in dishwashers and showers.
- Ventilate the attic and crawl spaces to prevent moisture build-up. Keeping humidity levels in these areas below 50% can help prevent water condensation on building materials.
- If cool mist or ultrasonic humidifiers are used, clean the appliances according to manufacturers' instructions and refill with fresh water daily. Because these humidifiers can become breeding grounds for biological contaminants, these humidifiers have the potential for spreading biological contaminants that cause such diseases as hypersensitivity pneumonitis and humidifier fever. Evaporation trays in air conditioners, dehumidifiers, and refrigerators should also be cleaned frequently.
- Thoroughly clean and dry water-damaged carpet and building materials (within 24 hours) or consider removal and replacement. Water-damaged carpets and building materials can harbor mold and bacteria, and ridding such materials of biological contaminants may be very difficult. Also, be sure to thoroughly dry carpet and building materials that have been cleaned with water or steam.
- Keep the building clean. Dust mites, pollens, animal dander, and other allergy-causing agents can be reduced, although not eliminated, through regular cleaning.
- Use allergen-proof mattress encasements, wash bedding in hot (130°F) water, and avoid room furnishings that accumulate dust, especially if these furnishings cannot be washed in hot water.
- Use central vacuum systems that are vented to the outdoors or vacuums with HEPA filters. Allergic individuals should also leave the house while it is being vacuumed because vacuuming can actually increase airborne mite allergens and other biological contaminant levels.
- Take steps to minimize biological pollutants in basements. Clean and disinfect the basement floor drain regularly. Do not finish a basement below ground level unless all water leaks are patched and outdoor ventilation and adequate heat are provided to prevent condensation. Operate a dehumidifier in the basement if needed to keep relative humidity levels between 30 and 50%.

6.8 HUMIDITY AND CONDENSATE EFFECTS: MANAGEMENT AND CONTROL

Molds and mildew are fungi that grow on object surfaces, within pores, and in deteriorated materials. These molds can cause discoloration and odor problems, deteriorate building materials, and lead to health problems. The following conditions are necessary for mold growth to occur on building surfaces:

- Temperature range above 40°F and below 100°F
- Mold spores
- Nutrient base (most surfaces contain nutrients)
- Moisture

Spores are almost always present in outdoor and indoor air, and almost all commonly used construction materials and furnishings can provide nutrients to support mold growth. Dirt on surfaces provides additional nutrients. Mold growth hot spots include damp basements and closets, bathrooms (especially shower stalls), places where fresh food is stored, refrigerator drip trays, house plants, air conditioners, humidifiers, garbage pails, mattresses, upholstered furniture, and old foam rubber pillows. Mold growth does not require standing water. Mold growth can occur when high relative humidity occurs or if the hygroscopic properties (the tendency to absorb and retain moisture) of building surfaces allow sufficient moisture to accumulate.

6.8.1 Relative Humidity, Vapor Pressure, and Condensation

Water enters buildings both as a liquid and as a gas (water vapor). Water, in liquid form, is introduced intentionally in bathrooms, kitchens, and laundries and accidentally via leaks and spills. Some of that water evaporates and joins the water vapor that is inhaled by building occupants or that is introduced by humidifiers. Water vapor also moves in and out of the building as part of the air that is mechanically introduced or that infiltrates and exfiltrates through openings in the building shell. A lesser amount of water vapor diffuses into and out of the building through the building materials themselves.

The ability of air to hold water vapor decreases as the air temperature is lowered. If an air unit contains half of the water vapor the air can hold, then 50% relative humidity (RH) is present. As the air cools, the relative humidity increases. If the air contains all of the water vapor the air can hold, then 100% RH is present, and the water vapor condenses, changing from a gas to a liquid. An RH of 100% can be reached without changing the water vapor amount in the air (its vapor pressure or absolute humidity). All that is required is for the air temperature to drop to the *dew point*.

Relative humidity and temperature often vary within a room, while the absolute humidity in the room air can usually be assumed to be uniform; therefore, if one side of the room is warm and the other side cool, the cool side has a higher RH than the warm side. The highest RH in a room is always next to the coldest surface. This is referred as the *first condensing surface*, as it will be the location where condensation first occurs if the relative humidity at the surface reaches 100%. When trying to understand why mold is growing on one patch of wall or only along the wall–ceiling joint, the condensing surfaces must be considered. The wall surface is probably cooler than the room air because a void exists in the insulation or because wind is blowing through cracks in the building exterior.

6.8.2 Taking Steps to Reduce Moisture

Mold and mildew growth can be reduced where relative humidity near surfaces can be maintained below the dew point. This can be accomplished by reducing the air moisture content (vapor pressure), increasing air movement at the surface, or increasing the air temperature (either the general space temperature or the temperature at building surfaces). Either surface temperature or vapor pressure can be the dominant factor in causing a mold problem. A surface-temperature-related mold problem may not respond very well to increasing ventilation, whereas a vapor-pressure-related mold problem may not respond well to increasing temperatures. Understanding which factor dominates will help in selecting an effective control strategy.

Consider an old, leaky, poorly insulated building. This building is in a heating climate and shows evidence of mold and mildew. Because the building is leaky, its high natural air exchange rate dilutes interior airborne moisture levels, maintaining a low absolute humidity during the heating season. Providing mechanical ventilation in this building in an attempt to control interior mold and mildew probably will not be effective in this case. Increasing surface temperatures by insulating the exterior walls and thereby reducing relative humidity next to the wall surfaces would be a better strategy to control mold and mildew.

Reduction of surface-temperature-dominated mold and mildew is best accomplished by increasing the surface temperature through either or both of the following approaches:

- Increase the air temperature near room surfaces either by raising the thermostat setting or by improving air circulation so that supply air is more effective at heating the room surface.
- Decrease the heat loss from room surfaces either by adding insulation or by closing cracks in the exterior wall to prevent wind-washing (air that enters a wall at one exterior location and exits through another exterior location without penetrating into the building).

Vapor-pressure-dominated mold and mildew can be reduced by one or more of the following strategies:

- Source control (e.g., direct venting of moisture generating activities such as showers) to the exterior
- Dilution of moisture-laden indoor air with outdoor air that is at a lower absolute humidity
- Dehumidification

Note that dilution is only useful as a control strategy during heating periods, when cold outdoor air tends to contain less moisture. During cooling periods, outdoor air often contains as much moisture as indoor air.

6.9 COMMON MOLD AND MILDEW AMPLIFICATION AREAS

6.9.1 Exterior Corners

Mold and mildew are commonly found on the exterior wall surfaces of corner rooms in heating climate locations. An exposed corner room is likely to be significantly colder than adjoining rooms. Exterior corners are common locations for mold and mildew growth in heating climates and in poorly insulated buildings in cooling climates. These corners tend to be closer to the outdoor temperature than other building surface parts for one or more of the following reasons:

- Poor air circulation (interior)
- Wind-washing (exterior)
- Low insulation levels
- Greater surface area of heat loss

Sometimes mold and mildew growth can be reduced by removing obstructions to airflow (e.g., rearranging furniture). Buildings with forced-air heating systems and/or room ceiling fans tend to have fewer mold and mildew problems than buildings with less air movement, other factors being equal.

A balance between the RH and the room temperature must be achieved. The essential question to be considered is "Is the RH above 70% at the surfaces because the room is too cold or because too much moisture is present (high water vapor pressure)?" The moisture in the room can be estimated by measuring temperature and RH at the same location and at the same time. For example, the following two cases illustrate rooms where correction must be made due to measured RH and temperature that are out of balance.

1. Assume that the RH is 30% and the temperature is 70°F in the middle of the room. The low RH at that temperature indicates that the water vapor pressure (or absolute humidity) is low. The high surface RH is probably due to room surfaces that are too cold. Temperature is the dominating factor, and control strategies should involve increasing the temperature at cold room surfaces.
2. Assume that the RH is 50% and the temperature is 70°F in the middle of the room. The higher RH at that temperature indicates that the water vapor pressure is high and a relatively large amount of moisture is present in the air. The high surface RH is probably due to air that is too moist. Humidity is the dominating factor, and control strategies should involve decreasing the indoor air moisture content.

6.9.2   Setback Thermostats

Mold and mildew can often be controlled in heating climate locations by increasing interior temperatures during heating periods. Unfortunately, this heating also increases energy consumption and reduces relative humidity in the breathing zone, which can create discomfort. Setback thermostats are used to reduce energy consumption during the heating season. Mold and mildew growth can occur when building temperatures are lowered during unoccupied periods. (*Note:* Maintaining a room at too low a temperature can have the same effect as a setback thermostat.)

6.9.3   Air Conditioned Spaces

Mold and mildew problems can be as extensive in cooling climates as in heating climates. The same principles apply: Either surfaces are too cold or moisture levels are too high, or both. A common mold growth example in cooling climates can be found in rooms where conditioned cold air blows against the interior surface of an exterior wall. This condition may be due to poor duct design, diffuser location, or diffuser performance; the cold air creates a cold spot on interior finish surfaces.

Rooms decorated with low-maintenance interior finishes such as impermeable wall coverings (vinyl wallpaper) can trap moisture between the interior finish and the gypsum board. Mold growth can be rampant when these interior finishes are coupled with cold spots and exterior moisture. Possible solutions for this problem include:

- Preventing hot, humid exterior air from contacting the cold interior finish (i.e., controlling the vapor pressure at the surface)
- Eliminating the cold spots (elevate the surface temperature) by relocating ducts and diffusers
- Ensuring that vapor barriers, facing sealants, and insulation are properly specified, installed, and maintained
- Increasing the room temperature to avoid overcooling

6.9.4   Concealed Condensation

A mold problem can occur within the wall cavity as outdoor air comes in contact with the cavity side of the cooled interior surface. The use of thermal insulation in wall cavities increases interior surface temperatures in heating climates, reducing the likelihood of interior surface mold, mildew, and condensation, and it reduces the heat loss from the conditioned space into the wall cavities, thus decreasing the temperature in the wall cavities and increasing the likelihood of concealed condensation.

The first condensing surface in a wall cavity in a heating climate is typically the inner surface of the exterior sheathing (i.e., the plywood or fiberboard backside). As the insulation value is increased in the wall cavities, so, too, is the potential for hidden condensation. Concealed condensation can be controlled by either or both of the following strategies:

1. Reduce the entry of moisture into the wall cavities (e.g., by controlling infiltration and/or exfiltration of moisture-laden air).
2. Elevate the first condensing surface temperature.

These changes can be made:

- In heating climate locations, by installing exterior insulation, assuming that no significant wind-washing is occurring
- In cooling climate locations, by installing insulating sheathing to the wall-framing interior and between the wall framing and interior gypsum board

6.9.5 Thermal Bridges

Localized surface cooling commonly occurs as a result of thermal bridges. Thermal bridges are building structure elements that are highly conductive of heat (e.g., steel studs in exterior frame walls, uninsulated window lintels, and the edges of concrete floor slabs). Dust particles sometimes mark the locations of thermal bridges, because dust tends to adhere to cold spots. The use of insulating sheathings significantly reduces the thermal bridge impacts in building envelopes.

6.9.6 Windows

In winter, windows are typically the coldest surfaces in a room, and the interior window surface is often the first condensing surface in a room. Condensation on window surfaces has historically been controlled by using storm windows or insulated glass (e.g., double-glazed windows or selective surface gas-filled windows) to raise interior window surface temperatures. Higher performance glazing systems have led to a greater incidence of moisture problems in heating climate building enclosures. The buildings can now be operated at higher interior vapor pressures (moisture levels) without visible surface condensation on windows. In older building enclosures with less advanced glazing systems, visible condensation on the windows often alerts occupants to the need for ventilation to flush out interior moisture (i.e., opening the windows).

6.10 INTERIOR ZONING

Buildings require outdoor air as make-up air. Often, heating or cooling of make-up air in association with the air currently within the building is also required. As outdoor air is drawn into the building, indoor air is exhausted or allowed to escape (passive relief), thus removing air contaminants. The term *HVAC system* is used to refer to the equipment that can provide heating, cooling, filtered outdoor air, and humidity control to maintain comfort conditions in a building. Not all HVAC systems are designed to accomplish all of these functions. Some buildings rely on only natural ventilation. Others lack mechanical air cooling (AC) equipment, and many function with little or no humidity control. The HVAC system features in a given building will depend on several variables, including:

- Design age
- Climate
- Building codes in effect
- Budget
- Planned use
- Owners' and designers' preferences
- Subsequent modifications

HVAC systems range in complexity from stand-alone units that serve individual rooms to large, centrally controlled systems serving multiple zones in a building. In large modern office buildings with heat gains from lighting, people, and equipment, interior spaces often require year-round cooling. Rooms at the perimeter of the same building (i.e., rooms with exterior walls, floors, or roof surfaces) may require variable heating and/or cooling as hourly or daily outdoor weather conditions change. In buildings over one story in height, perimeter areas at the lower levels also tend to experience the greatest uncontrolled air infiltration.

Some buildings use only natural ventilation or exhaust fans to remove odors and contaminants. In these buildings, thermal discomfort and unacceptable indoor air quality may occur if occupants keep the windows closed because of extreme hot or cold temperatures. Problems related to under-ventilation are also likely when infiltration forces are weakest (i.e., during the swing seasons and summer months).

Modern public and commercial buildings generally use mechanical ventilation systems to introduce outdoor air during the occupied mode. Thermal comfort is maintained by mechanically distributing conditioned (heated or cooled) air throughout the building. In some designs, air systems are supplemented by piping systems that carry steam or water to the building perimeter zones. Areas regulated by a common control (e.g., a single thermostat) are referred to as zones.

6.10.1 Single-Zone HVAC Systems

A single air-handling unit can serve more than one building area if the areas served have similar heating, cooling, and ventilation requirements or if control systems compensate for differences in heating, cooling, and ventilation needs among the spaces served. Thermal comfort problems can result if the design does not adequately account for differences in heating and cooling loads between rooms that are in the same zone. Such differences can easily occur if the cooling loads in some areas within a zone change due to increased occupant population or increased lighting or if new heat-producing equipment (e.g., computers, copiers) is introduced. Areas within a zone can have different solar exposures, which can produce radiant heat gains and losses, which, in turn, create unevenly distributed heating or cooling needs (e.g., as the sun angle changes daily and seasonally).

6.10.2 Multiple-Zone HVAC Systems

Multiple-zone systems can provide each zone with air at a different temperature by heating or cooling the airstream in each zone. Alternative design strategies involve delivering air at a constant temperature while varying the airflow volume or modulating room temperature with a supplementary system (e.g., perimeter hotwater piping).

6.10.3 Constant-Volume HVAC Systems

Constant-volume systems deliver a constant airflow to each space. Changes in space temperatures are made by heating or cooling the air or by switching the air-handling unit on and off. Changes are not made by modulating the supplied air volume. These systems often operate with a fixed minimum percentage of outdoor air or with an air economizer feature.

6.10.4 Variable Air Volume HVAC Systems

Variable air volume (VAV) systems maintain thermal comfort by varying the amount of heated or cooled air delivered to each space, rather than by changing the air temperature; however, many VAV systems also have provisions for resetting the delivery air temperature on a seasonal basis, depending on the weather severity. Overcooling or overheating can occur within a given zone if the system is not adjusted to respond to the load. Underventilation frequently occurs if the system is not designed to introduce at least a minimum quantity (as opposed to percentage) of outdoor air as the VAV system throttles back from full airflow or if the system supply air temperature is set too low for the loads present in the zone.

6.11 TESTING AND BALANCING

Modern HVAC systems typically use sophisticated automatic controls to supply the proper amounts of air for heating, cooling, and ventilation in commercial buildings. In addition to providing acceptable thermal conditions and ventilation air, a properly adjusted and balanced system can reduce operating costs and increase equipment service life. Testing and balancing involve the testing, adjusting, and balancing of HVAC system components so that the entire system provides airflows that are in accordance with the design specifications. Typical components and system parameters tested include:

- All supply, return, exhaust, and outdoor airflow rates
- Control settings and operation
- Air temperatures
- Fan speeds and power consumption
- Filter or collector resistance

The typical test and balance agency or contractor coordinates with the control contractor to accomplish three goals:

1. Verify and ensure the most effective system operation within the design specifications.
2. Identify and correct any problems.
3. Ensure the system safety.

A test and balance report should provide a complete record of the design, preliminary measurements, and final test data and include any discrepancies between the test data and the design specifications, along with reasons for those discrepancies. To facilitate future performance checks and adjustments, appropriate records should be kept on:

- All damper positions
- Equipment capacities
- Control types and locations
- Control settings and operating logic
- Airflow rates
- Static pressures
- Fan speeds; and horsepowers

Testing and balancing of existing building systems should be performed whenever the system is not functioning as designed or when current records do not accurately reflect the actual system operation. The following guidelines are recommended for determining whether testing and balancing are required:

- Space has been renovated or changed to provide for new occupancy.
- HVAC equipment has been replaced or modified.
- Control settings have been readjusted by maintenance or other personnel.
- The air conveyance system has been cleaned.
- Accurate records are required to conduct an IAQ investigation.
- The building owner is unable to obtain design documents or appropriate air exchange rates for compliance with IAQ standards or guidelines.

6.12 OUTDOOR AIR INTAKE

Building codes require the introduction of outdoor air for ventilation in most buildings. Most nonresidential air handlers are designed with an outdoor air intake on the ductwork return side. Outdoor air introduced through the air handler can be filtered and conditioned (heated or cooled) before distribution. Other designs may introduce outdoor air through air-to-air heat exchangers and operable windows.

Indoor air quality problems can be produced when contaminants enter a building with the outdoor air. Rooftop or wall-mounted air intakes are sometimes located adjacent to or downwind of building exhaust outlets or other contaminant sources. Problems can also result if debris (e.g., bird droppings) accumulates at the intake, obstructing airflow and potentially introducing micro-biological contaminants.

If more air is exhausted than is introduced through the outdoor air intake, then outdoor air will enter the building at any leakage sites in the shell. Indoor air quality problems can occur if the leakage site is a door to a loading dock, parking garage, or some other area associated with pollutants.

6.13 MIXED-AIR PLENUM AND OUTDOOR AIR CONTROLS

Outdoor air is mixed with return air (air that has already circulated through the HVAC system) in the air-handling unit mixed-air plenum. If outdoor air make-up and exhaust are balanced and the zones served by each air handler are separated and well defined, the minimum flow of outdoor air to each space may be estimated. This estimate can then be compared to ventilation standards (i.e., ASHRAE standards). Techniques used for this evaluation include:

- Direct measurement of the outdoor air at the intake
- Calculation of the outdoor air percentage by a temperature or CO_2 balance (carbon dioxide measured in an occupied space is also an indicator of ventilation adequacy)
- The use of tracer gases to assess ventilation quantities and airflow patterns

6.13.1 Outdoor Dampers

Indoor air quality problems frequently result if the outdoor air damper is not operating properly. Improper damper operation is defined as a system where the damper is not designed or adjusted to allow the introduction of sufficient outdoor air. The amount of outdoor air introduced in the occupied mode should be sufficient to meet needs for ventilation and exhaust make-up. Air intake may be fixed at a constant volume or may vary with the outdoor temperature. Modulating dampers that regulate the outdoor airflow bring in a minimum amount of outdoor air (in the occupied mode) under extreme outdoor temperature conditions and open further as outdoor temperatures approach the desired indoor temperature.

6.13.2 Air Economizer Cooling Systems

Systems that use outdoor air for cooling are referred to as air economizer cooling systems, which:

- Blend return air (typically at 74°F) with outdoor air to reach a mixed air temperature of 55 to 65°F. (*Note:* Mixed air temperature settings above 65°F may lead to the introduction of insufficient quantities of outdoor air for office space use.)
- Use a mixed air temperature controller and thermostat to control blending rates and volumes.
- Have a sensible/enthalpy control that signals the outdoor air damper to go to the minimum position when the outdoor air is too warm or humid. (*Note:* Economizer cycles that do not provide dehumidification may produce discomfort even when the indoor temperature is the same as the thermostat setting.)
- Further heat or cool the mixed air prior to delivery to occupied spaces.

6.13.3 Freezestat

Many HVAC designs protect the coils by closing the outdoor air damper if the airstream temperature falls below the freezestat setpoint. Inadequate ventilation can occur if the freezestat trips and is not reset or is set to trip at an excessively high temperature. Stratification of the cold outdoor air and warmer return air in the mixing plenums is a common situation that causes nuisance tripping of the freezestat. Unfortunately, the remedy often employed to prevent this problem is to close the outdoor air damper. Obviously, solving the problem in this way can quickly lead to inadequate outdoor air in occupied parts of the building.

6.14 AIR FILTERS

Proper air filtration can play an important role in protecting the HVAC system and in maintaining good indoor air quality in occupied spaces. Air filters should be selected and maintained to provide maximum filtration, while not overtaxing the supply fan capability or leading to blow-out situations with no air filtration. Filters are primarily used to remove particles from the air. The type and design of the filter determine its efficiency at removing particles of a given size and the amount of energy needed to pull or push air through the filter. Filters are rated by different standards and test methods, such as dust spot and arrestance, that measure various performance aspects.

6.14.1 Air Filter Efficacy

Air filters, whatever their design or efficiency rating, require regular maintenance (cleaning for some and replacement for most). As a filter loads up with particles, the filter material becomes more efficient at particle removal but increases the pressure drop through the system, thereby reducing airflow. Filter manufacturers can provide information on the pressure drop through their products under different conditions. Choosing an appropriate filter and proper maintenance are important to keeping the ductwork clean. If dirt accumulates in ductwork and if the relative humidity reaches the dew point (so that condensation occurs), then the nutrients and moisture may support microbiological amplification. Air handlers that are located in places that are difficult to access (e.g., places that require ladders for access, have inconvenient access doors to unbolt, or are located on roofs with no roof hatch access) will be more likely to suffer from poor air filter maintenance and overall poor maintenance. Quick release and hinged access doors for maintenance are more desirable than bolted access panels.

6.14.2 Low-Efficiency Filters

Low-efficiency filters (ASHRAE dust spot rating of 10 to 20% or less) are often used to keep lint and dust from clogging the system heating and cooling coils. Low-efficiency filters, if loaded to excess, will become deformed and even blow out of their filter rack. When filters blow out, bypassing of unfiltered air can lead to clogged coils and dirty ducts. Filtration efficiency can be seriously reduced if the filter cells are not properly sealed to prevent air from bypassing.

6.14.3 Medium-Efficiency Filters

Filters should be selected for their ability to protect both the HVAC system components and general indoor air quality. To maintain the proper airflow and minimize the amount of additional energy required to move air through these higher efficiency filters, pleated-type extended surface filters are used. These filters have a higher removal efficiency than low-efficiency filters, yet will not clog up as quickly as high-efficiency filters, and they can provide much better filtration than low-efficiency filters. In order to maintain clean air in occupied spaces, filters must also remove bacteria, pollens, insects, soot, dust, and dirt with efficiency suited to the building use (ASHRAE dust spot rating of 30 to 60%).

6.14.4 High-Efficiency Extended Surface Filters

Some manufacturers recommend high-efficiency extended surface filters (ASHRAE dust spot rating of 85%) without prefilters as the most cost-effective approach to minimizing energy consumption and maximizing air quality in modern VAV systems that serve office environments. In buildings that are designed to be exceptionally clean, the designers may specify that the equipment must utilize both a medium-efficiency prefilter and a high-efficiency extended surface filter (ASHRAE dust spot rating of 85 to 95%).

6.14.5 Gas and Volatile Organic Compound Removal Filters

Filters are available to remove gases and volatile organic contaminants from ventilation air. These systems are not generally used in normal occupancy buildings. In specially designed HVAC systems, permanganate oxidizers and activated charcoal may be used for gaseous removal filters. Some manufacturers offer partial bypass carbon filters and carbon-impregnated filters to reduce volatile organics in the ventilation air of office environments. Gaseous filters must be regularly maintained (replaced or regenerated) in order for the system to continue to operate effectively.

6.14.6 Acoustical Lining

Acoustical lining is used in air handler fan housings and supply ducts to reduce sound transmission and provide thermal insulation. This lining is often porous or consists of fiberglass that has lofted over time. The porous surface of fiberglass duct liner presents more surface area (which can trap dirt and subsequently collect water) than sheetmetal ductwork. Proper design, installation, filtration, humidity, and maintenance of ducts that contain porous materials are essential. Techniques developed for cleaning unlined metal ducts often are not suitable for use with fiberglass thermal liner or fiberboard. Such ducts may require a special type of cleaning to maintain the duct integrity. Attention to air filters is particularly important in HVAC systems with acoustical duct liner. Duct lining areas that have become contaminated with microbiological growth must be replaced. Sound reduction can also be accomplished with the use of special duct-mounted devices such as attenuators or with active electronic noise control.

6.15 DUCTS

Building owners and managers should take great precautions to prevent dirt, high humidity, or moisture from entering the ductwork. Special attention should be given to trying to find out if ducts are contaminated when specific problems are present, such as when water damage or biological growth is observed in ducts, debris is found in the ducts, or dust is discharging from supply diffusers. Problems with contamination in the ductwork are a function of filtration efficiency, HVAC system maintenance, the airflow rate, and good housekeeping practices in the occupied space. Problems with biological pollutants can be prevented by minimizing dust and dirt build-up, promptly repairing leaks and water damage, preventing moisture accumulation in the components that are supposed to be dry, and cleaning the components such as the drip pans that collect and drain water.

In cases where sheetmetal ductwork has become damaged or water soaked, building owners will need to undertake clean-up or repair procedures. These procedures should be scheduled and performed in a way that does not expose building occupants to increased pollutant levels and should be carried out by experienced workers.

Correcting the problems that allowed the ductwork to become contaminated in the first place is important; otherwise, the corrective action will be temporary. Workers who are doing the duct cleaning should be encouraged to look for other types of problems, such as holes or gaps in the ducts that could allow contaminants to enter the ventilation airstream.

6.16 DUCT LEAKAGE

Air leakage from ducts can cause or exacerbate air quality problems, in addition to wasting energy. Sealed duct systems specified with a leakage rate of less than 3% will have a superior life-cycle cost analysis and reduce the likelihood of problems associated with leaky ductwork. Examples of excessive duct leakage leading to problems include:

- Leakage of light-troffer-type diffusers installed in a return plenum at the diffuser/light fixture interface; such leakage has been known to cause gross shortcircuiting between the supply and return, wasting much of the conditioned air. If the room thermostat is located in the return plenum, the room can be very uncomfortable, while the temperature in the plenum is at the control setpoint.
- Supply ductwork leakage due to loose-fitting joints and connections.
- Blow outs of improperly fabricated seams.
- Leakage of return ducts located in crawl spaces or below slabs, allowing soil gases and molds to enter the ductwork.

6.17 HEATING AND COOLING COILS

Heating and cooling coils are placed in the airstream to regulate the air temperature delivered to the interior occupied space. A malfunctioning coil control can result in thermal discomfort. Condensation on inadequately insulated pipes and leakage in piped systems will create moist conditions conducive to the growth of molds, fungus, and bacteria. During the cooling mode (air conditioning), the cooling coil provides dehumidification as water condenses from the airstream. Dehumidification can only take place if the chilled fluid is maintained at a cold enough temperature (generally below 45°F for water).

Condensate collects in the drain pan under the cooling coil and exits via a deep seal trap. Standing water will accumulate if the drain pan system has not been designed to drain completely under all operating conditions (sloped toward the drain and properly trapped). Under these conditions, molds and bacteria will proliferate unless the pan is cleaned frequently.

Condensate lines must be properly trapped and charged with liquid. An improperly trapped line can be a contamination source, depending on where the line terminates. A properly installed trap could also be a source, if the water in the trap evaporates and allows air to flow through the trap into the conditioned air.

During the heating mode, problems can occur if the hotwater temperature in the heating coil has been set too low in an attempt to reduce energy consumption. If outdoor air is brought in to provide sufficient ventilation, air may not be heated sufficiently to maintain thermal comfort, or, to adequately condition the outdoor air, air intake may be reduced so that insufficient outdoor air is available to meet ventilation needs.

6.18 SUPPLY FANS

After passing through the coil section where heat is either added or extracted, air moves through the supply fan chamber and the distribution system. Air distribution systems commonly use ducts that are constructed to be relatively airtight. Building construction elements can also serve as part of the air distribution system. Such elements include pressurized supply plenums and return air plenums located in the cavity space above the ceiling tiles and below the deck of the floor above. Proper fan selection and duct layout coordination during the building design and construction phase and ongoing maintenance of mechanical components, filters, and controls are all necessary for effective air delivery.

Fan performance is expressed as the ability to move a given quantity of air (cubic feet per minute, or cfm) at a given resistance or static pressure (measured in inches of water column). Airflow in the ductwork is determined by:

- Duct opening size
- Duct configuration resistance
- Air velocity through the duct

The static pressure in a system is calculated using factors for duct length, air movement speed, and changes in the air movement direction.

The original duct design and the final installation often differ. Ductwork installation may be altered due to limited space that must be shared with structural members and other hidden elements of the building system (e.g., electrical conduit, plumbing pipes). If the friction in the system increases to a point that approaches the fan performance limits, air distribution problems can occur. These problems are particularly evident at the end of duct runs. Inappropriate use of long runs of flexible ducts with sharp bends causes excessive friction. Poor system balancing (adjustment) is another common cause of air distribution problems.

Dampers are used as controls to restrict airflow. Damper positions may be relatively fixed (e.g., set manually during system testing and balancing) or may change in response to signals from the control system. Fire and smoke dampers can be triggered to respond to indicators such as high temperatures or signals from smoke detectors. Modulating dampers should be checked during inspections for the proper settings. ASHRAE and the Associated Air Balance Council provide guidance on proper intervals for testing and balancing.

6.19 RETURN AIR SYSTEMS

Above-ceiling spaces may be utilized for the unducted passage of return air. This system approach reduces initial HVAC system costs but requires that the designer, maintenance personnel, and contractors obey strict guidelines. Life and safety codes (e.g., building codes) must be followed for materials and devices that are located in the plenum. If a ceiling plenum is used for the collection of return air, openings into the ceiling plenum created by the removal of ceiling tiles will disrupt airflow patterns. The ceiling and adjacent wall integrity must be maintained in areas that are designed to be exhausted, such as supply closets, bathrooms, and chemical storage areas.

Return air enters either a ducted return air grille or a ceiling plenum and then is returned to the air handlers. Systems may utilize return fans in addition to supply fans to properly control air distribution. When return and supply fans are utilized, especially in a VAV system, their operation must be coordinated to prevent under- or overpressurization of the occupied space or overpressurization of the mixing plenum in the air handler.

6.20 EXHAUSTS, EXHAUST FANS, AND PRESSURE RELIEF

Most buildings are required by law, including building or plumbing codes, to provide area exhaust where contaminant sources accumulate. Such areas are toilet facilities, janitorial closets, cooking facilities, and parking garages. Other areas where exhaust is frequently recommended but may not be legally required include reprographics areas, graphic arts facilities, beauty salons, smoking lounges, shops, and any area where contaminants are known to originate.

For successful confinement and exhaust of identifiable sources, the source area must be at a higher overall pressure as compared to the area receiving the exhaust. Any area designed to be exhausted must be isolated and, thus, disconnected from the return air system so that contaminants are not transported to other building areas. To prevent operating the building under negative pressures and to limit the amount of unconditioned air brought into the building by infiltration, make-up air from outdoors must be brought into the HVAC system. The amount of make-up air drawn in at the air handler should always be slightly greater than the total amount of relief air, exhaust air, and air exfiltrating through the building shell. This make-up air is typically drawn in at the mixed-air plenum and distributed within the building. The make-up air must have a clear path to the area that is being exhausted.

The total cubic feet per minute of powered exhaust should be compared to the minimum quantity of mechanically introduced outdoor air. Excess make-up air is generally relieved at an exhaust or relief outlet in the HVAC system, especially in air economizer systems.

In addition to reducing the effects of unwanted infiltration, designing and operating a building at slightly positive or neutral pressures will reduce the soil gas entry rate when the systems are operating. For a building to actually operate at a slight positive pressure, the building must be tightly constructed. Tightly constructed can be defined as permitting less than 0.5 air change per hour at 0.25 Pascals. Without this tight construction, unwanted exfiltration will prevent the building from achieving a neutral or slightly positive pressure.

6.21 TERMINAL DEVICES

Thermal comfort and effective contaminant removal demand that air delivered into a conditioned space be properly distributed within that space. Terminal devices that distribute and collect air include supply diffusers, return and exhaust grilles, and associated dampers and controls. The number, design, and location (ceiling, wall, floor) of terminal devices are very important. Improper placement can cause an HVAC system with adequate capacity to produce unsatisfactory results (i.e., drafts, odor transport, stagnant areas, uneven temperatures, or shortcircuiting).

Occupants who are uncomfortable because of distribution deficiencies often try to compensate by adjusting or blocking the airflow from supply outlets. Adjusting system flows without any knowledge of the proper design frequently disrupts the proper air supply to adjacent areas. Distribution problems can also be produced if the arrangement of movable partitions, shelving, or other furnishings interferes with airflow. Such problems often occur if walls are moved or added without evaluating the expected impact on airflows.

6.22 HUMIDIFICATION AND DEHUMIDIFICATION EQUIPMENT

In some buildings or zones within buildings, special needs warrant the strict control of humidity (e.g., operating rooms or computer rooms). This control is accomplished by adding humidification or dehumidification equipment and controls. In office facilities, relative humidity above 20 or 30% during the heating season and below 60% during the cooling season is preferable. ASHRAE Standard 55 provides guidance on acceptable temperature and humidity conditions. The use of a properly designed and operated air conditioning system will generally keep relative humidity below 60% during the cooling season in office facilities with normal densities and loads.

Office buildings in cool climates that have high interior heat gains, thermally efficient envelopes (e.g., insulation), and economizer cooling may require humidification to maintain relative humidity within the comfort zone. Humidification must be added in a manner that prevents the growth of microbiologicals within the ductwork and air handlers.

Steam humidifiers should utilize clean steam, rather than treated boiler water, so occupants will not be exposed to chemicals. Systems using media other than clean steam must be rigorously maintained in accordance with the manufacturers' recommended procedures to reduce the likelihood of microbiological growth.

Mold growth problems are more likely if the humidistat setpoint located in the occupied space is above 45%. The high-limit humidistat, typically located in the ductwork downstream of the point at which water vapor is added, is generally set at 70% to avoid condensation in the ductwork. Adding water vapor to a building that was not designed for humidification can have a negative impact on the building structure and the occupants' health if condensation occurs on cold surfaces or in wall or roof cavities.

6.23 SELF-CONTAINED UNITS

In some designs, small decentralized units are used to provide cooling or heating to interior or perimeter zones. With the exception of induction units, units of this type seldom supply outdoor air. These units are typically considered a low-priority maintenance item. Self-contained units that are overlooked during maintenance may become significant contaminant sources, especially for the occupants located nearby.

6.24 CONTROLS

Heating, ventilation, and air conditioning systems can be controlled manually or automatically. Most systems are controlled by some combination of manual and automatic controls. The control system can be used to:

- Switch fans on and off
- Regulate the air temperature within the conditioned space
- Modulate airflow and pressures by controlling fan speed and damper settings

Regular maintenance and calibration are required to keep controls in good operating order. All programmable timers and switches should have battery backup to reset the controls in the event of a power failure. Local controls such as room thermostats must be properly located in order to maintain thermal comfort. Problems can result from poorly designed temperature control zones, such as single zones that combine areas with very different heating or cooling loads. Other problems arise when thermostats are located outside of the occupied space or in a return plenum; are subject to drafts, radiant heat gain or loss, or direct sunlight; or are affected by heat from nearby equipment.

6.25 BOILERS

A boiler must be adequately maintained to operate properly. Combustion equipment must operate properly to avoid hazardous conditions such as explosions or carbon monoxide leaks, as well as to provide good energy efficiency. Codes often require boiler operators to be properly trained and licensed. Both the American Society of Mechanical Engineers (ASME) and ASHRAE have made recommendations as to the amount of combustion air needed for fuel-burning appliances. Boiler operation considerations particularly important to indoor air quality and thermal comfort include:

- Boiler and distribution loop operation at a high enough temperature to supply adequate heat in cold weather
- Proper gasket and breaching maintenance to prevent carbon monoxide from escaping into the building
- Proper fuel-line maintenance to prevent any leaks that could emit odors into the building
- Adequate outdoor air for combustion
- Boiler combustion exhaust designed to prevent reentrainment

Modern office buildings tend to have much smaller capacity boilers than older buildings because of advances in energy efficiency. In some buildings, the primary heat source is waste heat recovered from the chiller that operates year-round to cool the building core.

6.26 COOLING TOWERS

Cooling tower maintenance ensures proper operation and keeps the cooling tower from becoming a niche for breeding pathogenic bacteria, such as *Legionella* organisms. Cooling tower water quality must be properly monitored and chemical treatments used as necessary to minimize conditions that could support biological growth. Proper maintenance may entail physical cleaning to prevent sediment accumulation and installing drift eliminators.

6.27 WATER CHILLERS

Water chillers are frequently found in the air conditioning systems of large buildings. A water chiller must be maintained in proper working condition to perform its function of removing the heat from the building. Chilled water supply temperatures should operate in the range of 45°F or colder in order to provide proper moisture removal during humid weather. Piping should be insulated to prevent condensation. Other than thermal comfort, IAQ concerns associated with water chillers involve potential release of the working fluids from the chiller system. To control such IAQ concerns, the rupture disk (safety release) of the system should be piped to the outdoors, refrigerant leaks should be located and repaired, and waste oils and spent refrigerant should be disposed of properly.

RESOURCES

American Society of Heating, Refrigerating, and Air Conditioning Engineers (ASHRAE)

Guideline for the Commissioning of HVAC Systems, ASHRAE Guideline 1–1989.
Method of Testing Air-Cleaning Devices Used in General Ventilation for Removing Particulate Matter, ASHRAE Standard 55, 1992.
Method of Testing General Ventilation Air-Cleaning Devices for Removal Efficiency by Particle Size, ASHRAE Standard 52.2. 2000.
Ventilation for Acceptable Air Quality, ASHRAE Standard 62, 1999.

U.S. Environmental Protection Agency

IAQ Tools for Schools, EPA-402-K-95–001, May 1995.

EPA Internet Resources

An office building occupant's guide to IAQ: www.epa.gov/iaq/pubs/occupgd.html
Biological contaminants: www.epa.gov/iaq/pubs/bio_1.html
Building air quality action plan (for commercial buildings): www.epa.gov/iaq/base/actionpl.html
Floods/flooding: www.epa.gov/iaq/pubs/flood.html
Indoor air quality (IAQ) home page: www.epa.gov/iaq
IAQ in large buildings/commercial buildings: www.epa.gov/iaq/base/index.html
IAQ in schools: www.epa.gov/iaq/schools/index.html
Mold remediation in schools and commercial buildings: www.epa.gov/iaq/pubs/molds.html
Mold resources: www.epa.gov/iaq/pubs/moldresources.html
U.S. EPA indoor air quality (IAQ) information clearinghouse: phone, (800) 438-4318 or (703) 356-4020; fax, (703) 821-8236; e-mail, iaqinfo@aol.com; includes indoor air-related documents, answers to indoor air quality (IAQ) questions, listing of state IAQ and regional EPA contacts

Resources List

American Academy of Allergy, Asthma & Immunology (AAAAI); (800) 822-2762; www.aaaai.org
American College of Occupational and Environmental Medicine (ACOEM); (847) 818-1800; www.sioux-land.com/acoem/
American Society of Heating, Refrigerating, and Air-Conditioning Engineers (ASHRAE); (800) 527-4723; www.ashrae.org (physician referral directory and information on allergies and asthma)
American Conference of Governmental Industrial Hygienists (ACGIH); (513) 742-2020; www.acgih.org (occupational and environmental health and safety information)
American Industrial Hygiene Association (AIHA); (703) 849-8888; www.aiha.org (information on industrial hygiene and indoor air quality issues including mold hazards and legal issues)
American Lung Association (ALA); (800) LUNG-USA (800-586-4872); www.lungusa.org (information on engineering issues and indoor air quality)
Association of Occupational and Environmental Clinics (AOEC); (202) 347-4976; www.aoec.org (referrals to clinics with physicians who have experience with environmental exposures, including exposures to mold; databases of occupational and environmental cases maintained)
Association of Specialists in Cleaning and Restoration (ASCR); (800) 272-7012; www.ascr.org (disaster recovery, water and fire damage, emergency tips, referrals to professionals)
Asthma and Allergy Foundation of America (AAFA); (800) 7-ASTHMA (800-727-8462); www.aafa.org (referrals to physicians having experience with environmental exposures)
Asthma and Allergy Network/Mothers of Asthmatics (AAN-MA); (800) 878-4403 or (703) 641-9595; www.aanma.org (information on allergies and asthma)
Canada Mortgage and Housing Corporation (CMHC); (613) 748-2003 (international); www.cmhc-schl.gc.ca/cmhc.html (several documents on mold-related topics available)
Carpet and Rug Institute (CRI); (800) 882-8846; www.carpet-rug.com (carpet maintenance, restoration guidelines for water-damaged carpet, other carpet-related issues)
Centers for Disease Control and Prevention (CDC); (800) 311-3435; www.cdc.gov (information on health-related topics including asthma, molds in the environment, and occupational health)
CDC's National Center for Environmental Health (NCEH); (888) 232-6789; www.cdc.gov/nceh/asthma/factsheets/molds/default.htm (questions and answers on *Stachybotrys chartarum* and other molds)
Energy and Environmental Building Association; (952) 881-1098; www.eeba.org (information on energy-efficient and environmentally responsible buildings, humidity/moisture control/vapor barriers)
Federal Emergency Management Agency (FEMA); (800) 480-2520; www.fema.gov/mit (publications on floods, flood proofing, etc.)

Health Canada, Health Protection Branch, Laboratory Centre for Disease Control, Office of Biosafety; (613) 957-1779; www.hc-sc.gc.ca/main/lcdc/web/biosafety/msds/index.html (Material Safety Data Sheets with health and safety information on infectious microorganisms, including *Aspergillus* and other molds and airborne biologicals)

Indoor Environmental Remediation Board (IERB); (215) 387-4097; www.ierb.org (information on best practices in building remediation)

Institute of Inspection, Cleaning and Restoration Certification (IICRC); (360) 693-5675; www.iicrc.org (information on and standards for the inspection, cleaning, and restoration industry)

International Sanitary Supply Association (ISSA); (800) 225-4772; www.issa.com (education and training on cleaning and maintenance)

International Society of Cleaning Technicians (ISCT); (800) WHY-ISCT (800-949-4728); www.isct.com (information on cleaning such as stain removal guide for carpets)

Material Safety Data Sheets (MSDSs), Cornell University; http://msds.pdc.cornell.edu/msdssrch.asp (MSDSs contain information on chemicals or compounds including topics such as health effects, first aid, and protective equipment for people who work with or handle these chemicals)

MidAtlantic Environmental Hygiene Resource Center (MEHRC); (215) 387-4096; www.mehrc.org (indoor environmental quality training on including topics such as mold remediation)

National Air Duct Cleaners Association (NADCA); (202) 737-2926; www.nadca.com (duct cleaning information)

National Antimicrobial Information Network (NAIN); (800) 447-6349 http://ace.orst.edu/info/nain/ (regulatory information, safety information, and product information on antimicrobials)

National Association of the Remodeling Industry (NARI); (847) 298-9200; www.nari.org (consumer information on remodeling, including help finding a professional remodeling contractor)

National Institute for Occupational Safety and Health (NIOSH); (800) 35-NIOSH (800-356-4674); www.cdc.gov/niosh (health and safety information with a workplace orientation)

National Institute of Allergy and Infectious Diseases (NIAID); (301) 496-5717; www.niaid.nih.gov (information on allergies and asthma)

National Institute of Building Sciences (NIBS); (202) 289-7800; http://nibs.org (information on building regulations, science, and technology)

National Jewish Medical and Research Center; (800) 222-LUNG (800-222-5864); www.njc.org (information on allergies and asthma)

National Pesticide Telecommunications Network (NPTN); (800) 858-7378; http://ace.orst.edu/info/nptn (information on pesticides/antimicrobial chemicals, including safety and disposal information)

New York City Department of Health, Bureau of Environmental & Occupational Disease Epidemiology; (212) 788–4290; www.ci.nyc.ny.us/html/doh/html/epi/moldrpt1.html (guidelines on assessment and remediation of fungi in indoor environments)

Occupational Safety and Health Administration (OSHA); (800) 321-OSHA (800-321-6742); www.osha.gov (information on worker safety, includes topics such as respirator use and safety in the workplace)

Sheet Metal and Air Conditioning Contractors' National Association (SMACNA); (703) 803-2980; www.smacna.org (technical information on topics such as air conditioning and air ducts)

Smithsonian Center for Materials Research and Education (SCMRE); (301) 238-3700; www.siedu/scmre (guidelines for caring for and preserving furniture and wooden objects, paper-based materials; preservation studies)

University of Michigan Herbarium; (734) 764-2407; www.herb.lsa.umich.edu (specimen-based information on fungi; information on fungal ecology)

University of Minnesota, Department of Environmental Health & Safety; (612) 626-5804; www.dehs.umn.edu/remanagi.html (managing water infiltration into buildings)

University of Tulsa Indoor Air Program; (918) 631-5246; www.utulsa.edu/iaqprogram (courses, classes, and continuing education on indoor air quality)

University of Wisconsin-Extension, *The Disaster Handbook*; (608) 262-3980; www.uwex.edu/ces/news/handbook.html (information on floods and other natural disasters)

Water Loss Institute, Association of Specialists in Cleaning and Restoration; (800) 272-7012 or (410) 729-9900; www.ascr.org/wli.asp (information on water and sewage damage restoration)

CHAPTER 7

Maintenance

Martha J. Boss, Dennis W. Day, and Marwan Bader

CONTENTS

1-56670-606-8/03/$0.00+$1.50
© 2003 by CRC Press LLC

Maintenance may involve sterilization, disinfection, decontamination, or dilution. Of these, only sterilization attempts to kill all biological contaminants. Disinfection attempts to kill sufficient numbers in order to lessen the infective potential of contaminants. Decontamination and dilution seek to lessen the numbers of biological contaminants to some defined limit. For molds, fungi, and yeasts, various limits are considered acceptable given the use of the buildings or areas and the health status of the people potentially exposed.

7.1 FUNGI CONTROL

Mold growth can eventually cause structural damage if a mold or moisture problem remains unaddressed for a long time. In the case of a long-term roof leak, for example, molds can weaken floors and walls as the molds feed on wet wood (Figures 7.1 and 7.2). Indoor mold growth may not be obvious. Mold may be growing on hidden surfaces, such as the back side of drywall (Figures 7.3 and 7.4), wallpaper (Figure 7.5), or paneling; the top of ceiling tiles (Figure 7.6); the underside of carpets and pads; pipe chases and utility tunnels with leaking or condensing pipes; walls behind furniture where condensation forms; condensate drain pans inside air-handling units; porous thermal

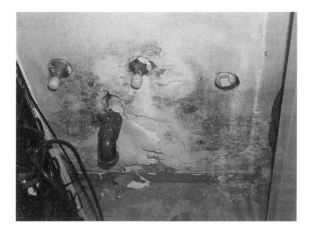

Figure 7.1   Typical mold patterns on walls. (Courtesy of Aerotech Laboratories, Phoenix, AZ.)

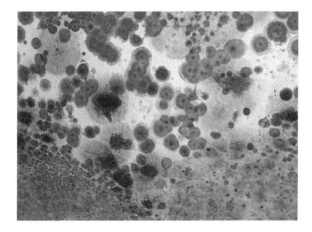

Figure 7.2   Typical mold patterns on walls. (Courtesy of Aerotech Laboratories, Phoenix, AZ.)

Figure 7.3   Mold growing on and into wood. (Courtesy of Aerotech Laboratories, Phoenix, AZ.)

Figure 7.4   Mold growing on and into wood. (Courtesy of Aerotech Laboratories, Phoenix, AZ.)

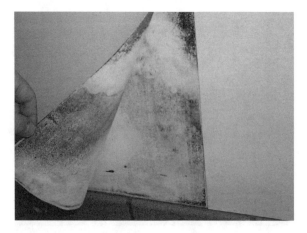

Figure 7.5   Mold under wallpaper. (Courtesy of Aerotech Laboratories, Phoenix, AZ.)

Figure 7.6   Mold on ceiling (be sure to check electrical fixtures also). (Courtesy of Aerotech Laboratories, Phoenix, AZ.)

or acoustic liners inside ductwork; or roof materials above ceiling tiles due to roof leaks or insufficient insulation.

Assess the size of the mold and/or moisture problem and the type of damaged materials before planning the remediation work. The remediation plan should: include steps to fix the water or moisture problem to prevent the problem from returning, cover the use of appropriate personal protective equipment (PPE), and include steps to carefully contain and remove moldy building materials to avoid spreading the mold.

The highest priority must be to protect the health and safety of the building occupants and remediators. Communication must be established with building occupants when mold problems are identified. Temporary relocation of some or all of the building occupants may be required. The decision to relocate occupants should consider the size and type of the area affected by mold growth, the type and extent of health effects reported by the occupants, the potential health risks that could be associated with debris, and the amount of disruption likely to be caused by remediation activities. If possible, remediation activities should be scheduled during off-hours when building occupants are less likely to be affected.

Some building materials, such as drywall with vinyl wallpaper or wood paneling, may act as vapor barriers, trapping moisture underneath their surfaces and thereby providing a moist environment where mold can grow. Removal of this wallpaper can lead to a massive release of spores from mold growing on the underside of the paper.

The standing rules are:

- Bleach what you can bleach.
- Use biocides with caution.
- Throw out what you can throw out.

If you are unsure as to any of these protocols, get help.

Moisture, heat, and dirt or dusts are the ingredients needed to grow fungi. As part of routine building maintenance, buildings should be inspected for evidence of water damage and visible mold. Conditions causing mold (such as water leaks, condensation, infiltration, or flooding) should be corrected.

Good preventive maintenance can reduce the risk of a problem with molds growing inside the home and other buildings. Homes and buildings with water damage should be repaired, and all moldy material should be removed. Avoiding or diminishing other contributors of humidity may help. Some causes and contributors of high humidity may include leaking pipes, water-damaged drywall and ceiling tile, leaking roofs, flooding, faulty or obstructed dryer vent connections, use of steaming hot water in washing machines, many showers, faulty or obstructed bathroom/kitchen ventilation fans, boiling water for long periods of time, canning or pressure cooking, hand washing and rack drying laundry, use of humidifiers, and excessive sealing of homes so inadequate air exchange occurs.

7.2 STACHYBOTRYS

Some molds can be killed by cleaning the moldy surface with chlorine; however, *Stachybotrys* often has a germ mycelium that is buried inside the water-damaged surface and may be inaccessible to chlorine. Changing the humidity may lead to death of the *Stachybotrys* colony; however, changing the humidity can also induce heavy sporulation. While the spores may die quickly, these spores can remain toxic and continue to cause allergic reactions; therefore, it is best to remove all of the water-damaged material. Visual identification of black mold in a chronically wet area is considered to be a possible indicator of mold amplification in interiors. The New York City Department of Health (NYCDOH) convened an expert panel on *Stachybotrys chartarum* (SC) in

1993, which recommended different methods of mold removal depending on the size of the mold problem. Their recommendations based on mold surface area are presented as an example of response decision logic.

7.2.1   Level I: 2 Square Feet or Less

The area can be cleaned by individuals who have received training on proper clean-up methods, protection, and potential health hazards. These individuals should be free from asthma, allergy, and immune disorders. Gloves and a half-face respirator should be worn. Contaminated material should be placed in a sealed plastic bag before being taken out of the building to prevent contamination of other parts of the building. Surrounding areas should be cleaned with household bleach.

7.2.2   Level II: More than 2 Square Feet but Less than 30 Square Feet

The recommendations are the same as Level I, with the added precaution that moldy materials should be covered with plastic sheets and taped before any handling or removal is done. For instance, in the case of a moldy panel of gypsum board (measuring 4 × 8 ft), plastic sheeting should be taped on the wall over the affected area before the wallboard is cut to remove the contaminated section. Once cut from the wall, that section should be placed within another layer of plastic before being carried through the building for disposal.

7.2.3   Level III: More than 30 Square Feet

Personnel conducting decontamination efforts must be trained in the handling of hazardous materials. Decontamination planning must assume hazardous materials may be present.

7.2.4   Level IV

The Level IV designation indicates that *Stachybotrys* is present in the HVAC system. Precautions are the same as those for Level III (NYCDOH, 2000).

7.3 IMMEDIATE WORKER PROTECTION

Whenever possible, use remote methods for clean-up. At a minimum, entry where any invasive activities will occur requires use of respirators with high-efficiency particulate air (HEPA) filters and dermal protection for hands. All material worn or used must be either decontaminated or properly disposed. If the remediation job disturbs mold and mold spores become airborne, then the risk of respiratory exposure increases. Actions that are likely to disturb mold include break-up of moldy porous materials, such as wallboard; invasive procedures used to examine or remediate mold growth in a wall cavity; active stripping or peeling of wallpaper; and the use of fans to dry items.

The primary function of full-face respirators is to avoid inhaling mold and mold spores and to avoid mold contact with the skin or eyes. The following sections discuss the different types of personal protective equipment that can be used during remediation activities. Please note that all individuals using certain PPE, such as half-face or full-face respirators, must be trained, have medical clearance, and be fit-tested by a trained professional. In addition, the use of respirators must follow a complete respiratory protection program as specified by the Occupational Safety and Health Administration (OSHA).

7.3.1   Skin and Eye

Gloves are required to protect the skin from contact with mold allergens (and in some cases mold toxins) and from potentially irritating cleaning solutions. Long gloves that extend to the middle of the forearm are recommended. The glove material should be selected based on the type of materials being handled. If a biocide (such as chlorine bleach) or a strong cleaning solution is used, select gloves made from natural rubber, neoprene, nitrile, polyurethane, or polyvinylchloride (PVC). If you are using a mild detergent or plain water, ordinary household rubber gloves may be used. To protect your eyes, use properly fitted goggles or a full-face respirator with a HEPA filter. Goggles must be designed to prevent the entry of dust and small particles. Safety glasses or goggles with open vent holes are not acceptable.

7.3.2   Respiratory Protection

Respirators protect clean-up workers from inhaling airborne mold, mold spores, and dust. Their use is classified as follows:

- *Minimum* — When cleaning up a small area affected by mold, you should use an N-95 respirator. This device covers the nose and mouth, filters out 95% of the particulates in the air, and is available in most hardware stores. In situations where a full-face respirator is used, additional eye protection is not required.
- *Limited* — Limited PPE includes use of a half-face or full-face air-purifying respirator (APR) equipped with a HEPA filter cartridge. These respirators contain both inhalation and exhalation valves that filter the air. Half-face APRs do not provide eye protection. HEPA filters do not remove vapors or gases.
- *Full* — In situations where high levels of airborne dust or mold spores are likely or intense or long-term exposures are expected (e.g., the clean-up of large areas of contamination), a full-face, tight-fitting, powered air-purifying respirator (PAPR) is recommended. Full-face PAPRs use a blower to force air through a HEPA filter. The HEPA-filtered air is supplied to a mask that covers the entire face. The positive pressure within the hood prevents unfiltered air from entering through penetrations or gaps. Individuals must be trained to use their respirators before remediation begins.

7.3.3   Disposable Clothing

Disposable clothing is recommended during a medium or large remediation project to prevent the transfer and spread of mold to clothing and to eliminate skin contact with mold. Their use is classified as follows:

- *Limited* — Disposable paper overalls
- *Full* — Mold-impervious disposable head and foot coverings and a body suit made of a plastic-coated material with all gaps, such as those around ankles and wrists, sealed with duct tape

7.4 DECONTAMINATION

7.4.1   Method 1: Wet Vacuum

Wet vacuums are vacuum cleaners designed to collect water. Wet vacuums can be used to remove water from floors, carpets, and hard surfaces where water has accumulated. They should not be used to vacuum porous materials, such as gypsum board and should be used only when materials are still wet, as wet vacuums may spread spores if sufficient liquid is not present. The tanks, hoses, and attachments of these vacuums should be thoroughly cleaned and dried after use because mold and mold spores may stick to the surfaces.

7.4.2 Method 2: Damp Wipe

Whether dead or alive, mold is allergenic, and some molds may be toxic. Mold can generally be removed from nonporous (hard) surfaces by wiping or scrubbing with water or water and detergent. These surfaces should be dried quickly and thoroughly to discourage further mold growth. Instructions for cleaning surfaces, as listed on product labels, should always be read and followed. Porous materials that are wet and have mold growing on them may have to be discarded. Because molds will infiltrate porous substances and grow on or fill in empty spaces or crevices, the mold can be difficult or impossible to remove completely.

Do not paint or caulk moldy surfaces. Clean and dry surfaces before painting. Paint applied over moldy surfaces is likely to peel. Decontamination may consist of washing with chlorinated or other oxidizing chemicals (e.g., bleach, oxidizing color-safe bleach, or ozone). Biocides may also be used; however, the biocide used should be proven effective for the particular biologicals present. All of these decontamination chemicals pose some risk to workers. At a minimum, Material Safety Data Sheets (MSDS) should be obtained to communicate this risk to workers.

For porous surfaces, including fiberglass liners inside ducts, encapsulation of the porous surface may be required prior to removal. Contaminated fiberglass liners cannot be cleaned. The HVAC system should not be operating during mold removal. The system may be contaminated or may spread contamination. In many buildings, fiberglass-lined ductwork lofts due to continual airflow. These lofted spaces collect dirt and become microbial nests. The microbes grow and multiply and then are blown all over the building to infest other areas.

Furnace filters may be subject to breakthrough, where mold spores pass though filter sections and reenter the airstream. As the filters become dirtier, the filter material may catch the microbes, provide a growth location, and transfer the microbial contamination into the airstream. Another hot spot for microbial growth is the humidifier assembly on furnaces. Typical reservoir humidifiers contain pools of standing, stagnant water throughout much of the year that allow mold to grow and infiltrate surrounding ductwork.

7.5 ABATEMENT

Biologicals grow and reproduce. In that regard, concentrations of biologicals are not equivalent to such things as chemical concentrations. A dilute concentration of biologicals in a good growth environment will result in a concentrated level of contamination over time. All abated buildings must be sampled and certified as suitable for reentry prior to normal building usage. This certification states that the biological contamination of the building has been diminished through abatement activities and the level is now equivalent or below the ambient exterior conditions or interior baseline conditions previously agreed upon. Certification does not state that the building cannot be recontaminated in the future. Always ask for recommendations as to how to prevent future contamination. Any porous materials that have been contaminated and removed from the facility and will be returned at a later date must be decontaminated prior to the return, or facility users must be advised that recontamination may be inevitable.

7.6 CLEANING SEQUENCE AND HAZARDOUS MATERIALS CONSIDERATIONS

Health, safety, and environmental factors are of primary concern in any undertaking. Where hazardous materials are known to exist, administrative procedures that address these hazards must be considered at the outset of any decontamination process. These administrative procedures assure that compliance with all federal, state, and local laws and regulations are met; appropriate permits are obtained; and the entire project is adequately documented and reviewed. Workers must be specially trained prior to doing the work. A walk-through of the work area as well as a review of hazards and work practices has to be conducted at each project site prior to the start of work each day. Work practices must also protect those not involved in the decontamination from hazards. Appropriate signage and the use of roped-off work areas are required. Work is performed during hours when the usual work force is absent or at a minimum.

7.6.1 Double Containment

As a general precaution, double containment is always employed. When working in areas with concrete floors, cover the floor with plastic sheeting. Although this increases the amount of waste generated, a spill may penetrate the concrete, and clean-up then becomes very difficult. Take precautions while working to prevent the spread of contamination of any kind.

7.6.2 HVAC System Wetting

Some HVAC systems must be made wet during cleaning. Aggressive penetration of any system is always preceded and accompanied by thorough wetting with water or steam. When opening seams, a spray is continuously directed into the fold. Screws and bolts are often rusted, so sufficient time must be allowed following wetting to permit penetration of the water behind the screw heads and nuts. Cutting sheetmetal is preferred in some cases, rather than attempting to remove screws and bolts that are severely corroded. Contaminated parts are bagged in plastic to maintain a moist environment. Parts to be discarded as well as those that may be scheduled for reinstallation are decontaminated.

7.6.3 System Steam Cleaning and Disassembly

Steaming as a decontamination procedure is useful only when a system is known to be intact. Systems appearing to be intact by visual inspection often are not. Where the entire system has no leaks, steaming the system without dismantling is practical; however, steaming is limited to systems where sections are joined by smooth welds. In most of the systems encountered, the joints are fastened by draw bands. In these cases, even if steam had penetrated the joints, much of the contamination would be left in place. Disassembly of the system for decontamination has been found to be cost effective in most situations. Considerable amounts of contamination may be found on the tops of chemical fume hoods. The tops of hoods are defined as the hood interior casement tops that surround the portal to the exhaust ductwork. Accumulated contamination penetrates improperly installed or damaged ductwork connections. For aesthetic reasons, many fume hood systems are installed with faces that extend to the ceiling. Concealed contamination may be present behind this paneling and must be considered.

7.7 VENTILATION TROUBLESHOOTING

Ventilation systems are designed and installed in buildings to replace stale, contaminated air with fresh air from the outside. Air also enters and leaves in other ways. Air can enter by infiltration, through construction joints and cracks around windows and doors and through the foundation and crawl spaces. Air also enters through natural ventilation openings, such as open doors and windows. An approach to lowering the concentrations of any indoor air pollutants is to increase the amount of outdoor air coming indoors. Opening windows and doors, operating window or attic fans (when the weather permits), or running a window air conditioner with the vent control open increases the outdoor ventilation rate. Local bathroom or kitchen fans that exhaust outdoors remove contaminants directly from the room where the fan is located and also increase the outdoor air ventilation rate.

Often the essence of indoor air quality (IAQ) problems lies in a singular phenomenon: lack of sanitation. Because air-handling systems often seem to be invisible, we tend to forget that the air we breathe did not magically appear in the room. Most systems, even if inspected by normal maintenance means, are not accessible for visual observation along many ductwork and plenum runs. Telescopic, fiberoptic, or camera scoping of these systems is sometimes the only means to determine levels of contamination.

The following are some of the problems often identified in air-handling systems:

- Adequate spaces are often not provided to access items that require maintenance such as filters, coils and drain pans, and strainers. These areas, in addition to accumulating the normal dust and dirt associated with airstreams, may accumulate water through condensation events and become hospitable areas for mold and bacterial amplification (i.e., rapid growth). Biocide application is not effective without thorough mechanical cleaning of these areas and system alteration to prevent moisture accumulation. Biological amplification may then migrate to ductwork, plenums, and habitable spaces served by the system.
- Renovation activities may lead to unbalanced HVAC systems as ductwork and other air-handling appurtenances are added. Positive and negative pressure areas within a building and within HVAC systems may become different from the design intent. This problem is especially dangerous where infection control or chemical source reduction is required.
- Permanently sealing fresh-air intakes by welding plates to the intake faces, permanently closing dampers within ductwork, or programming control systems eliminate the fresh air intake and cause carbon dioxide to build up. Thus, buildings receive make-up air only through doors being opened or seepage through construction materials (e.g., cracks in walls, up the sides of foundations to basement window casings with gaps). In addition to the potential for excess carbon dioxide, this situation encourages the infiltration of radon gas up the sides of foundations and into building structural gaps or openings.
- Intake air for the general structure interior is usually of poor quality if supplied by attic space air and/or crawl space air. Obvious problems with air quality occur when air is supplied from damp crawl spaces frequented by rodents or from attic spaces frequented by birds and sometimes even bats. Even without animal habitation, degradation of building materials and water intrusion into these spaces surrounding habitable spaces can lead to indoor air problems associated with both particulates and biological growth (molds, bacteria). These problems can be present even if the only make-up air supplied from the crawl spaces or attics is through cracks or other openings to these spaces.
- Dirty air plenums and building spaces are also used as make-up air sources and supply dirty air to HVAC systems. Vertical or horizontal air plenums may contain paint chips (lead, cadmium), accumulated dusts, and biological risk factors (spores, bacteria, animal droppings). Some plenums and building spaces are used as overflow storage for a wide variety of real property. This real property may be in various stages of degradation and is often laden with soil particulate (dirt and dust). Stored paper materials and boxes may contribute to biological growth and dispersal of cellulose fibers or general particulate laden with spores and mold vegetative structures.
- Air intakes that are sometimes located in tunnels or crawl spaces where sewage and water pipes are also located may supply aerosolized components from the pipes to the HVAC system. In a worst-case scenario, these spaces may also contain delaminated asbestos mixed with leakage from sewers. Given that hepatitis B and other biological organisms can remain pathogenic for weeks, these air supply areas may provide air laden with human pathogens.
- Air intakes for furnace blower systems that obtain make-up air at the base of the system (floor) are sometimes located near condensate pans, stored chemicals, maintenance shops with wood and paint dusts, and areas of debris. These intakes will ultimately provide poor-quality make-up air.

More complicated sanitation methods may require chemical and mechanical cleaning and/or mechanical encapsulation of dirty system. Chemical and particulate outfall from these sanitation events may also be of concern if the HVAC system being treated cannot be partitioned (i.e., shut down, separated by critical barriers) during the chemical application or mechanical cleaning event.

7.8 DUCTWORK MAINTENANCE AND CLEANING

The original engineering concept for any HVAC system relies upon sufficient filtration and airflow velocity to keep the system dry and, with the exception of the filter beds, free of contamination. When this concept fails, either through intended changes in the system or system breakdown, increased vigilance in maintenance activities may be required. Unfortunately, some systems cannot be rendered dry and other systems cannot be adequately cleaned. However, for those systems where debris and stored extraneous materials can be easily removed, removing them may substantially improve air quality and reduce the necessity for endpoint filtration as the sole guarantor of a clean airstream. Duct cleaning generally refers to the cleaning of these HVAC system components:

- Supply and return air ducts and registers — The supply ductwork must be cleaned starting with the supply fan and ending at the supply diffuser. Clean return air ducts starting at the outer ends of the return air system and concluding at the mixed-air chamber and the exhaust stack.
- Grilles and diffusers — Clean outside air-intake grill and shaft. Vacuum diffusers, grills, and registers. Reset as needed.
- Heat exchangers and heating and cooling coils.
- Condensate drain pans (drip pans).
- Fan motor and fan housing.
- Air-handling unit housing — Clean the interior of the air-handling unit. Remove interior insulation. Reinstall according to manufacturer's directions.

Failure to clean a component of a contaminated system can result in recontamination of the entire system, thus negating any potential benefits of decontamination.

7.8.1 Duct-Cleaning Industrial Standards and Limitations

Duct-cleaning industry standards referenced by the Environmental Protection Agency (EPA) include those developed by:

- National Air Duct Cleaners Association (NADCA)
- North American Insulation Manufacturers Association (NAIMA)
- Underwriters Laboratories (UL), the American Society for Testing and Materials (ASTM), and the National Fire Protection Association (NFPA)

Fiberglass duct board, sheetmetal ducts internally lined with fiberglass duct liner, and flexible duct engineered for use in HVAC systems are tested in accordance with standards established by UL, ASTM, and NFPA. Chemical biocides used on inanimate surfaces are regulated by the EPA under the Federal Insecticide, Fungicide, and Rodenticide Act (FIFRA). A product must be registered by EPA for a specific use in order to be legally used for that purpose. The specific uses must appear on the pesticide (e.g., biocide) label. Using a pesticide product in any manner inconsistent with the label directions is a violation of FIFRA.

A small number of products are currently registered by the EPA specifically for use on the inside of bare sheetmetal air ducts. A number of products are also registered for use as sanitizers on hard surfaces, which could include the interior of bare sheetmetal ducts. While many such products may be used legally inside of unlined ducts if all label directions are followed, some of the directions on the label may be inappropriate for use in ducts. For example, if the directions indicate rinse with water, the added moisture could stimulate mold growth. No products are currently registered as biocides for use on fiberglass duct board or fiberglass-lined ducts. (USEPA, 1997)

7.8.2 Dry Preventive Maintenance

Use the highest efficiency air filter recommended by the manufacturer of the heating and cooling system. Change filters regularly. If the filters become clogged, change them more frequently. Be sure that no filters are missing and that air cannot bypass filters through gaps around the filter housing. When the heating and cooling system is maintained or checked for other reasons, the service provider should be asked to clean cooling coils and drain pans. During construction or renovation work that produces dust, seal off supply and return registers and do not operate the heating and cooling system until after the dust is cleaned up. Remove dust and vacuum regularly. Use a vacuum cleaner equipped with a HEPA filter or the highest efficiency filter bags the vacuum cleaner can take. Vacuuming can increase the amount of dust in the air during and after vacuuming as well as in your ducts. (USEPA, 1997)

7.8.3 Wet Preventive Maintenance

Moisture should not be present in ducts. The presence of condensation or high relative humidity is an important indicator of the potential for mold growth on any type of duct. Controlling moisture is the most effective way to prevent biological growth in air ducts. Moisture can enter the duct system through leaks or if the system has been improperly installed or serviced. Properly seal and insulate ducts in all spaces (e.g., attics and crawl spaces) that are not air conditioned. (*Note:* Sealing will help prevent moisture due to condensation from entering the system.) Properly insulate the HVAC system to prevent water condensation. Promptly and properly repair any leaks or water damage.

Condensation on or near cooling coils of air-conditioning units is a major factor in moisture contamination of the system. Condensation occurs when a surface temperature is lower than the dew point temperature of the surrounding air; consequently, improper cycling of an HVAC system between heating and cooling phases contributes to unwanted condensation events.

Maintain and clean cooling coils. Cooling coils are designed to remove water from the air and can be a major source of moisture contamination that can lead to mold growth. Remove standing water under cooling coils of air-handling units by making sure that drain pans slope toward the drain. Check insulation near cooling coils for wet spots. Correct any water leaks or standing water. If the heating system includes in-duct humidification equipment, be sure to operate and maintain the humidifier strictly as recommended by the manufacturer.

Remove fiberglass or any other insulation material that is wet or visibly moldy (or if an unacceptable odor is present). (*Note:* Replacement should be done only by a qualified contractor.) Do not use steam cleaning or any other methods involving moisture without adequate biocide application and/or drying. Use steam cleaning with extreme caution, as the outer core of the steam application will not be of sufficient temperature to kill biologicals, and the steam adds to the moisture loading.

Proper design makes all of the maintenance tasks much easier. Air-handling units should be constructed so that maintenance personnel have easy, direct access to heat exchange components and drain pans for proper cleaning and maintenance. (USEPA, 1997)

7.8.4 Mold Detection and Cleaning Confirmation

Mold detection and subsequent cleaning confirmation in the HVAC systems may be complicated by several factors:

- Many sections of the heating and cooling system may not be accessible for a visible inspection. Some service providers use remote photography to document conditions inside ducts.
- Although a substance may look like mold, a positive mold determination can only be accomplished through laboratory analysis or microscopic inspection.

- Wet insulated air ducts and the insulation cannot be effectively cleaned. The insulation should be removed and replaced.
- Ducts infested with vermin (e.g., rodents or insects) constitute an additional risk.

If the conditions causing the mold growth in the first place are not corrected, mold growth will recur.

The EPA does not recommend that air ducts be cleaned except on an as-needed basis because of the continuing uncertainty about the benefits of duct cleaning under most circumstances. If a service provider or advertiser asserts that the EPA recommends routine duct cleaning or makes claims about its health benefits, notify the EPA in writing. The EPA neither establishes duct-cleaning standards nor certifies, endorses, or approves duct-cleaning companies.

7.8.5 Cleaning Tasks

Typically, a service provider will:

- Open access ports or doors to allow the entire system to be cleaned and inspected; smaller ducts or turns may require the use of a fiberoptic borescope.
- Inspect the system before cleaning to be sure that no asbestos-containing materials are present in the heating and cooling system; asbestos-containing materials require specialized procedures and should not be disturbed or removed except by specially trained and equipped contractors.
- Protect carpet and household furnishings during cleaning.
- Use specialized tools to dislodge dirt and other debris in ducts, then vacuum with a high-powered vacuum cleaner. These tools include a vacuum collector unit, omnidirectional air nozzles, rotary brushes and vacuum, and air pressure wash and vacuum.
- Remove exterior insulation to gain access to ductwork; reinstallation may be required.
- Design sound attenuation modification, if needed.
- Wash and vacuum clean each duct section using well-controlled brushing of duct surfaces in conjunction with contact vacuum cleaning to dislodge dust and other particles; for fiberglass duct board and sheetmetal ducts internally lined with fiberglass, only soft-bristled brushes are used.
- Remove and replace any flex ductwork, replacing it with ductwork that has a smooth surface interior, if possible.
- Use vacuum equipment that exhausts particles outside of the building or, if the vacuum exhausts inside the home, use only vacuuming equipment equipped with HEPA filtration.
- Plug the access openings; round access holes have airtight plugs with plastic caps, while rectangular access openings may require overlapping material and a cover of the same gauge thickness as the existing duct. Rectangular covers are fastened using self-tapping metal screws, silicone bead sealing gaskets, and/or sealant tape.
- Seal all joints airtight.
- Install new test holes and access doors in ducts as needed — on the side of the duct where adequate clearance is available, and at other locations requiring access for inspection, cleaning, adjusting, maintenance, and operation.
- Verify that the air supply and return system of the building are properly balanced (tested and adjusted) as needed.
- Secure all manual dampers at full open position.
- Set splitters straight with the main duct.
- Purge the HVAC system.
- Cover all terminal air outlets (diffusers, registers, grilles, etc.) with synthetic filter media that is at least 30% efficient.
- Tape filter to terminal device frame to eliminate air leakage.
- Start HVAC unit and, in the event of variable speed/volume systems, operate unit up and down between low and high speed to dislodge dirt and debris for minimum of 1 hour.

- Propose applying chemical biocides to the inside of the ductwork and to other system components, if necessary.
- Suggest applying sealants or other encapsulants to seal or cover the inside surfaces of the air ducts and equipment housings, if necessary; the rationale is that these chemicals will control mold growth or prevent the release of dirt particles or fibers from ducts.

At the conclusion of cleaning activities, any synthetic filter media applied to air terminal outlets must be washed with an approved cleaning solution. The media and outlet frame must then be air dried prior to reinstallation. Residual chemical treatment should be applied only after the system has been properly cleaned of all visible dust or debris.

If a service provider fails to follow proper duct-cleaning procedures, duct cleaning can cause indoor air problems. For example, an inadequate vacuum collection system can release more dust, dirt, and other contaminants than if the ducts had been left alone. A careless or inadequately trained service provider can damage the ducts or HVAC system, possibly increasing heating and air conditioning costs or creating a need for difficult and costly repairs or replacements.

7.8.6 Example of Cleaning Sequence (USEPA, 1997)

1. Isolate HVAC unit housing with polyethylene sheeting
2. Protect all motors, bearing assemblies, and belt-drive assemblies within the HVAC unit housing with taped-on polyethylene sheeting
3. Carefully remove filters/filter media from holding frames and/or spools. Store filters and protect them from damage. Reinstall filter media after the cleaning operation is complete, in accordance with the filter manufacturer's instructions to ensure a leak-free installation. Do not restart fans until all filters have been reinstalled and inspected.
4. Remove insulation on the supply fan interior. Replace with exterior insulation. Alternatively, closed-cell foam insulation may be applied to the interior of the supply fan.
5. Vacuum clean entire internal space of HVAC unit, including each component and component supports, frames, and mounts.
6. Pressure wash and hand scrub each HVAC unit as needed, using EPA-approved cleansing agents.
7. Clean all the internal surfaces of the HVAC unit housing; all the internal components of the HVAC unit; all fan surfaces, inside and outside; cooling and heating coil banks, including both upstream and downstream coil faces; filter bank support frames; and contiguous control damper assemblies.
8. Rinse thoroughly with clear water.
9. Vacuum clean and dry all washed surfaces.

7.9 CHEMICAL TREATMENT

If the interior of a variable air volume (VAV) box is encapsulated, the controls on the box must be protected from the encapsulant. The performance of the box must not be compromised by the duct cleaning.

7.9.1 Biocides and Ozone

Biocides and ozone treatment may be duct-cleaning options, but the hazards of this treatment must be considered. Ozone is a highly reactive gas that is regulated in the outside air as a lung irritant, and some people may react negatively to the biocide or ozone and experience adverse health reactions.

7.9.2   Sealants

Sealants should never be used on wet duct liner, to cover actively growing mold, or to cover debris in the ducts and should only be applied after cleaning according to NADCA or other appropriate guidelines or standards. Most organizations concerned with duct cleaning, including the EPA, NADCA, NAIMA, and the Sheet Metal and Air Conditioning Contractors' National Association (SMACNA) do not currently recommend the routine use of sealants in any type of duct. Sealants may prevent dust and dirt particles inside air ducts from being released into the air. The following issues with sealant use should be considered:

- A sealant is often applied by spraying it into the operating duct system; laboratory tests indicate that materials introduced in this manner tend not to completely coat the duct surface.
- Application of sealants may also affect the acoustical (noise) and fire-retarding characteristics of fiberglass-lined or -constructed ducts and may invalidate the manufacturer's warranty.
- Many questions remain unanswered about the safety, effectiveness, and overall desirability of sealants.
- Little is known about the potential toxicity of these products under typical use conditions or in the event of fire.
- Sealants have yet to be evaluated for their resistance to deterioration over time. Deterioration could add particulates to the air traveling through ductwork.

7.10 INSULATED DUCTS

Many air duct systems are constructed of fiberglass duct board, sheetmetal ducts internally lined with fiberglass, and flexible ducts internally lined with plastic. The use of insulated duct material has increased due to the contribution of the insulation to improved temperature control, energy conservation, acoustical control, and reduced condensation. Porous insulation materials (e.g., fiberglass) are more prone to microbial contamination than bare sheetmetal ducts. Once fiberglass duct liner is contaminated with mold, cleaning is not sufficient to prevent regrowth. The EPA has not approved or registered biocides for the treatment of porous duct materials. The replacement of wet or moldy fiberglass duct material is the only safe option. (USEPA, 1997)

7.11 CLEAN CONFIRMATION CHECKLIST

See Table 7.1.

7.12 DUCTWORK ACCESS AND ZONING

The interior of ductwork can be entered for cleaning using access panels. These panels and other planned interior duct entry options should be shown on preplanning documents. Ductwork entry must be preceded by analysis of structural integrity impacted by entry, electrical safety and confined space protocol development, and investigation of the current conditions. Disassembly of ductwork may be required. Fan sequencing may have to be altered or the entire system may have to be shut down. The zones within the ductwork that are to be cleaned must be partitioned off and sealed both from other components of the HVAC system and the ambient room airstream. Synthetic filter media (one-inch thick and 30% efficiency) or layers of plastic may be temporarily fitted over each register, grille, and diffuser in the duct system to intercept any migrating loose dirt and debris.

Table 7.1   Clean Confirmation Checklist

System access and cleaning	Did the service provider obtain access to and clean the entire heating and cooling system, including ductwork and all components (drain pans, humidifiers, coils, and fans)? Has the service provider adequately demonstrated that ductwork and plenums are clean? (Plenum is a space in which supply or return air is mixed or moves; it can be a duct, joist space, attic or crawl space, or wall cavity.)
Heating	Is the heat exchanger surface visibly clean?
Cooling components	Are both sides of the cooling coil visibly clean? If you point a flashlight into the cooling coil, does light shine through the other side? It should if the coil is clean. Are the coil fins straight and evenly spaced (as opposed to being bent over and smashed together)? Is the coil drain pan completely clean and draining properly?
Blower	Are the blower blades clean and free of oil and debris? Is the blower compartment free of visible dust or debris?
Plenums	Is the return air plenum free of visible dust or debris? Do filters fit properly and are the filters the proper efficiency as recommended by HVAC system manufacturer? Is the supply air plenum (directly downstream of the air-handling unit) free of moisture stains and contaminants?
Metal ducts	Are interior ductwork surfaces free of visible debris? (Select several sites at random in both the return and supply sides of the system.)
Fiberglass	Is all fiberglass material in good condition (e.g., free of tears and abrasions; well adhered to underlying materials)?
Access doors	Are newly installed access doors in sheet metal ducts attached with more than just duct tape (e.g., screws, rivets, mastic, etc.)? With the system running, is air leakage through access doors or covers very slight or nonexistent?
Air vents	Have all registers, grilles, and diffusers been firmly reattached to the walls, floors, or ceilings? Are the registers, grilles, and diffusers visibly clean?
System operation	Does the system function properly in both the heating and cooling modes after cleaning?

7.13 AIR CLEANERS

Air cleaning is one of three methods of reducing pollutants in indoor air. In order of effectiveness, the three methods are:

1. Removal of the source or control of its emissions
2. Ventilation
3. Air cleaning

Air cleaning can be used as an adjunct to source control and ventilation; however, air cleaning alone cannot adequately remove all of the pollutants typically found in indoor air.

7.13.1 Contaminant Removal

Air cleaners may reduce the health effects from some particles — small solid or liquid substances suspended in air, such as dust or light spray mists. Some air cleaners, under certain conditions, can effectively remove some respirable-size particles (for example, tobacco smoke). These invisible particles can be inhaled deeply into the lungs and therefore are of concern. Removing such particles may reduce associated health effects in exposed people. These health effects may range from eye and lung irritation to more serious effects such as cancer and decreased lung function. Some controversy exists about whether air cleaners can reduce the allergic reactions produced by larger particles such as pollen, house dust allergens, some molds, and animal dander. Most of these particles settle on surfaces in the home, rather than staying indefinitely suspended in air. Consequently, these particles cannot be removed by an air cleaner unless disturbed and resuspended in the air. Air cleaners that do not contain special media, such as activated carbon or alumina, will not remove gaseous pollutants, including radon, or reduce their associated health effects. Whether air cleaners that contain these media are effective in reducing health risks from gaseous pollutants cannot be adequately assessed at this time.

7.13.2 Types

Some air cleaners may be installed in the ducts that are part of central heating or air-conditioning systems in homes. Portable air cleaners stand alone in a room. Types of air cleaners include:

* Mechanical filters similar to, and including, the typical furnace filters
* Electronic air cleaners (e.g., electrostatic precipitators) that trap charged particles using an electrical field
* Ion generators that charge the particles, which are then attracted to walls, floors, draperies, etc. or a charged collector
* Hybrid devices, which contain two or more of the particle removal devices discussed above

7.13.3 Performance Assessment

At a minimum, the following major factors should be considered: (1) the percentages of the particles removed as these particles go through the device (that is, the efficiency); (2) the amount of air handled by the device (in some cases, pollutants may be generated more quickly than the pollutants can be removed); the effective volume of the air to be cleaned (a single portable unit used in a room within a large building in which the air flows among several apartments or offices would be of little or no value); and (3) the decrease in performance between maintenance periods and the impact if periodic maintenance is not performed on schedule.

7.13.4 Additional Factors

Without proper installation and maintenance, ion generators and electronic air cleaners may produce ozone, which can be a lung irritant. Ion generators, especially those that do not contain a collector, may cause soiling of walls and other surfaces. Gases and odors from particles collected by the devices may be redispersed into the air. The odor of tobacco smoke is largely due to gases in the smoke rather than particles. Thus, a tobacco odor can be detected even when the smoke particles have been removed. Some devices scent the air to mask odors, which gives the impression that the odor-causing pollutants have been removed. Noise from portable air cleaners may be of concern, even at low speeds. Maintenance costs, such as costs for the replacement of filters, may be significant. In general, the most effective units are also the most costly.

7.13.5 Installation and Maintenance

Follow the manufacturer's directions to ensure that the air cleaner works properly. To avoid any electrical or mechanical hazards, be sure the unit is listed with Underwriters Laboratories or another recognized independent safety testing laboratory. Perform routine maintenance, as required, and frequently clean and replace filters. The air cleaner must be placed near a specific pollutant source and in such a position as it forces the cleaned air into occupied areas. The inlet and outlet must not be obstructed by walls, furniture, or other building components.

7.13.6 Efficiency Ratings

Air cleaner effectiveness is expressed as a percentage efficiency rate. The rate of airflow through the cleaning or filtering element is expressed in cubic feet per minute (cfm). Efficiency is lowered by a low air circulation rate with a very efficient collector or high air circulation rate with a less efficient collector. One common method of rating high-efficiency filters uses a procedure detailed in Military Standard 282. This procedure measures how well small particles of a specific chemical are removed by the filter. The federal government has not published guidelines or standards that can be used to determine how well low- to medium-efficiency air cleaners work; however, standards have been developed by private standard-setting trade associations. These standards may be useful in comparing air cleaners.

7.14 FILTRATION

The standing assumptions used to be that endpoint filtration took care of all of the above problems. Filters initially were flat filter media. Over time these filtration systems have become more geometric with pleating and layering of filters. The current accepted filtration dynamics assumption is that filter loading increases the filter efficiency up to the failure point; however, as the failure point is approached, the weight of the filter media causes the filter to pull away from the supportive framework and breakthrough of particulate then occurs. Thus, the timing of filter replacement becomes imperative.

When properly maintained and monitored, filter systems do a very good job of filtering most air particulates 1 μm and larger. Smaller particulates, gases, fumes, vapors, and mists may go readily through the filter media, or partially through, and can contribute to filter failure. If standing moisture is an issue, biological and chemical reactions may ensue within the filter membrane and throughout lined ductwork. Filter systems may become the final resting places for rodents, birds, various insects, and spiders. Initial entry may be through broken air intake systems or failed ductwork. Often residual moisture in condensate pans and near coils is an attractant for rodents, and mice or rat populations may flourish in these environments. Upon death, the rotting tissues of these animals contribute a host of unpleasant and potentially toxic gases to the airstream, as well as providing biological amplification sites.

7.15 HOUSE DUST

House dust is not dust that blows in from the outside. House dust is produced indoors from several sources:

- The breakdown and release of plant and animal materials used in the home. These contaminants include such items as feathers, cotton, wool, jute, hemp, and animal hairs. These materials come from clothing, carpets, rugs and furniture.
- The disintegrated stuffing material from mattresses, pillows, quilts, and upholstered furniture. Prolonged use seems to cause these resilient fibers to weaken and eventually break down into particles small enough to be inhaled.

- Human skin scales, animal dander, body parts from insects such as cockroaches and dust mites, saliva, molds and mildew, bacteria, viruses, and pollen. As people go through their daily activities, particles that have settled onto the floor and other surfaces are stirred into the air.

7.16 DUST MITES

Dust mites (*Dermatophagoides farinae*) are microscopic spider-like insects found everywhere and are thought to be the principle irritant found in house dust when inhaled by sensitive people. The mites live for about 30 days, and the female lays approximately one egg each day. During warm weather when the humidity is above 50%, mites thrive and produce waste pellets. In less than ideal conditions, mites can go into dormancy. Upon death, their bodies disintegrate into small fragments that can be stirred into the air and inhaled.

7.17 MOLDS IN THE AIR

Molds are persistent and eventually land on surfaces and settle into the tiniest cracks and crevices of carpets, furniture, draperies, insulation, rough textures and smooth surfaces. Dirty heating and cooling ducts, wet carpets, damp upholstery, and dirty air filters on air conditioners and furnaces become hiding places for molds, making it difficult to completely rid a home of mold spores.

7.18 CONTAINMENT

If removal of contaminated items is the chosen remediation option, workers and building occupants must be protected during the abatement process. Control of airstreams in and out of the contaminated area is a requirement in order to limit contamination in other areas of the building. The purpose of containment during maintenance or remediation activities is to limit the release of mold into the air and surroundings. In an effort to minimize the exposure of remediators and building occupants to mold, mold and moldy debris should not be allowed to spread to areas in the building beyond the contaminated site being remediated.

The two types of containment are limited and full. The larger the area of moldy material, the greater the possibility of human exposure and the greater the need for containment. In general, the size of the area helps determine the level of containment; however, a heavy growth of mold in a relatively small area could release more spores than a lighter growth of mold in a relatively large area.

The primary objective of containment should be to prevent occupant and remediation personnel exposure to mold. To accomplish this:

- Always maintain the containment area under negative pressure.
- Exhaust fans to the outdoors and ensure that adequate make-up air is provided.
- Routinely check if the containment is working. The polyethylene sheeting should billow inward on all surfaces. If the sheeting flutters or billows outward, containment has been lost. This problem must be corrected before continuing remediation activities.

7.18.1 Limited Containment

Limited containment is generally recommended for areas involving between 10 and 100 square feet (ft^2) of mold contamination. The enclosure around the moldy area should consist of a single layer of 6-mil, fire-retardant polyethylene sheeting and should have a slit entry and covering flap on the outside of the containment area. For small areas, the polyethylene sheeting can be affixed to floors and ceilings with duct tape. For larger areas, a steel, pvc, or wooden stud frame can be erected and the polyethylene sheeting can be anchored to the frame.

All supply and air vents, doors, chases, and risers within the containment area must be sealed with polyethylene sheeting to minimize the migration of contaminants to other parts of the building. Heavy mold growth on ceiling tiles may impact HVAC systems if the space above the ceiling is used as a return air plenum. In this case, containment should be installed from the floor to the ceiling deck (i.e., slab-to-slab), and the filters in the air-handling units serving the affected area may have to be replaced once remediation is finished.

To ensure that contaminated air does not flow into adjacent areas, the containment area must be maintained under negative pressure relative to surrounding areas. A HEPA-filtered fan unit exhausted outside of the building can be used. For small, easily contained areas, an exhaust fan ducted to the outdoors can also be used. The surfaces of all objects removed from the containment area should be remediated or cleaned prior to removal.

7.18.2 Full Containment

Full containment is recommended for the clean-up of mold-contaminated surface areas > 100 ft^2 or in any situation in which it appears likely that the occupant space would be further contaminated without full containment. Double layers of polyethylene should be used to create a barrier between the moldy area and other parts of the building. A decontamination airlock should be constructed for entry into and exit from the remediation area. The entryways to the airlock from the outside and from the airlock to the main containment area should consist of a slit entry with covering flaps on the outside surface of each slit entry. The airlock should be large enough to hold a waste container and allow a person to put on and remove PPE. PPE must be worn throughout the final stages of HEPA vacuuming and damp-wiping of the contained area. PPE must also be worn during HEPA vacuum filter changes or clean-up of the HEPA vacuum. All contaminated PPE, except respirators, should be placed in a sealed bag for removal through the airlock. Respirators should be worn until remediators are outside the airlock. Indoor houseplants should not be over-watered because overly damp soil may promote the growth of microorganisms that can affect allergic individuals.

7.19 RESIDENTIAL SANITATION AND PREVENTION

- Wear rubber gloves, long pants, a long-sleeved shirt, and a dust mask when conducting an activity that will stir up dust.
- Properly maintain flues and the chimneys of furnaces, water heaters, and fireplaces to keep smoke from entering the living areas of homes.
- Check and clean refrigeration equipment annually to be sure the air intake and exhaust systems are working correctly. In addition to changing filters and cleaning coils, remember the drip pan that is often hidden at the base.
- Avoid stove fans and other filter systems that recirculate the air rather than exhaust it to the outside of the home. Replace them with models that exhaust to the outside. If that is not possible, keep filters of recirculating models very clean.
- Do not allow furnace and air conditioner air filters to become clogged; change them regularly. Within 24 hours, clean and dry areas that become damp or wet, to prevent molds from establishing themselves.
- Soaked carpeting should be professionally cleaned within 24 hours, as mold and bacteria will grow in the fibers and backing and under the carpeting. If that happens, remove and throw away the carpeting and pad. Also, clean and disinfect the floor.
- Tear out soaked wall materials, ceiling tiles, and wet insulation and disinfect the area. Wear protective eyewear, gloves, and a mask. In the case of flooding, look above the flood line. Water may have run through materials and/or been absorbed to a higher level. Dry the area as quickly as possible.

- Vacuum the entire house and immediately discard the disposable bags. Steam clean all carpets and dry quickly. Discard potential sources of molds such as accumulations of old newspapers and boxes. Clothing that can be laundered with bleach or drycleaned may be salvageable.
- Mechanical filters that use standard disposal fiberglass filters should be changed monthly. Permanent filters with baffles should be cleaned periodically. The most effective mechanical filter is a HEPA filter. An electrostatic precipitator is an electric filter. These must be cleaned frequently to prevent the production of irritating ozone.
- Encase mattresses in airtight covers. After vacuuming pillows, mattresses, and box springs, encase them in zippered, airtight plastic or special allergen-proof fabric covers.
- Replace comforters and pillows made with down feathers, kapok, and cotton with ones made with synthetic fibers such as Dacron and Orlon.
- Wash bedding weekly in hot water (130°F) to kill dust mites. Wash comforters and pillows regularly. Replace synthetic pillows every two to three years.
- When possible, remove carpeting where dust mites, mold spores, animal dander, and other particulates can accumulate. Carpeting laid over concrete floors tends to have more dust mites because of increased humidity. Replace carpeted floors with hardwood or linoleum. Wash scatter rugs and furniture covers regularly.
- Vacuuming can stir dust into the air. Use high-quality vacuum bags and change them frequently.
- Wet mop or wet-wipe hard surfaces such as floors, walls, and ceilings.

7.20 FLOOD EVENT SANITATION

7.20.1 Remove Standing Water

Standing water is a breeding ground for microorganisms, which can become airborne and be inhaled. Where floodwater contains sewage or decaying animal carcasses, infectious disease is of concern. Even when flooding is due to rainwater, the growth of microorganisms can cause allergic reactions in sensitive individuals. For these health reasons, and to lessen structural damage, all standing water should be removed as quickly as possible.

7.20.2 Dry

Excess moisture in the home is an indoor air quality concern for three reasons:

1. Microorganisms brought into the home during flooding may present a health hazard; these organisms can penetrate deep into soaked, porous materials and later be released into air or water.
2. High humidity and moist materials provide ideal environments for the excessive growth of microorganisms that are always present in the home.
3. Long-term increases in humidity in the home can also foster the growth of dust mites. Dust mites are a major cause of allergic reactions and asthma.

Be patient. The drying out process could take several weeks, and growth of microorganisms will continue as long as humidity is high. If the house is not dried out properly, a musty odor, signifying growth of microorganisms, can remain long after the flood.

7.20.3 Remove Wet Materials

Throwing away items in a home, particularly those with sentimental value, may be difficult; however, keeping certain items that were soaked by water may be unhealthy, as mold and bacteria amplification can occur in these items and residual spores will remain. Materials that are wet and cannot be thoroughly cleaned and dried should be discarded. In cases where items must be saved, a special restoration method may be required. Certain building materials take a long time to dry and may be an ongoing source of microbial growth. These materials include wallboard, fiberglass insulation, wall-to-wall carpeting, and plaster. Fiberboard, fibrous insulation, and disposable filters present in a heating and air conditioning system should be replaced after contact with water. If a filter was designed to be cleaned with water and was in contact with clean rainwater only, ensure that it is thoroughly cleaned before reinstalling.

7.20.4 Clean-Up

The clean-up process involves thorough washing and disinfecting of the walls, floors, closets, shelves, and contents of the house. In most cases, common household cleaning products and disinfectants are used for this task. Disinfectants and sanitizers may be applied to solid-surface ductwork. Additional PPE may be required if human pathogens are suspected in the floodwaters. Human pathogens should be expected if raw sewage is a component of the floodwaters, or if the flood involved compromising the current sewage systems.

7.21 GOOD MAINTENANCE PROGRAMS

Establish a preventive maintenance program. Certain elements of any ventilation system should be checked on a regular schedule and replaced if found to be defective. Keep written records. Maintain written documentation not only of original installations but also of all modifications as well as problems and their resolution. Establish a safe place to file drawings, specifications, fan curves, operating instructions, and other papers generated during design, construction, and testing.

7.21.1 Periodic Inspections

The type and frequency of inspection depend on the operation of the system and other factors:

- Daily — Visual inspection of hoods, ductwork, access and clean-out doors, blast gate positions, hood static pressure, pressure drop across air cleaner, and verbal contact with users (e.g., "How is the system performing today?")
- Weekly — Air cleaner capacity, fan housing, pulley belts
- Monthly — Air cleaner components

A quick way to check for settled material in a duct is to tap the underside of all horizontal ducts with a broomstick. If the tapping produces a clean sheetmetal sound, the duct is clear. If the tapping produces heavy, thudding sounds and no sheetmetal vibration, liquids or settled dust may be in the duct.

7.21.2 Training

Workers need to be trained in the purpose and functions of the ventilation system. For example, workers need to know how to work safely and how best to utilize the ventilation system. Prohibitions against welding shut or otherwise permanently disabling the outside air intakes for make-up air should be a top-priority training item.

7.21.3 Microorganisms

Check for stagnant water in the ventilation system. The presence of mold or slime is a possible sign of trouble. The following are general preventive measures for controlling microbial problems in ventilation systems:

- Prevent buildup of moisture in occupied spaces (relative humidity of 60% or less).
- Prevent moisture collection in HVAC components.
- Remove stagnant water and slime from mechanical equipment.
- Use steam (not hot water) for humidifying.
- Avoid use of water sprays in HVAC systems.
- Use filters with a 50 to 70% collection efficiency rating.
- Find and discard microbe-damaged furnishings and equipment.
- Provide regular preventive maintenance.

7.22 CARPET

Carpet, while an effective floor covering in many applications, does present some maintenance and IAQ challenges. The costs associated with the increased maintenance must be factored into any decision to use carpet vs. hard, nonporous flooring materials. Carpets as originally installed may vary greatly from the current status of carpets used in schools and offices. Carpet is porous material that will absorb and adsorb other contaminants of concern. These contaminants may include biological risk agents such as mold spores and bacteria, heavy-metal particulates such as lead and cadmium, and volatile organics adsorbed during painting or other maintenance events. Carpet padding and adhesives will contribute to IAQ issues both through chemical changes that occur during installation and the ongoing phenomena of adsorption and absorption of other contaminants.

Cleaning events may impart additional residual chemicals on carpeting and pad. Pesticide residues may also be present as both volatile and particulate components. Vacuuming may resuspend both particulate and semivolatile solid residuals into the surrounding environs. Regular carpet maintenance must involve the use of HEPA vacuuming and vigilance to prevent either moisture or contaminant entrapment into carpet matrices. Steam or hotwater cleaning should only be considered with sufficient dwell times of biocide or detergent agents and thorough drying prior to any reoccupancy.

If carpet is installed in high-traffic areas or in locations where contaminant entrapment cannot be avoided, maintenance may involve the use of even stronger chemical stripping agents that include more concentrated biocides. If mold or bacteria amplification has occurred, removal of the contaminated carpet may be more cost effective than cleaning. Removal of contaminated carpet must be done in a controlled manner so as not to contaminate surrounding environs and potentiate exposure scenarios for building occupants.

7.22.1 Carpet Studies

In 1988, the installation of new carpeting at the EPA headquarters in Washington, D.C. led to a rash of health problems and complaints from the staff. This incident became the first highly publicized case of what has been called *sick building syndrome* (SBS). Although the cause of the problem was never verified, speculation has focused on the adhesives used to install the carpet and on a chemical by-product known as 4-PC (4-phenylcyclohexene), which is released from the backing material of the carpet.

Of the chemicals released from carpets, most notable are styrene and 4-PC, both of which come from the styrene butadiene (SB) latex backing used on 95% of carpets. Styrene is a known toxin and suspected carcinogen. 4-PC has not been shown to be toxic, has a detectable odor even at levels below one part per billion, and is the chemical most responsible for the distinctive smell associated with new carpets. 4-PC is less volatile than many of the other chemicals measured and continues to be emitted at measurable levels for a longer time.

Vinyl-backed carpet tiles used in some commercial installations emit a distinctly different set of chemicals, notably vinyl acetate and formaldehyde. These were determined by a Consumer

Product Safety Commission (CPSC) study to be at levels far below those likely to contribute to adverse health effects. Other issues to consider regarding volatile emissions from carpeting include:

- Interactive effects between two or more of the many chemicals involved may impact the effect of any one substance.
- Periodic fluctuations in the manufacturing process may generate occasional batches of particularly volatile carpet with far higher emissions.

One component of the problem appears to be that some people are simply much more sensitized to the effects of chemicals that have no noticeable effect on most. These individuals, with multiple chemical sensitivity (MCS) or environmental illness (EI), appear to be severely affected by conditions that most people consider normal. The syndrome may result from the cumulative effects of low-level chemical exposure and everyone is potentially at risk.

In an agreement negotiated with the EPA, the Carpet and Rug Institute (CRI) began a testing and labeling program in 1992. The program was widely criticized by consumer advocates and the attorneys general of several states for failing to warn consumers about potential hazards and for inadequate testing. The program was strengthened in 1994.

At present, each carpet line is tested four times a year for four categories of emissions: total volatile organic chemicals (tVOCs), styrene, 4-PC, and formaldehyde. As no industry standards exist, somewhat arbitrary maximum emission levels, measured 24 hours after manufacture, have been established. Products that pass the test can carry the new label. The Canadian Carpet Institute (CCI) recently joined with the CRI in offering this labeling program. All testing for the program and most of the industry-sponsored research on carpets has been done by Air Quality Sciences, Inc., of Atlanta, GA. The State of Washington has proposed a program establishing maximum emissions for new carpet used in state projects that are stricter than the industry's voluntary standards (Nalin, 1994).

7.22.2 Carpet: Special Considerations

Carpet fibers, cushions, pads, adhesives, and seam sealants emit volatile organic chemicals as gases after installation or heating events. Heating events, called bake-outs, may be used to force volatile organic chemicals from carpets. Bake-outs require air temperatures in the room to be raised to approximately 95°F. The chemicals emitted may be from integral chemicals or from biocides intentionally applied to carpet materials during manufacture. Formaldehyde is a common biocide chemical applied to preserve material, including the cloth materials, that constitute carpet fibers. Total VOC emissions should be below 100 $\mu g/m^2$ per hour, measured after 24 hours. The manufacturer should specify the adhesive and provide a warranty of tVOC emissions for the installation that includes VOC emission from carpet, adhesive, and sealant as the carpet is used after the 24-hour installation interval. (Nalin, 1994)

7.22.3 Old Carpets

Installed carpet fibers filter materials from the air and retain materials when it is walked on. The materials trapped include particulates, some with adsorbed or absorbed semivolatile or volatile chemicals. Adsorption is a surficial adherence to the particulate nucleus, while absorption implies that the contamination has penetrated throughout the particulate solid matrix. The carpet fibers can trap particulates and adsorb or absorb contaminants from the particulates, in addition to liquid spillage and ambient air gases. Wool fibers may trap more volatiles (e.g., formaldehyde and nitrogen oxides) than do synthetics. The potential for the wool to subsequently release these contaminants at room temperature or through bake-out varies.

Carpet can act as a sink for chemical and biological pollutants including pesticides, dust mites, and fungi. Dust mites, which consume flakes of dead human skin, leave highly allergenic excrement. Upon wetting, the carpet quickly becomes an amplification site for biologicals. Biocides may limit this amplification but will not totally eliminate biologicals. Recycled-content polyester (PET), or polyester, is the only carpet fiber made with significant recycled content. Polyester is generally considered to be a less durable fiber than nylon, so polyester carpets are usually used only in residential applications. (Nalin, 1994)

7.22.4 Installation of Carpet

Clean old carpet before removal. Dirty carpet, when removed, will release particulates. These particulates will circulate in the air and can be trapped in the ventilation system. Even with cleaning, ventilation shutdown should be considered. The following procedure should be used when installing carpet (Nalin, 1994):

- Clean the area thoroughly. Pad and attachment materials under the carpet may remain soiled even after carpet cleaning. The substrate wood or concrete surface may also remain soiled. Prior to installation of new pad and carpet, the substrate and attachment surfaces must be cleaned and thoroughly dried. Sealants should be considered for porous substrates.
- Preventilate. To remove volatiles, unroll the carpet in an unoccupied area with the ventilation system disabled in that area. A bake-out can also be performed during this interval.
- Use adhesives with low volatile organic levels and, thus, low potential for off gassing.
- Ventilate the area while monitoring volatile off-gassing; ventilate directly to the building exterior if possible.
- Vacuum using a HEPA filtration vacuum; vacuum the carpet during installation and thereafter.

7.22.5 Maintenance

Vacuums equipped with HEPA filters should be used. Steam cleaning should only be conducted with the concurrent application of carpet cleaning chemicals with biocide properties. Steam cleaning does not generate sufficient heat to act as an effective biocide. Carpets must be totally dried as quickly as possible. The process of steam cleaning reaerosolizes all materials previously trapped in carpets. Until this aerosolized material settles out of the air, a vector has been created for inhalation exposure to all formerly trapped materials. Consequently, rehabitation of the carpeted areas should be delayed for 4 to 6 hours after the carpet has been determined to be dry by touch. Small children in particular must not be allowed to crawl or walk across wet or newly dried carpet.

Carpet wetted with unsanitary waters through sewer back-up, floods, or spillage may be very difficult to clean. Mold spores and bacterial endospores will be activated by moisture and fed by the deposition of organic residues. Cleaning of substrate and padding may be even more difficult. Steam cleaning of these carpets is more dangerous than steam cleaning of carpets soiled through intended usage (Figure 7.7). The usual recommendation is to dispose of all carpet that has remained wet for more than 24 hours or has been soiled with sewage or other biocontaminated materials. The installation or reinstallation of carpet into areas where biocontamination is likely is not advisable.

Figure 7.7   Material under carpet is difficult to clean. (Courtesy of Aerotech Laboratories, Phoenix, AZ.)

RESOURCES

ASHRAE, Method of Testing General Ventilation Air-Cleaning Devices for Removal Efficiency by Particle Size, Standard 52.2, American Society of Heating, Refrigerating, and Air Conditioning Engineers, Atlanta, GA, 2000.

Canada Mortgage and Housing Corporation, Clean-Up Procedures for Mold in Houses, CHMC publication NHA6789, Ottawa, Canada, 1993.

Dillon, H.K., Heinsohn, H.K., and Miller, J.D., *Field Guide for the Determination of Biological Contaminants in Environmental Samples*, American Industrial Hygiene Association, Fairfax, VA, 1996.

Hurst, C., Ed., *Manual of Environmental Microbiology*, ASM Press. Washington, D.C., 1997.

IICRC, *IICRC S500, Standard and Reference Guide for Professional Water Damage Restoration,* 2nd ed., Institute of Inspection, Cleaning and Restoration Certification, Vancouver, WA, 1999.

Lstiburek, J., *Building Science Corporation Builder's Guide, Mixed-Humid Climates*, Building Science Corporation and the Energy Efficient Building Association, Westford, MA, 1999.

Malin, N., Carpeting, indoor air quality, and the environment, *Environmental Building News*, 3, 6, 1994.

National Academy of Sciences, Committee on the Assessment of Asthma and Indoor Air. *Clearing the Air: Asthma and Indoor Air Exposures*, National Academy Press, Washington, D.C., 2000.

NIOSH, Guide to the Selection and Use of Particulate Respirators Certified Under 42 CFR 84, DHHS (NIOSH) Publ. No. 96–101, National Institute for Occupational Safety and Health, Washington, D.C., 1996.

NYCDOH, Guidelines on Assessment and Remediation of Fungi in Indoor Environments, New York City Department of Health, Bureau of Environmental and Occupational Disease Epidemiology, New York, NY, 2000.

OSHA, Respiratory Protection Standard, 29 CFR 1910.134, 63 FR 1152, Occupational Safety & Health Administration, Washington, D.C., 1998.

Pope, A.M., Patterson, R., and Burge, H., Eds., National Academy of Sciences, *Indoor Allergens: Assessing and Controlling Adverse Health Effects*, National Academy Press, Washington, D.C., 1993.

USEPA, IAQ Tools for Schools, EPA-402-K-95–001, U.S. Environmental Protection Agency, Washington, D.C., 1995.

USEPA, Should You Have the Air Ducts in Your Home Cleaned?, EPA-402-K-97–002, U.S. Environmental Protection Agency, Washington, D.C., 1997.

CHAPTER 8

General Infection Control

Renee Dufault, Martha J. Boss, and Edward Rau

CONTENTS

1-56670-606-8/03/$0.00+$1.50
© 2003 by CRC Press LLC

This chapter deals in general terms with infection control. Emphasis is placed on infection control in laboratory settings where forced amplification of biological organisms may occur and such organisms may be handled repetitively. This chapter is intended to give those individuals who use laboratory services a general idea of the types of protocols required in these labs. Protocols recommended for handling specific microorganisms are published by the Centers for Disease Control and Prevention (CDC). Scientists develop these protocols through trial and error as they learn more about the safe containment and handling of infectious microorganisms and other materials.

Managing infectious materials safely in the laboratory setting is known as practicing "biosafety." The level of safety required for a specific biological agent is dependent on the risk or probability that disease will occur if laboratory personnel become exposed to the agent. There are four biosafety levels (BSLs) described in the CDC publication entitled "Biosafety in Microbiological and Biomedical Laboratories (BMBL)." Each BSL consists of laboratory practices and techniques (containment and handling protocols), safety equipment, and laboratory facilities. The levels are identified in ascending order, by degree of protection provided to laboratory personnel, the environment, and the community, with BSL 1 being the least restrictive and BSL 4, the most restrictive work environment. There is a recommended BSL for each known agent. For example, a laboratory receiving mostly mold samples for incubation will be one that requires BSL 2 containment protocols and the mold samples will be incubated in Class 2 Biosafety Cabinets (BSLs).

Most laboratory risk assessments or evaluations are based on the potential level of infection from an organism to be evaluated in that laboratory. Infection requires that all three of the following *chain of infection* links are present:

- A susceptible host
- A pathogen with sufficient infectivity and numbers to cause infection
- A portal through which the pathogen may enter the host.

Effective infection-control strategies are intended to break one or more of these links in the *chain of infection*, thereby preventing infection. Many of the infection control procedures required in laboratories are also required in medical and healthcare facilities where human pathogens may be present. The term *facility* used herein should be considered an all-inclusive term that denotes laboratories, medical facilities, healthcare facilities, and other locations where contact with pathogens may occur.

Pathogenic risk judgements are based on:

- Disease severity potential for infected individuals
- Infection route
- Presence or absence of transmittal vectors
- Virulence
- Infectivity
- Quantity of agent
- Effective therapies
- Immunization
- Indigenous aspects given where the agent was discovered
- Possible effects on other species, including plants and animals

Some microorganisms (viruses, bacteria, fungi) are species specific, selectively infecting and causing disease in a limited number of, or only one, host species. Unrelated and distantly related species may not be similarly affected by the same infectious microorganism due to differences in physiology, metabolism, or biochemistry. In general, human pathogens from human sources are the most dangerous to humans.

Although a single mode of transmission may predominate, disease-causing microorganisms can be spread or transmitted from one host to the next, directly or indirectly, by a number of methods, including:

- Aerosol generation and inhalation
- Ingestion of contaminated food and water
- Skin and mucous membrane contact with contaminated surfaces
- Contact contamination of an open wound or lesion
- Self inoculation via a cut, laceration or puncture with a contaminated instrument

Containment for pathogenic microorganisms is defined as either primary or secondary. The term *Containment Level* may be used to describe the Risk Group and corresponds to a particular group of microorganisms that can be safely contained within a particular Containment Level given its respective safety requirements. The assumption is that activities within a given Containment Level area will be ordinary and that any growth of agent will only produce a small volume of the agent. BSLs are analogous to Containment Levels and include, in addition to containment and handling protocols, general safety requirements. In all cases, a site-specific hazard assessment should be conducted to determine the required safety criteria.

The safety criteria as defined by Biosafety and Containment Levels described herein are for guidance only and do not take the place of the site-specific hazard assessment.

8.1 GENERAL SAFETY REQUIREMENTS

The following general safety requirements lead to more specific discussions of safety criteria for the various agent Risk Groups and their associated Biosafety and Containment Levels. The Risk Group discussions presented contain information from the BMBL. Site-specific facility programs must be maintained with a clear designation of authority as to the assigned responsibility for program implementation. General safety requirements include infection control procedures to ensure that the facility and all work surfaces therein are kept clean and free of microbial contamination. Minimizing extraneous materials is important and this requirement extends to porous materials such as papers and boxes that cannot be readily decontaminated. The following are examples of general safety requirements.

- Food preparation, eating, applying cosmetics, and inserting contact lenses are not permitted in work areas. The term *eating* should be taken in the broadest definition to include drinking.
- Long hair, jewelry, shoe ornaments, and body piercing should all be addressed programmatically to ensure additional risk is not engendered. In particular, body ornaments that react to disinfecting chemicals or may mechanically degrade personal protective equipment (PPE) must not be worn.
- All personnel must wash their hands after handling infectious materials, doffing gloves, or upon leaving the laboratory. Hand washing soaps must be formulated to prevent chapped hands. Bar soaps are not to be used as sufficient drying intervals will probably not be present to ensure desiccation of bacteria on the soap surface.
- Mouth (oral) pipetting is prohibited in any laboratory. Mechanical pipetting devices must be used instead.
- Plastic ware should be substituted for glassware whenever possible and properly disposed.
- Animals not involved in the work being performed must not be permitted in the lab.
- Illumination must be adequate for all activities, avoiding reflections and glare that could impede vision.
- Eye wash stations are required if caustics or corrosives are used.
- Laboratory equipment and work surfaces must be decontaminated routinely with an effective disinfectant, after work with infectious materials is finished, and especially after overt spills, splashes, or other contamination. Bench tops are impervious to water and are resistant to moderate heat, organic solvents, and decontamination chemicals.

8.2 LABORATORY DIRECTOR

The laboratory director ensures that laboratory personnel are proficient in standard microbiological practices and techniques and safely conduct operations specific to the laboratory facility. Laboratory personnel must receive training on the following topics:

- Occupational safety and health hazards in their workplace
- Safe handling and containment of pathogenic and potentially infectious materials
- Safe handling procedures for hazardous materials (chemicals, radioactive isotopes)
- Use of safety equipment including PPE
- Proper disposal techniques and handling of contaminated waste
- Emergency response procedures

In addition, the laboratory director has sole responsibility for directing downgrades from BSL 3 protocols and provides written procedures that are posted or easily available within the facility outlining requirements for general safety, waste disposal, decontamination, security, laboratory access, emergency response, and occupational health.

8.3 OCCUPATIONAL HAZARDS OF LABORATORY PERSONNEL

8.3.1 Human Bloodborne Pathogens

The OSHA Occupational Exposure to Bloodborne Pathogens Standard found in 29 CFR 1910.1030 was originally intended to provide measures to protect laboratory personnel against the untoward effects of exposure to bloodborne pathogens. The covered population was broadened as this standard was finalized to include any workplace where exposures to bloodborne pathogens may occur. While the hepatitis B virus (HBV) and the human immunodeficiency virus (HIV) are often thought of as the only microorganisms covered by 29 CFR 1910.1030; a *bloodborne pathogen* is by OSHA definition "any pathogenic micro-organism that is present in human blood or other potentially infectious materials that can infect and cause disease in persons who are exposed to material containing this pathogen." OSHA includes both human blood and fluid derived from blood or potentially contaminated by proximity to blood in the body as potentially infectious. The term *OPIM* refers to other potentially infectious materials and includes infected human tissues, semen, vaginal secretions, cerebrospinal fluid, synovial fluid, pleural fluid, peritoneal fluid, pericardial fluid, amniotic fluid, saliva in dental procedures, and any other body fluid that is visibly contaminated with blood.

8.3.1.1 Animal Cells, Blood and Body Fluids, and Fixed Tissues

Occupational hazards associated with cells, fluids, and fixed tissues may include the potential for infection as these items are tested, handled, and disposed of. If the infective agent is known, the cells, primary cultures, and tissues are assigned the Risk Group (1–4) for that agent and are maintained in the appropriate Containment Levels (BSL 1–4). All cell lines should be considered potentially infectious until proven otherwise. Culturing should be assumed to successfully increase the quantity of infective agents and thus may necessitate heightened Containment Levels.

Primate cell lines are of particular concern given their genetic similarity to human cell lines. Risk Group classification may change and in this case be increased to Risk Group III or IV depending on other factors. Some agents may be diluted or inactivated by chemical treatments (including tissue fixing). Empirical evidence should be collected to document when lessened levels of protection are appropriate.

Non-primate cells, blood, and tissue may also be of concern. The level of potential infectivity and resultant Containment Level needs must be evaluated, with the starting point often being the infective agent's Risk Grouping.

Transfer of infective agents from any source, including animal bites, scratches, and insect bites, should be considered whenever animals are housed for research purposes. Exposure to animal waste products, bedding, and dead animals should also be evaluated.

Cells transformed with viral agents, such as SV-40, EBV, or HBV, as well as cells carrying viral genomic material, present potential hazards to laboratory workers. Tumorigenic human cells also are potential hazards as a result of self-inoculation.

8.3.1.2 Universal Precautions

The Centers for Disease Control and Prevention (CDC) has published a series of recommendations, *Universal Blood and Body Fluid Precautions* or simply, *Universal Precautions* that form the basis for the OSHA 29 CFR 1910.1030 requirements. These precautions are intended to prevent parenteral, mucous membrane, and nonintact skin exposure of workers to bloodborne pathogens. Universal precautions in the laboratory setting include the following:

- Not eating, drinking, smoking, applying cosmetics or lip balm, or handling contact lenses in the laboratory.
- Using appropriate barrier precautions to prevent dermal exposure. Gloves must be worn when blood or OPIM will otherwise be in direct contact with the skin. Soiled equipment and materials with blood and OPIM residues must not be handled without the use of gloves. Torn or damaged impermeable gloves should be discarded and replaced.
- Using impermeable gowns, aprons, or coverall suits to prevent contamination of clothing.
- Wearing safety eyewear and mask to protect the mucous membrane around and in the mouth, nose, and eyes.
- Washing immediately any exposed skin or mucous membrane surfaces if contact with blood or OPIM occurs. Because no impermeable clothing, including gloves, is 100% effective, any body surfaces beneath the impermeable clothing and positioned so as to be immediately beneath areas where the blood or OPIM was contacted should be washed when the clothing is doffed.
- Using caution when handling sharps. Needle stick precautions are particularly important as a needle stick may directly inject biological agents into the circulatory system or surrounding tissues. The development and use of alternative methods that do not require the use of needles and other sharps must be encouraged. All sharps must be carefully used and containerized. Reusable sharps must be carefully disinfected, and disinfection may also be required for disposable sharps. While medical sharps are the most often thought of regarding universal precautions, in reality any cutting surface should be considered a "sharp." Thus, in workplaces where cutting injuries can be expected either through the use of tools or other equipment with sharp edges, precautions must also be taken to prevent first aid responses from becoming bloodborne pathogen incidents.
- Clean up spills of blood or OPIM promptly using appropriate PPE and a disinfectant. Properly dispose of all contaminated cleanup materials, including PPE treating all waste as potentially infectious. All waste containers must be puncture and leak resistant, and have appropriate warning signs or labels to ensure that workers do not inadvertently come in contact with their contents. Infectious waste disposal may present additional challenges in the laboratory setting as the waste is recontainerized and transferred at the loading dock.
- When first aid response may involve contact with blood or OPIM, the first aid responders must be trained as to the hazards of bloodborne pathogens and correct first aid response. A valve cardio-pulmonary resuscitation (CPR) shield must be available and at least two CPR shields should be available at all first aid stations, as the first one may become fouled with OPIM or blood and be rendered unusable should a vomiting event occur.

8.3.2 Toxins of Biological Origin

Given the growing number of microbiological and biomedical laboratories working with toxins of biological origin, the *Biological Defense Safety Program, Technical Safety Requirements (DA Pamphlet 385-69)* and Appendix A of the United States Department of Labor Occupational Safety and Health Association rule *Occupational Exposure to Hazardous Chemicals in Laboratories* have been used to provide safety guidelines.

Laboratory managers and facility safety officials are encouraged to consult with subject matter experts before using any toxin to ensure that appropriate facilities, containment equipment, policies and procedures, personnel training programs, and medical surveillance protocols specific to the toxin and the laboratory are in place.

All high-risk operations should be conducted with two knowledgeable individuals present. Each must be familiar with the applicable procedures, maintain visual contact with the other, and be ready to assist in the event of an accident.

8.3.2.1 General

The laboratory facilities, equipment, and procedures appropriate for work with toxins of biological origin must reflect the intrinsic level of hazard posed by a particular toxin as well as the potential risks inherent in the operations performed. If toxins and infectious agents are used, both must be considered when containment equipment is selected and policies and procedures are written. If animals are used, animal safety practices must also be considered. When vacuum lines are used with systems containing toxins, the lines should be protected with a high-efficiency particulate air (HEPA) filter to prevent toxin entry into the lines. Sink drains should be similarly protected when water aspirators are used.

8.3.2.2 Special Practices

Special practices and safety equipment requirements listed under the applicable Risk Group physical and operational requirements should be reviewed and incorporated into protocols for work with toxins. Training specific to the toxin(s) used should be required and documented for all laboratory personnel working with toxins, before starting work with the toxin and at intervals thereafter.

Each laboratory should develop a chemical hygiene plan specific to the toxin(s) used in that laboratory. The chemical hygiene plan should (1) identify the hazards that will be encountered in normal use of the toxin, and those that could be encountered in case of a spill or other accidents, and (2) specify the policies and practices to be used to minimize risks (e.g., containment and personal protective equipment, management of spills and accidental exposures, medical surveillance). Suggested policies include:

- An inventory control system.
- Storage of toxins in locked storage rooms, cabinets, or freezers when not in use.
- Restricted access to areas containing toxins. Any special entry requirements should be posted on the entrance(s) to the room. When toxins are in use, the room should be posted to indicate *Toxins in Use — Authorized Personnel Only*. Only personnel whose presence is required should be permitted in the room while toxins are in use.
- When using an open-fronted fume hood or BSC, protective clothing, including gloves and a disposable long-sleeved body covering (gown, laboratory coat, smock, coverall, or similar garment), should be worn so that hands and arms are completely covered.
- Eye protection should be worn if an open-fronted containment system is used.

Other protective equipment may be required, depending on the characteristics of the toxin and the containment system. For example, use additional respiratory protection if aerosols may be generated and it is not possible to use containment equipment or other engineering controls.

8.3.2.3 Toxin Containers

Preparation of primary containers of toxin stock solutions and manipulations of primary containers of dry forms of toxins should be conducted in a chemical fume hood, a glove box, or a BSC or equivalent containment system. HEPA and/or charcoal filtration of the exhaust air may be required, depending on the toxin. The user should verify inward airflow of the hood or BSC before initiating work and all work should be done within the operationally effective zone of the hood or BSC. Before containers are removed from the hood, cabinet, or glove box, the exterior of the closed primary container should be decontaminated and placed in a clean secondary container. Toxins should be transported only in leak /spill-proof secondary containers.

8.3.2.4 Decontamination

Contaminated and potentially contaminated protective clothing and equipment should be decontaminated using methods known to be effective against the toxin before removal from the laboratory for disposal, cleaning, or repair. If decontamination is not possible/practical, materials (e.g., used gloves) should be disposed of as toxic waste. Materials contaminated with infectious agents as well as toxins should also be autoclaved or otherwise rendered noninfectious before leaving the laboratory. The interior of the hood, glove box, or cabinet should be decontaminated periodically, for example, at the end of a series of related experiments. Until decontaminated, the hood, box, or cabinet should be posted to indicate that toxins are in use and that access to the equipment and apparatus is restricted to necessary, authorized personnel.

8.3.2.5 Electrostatic Toxins

When handling dry forms of toxins that are electrostatic do not wear gloves (such as latex) that help to generate static electricity and use a glove bag within a hood or BSC, a glove box, or a class III BSC. When handling toxins that are percutaneous hazards (irritants, necrotic to tissue, or extremely toxic upon dermal exposure), select gloves that are known to be impervious to the toxin. Make sure to consider both the toxin and the diluent when selecting gloves and other protective clothing. If infectious agents and toxins are used together in an experimental system, consider both when selecting protective clothing and equipment

8.4 MEDICAL SURVEILLANCE

A medical surveillance program must be programmatically defined and must include all appropriate testing given the agents being handled. Baseline serum sample requirements, immunization schedules, and incident reporting follow-up must all be defined in this program element. Medical tests may include tests for HIV, Epstein Barr, Hepatitis C, Mantoux (tuberculosis screening), Q Fever serological, and others as needed.

8.4.1 Baseline Sera

Laboratory personnel must receive appropriate immunizations or tests for the agents potentially present. When appropriate, considering the agent(s) handled, baseline serum samples for laboratory and other at-risk personnel are collected and stored. Additional serum specimens may be collected periodically.

8.4.2 Sharps

All personnel sustaining sharps injuries must be reported and referred to the facility infection control or occupational health staff for medical surveillance. The use of needles and syringes and other sharp objects must be strictly limited to prevent injuries. Extreme caution must be used when handling needles and syringes to avoid self-inoculation and the generation of aerosols during use and disposal. A high degree of precaution must be taken with any contaminated sharp items. Sharp instruments should be restricted in the laboratory and the following procedures should be in place:

- Syringes that re-sheathe the needle, needleless systems, and other safety devices are used when appropriate.
- Only needle-locking syringes or disposable syringe-needle units (i.e., needle is integral to the syringe) are used for injection or aspiration of infectious materials.
- Disposable needles and syringes must not be replaced in their sheath or guard. These sharps must be placed into a puncture-resistant container and autoclaved, if contaminated, before disposal, or incinerated.
- Disposable sharp waste must be collected in a designated puncture-resistant container. Broken glassware must not be handled directly by hand, but must be removed by mechanical means such as a brush and dustpan, tongs, or forceps. Used disposable needles are not be bent, sheared, broken,

recapped, removed from disposable syringes, or otherwise manipulated by hand before disposal; rather, these items are to be carefully placed in conveniently located puncture-resistant containers used for sharps disposal.

- Nondisposable sharps must be placed in a hard-walled container for transport to a processing area for decontamination, preferably by autoclaving.

8.4.3 Dermal Hazards

If the outer skin surface is breached or mucous membrane is exposed to biological contaminants, emergency procedures are required. These procedures include washing the site. In the case of puncture wounds (including animal bites), sharps injuries, and scratches, the wound should often be encouraged to bleed freely.

When mucous membrane or nonintact skin due to skin lesions has been exposed, water flushing of the surface may be needed.

In all such cases, a medical doctor with experience dealing with the biological agents of concern should be contacted. If bloodborne pathogen exposure is expected, the individual should receive counseling to determine the upcoming course of action (including in the case of HBV the advisability of HBV vaccination and worker counseling required by 29 CFR 1910.1030).

If exposure presents a risk both to the exposed individual and to those who may be in contact with the exposed individual, all parties must be notified of the risk potential. Medical and life style alternatives should be discussed.

As with any incident, a written record should be kept of the incident, how exposure occurred, what the responses were, and the current status of counseling efforts.

8.4.4 Immunoprophylaxis Programs

Immunization programs should be offered to workers as needed. When bloodborne pathogen exposures are anticipated, HBV immunizations are to be offered to workers per 29 CFR 1910.1030; however, workers may decline the immunization using a written declination statement. These same workers may change their minds and later request and be given the HBV vaccination.

Other vaccination programs may include immunizations for Anthrax, Botulism, Cholera, Diphtheria, Hemophilus influenzae type b, Hepatitis A Virus, Influenza A, Influenza Virus, Japanese encephalitis, Lyme disease, Measles, Meningococcus, Mumps, Pertussis, Plague, Pneumococcus, Polio, Rabies Virus, Rubella, Tetanus, Tuberculosis (BCG), Typhoid, Vaccinia, Varicella, and Yellow fever.

8.5 PERSONAL PROTECTIVE EQUIPMENT (PPE)

The PPE used must be chosen to adequately protect laboratory personnel from exposure to biological agents, and also the decontamination solutions used on these agents. The level of permeability required for PPE will in part be determined by the dermal hazard potential. When splash protection is also required, splash aprons, shields, goggles, and safety glasses with shields should all be considered.

Glove use should be carefully controlled to ensure that both the person wearing the gloves and those individuals this person may touch are protected. The reuse of gloves presents many challenges with regard to proper donning and doffing and the level of decontamination needed during repetitive use. Obviously, reaching under the glove to remove the glove may in one step contaminate the formerly gloved hand. Issues associated with perspiration build up within gloves and glove material sensitivities must also be considered.

Footwear must not be neglected in these assessments, particularly since leather is a permeable material with pores that can retain contamination. Closed toe and heel shoes are required in laboratory or other infection control areas.

8.6 ACCESS RESTRICTIONS

Laboratory access is limited or restricted when experiments or work with cultures and specimens are in progress. The general rule in any workplace is that only those who need to be in hazardous locations should in fact be in those locations. Special restrictions may be needed for laboratories depending on the agents studied in those laboratories, security clearances required to work on the laboratory projects, and the level of confidentiality required. As with any workplace, hazard communication and training are required to ensure that work proceeds safely. The special risks that may be present for immunocompromised individuals and pregnant women must be communicated to these individuals and appropriate decisions must be made as to their access to various facility locations. Immunization requirements must also be programmatically defined.

Containment and Biosafety Levels 2, 3, and 4 require various access restrictions including signage on entry doors that states the Containment/Biosafety Level and/or the Risk Group information. Infectious agents if present may require that special entry instructions or conditions be included on all signage. The names of all responsible parties who may need to be contacted per the entry restrictions must also be listed.

8.7 BIOHAZARD SIGNAGE

Appropriate information to be posted includes:

- Investigator's and laboratory director's name(s) and phone number(s)
- Agent(s) in use
- Biosafety Level
- Required immunizations
- Required PPE
- Exit and emergency procedures

8.8 EMERGENCY PROCEDURES

Emergency procedures must be programmatically defined and the resultant emergency plan must be communicated through training to all laboratory personnel and potential emergency responders. Security and medical responses must be clearly defined. All personnel who are required to be notified must be listed, and their phone numbers provided.

In accordance with applicable regulations, contaminated equipment (including that used in the spill response) must be decontaminated before removal from the facility for repair or maintenance or packaging for transport.

8.9 BIOLOGICAL HAZARDS PHASE ANALYSIS

For biological agents, inherent risk may increase if the agent reproduces and thus increases in volume, and possibly dispersal potential. This increase in volume may also be an increase in the numbers of infective agents in a unit area. Culturing *in vitro* and infection *in vivo* can cause infective agents to multiply.

Biological agents can be spread in many ways, including as aerosols or effluents.

8.9.1 Aerosols and Effluents

Aerosols can be produced by procedures that cause solids or liquids to be dispersed. If after the initial dispersal, room air currents pick up the aerosols, more extensive dispersal will occur. Under a worst-case scenario, the building ventilation system may transmit the aerosols to areas far from the original source. For this reason, aerosols must be strictly controlled.

The obvious means of control is to keep all solids and liquids in closed containers. Unfortunately, procedures to transfer, change, resuspend, and mix phase components often require opening vessels and containers. Precautions may include the use of appropriate BSCs. Some of the procedures of concern include sonicating, centrifugation, streaking of cultures, flaming of inoculation loops, and releasing vacuum on a freeze dryer.

Equipment manufacturers should be consulted to determine if equipment safeguards against aerosol dispersal are available and have been tested. Copies of these test results, including the conditions under which the test was held, should be requested and kept on file.

The first control to be considered should be source reduction. Source reduction involves either use of less hazardous materials or, in this case, use in such a way that aerosol dispersal is precluded. End-point filtration of aerosols that have been released to atmosphere should be considered as a control method only when source reduction cannot achieve infection control goals. All filtration equipment must be tested and maintained to ensure adequacy of the initial aerosol capture rate and effective transfer of the aerosol to the desired location. This location may be a filter unit, or ultimately, destruction or dispersal to the atmosphere (outdoors).

In cases were source reduction and engineering controls both prove ineffective, personal protective equipment and entry restrictions must be used.

Effluents may include all phases of matter — gases from air exhaust, fluids that include both gas and liquid phase materials, and solids. Control and management of effluents from a facility require effective capture mechanisms for all potential effluents and correct laboratory protocols.

8.9.2 Gases

Air within a facility is captured by ventilation units for either recirculation or exhaust. The type of infective agent will determine the proper ventilation unit and air transfer method. These ventilation systems must meet criteria for both contaminant capture and safe release. Safe release can be interpreted as discharge of the captured air to areas appropriate for the level of infectivity of the aerosols that may be in the air stream. This air movement must be within an enclosed ventilation system that is frequently tested for gaps, corrosion, and system integrity. Such testing can be very difficult particularly if ductwork is hidden and not readily accessible for inspection.

Airflow patterns within a room must be considered. Much emphasis is placed upon the use of pressure differentials within a room to control contaminant spread. The idea is that a negative air-pressure area will receive air from a more positively pressured area. Thus areas where the influx of contaminants or other airstreams is not desired must be under positive pressures. The flip side of this use of positive pressure is that whatever is in the positively pressured area will move out to other areas.

When design intent is translated into actual building construction and use, the original designs for pressure differential often may not be consistent throughout the facility. In particular, the retrofitting of spaces may change the pressure dynamics.

Vacuum lines and compressed air lines may also uptake contaminants and move them to other locations. All supply lines with valves must be checked for valve integrity upon shut down and also traced to potential routes for contaminant movement from a facility. An added problem with compressed air lines is the potential for condensate to build up in unused lines that are not regularly flushed.

Gases as carriers of liquid and solid contaminants are usually controlled by source reduction, engineering controls, or filtration.

8.9.2.1 HEPA Filters

Filtration of contaminants must be consistent with the contaminant of concern. HEPA filters will capture particulates as small as 0.3 μm at 97% efficiency. However, these filters — if loaded with contaminant, covered with mold, or wetted so as to pull away from their framework — will not function correctly. All filtration units must be continually checked on a defined preventative maintenance schedule to ensure both filter structural integrity and correct airflow through the filter. In some cases, downwind (from the filter point of exhaust) locations should be routinely sampled to ensure that filters are operating correctly. Gases absorbed or adsorbed to particulates will be trapped by HEPA filtration as the particulates are trapped; however, HEPA filtration will not inactivate these gases and off-gassing may occur, with the gases emitted unchanged after desorbing from their particulate carriers.

Dioctylphthalate (DOP) aerosol may be used to test HEPA filters in ventilation and exhaust systems. The DOP challenge has been used for many years by the National Institute of Occupational Health and Safety to test HEPA filters in respirators. This test relies on observing the amount of DOP that penetrates the filter. Correctly placed HEPA filters should capture > 99.99% of the DOP challenge aerosol penetration applied upstream of the filter. In other words, < 0.01% of DOP should be present downstream of the HEPA filter.

8.9.2.2 Chemical Cartridge

The addition of a chemical cartridge unit to the filter serves to promote certain chemical's absorption and adsorption to the chemically treated cartridge media. The cartridges will work effectively up until the point where the chemical loading on the cartridge exceeds the absorptive and adsorptive potential of the cartridge. After that point, some of the chemical may continue to react in the cartridge interstitial spaces with the chemical treatment material; however, the chemical will not be molecularly affixed to the base media (activated charcoal, silica gel). Breakthrough of the chemical occurs at that time, usually through capillary action phenomena at the downstream planar surface of the cartridge. At this time, any gaseous or other phase agents will be transmitted downstream of the cartridge, sometimes in very high concentrations as breakthrough continues. For this reason, if used, chemical cartridges must be subject to very strict change out policies as directed by the manufacturer's information and empirical efficacy studies.

8.9.2.3 Ventilation Testing and Airflow Smoke Patterns

Smoke testing for ventilation adequacy is one way to visibly show airflow patterns. Other methods that rely on ventilation calculations and measurements of airflow and pressure differential may be used. Most smoke tubes, bottles, and sticks use titanium chloride (TiCl4) to produce a visible fume. No OSHA PEL or ACGIH TLV has been developed for this chemical, although it is a recognized inhalation irritant. Caution should therefore be utilized when using smoke testing, and nonirritant smoke is available and preferred.

In testing BSC integrity, visible smoke should flow evenly when delivered to the various cabinet zones. Even exhaust is defined as smoke proceeding in an even band without gaps, swirling, or flow away from the exhaust direction. Smoke delivered within the cabinet should remain within the cabinet or be exhausted only in the direction of intended initial cabinet exhaust. For a Class II BSC, no smoke should pass over the work surface or penetrate the work zone. The manufacturer's instructions should be followed to determine the correct sequence and delivery system for smoke testing.

8.9.3 Liquids

Liquids and semi-solids (agar) that may contain biological contaminants must be treated before general disposal. Treatment usually involves sterilization or some level of disinfection.

The manufacturer's information on the biocides used for disinfection must be checked to ensure that their usage is indeed intended to disinfect the biological agents of concern. Autoclaving is an effective means of sterilization against many liquid agents.

8.9.4 Solids

The term *solids* may include solidified biological material or contaminated equipment with traces of biological contamination. Disposable equipment also constitutes a troubling waste stream; especially when in an effort to save money, such equipment is maintained for reuse. Disposable equipment, because of equipment design or material components, may not be amenable to the disinfecting and sterilizing methods that would guarantee safe reuse.

8.10 DECONTAMINATION: STERILIZATION VS. DISINFECTION

Biological contamination refers to the presence of infectious microorganisms upon surfaces or in materials such as aerosols, gases, solids, and liquids. Decontamination is the process of inactivating the infectious microorganisms wherever they are located so that they are rendered harmless and no longer capable of causing disease.   Inactivation or treatment of media contaminated with infectious or potentially infectious microorganisms can be accomplished through either sterilization or disinfection depending upon the agent of concern and the desired level of antimicrobial activity.

8.10.1 Sterilization

Antimicrobial pesticides identified by the term *sterilizer* or *sterilant* are intended to inactivate all viruses and all living bacteria, fungi, and their spores on inanimate surfaces. Sterilization represents the highest level of antimicrobial activity. Manufacturers or registrants of sterilizing antimicrobial pesticides are required to submit to the EPA specific effectiveness data using resistant bacterial spores in testing to support sterilization claims on their product labels.

8.10.2 Disinfection

Antimicrobial pesticides identified by the term *disinfectant* are intended to provide a lower level of antimicrobial activity than sterilization and destroy or irreversibly inactivate *specific* viruses, bacteria, or pathogenic fungi, but not necessarily their spores, on inanimate surfaces.

Most disinfectants, even with prolonged contact times, are not as effective as sterilants.   To support effectiveness as a disinfectant, specific data relative to each bacterium, pathogenic fungus, or virus against which a product is claimed to be effective must be submitted to EPA by the manufacturer or registrant.

To be registered as a hospital disinfectant, a product must be shown to be effective against *Staphylococcus aureus*, *Salmonella choleraesuis*, and *Pseudomonas aeruginosa*.

A registrant may optionally claim effectiveness against additional microorganisms such as *Mycobacterium tuberculosis*, pathogenic fungi, or certain specific viruses, provided such efficacy is documented with appropriate data.

8.11 HOT STERILIZATION

Heating processes capable of equipment sterilization include steam under pressure {autoclaving}, dry heat, and heat/chemical vapor application.

Heat-sensitive chemical indicators that change color after exposure to heat do not ensure the adequacy of a sterilization cycle. These indicators may be used to identify packs or equipment that have been processed through the heating cycle.

Disposable nonsharp items contaminated with biological agents must be collected in autoclave bags. After autoclaving and cooling, these bags of waste must be placed into containers for disposal.

8.12 COLD STERILIZATION — LIQUID CHEMICAL GERMICIDES

In all healthcare settings, indications for the use of liquid chemical germicides to sterilize instruments are limited. For heat-sensitive instruments, this procedure may require up to 10 hours of exposure to a liquid chemical agent registered with the EPA as a *sterilant/disinfectant*. This sterilization process should be followed by septic rinsing with sterile water, drying, and placement in a sterile container.

8.12.1 EPA-Registered Sterilant/Disinfectant

EPA-registered sterilant/disinfectant chemicals are used to attain high-level disinfection of heat-sensitive semicritical instruments. The product manufacturers' directions regarding appropriate concentration and exposure time should be followed closely. The EPA classification of the liquid chemical agent will be shown on the chemical label.

Liquid chemical agents that are less potent than the sterilant/disinfectant category are not appropriate for reprocessing critical or semicritical instruments.

8.12.2 Intermediate Chemical Germicides

Intermediate disinfectants are not recommended for reprocessing critical or semicritical dental instruments. Countertops and surfaces that may have become contaminated with patient material may be cleaned with intermediate disinfectants. Surfaces then should be disinfected with a suitable chemical germicide. Chemical germicides should be registered with the EPA as a hospital disinfectant and labeled for tuberculocidal and mycobactericidal activity. Because mycobacteria are among the most resistant groups of microorganisms, germicides effective against mycobacteria should be effective against many other bacterial and viral pathogens.

Intermediate-level disinfectants include phenolics, iodophors, chlorine-containing compounds. A fresh solution of sodium hypochlorite (household bleach) prepared daily is an inexpensive and effective intermediate-level germicide. Concentrations ranging from 500 to 800 ppm of chlorine (1:100 dilution of bleach and tap water or 1/4 cup of bleach to 1 gallon of water) are effective on environmental surfaces that have been cleaned of visible contamination. Caution should be exercised, since chlorine solutions are corrosive to metals, especially aluminum.

8.13 CONTAINMENT

The term *containment* describes methods to manage infectious materials and reduce or eliminate exposure.

8.13.1 Primary Containment and Barriers

Primary containment includes good microbiological technique and appropriate safety equipment. Vaccines may also be used to supplement primary containment goals.

BIOLOGICAL SAFETY CABINETS (BSC)

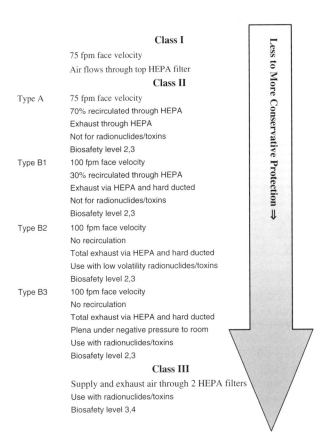

Figure 8.1 Various biological safety cabinets are used to mitigate hazards.

Safety equipment includes BSCs (Figure 8.1); enclosed containers, such as the safety centrifuge cup which is designed to prevent aerosols from being released during centrifugation; other equipment and engineering controls; and PPE, such as gloves, coats, gowns, shoe covers, boots, respirators, face shields, safety glasses, and goggles

BSC Classes are denoted by the agent numerators for the Risk Groups that can be stored, worked on, and/or incubated within these cabinets. Thus, Class II BSCs can be used for Risk Group II agents, and so forth. BSCs are designed to minimize the potential for infectious splashes and exposure to aerosols generated by microbiological procedures.

When used with good microbiological techniques, open-fronted Class I and Class II BSCs are primary barriers that offer significant levels of protection to laboratory personnel and to the environment. BSCs should be used when performing procedures that might cause splashing, spraying, or splattering of droplets and for the initial processing of clinical specimens when the nature of the test suggests the likely presence of an agent readily transmissible by infectious aerosols (e.g., *M. tuberculosis*) or use of a BSC (Class II) is indicated to protect the integrity of the specimen.

The Class II BSC also provides protection from external contamination of the materials (e.g., cell cultures, microbiological stocks) being manipulated inside the cabinet. The gas-tight Class III BSC provides the highest attainable Containment Level of protection to personnel and the environment.

All BSCs must be inspected and tested at least annually to ensure that the cabinet is functioning correctly. The qualifications of the inspecting individuals, the dates and results, and the certification statements must all be retained as required laboratory documentation papers.

BSCs must be installed so that fluctuations of the room supply of air and exhaust air do not cause the BSCs to operate outside their parameters for containment. Thus, BSCs should be located away from doors, from windows that can be opened, from heavily traveled laboratory areas, and from other potentially disruptive equipment. If the BSCs are relocated, reinspection and retesting must occur prior to BSC use.

8.13.1.1 Paraformaldehyde Vapor Decontamination

Cabinet decontamination with paraformaldehyde vapor should be conducted prior to the testing of BSCs that have been used for activities involving biological agents. The BSC should be sealed and decontaminated using the paraformaldehyde vapor technique. The paraformaldehyde holding/contact time should be a minimum of 2 hours, after which the paraformaldehyde vapor should be neutralized or vented to the building exterior.

Biological monitors such as *Bacillus subtilis globigii* or a combination of *B. globigii* with *B. stearothermophilus* are used to determine paraformaldehyde decontamination success. The steps are:

- Spore strips are removed from their protective envelopes and suspended in the work compartment where these spore strips will be exposed to filtered air during the purge cycle.
- After final BSC decontamination, these spore strips are aseptically transferred to test tubes of Trypticase Soy Broth.
- Test tubes are incubated at 37°C for *B. globigii* and 55°C for *B. stearothermophilus*. BSC decontamination clearance will require that no growth above control levels occurs during a 72-hour incubation period. Controls will also be incubated and will be positive controls since growth at some level will be assumed.

8.13.2 Secondary Containment and Barriers

Secondary containment includes operational methods and facility design. The recommended secondary barrier(s) will depend on the risk of transmission of specific agents. For example, the exposure risks for most laboratory work in Containment Level 1 and 2 facilities will be direct contact with the agents or inadvertent contact exposures through contaminated work environments. Secondary barriers in these laboratories may include separation of the laboratory work area from public access, availability of a decontamination facility (e.g., autoclave), and hand washing facilities.

When the risk of infection by exposure to an infectious aerosol is present, higher levels of primary containment and multiple secondary barriers may become necessary to prevent infectious agents from escaping into the environment. Design features include:

- Specialized ventilation systems to ensure directional air flow
- Air treatment systems to decontaminate or remove agents from exhaust air
- Controlled access zones, airlocks at laboratory entrances, or separate buildings or modules to isolate the laboratory.

Design engineers for laboratories may refer to specific ventilation recommendations as found in the *ASHRAE HVAC Applications Handbook 1999* (1-P) published by the American Society of Heating, Refrigerating, and Air-Conditioning Engineers (ASHRAE).

8.14 RISK GROUP 1: BIOSAFETY AND CONTAINMENT LEVEL 1

Risk Group 1 includes:

Defined and characterized strains of viable microorganisms not known to consistently cause disease in healthy adult humans. *Bacillus subtilis, N. aegleria gruberi,* infectious canine hepatitis virus, and exempt organisms under the NIH Recombinant DNA Guidelines are representative of microorganisms meeting these criteria. Many agents not ordinarily associated with disease processes in humans are, however, opportunistic pathogens and may cause infection in the young, the aged, and immunocompromised individuals. Vaccine strains that have undergone multiple *in vivo* passages should not be considered avirulent simply because they are vaccine strains. (DHHS, 1999)

8.14.1 Physical Requirements

No special design features beyond those suitable for a well-designed and functional facility are required. BSCs are not required nor is special ventilation and work may be done on an open bench top using standard techniques. Containment is achieved through the use of practices normally employed in a basic microbiology laboratory.

The following must be provided for Risk Group 1:

- A room separated from public areas with a door for access control. This door should remain closed.
- Fly screens over windows that can be opened near working areas or containment equipment.
- Coat hooks for laboratory clothing near the exit. Note: Separate areas for street clothes vs. laboratory clothes.
- Laboratory furniture capable of supporting anticipated loading.
- Facility furnishings and work surfaces impervious to liquids and readily cleanable:
  - No carpets and rugs
  - Cleanable coatings on walls, ceilings, furniture, and floors
  - Accessible spaces between benches, cabinets, and equipment
  - Bench tops impervious to liquids and decontamination chemicals and resistant to moderate heat, organic solvents, acids, alkalis
  - Hand washing sink in each laboratory and near point of exit to public areas

8.14.2 Operational Requirements

Basic facility safety practices must be followed.

- Personnel wash their hands:
  - After handling viable materials
  - After removing gloves
  - Before leaving
- Persons who wear contact lenses in laboratories also wear goggles or a face shield.
- Food is stored outside work area in cabinets or refrigerators designated and used for this purpose only.
- Policies for the safe handling of sharps are instituted.
- All procedures are performed carefully to minimize the creation of splashes or aerosols.
- A biohazard sign is posted at the entrance to the laboratory whenever infectious agents are present. The sign includes the name of:
  - The agent(s) in use; and
  - Investigator's name and phone number.
- An insect and rodent control program is in effect.

8.15 RISK GROUP 2: BIOSAFETY AND CONTAINMENT LEVEL 2

Except in extraordinary circumstances, the initial processing of clinical specimens and serological identification of isolates can be done safely at Biosafety Level 2, the recommended level for work with bloodborne pathogens such as HBV and HIV.

The containment elements described in Biosafety Level 2 are consistent with the OSHA standard *Occupational Exposure to Bloodborne Pathogens*. Other recommendations specific for clinical laboratories may be obtained from the National Committee for Clinical Laboratory Standards. These recommendations and OSHA requirements focus on the prevention of percutaneous and mucous membrane exposures to clinical material. Primary hazards relate to percutaneous or mucous membrane exposures and ingestion of infectious materials. Extreme caution should be taken with contaminated needles or sharp instruments.

With good microbiological techniques, these agents can be used safely in activities conducted on an open bench, provided the potential for producing splashes or aerosols is low.

Even though organisms routinely manipulated at Biosafety Level 2 are not known to be transmissible by the aerosol route, procedures with aerosol or high splash potential that may increase the risk of such personnel exposure must be conducted in primary containment equipment or in devices such as a BSC or safety centrifuge cups, or by using other primary barriers as appropriate, such as splash shields, face protection, gowns, and gloves.

Hepatitis B virus, HIV, the salmonellae, and *Toxoplasma* spp. are representative of microorganisms assigned to this Containment Level.

8.15.1 Risk Group 2 Agents

Examples of Risk Group 2 agents are as follows:

BACTERIA
Clostridium botulinum
Escherichia coli: enterotoxigenic/invasive/hemorrhagic
strains
Haemophilus influenzae, *H. ducreyi*
Helicobacter pylori
Legionella spp.
Leptospira interrogans: all serovars
Listeria monocytogenes
Staphylococcus aureus
Streptobacillus moniliformis
Streptococcus spp: Lancefield Groups A, B, C, D, G
FUNGI
Candida albicans
Cryptococcus neoformans
Aspergillus flavus
Aspergillus fumigatus
Sporothrix schenckii
Trichophyton spp.
VIRUSES
Lymphocytic choriomeningitis virus: laboratory adapted
  strains
Hepatitis C virus, Hepatitis B virus, including Delta agent
Genus cytomegalovirus: all isolates including CMV (HHV 5)
Rubella virus

Containment Level 2 is suitable for work with agents in Risk Group 2. In addition to the requirements of Containment Level 1, the following are required.

8.15.2 Physical Requirements

The facility should be located away from public areas, offices, and patient use areas. Facility doors are to be self-closing and lockable doors should be provided for facilities that house restricted agents (as defined in 42 CFR 72.6). An autoclave should be available in or near the facility. Vacuum lines used for work involving the agent are to be protected from contamination by HEPA filters or equivalent equipment.

8.15.3 Operational Requirements

Centrifugation must be conducted with closed containers or aerosol-proof safety heads or cups. These containers should be opened only in the BSC.

8.15.3.1 BSCs

Properly maintained BSCs, preferably Class II, appropriate PPE, or physical containment devices are used whenever (Figure 8.2) procedures with a potential for creating infectious aerosols or splashes are conducted. These procedures may include centrifuging, grinding, blending, vigorous shaking or mixing, sonic disruption, opening containers of infectious materials whose internal pressures may be different from ambient pressures, inoculating animals intranasally, and harvesting infected tissues from animals or embryonate eggs.

BSCs are also needed when high concentrations or large volumes of infectious agents are used. (*Note*: Such materials may be centrifuged in the open laboratory if sealed rotor heads or centrifuge safety cups are used, and if these rotors or safety cups are opened only in a BSC.)

Air from BSCs may be recirculated to the room only after passage through an HEPA filter.

8.15.3.2 PPE

Special care should be taken to avoid contamination of the skin with infectious materials. Clothing designed to be used in a laboratory is required; coats that fasten on the front are permissible. Laboratory clothing is not to be worn outside the containment laboratory.

Gloves are worn when hands may contact potentially infectious material or contaminated surfaces or equipment. Wearing two pairs of gloves may sometimes be appropriate. Gloves are disposed of when overtly contaminated, and removed when work with infectious materials is completed or when the integrity of the glove is compromised. Disposable gloves are not washed, reused, used for touching *clean* surfaces, or worn outside the lab. Alternatives to powdered latex gloves are available. Hands are washed following removal of gloves.

Face protection (goggles, mask, face shield or other splatter guard) is used for anticipated splashes or sprays of infectious or other hazardous materials to the face when the microorganisms must be manipulated outside the BSC.

Protective clothing is removed and left in the laboratory before leaving for nonlaboratory areas. All protective clothing is either disposed of in the laboratory or laundered by the institution — never to be taken home.

8.15.3.3 Contaminated Materials and Spills

Contaminated glassware is decontaminated before leaving the facility. Using a proven method, potentially infectious wastes (cultures, tissues, specimens, wastes) are decontaminated before disposal. Material to be decontaminated outside the immediate laboratory are placed in a durable, leak-proof container and closed for transport outside the laboratory. Materials to be decontaminated off site must be containerized and transferred as required by applicable local, state, and federal regulations.

CLASS II BSC

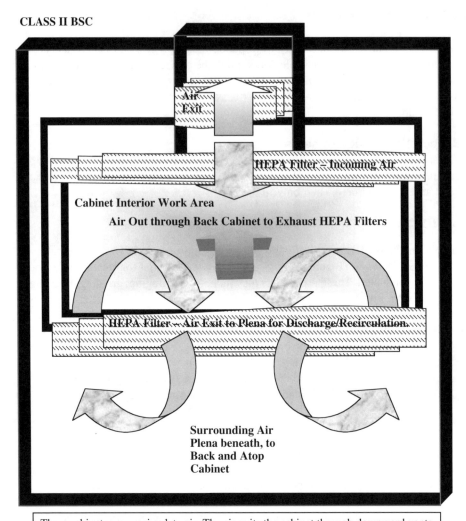

These cabinets may recirculate air. The air exits the cabinet through downward vents that go through a lower HEPA filter unit. This air goes around the cabinet in a plena system and reenters through another HEPA filter unit.

The air that does not recirculate exits from another portal and ultimately exhausts through a different HEPA filter unit.

As the need for more protection increases, the cabinets' design varies and less air is recirculated. The most protective Class II units recirculate no air and the cabinet interior air spaces (plena) are maintained at negative pressure as compared to the room.

For Class II B1, B2, and B3: All units exhaust and intake makeup air through fixed ductwork. In some B1 recirculating units, the recirculating air and the new intake air may temporarily mix prior to resupplying the cabinet interior.

Figure 8.2   Biological safety cabinets (BSCs) for Biosafety Level 2.

8.16 RISK GROUP 3: BIOSAFETY AND CONTAINMENT LEVEL 3

All Risk Group 3 agents must be stored within the Containment Level 3 facility. Level 3 is applicable to clinical, diagnostic, teaching, research, or production facilities and those where work is done with indigenous or exotic agents. These agents may cause serious or potentially lethal disease as a result of exposure by the inhalation route. *Mycobacterium tuberculosis*, St. Louis encephalitis virus, and *Coxiella burnetii* are representative microorganisms assigned to this level.

Examples of Risk Group 3 Agents

BACTERIA

Bacillus anthracis
Mycobacterium tuberculosis 1

FUNGI

Moniliaceae
Ajellomyces capsulatum (*Histoplasma capsulatum*,
  including var. *duboisii*)
Coccidioides immitis

VIRUSES

Lymphocytic choriomeningitis virus: neurotropic strains
Hantaan
Rift Valley fever virus
Yellow fever virus: wild type
Japanese encephalitis virus
Murray Valley encephalitis virus
Powassan encephalitis virus
Chikungunya virus
Venezuelan equine encephalitis virus (except Strain TC-83)
Unclassified viruses

PARASITES

None

The primary hazards to personnel working with these agents are related to self-inoculation, ingestion, and exposure to infectious aerosols.

At Biosafety Level 3, more emphasis is placed on primary and secondary barriers to protect personnel in contiguous areas, the community, and the environment from exposure to potentially infectious aerosols.

All laboratory manipulations should be performed in BSCs or other enclosed equipment, such as a gas-tight aerosol generation chamber. Secondary barriers for this level include controlled access to the laboratory and ventilation requirements that minimize the release of infectious aerosols

Laboratories engaging in primary isolation and identification of HTLV or HIV may perform these activities in Containment Level 2 laboratories (physical conditions) using Containment Level 3 operational procedures and conditions. All research and production activities require Containment Level 3 physical and operational conditions.

Containment Level 3 is suitable for work with agents in Risk Group 3. The operational requirements for the Level 3 facility are substantially greater than those for Levels 1 and 2 and the facility staff must receive specific training in the safe handling and manipulation of the agents used in this facility.

The Containment Level 3 facility is designed to minimize environmental release of hazardous materials and provide enhanced worker protection. It must undergo annual performance testing and verification.

A Containment Level 3 facility requires specialized design and construction. Those responsible for biosafety in an institution should maintain close control and seek expert advice in, and remain in close communication throughout, all phases of design, construction, initial and annual performance testing and verification, and operation and maintenance.

8.16.1 Physical Requirements

The laboratory is separated from areas that are open to unrestricted traffic flow within the building. Access to the laboratory is restricted. The basic requirement for entry from access corridors is passage through a series of two self-closing doors. A clothes-changing room may be included in the passageway.

- Floors should be monolithic and slip-resistant. Consideration should be given to the use of coved floor coverings.
- Penetrations in floors, walls, and ceiling surfaces are sealed.
- Openings, such as those around ducts and spaces between doors and frames, are capable of being sealed to facilitate decontamination.
- Facility windows must be sealed and unbreakable.
- Doors are lockable.

Each laboratory room contains a sink for hand washing that is hands-free with foot or knee controls, or automatically operated, and is located near the room exit door. In addition, a body shower should be provided within the containment perimeter. The facility must have a pass-through or stand-alone autoclave located in the work zone. Where physical constraints preclude the installation of an autoclave in an existing Level 3 facility, alternative technologies may be used for sterilization of contaminated materials.

Facility furnishings should be kept to a minimum and backup power provided to critical items such as BSCs, fume hoods, and freezers.

Respiratory and face protection should be used when personnel are in rooms containing infected animals.

The biohazard sign posted on the entrance should include (in addition to those things listed above for Class 2) requirements for entering the laboratory and needed immunizations, respirators, or other personal protective measures.

8.16.1.1 BSCs

All manipulations of infectious materials are conducted in Class II or III BSCs. No work in open vessels is conducted on the open bench. All procedures involving the manipulation of infectious materials are conducted (Figure 8.3):

- Within BSCs. All open manipulations involving infectious materials are conducted in BSCs or other physical containment devices within the containment module.
- Using other physical containment devices (e.g., centrifuge safety cups or sealed rotors). Centrifugation must be conducted with closed containers using aerosol-proof safety heads or cups that are opened and loaded and unloaded in the BSC.
- By personnel wearing PPE.

Class III BSCs must be installed in a manner that does not interfere with the air balance of the cabinet or the room and should be directly connected to the exhaust system. HEPA-filtered exhaust air from a Class II BSC can be recirculated into the laboratory, if the cabinet is tested and certified at least annually.

When exhaust air from Class II safety cabinets is to be discharged to the outside through the building exhaust air system, the cabinets must be connected in a manner that avoids any interference with the air balance of the cabinets or the building's exhaust systems.

If connected to the supply system, BSCs must be connected in a manner that prevents positive pressurization of the cabinets and should be located away from doors, room supply louvers, and heavily traveled laboratory areas.

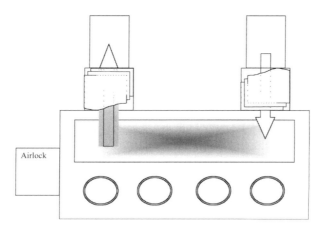

Figure 8.3   Biological safety cabinets (BSCs) for Biosafety Level 3.

8.16.1.2 Plumbing

Water supplied to the facility must be provided with reduced-pressure backflow preventers. All vent lines should be provided with HEPA filters or equivalent protection. Dunk tanks may be provided at the containment perimeter. Sink and floor drains should be piped separately to the main building drain and be appropriately labeled; floor drains are not generally recommended. Infectious materials must never be placed in sinks or floor drains.

Autoclave condensate drains should have closed connections and go directly to the sanitary sewer.

In care facilities for small animals, the disposal of wastes will not differ from other contaminated laboratory materials. Large animals producing large quantities of infectious wastes require special facilities that must be designed accordingly.

8.16.1.3 Vacuum System

Vacuum lines are protected with liquid disinfectant traps and HEPA filters, or their equivalent. Filters must be replaced as needed. No vacuum lines may exit through the containment perimeter. An alternative is to use portable vacuum pumps (also properly protected with traps and filters). Portable vacuum pumps must be fitted with in-line HEPA filters or equivalent equipment.

8.16.1.4 General Ventilation

Laboratory personnel must verify that the airflow direction (into the laboratory) is proper. A visual monitoring device that indicates and confirms directional inward airflow should be provided at the laboratory entry. An HVAC control system must be provided to ensure that the Level 3 facility does not become positively pressurized relative to the surrounding area. Audible alarms should be considered to notify personnel of HVAC system failure. The air supply system must be interlocked with the air exhaust system. The air supply and exhaust system should be equipped with manual dampers at the room perimeter, which may be closed as required to permit gas decontamination.

The facility is held at a negative pressure relative to the surrounding area at all times such that laboratory doors are kept closed when experiments are in progress, windows in the laboratory are closed and sealed, directional airflow is created, and air ingresses from both entry and exit areas. A ducted exhaust air ventilation system is provided into the laboratory from *clean* areas and toward *contaminated* areas.

The facility should be provided with a sealed dedicated air supply and exhaust system. The air discharged from the facility is prevented from recirculating back into the air supply system of the laboratory, the building, or adjacent buildings.The exhaust air is not recirculated to any other building area. The outside exhaust must be dispersed away from occupied areas and air intakes or HEPA-filtered. With a dedicated sealed exhaust system, air may be exhausted from the facility to the building exterior without HEPA filtration.The sealed exhaust system may draw part of the air from an adjacent Level 2 area, if this Level 2 area serves a support function for the Level 3 work. At the discharge point, the exhausted air must be dispersed away from air intakes or populated areas. When the air is not exhausted through a dedicated exhaust system, exhaust air must be HEPA filtered before discharge into the main building air exhaust system. The exhaust HEPA filter plenum must be designed to allow *in situ* decontamination and must pass annual testing and certification by aerosol challenge and scan testing techniques. When the supply air is not provided by a dedicated system, airtight back draught dampers or HEPA filters must be installed in the air supply system.

8.16.1.5 Centrifuge or Other Equipment Discharge

Continuous flow centrifuges or other equipment that may produce aerosols are contained in devices that exhaust air through HEPA filters before discharge into the laboratory. These HEPA systems are tested at least annually. Alternatively, the exhaust from such equipment may be vented to the outside if this exhaust is dispersed away from occupied areas and air intakes.

8.16.2 Operational Requirements

The following are the minimum operational requirements.

8.16.2.1 PPE

Staff are required to change into dedicated, solid-front laboratory clothing on entry to the facility. Protective laboratory clothing such as solid-front or wrap-around gowns, scrub suits, or coveralls are worn by workers when in the laboratory. Personal protective clothing may include head covers, and dedicated shoes or foot covers must be used while in the containment facility and removed on leaving. Reusable clothing is decontaminated before being laundered. Clothing is changed when overtly contaminated and protective clothing is not worn outside the laboratory. Gloves are worn when handling infectious or potentially infectious materials including animals or waste, or contaminated equipment. Gloves must be frequently changed and glove changes should be accompanied by hand washing. Disposable gloves are not reused. Appropriate respiratory protection should be considered depending on the infectious agents in use if infected animals are in the lab.

8.16.2.2 Decontamination

All potentially contaminated waste materials from laboratories are decontaminated before disposal or reuse. A decontamination method should be available in the facility and utilized, preferably within the laboratory (i.e., autoclave, chemical disinfection, incineration). Clean up is facilitated by using plastic-backed paper toweling on nonperforated work surfaces within BSCs. Effective disinfectants must be available at all times in the facility. Laboratory clothing must be removed upon work completion and autoclaved prior to laundering. Clothing is changed when overtly contaminated. Reusable clothing is decontaminated before being laundered. Body showering may be required depending on infectious agents used and manipulations involved.

8.16.2.3 Contaminated Material

Any contaminated material including wastes removed from the laboratory must be completely sealed in a leak-proof container. This material must not be transported into public access areas.

8.17 RISK GROUP 4: BIOSAFETY AND CONTAINMENT LEVEL 4

Level 4 is required for work with dangerous and exotic agents that pose a high individual risk of aerosol-transmitted laboratory infections and life- threatening disease. Agents with a close or identical antigenic relationship to Biosafety Level 4 agents are handled at this level until sufficient data are obtained either to confirm continued work at this level or to work with them at a lower level.

Biosafety Level 4 exposure usually produces very serious human or animal disease and has a high individual and community risk of causing life-threatening disease. All manipulations of potentially infectious diagnostic materials, isolates, and naturally or experimentally infected animals pose a high risk of exposure and infection to laboratory personnel, the community, and the environment. This level of agents includes dangerous and exotic biologicals and may be readily transmitted from one individual to another, or from animal to human or vice versa, directly or indirectly, or by casual contact; they also may be transmitted via the aerosol route. No available vaccine or therapy is currently available. Viruses such as Marburg or Congo-Crimean hemorrhagic fever are manipulated at Biosafety Level 4.

The primary hazards are respiratory exposure to infectious aerosols, mucous membrane or broken skin exposure to infectious droplets; and self-inoculation. The laboratory worker's complete isolation from aerosolized infectious materials is accomplished primarily by working in class III BSCs or by wearing a full-body, air-supplied positive-pressure personnel suit.

Examples of Level 4 agents are:

BACTERIA
None
FUNGI
None
VIRUSES
Lassa, Junin, Machupo, Sabia
Congo Crimean hemorrhagic fever virus
Marburg virus
Ebola virus
Tick-borne encephalitis complex including Russian Spring-
    Summer, Central European
Kyasanur forest virus
Omsk hemorrhagic fever virus
PARASITES
None

8.17.1 Physical Requirements

The Biosafety Level 4 facility consists of either a separate building or a clearly demarcated and isolated zone within a building and is generally a separate building or completely isolated zone with complex, specialized ventilation requirements and waste management systems to prevent release of viable agents. The Biosafety Level 4 laboratory has special engineering and design features to prevent microorganisms from being disseminated into the environment. Extraneous materials not related to the laboratory work are not permitted in the facility. Access to the facility is limited by means of secure, locked door(s) that are self-closing. Any windows are breakage-resistant and are sealed. Appropriate communication systems should be provided between the laboratory and the outside. At a minimum, an automatically starting emergency power source is provided for the exhaust system, life-support systems, alarms, lighting, entry and exit controls, and BSCs. Within facility work areas, all activities are confined to either cabinet or suit lab scenarios or a combination of both. Cabinet Labs use Class III BSCs. The facility rooms are arranged to ensure passage through a minimum of two doors prior to entering the room(s) containing the Class III BSC. Suit Labs use Class II BSCs used with one-piece positive pressure personnel suits ventilated by a life-support system. The suit pressure is maintained to be in excess of the surrounding laboratory air pressure.

Biosafety Level 4 facility design and operational procedures must be documented. A daily inspection of all containment parameters (e.g., directional airflow, chemical showers) and life-support systems must be completed before laboratory work is initiated. This inspection ensures that the laboratory is operating according to its operating parameters.

8.17.1.1 BSCs

The treated exhaust air from Class II BSCs, located in a facility where workers wear a positive pressure suit, may be discharged into the room environment or to the outside through the facility air exhaust system.The treated exhaust discharged to the outside through the facility exhaust system is connected so as to avoid any interference with cabinet air balance or facility exhaust systems. The supply systems of class III cabinets should prevent positive pressurization of the cabinet and should be directly connected to the exhaust system.

8.17.1.2 General

Internal facility appurtenances in the suit area, such as light fixtures, air ducts, and utility pipes, are arranged to minimize the horizontal surface area. Bench tops have seamless surfaces. Laboratory furniture is of simple open construction. Walls, floors, and ceilings of the cabinet room, suit area, chemical shower, inner change room, and airlocks are constructed to form a sealed internal shell to facilitate fumigation and to prohibit entry by insects, rodents, and other animals. Floors are integrally sealed and covered. The internal shell surfaces are resistant to liquids and chemicals to facilitate cleaning and decontamination of the area. All penetrations in these structures and surfaces are sealed. Openings around doors into the cabinet room, suit room, and inner change room are minimized and are capable of being sealed to facilitate decontamination.

8.17.1.3 Plumbing and Service Lines

Water fountains are automatically or foot operated and should be located in the facility corridors outside the laboratory. The water service to the fountain is isolated from the distribution system laboratory waters and equipped with a backflow preventer. A hands-free or automatically operated handwashing sink is provided in the suit area(s). Handwashing sinks in the outer and inner change rooms should be considered.

The process used for liquid waste decontamination must be validated physically and biologically. Liquid effluents from the dirty side inner change room (including toilets), sinks, floor drains, autoclave chambers, and other sources within the containment barrier are decontaminated by a proven method, preferably heat treatment, before being discharged to the sanitary sewer. Other liquid and gas services to the suit area are protected by devices that prevent backflow. Any drains in the cabinet or suit area contain traps filled with a chemical disinfectant of demonstrated efficacy against the target agent anare connected directly to the liquid waste decontamination system. Sewer vents and other service lines contain HEPA filters and protection against vermin intrusion.

8.17.1.4 Ventilation System

A dedicated nonrecirculating ventilation system is provided. The system supply and exhaust components are balanced to ensure directional airflow from the area of least hazard to the area(s) of greatest potential hazard. The airflow in the supply and exhaust components is monitored and the HVAC control system is designed to prevent sustained positive pressurization of the laboratory. The differential pressure/directional airflow between adjacent areas is monitored and alarmed to indicate any system malfunction. An appropriate visual pressure-monitoring device that indicates and confirms the pressure differential of the cabinet or suit room is located at the entry to the clean change room. Redundant supply fans are recommended, and redundant exhaust fans are required. Positioning of the supply and exhaust points is such that dead air space in the cabinet or suit room is minimized.

The supply air to and exhaust air from the cabinet room, suit room, inner change room, and anteroom pass through HEPA filter(s). The general room exhaust air from the suit area, decontamination showers, and decontamination airlock is treated by a passage through two HEPA filters in series prior to discharge to the outside. (*Note*: The exhaust HEPA filters' service life can be extended through adequate supply air prefiltration.) The air is discharged away from occupied spaces and air intakes. The HEPA filter(s) are located as near as practicable to the source in order to minimize the potentially contaminated ductwork.

The HEPA filter housing design should facilitate filter installation validation. The use of pre-certified HEPA filters could be an advantage. All HEPA filters must be tested and certified annually. The HEPA filter housings are designed to allow for *in situ* filter decontamination prior to removal or removal of the filter in a sealed, gas-tight primary container for subsequent decontamination and/or destruction by incineration.

8.17.1.5 Central Vacuum Systems

Central vacuum systems do not serve areas outside the suit area. In-line HEPA filters are placed as near as practicable to each use point or service cock. Filters are installed to permit in-place decontamination and replacement.

8.17.1.6 Decontamination Equipment

Double-door autoclaves are provided to decontaminate materials from class III BSC(s), cabinet room(s), and suit room(s). Autoclaves that open outside of the containment barrier must be sealed to the containment barrier wall. The autoclave doors are automatically controlled so that the outside door can only be opened after the autoclave *sterilization* cycle has been completed. Pass-through dunk tanks, fumigation chambers, or equivalent decontamination methods are provided for transfer of materials from room to room so that materials and equipment that cannot be decontaminated in the autoclave can be removed from both the class III BSC(s) and the cabinet room(s).

8.17.2 Operational Requirements

Laboratory personnel have specific and thorough training in handling extremely hazardous infectious agents and understand the primary and secondary containment functions of standard and special practices, containment equipment, and laboratory design characteristics. A specific facility operations manual is prepared or adopted. Only persons whose presence in the facility or individual laboratory rooms is required for program or support purposes are authorized to enter. The laboratory director has the final responsibility for assessing each circumstance and determining who may enter or work in the laboratory. Before entering, persons are advised of potential biohazards and instructed as to the safeguards. Authorized persons must comply with all instructions.

Persons who are immunocompromised or immunosuppressed may be at greater risk of acquiring infections. Therefore, persons who may be at increased risk of acquiring infection or for whom infection may be unusually hazardous, such as children or pregnant women, are not allowed in the laboratory or animal rooms. A logbook, signed by all personnel, indicates the date and time of each entry and exit. Practical and effective protocols for emergency situations are established.

8.17.2.1 Decontamination

Personnel enter and leave the laboratory only through the clothing-change and shower rooms. Entry is through an airlock fitted with airtight doors and inner and outer doors to the chemical shower and inner and outer doors to airlocks are interlocked to prevent both doors from being opened simultaneously. The rooms are arranged to ensure passage through the changing and decontamination areas prior to entering the room(s) where work is done with Risk Group 4 agents. Outer and inner change rooms separated by a shower are provided for personnel entering and exiting. A specially designed suit area is maintained in the facility to provide personnel protection equivalent to that provided by class III BSCs. Personnel who enter the facility wear a one-piece positive pressure suit that is ventilated by a life-support system protected by HEPA filtration and contains a life-support system that includes redundant breathing air compressors, alarms, and emergency backup breathing air tanks. For entry, personnel remove and store personal clothing in the outer clothing change room. Complete laboratory clothing, including undergarments, pants and shirts or jumpsuits, shoes, and gloves, are provided and used by all personnel entering the laboratory.

When exiting the laboratory and before proceeding into the shower area, personnel remove their laboratory clothing in the inner change room. Soiled clothing is autoclaved before laundering. Personnel take a decontaminating chemical shower before leaving the laboratory.

8.17.2.2 Medical Surveillance

The decision to establish a serologic surveillance program takes into account the method availability for the assessment of antibodies to the agent(s) of concern. The program provides for serum sample testing at each collection interval and the communication of results to the participants. Laboratory personnel receive available immunizations for the agents handled or potentially present. A system is established for reporting accidents, exposures, and employee absenteeism and medical surveillance of potential laboratory-associated illnesses. Written records are prepared and maintained.

(*Note*: An essential adjunct to such a reporting medical surveillance system is the availability of space in the facility for the quarantine, isolation, and medical care of personnel with potential or known laboratory-associated illnesses.)

8.17.2.3 Supplies and Materials

Supplies and materials needed in the facility are brought in by way of the double-door autoclave, fumigation chamber, or airlock, which is appropriately decontaminated between each use. After securing the outer doors, personnel within the facility retrieve the materials by opening the interior doors of the autoclave, fumigation chamber, or airlock. These doors are secured after materials are delivered.

Biological materials to be removed from the Class III BSC or from the Biosafety Level 4 laboratory in a viable or intact state are transferred to a nonbreakable, sealed primary container and enclosed in a nonbreakable, sealed secondary container. The material is removed from the facility through a disinfectant dunk tank, fumigation chamber, or an airlock designed for this purpose.

Materials (with the exception of biological materials that are to remain in a viable or intact state) are removed from the Biosafety Level 4 laboratory only after being autoclaved or decontaminated. Equipment or material that might be damaged by high temperatures or steam may be decontaminated by gaseous or vapor methods in an airlock or chamber designed for this purpose. Laboratory equipment is decontaminated routinely after work with infectious materials is finished, and especially after overt spills, splashes, or other contamination with infectious materials. Equipment is decontaminated before being sent for repair or maintenance.

8.18 DISPOSAL

All cultures, stocks, and other regulated wastes are decontaminated before disposal by an approved decontamination method such as autoclaving. Materials to be decontaminated outside of the immediate laboratory are placed in a durable, leak proof container and closed for transport and packaged in accordance with applicable regulations before removal from the facility.

If the material collected in the sharp waste container is contaminated with viable hazardous biological agents, the waste must be decontaminated, preferably by autoclaving, to inactivate the biological agents. Chemical disinfection of sharp waste is generally not recommended since this decontamination requires additional handling.

Before removal from a facility, hazardous chemical and radioactive solid wastes may require an additional procedure to inactivate viable biological agents that may be present. Autoclaving is generally not recommended in all situations involving such wastes, since the high temperature, steam, and pressure may contribute to potentially hazardous reactions.

8.19 DECONTAMINATION OF PATHOGENIC PRIONS IN MATERIALS AND WASTES: AN EMERGING ISSUE

8.19.1 Prion Disease Agents

Prions, or *infectious protein particles* lack nucleic acids and are unprecedented infectious agents that apparently cause a group of similar human and animal neurodegenerative diseases referred to as transmissible spongiform encephalopathies (TSEs) or transmissible degenerative encephalopathies (TDEs).

Creutzfeldt–Jakob disease (CJD) is the most well known of the neurodegenerative disorders in humans. The occurrence of CJD is generally sporadic, and a subset of the disease occurs in familial clusters, suggesting that CJD may be a genetic disease. TSEs may be considered transmissible, hereditary, or both.

Mad cow disease or bovine spongiform encephalopathy (BSE) affects cattle and certain other ruminants, and may have been passed to humans as a new variant of CJD (nvCJD). Several TSEs affect other animal species:

- Chronic wasting disease of elk and mule deer (CWD)
- Feline spongiform encephalopathy (FSE) of domestic cats and captive exotic felines
- Transmissible mink encephalopathy (TME)
- Scrapie, a disease of sheep and goats

The diseases in this group share several unusual characteristics including long latency times, without any characteristic clinical signs; no demonstrable immune or inflammatory response by the host; a disease process that includes conversion of normal cellular prion proteins (PrPᶜ) in the brain and other tissues of the central nervous system of the host to an abnormal, proteinase-resistant isoform (PRP^{Sc}); and relatively rapid progression of disease after onset of symptoms with an invariably fatal outcome

The etiologic agents responsible for TSEs have several unique properties including having no nucleic acids and being impossible to filter. They also cannot be cultured or observed at electron microscopy and exhibit high resistance to chemical and physical sterilization procedures that are effective against conventional microorganisms.

Some controversy remains as to the factors that cause the TSEs. None of the present concepts are explanatory for all aspects of these diseases in humans. According to prevailing theories, the infectious agent or *template* responsible for the conversion of PrP^c and the resulting degenerative disorders is the pathogenic form of the prion protein, or a component of the protein. A host-specific accessory protein may also accompany the agent.

Primary routes of transmission are by ingestion, inoculation, and implantation. Direct contact, droplet and airborne spread have not been implicated but cannot be ruled out.

According to the World Health Organization (WHO):

Although there have been no confirmed cases of occupational transmission of TSE to humans, cases of CJD in healthcare workers have been reported in which a link to occupational exposure is suggested. Therefore, it is prudent to take a precautionary approach. In the context of occupational exposure, the highest potential risk is from exposure to high infectivity tissues through needle-stick injuries with inoculation; however exposure to either high or low infectivity tissues through direct inoculation (e.g., needle-sticks, puncture wounds, sharps. injuries, or contamination of broken skin) must be avoided. Exposure by splashing of the mucous membranes (notably the conjunctiva) or unintentional ingestion may be considered a hypothetical risk and must also be avoided. Healthcare personnel who work with patients with confirmed or suspected TSEs, or with their high or low infectivity tissues, should be appropriately informed about the nature of the hazard, relevant safety procedures ...[39]

Existing information on the biosafety aspects of prions is not sufficient. Accordingly, recommendations issued by regulatory authorities tend to include highly conservative, precautionary measures aimed at theoretical risk.[19] These concerns are balanced against the concepts associated with risk prevalence. According to some studies "Prions are rare and thus do not constitute a major infection control risk."[7]

8.19.2 Inactivation Issues

Prions are notoriously resistant to most of the physical and chemical methods used for inactivation of pathogens that are conventional microorganisms. Surgical instruments, medical diagnostic laboratory equipment and other nondiscardable items that may be exposed to prion-contaminated materials present difficult problems. The treatment processes shown to be effective in inactivating these agents may damage or destroy these items rendering them unfit for reuse.

Prion-contaminated items should be segregated from all other wastes to assure that prion-contaminated items are treated by effective processes, and to minimize the amount of materials that must be subjected to these more difficult and potentially costly processes. Contact between materials or equipment used and infectious prions must be minimized. Where possible use disposable items that can be incinerated or treated by alkaline hydrolysis should be used.

8.19.3 Prion Detection

The lack of rapid and sensitive techniques for detection of prion contamination has made assessment of various inactivation methods difficult. Immunoassay techniques such as Western blotting lack sufficient sensitivity to detect the very low levels of PRP^{Sc} present in asymptomatic subjects and some contaminated materials that have been incompletely inactivated. These very low levels are currently detected by animal bioassay methods usually involving intracerebral injection of a dilution series of the specimen material into small laboratory animals such as hamsters. The animals must then be maintained and observed for symptoms of TSE, which may take 10 months or more to develop. Clinically positive animals are then tested by Western blot to verify disease transmission. These procedures require prolonged incubation times, use of large numbers of animals, and are costly to perform.[2,19] Recent advances have been made in animal bioassay and immunoassay techniques that will improve the capability to detect low levels of prion contamination.[19] Cyclic amplification of protein misfolding, a process conceptually analogous to the polymerase chain reaction (PCR) cycling may provide a sensitive method to detect the presence of currently undetectable prion agents in tissues and biological fluids.[8] Scrapie is most commonly used in prion inactivation studies because it is considered a nonhuman pathogen and is easily maintained in mice, hamsters, and other small laboratory animals. In studies of the efficacy of hyperbaric rendering procedures in inactivating prions, the BSE agent appeared to be more resistant to heat inactivation procedures than the scrapie agent, particularly at lower temperatures and shorter treatment times.

8.19.4 Reference Treatments

Because TSEs are insidious diseases of low incidence and long latency, cases may be unrecognized. This increases the potential for contamination of instruments and medical supplies by contact with tissues from undiagnosed patients. Reference treatments recommended by the World Health Organization include soaking in $1N$ sodium hydroxide (1 hour at 20°C), soaking in 12.5% bleach (1 hour at 20°C), and autoclaving (18 minutes at 134–138°C). Where possible, two or more different methods of inactivation should be combined in any sterilization procedure for these agents.

8.19.4.1 Inactivation by Autoclaving

Treatment of scrapie-infected brain tissue with formaldehyde before autoclaving stabilized infectivity. Treatment after autoclaving was either ineffective or further reduced infectivity. The presence of elemental carbon protects infectivity during autoclaving. Infectivity survives autoclaving at 132–138°C and the effectiveness of the process may actually decline at higher temperatures. Incomplete inactivation by autoclaving may result in the selection for or formation of subpopulations of prions that have biological characteristics different from the main population and cannot be inactivated by re-autoclaving. Stabilization of thermostable subpopulations of prions may occur through the smearing and drying of infected tissue on surfaces that can occur prior to autoclaving.[14] Combining sodium hydroxide treatment with autoclaving is extremely effective even at 121°C.[11]

8.19.4.2 Inactivation by Chemical Sterilants

Chemical inactivation methods are often favored for large or fixed surfaces and equipment; however, prions are extremely resistant to most commonly used disinfectants.[19]

Studies in the review by Taylor (1999) have shown that prions can remain infective after 24 hours of treatment with 1 M sodium hydroxide. Infectivity can survive exposure to 2 M sodium hydroxide for up to 2 hours.[11] The following chemical methods of decontamination reviewed by WHO (1999) and Taylor (2000) were found to be generally ineffective:

- Acids, except for formic acid
- Alcohol
- Alkylating agents — formalin, formaldehyde, acetylethleneimine, β-propiolactone, ethylene oxide
- Detergents — sodium dodecyl sulphate
- Guanidine hydrochloride
- Iodine 2% solution

- Phenolics
- Organic solvents — chloroform, heptane, hexane, perchloroethylene, petroleum
- Oxidizers — hydrogen peroxide 3%, peracetic acid 18%
- Proteolytic enzymes — trypsin, some proteases

Those rated as variably or partially effective are:

- Chlorine dioxide
- Glutaraldehyde
- Urea
- Iodophores
- Sodium dichloroisocyanurate
- Sodium metaperiodiate

Those considered effective include:

- Strong sodium hydroxide solutions (2 M) for at least 2 hours
- Strong sodium hypochlorite solutions
- Some proteases — proteinase K, pronase with prolonged digestion

Glycidol may be a potential prion disinfectant.[16]

Prions are more easily inactivated under alkaline conditions. Higher pH and increasing concentrations of urea or guanidine thiocyanate decrease the protease resistance of prion proteins.[22] Disinfectants such as ethanol and formalin that act by denaturing proteins are ineffective and probably enhance the resistance of prions to thermal inactivation.[11] Phenolic disinfectants are not effective[11] and soaking in aldehydes is generally ineffective.[28] Exposure to hypochlorite solutions containing 20,000 ppm of available chlorine is believed to be a reliable inactivating procedure.[11,38] Guanidine thiocyanate (GdnSCN) solutions have been found to be highly effective in inactivating prions, even in complex materials such as whole brain tissue.[23]

The loss of infectivity resulting from treatment of scrapie with guanidine thiocyanate (GdnSCN) solutions may be restored upon dilution of the reagent and restoration may be enhanced by the presence of copper.[5]

The inclusion of a formic acid step in routine formaldehyde fixation of prion-infected tissues provides histological sections of excellent quality and virtually eliminates the risk of handling infectious material.[30,31]

Organic solvents are generally not effective as inactivation agents. In one study, treatment with hot solvents including hexane, heptane, petroleum spirits, or perchloroethylene used in past meat rendering processes produced only slight inactivation of BSE and scrapie agents.[15]

8.19.4.3 Inactivation by Dry Heating

Thermal inactivation is most efficient with steam under pressure. Under dry conditions, TSEs exhibit extraordinary resistance to thermal inactivation. Specimens of scrapie-infected brain tissue subjected to dry heat at 360°C for 1 hour still retain a small amount of infectivity. Limited infectivity has also been found in ash from similar tissues heated to 600°C, approaching the operating temperature of medical waste incinerators. A thermally resilient inorganic template of replication has been suggested to explain this pattern of infectivity.[2,29]

8.19.4.4 Inactivation by Alkaline Hydrolysis

Treatment of prion-contaminated materials with 2 M sodium hydroxide leads to substantial but incomplete inactivation.[11] Systems employing heating and alkaline hydrolysis that are suitable for treatment and disposal of animal carcasses and other solid wastes contaminated with prions are now available commercially. These systems are scalable and may be used to treat carcasses ranging in size from small laboratory animals to livestock.[27] By treating all animal carcasses and potentially infected tissues by alkaline hydrolysis the need for segregated wastes known or suspected to be contaminated with prions is eliminated. A disadvantage of alkaline hydrolysis is that treated wastes may be subject to additional management requirements. Treated wastes that meet the Department of Transportation (DOT) definition of corrosive material (Class 8) must comply with hazardous material shipping requirements.[34] Wastes containing free liquids with a pH ≥ 12.5 are regulated as unlisted hazardous wastes (corrosivity characteristic) under EPA regulations.[33] Such wastes may not be acceptable to medical waste disposal facilities, which are usually not permitted to accept hazardous waste. Effluents from alkaline hydrolysis may exceed local discharge limits for pH, biological oxygen demand (BOD), chemical oxygen demand (COD), and other criteria. This may preclude disposal of process effluents and residues to the sanitary sewer. Alternative disposal methods such as drying and landfilling may be required.

8.19.5 Biosafety Classification

Human and BSE prions, and those propagated in apes and monkeys, are manipulated at Biosafety Level 2 or 3, depending on the studies being conducted. All other animal prions are considered Biosafety Level 2.[26]

8.19.6 Decontamination — WHO Policy Guidelines

The WHO *Infection Control Guidelines for Transmittable Spongiform Encephalopathies* advised policy makers to be guided by the tissue infectivity level and instrument reuse issues for instruments potentially contaminated by the infectious agent. Thus, instruments used on a person with a known TSE that are subsequently slated for reuse in the CNS or spinal column of another individual would require the most stringent decontamination methods.

While appropriate disposal of all used instruments would be preferable to reuse, economic considerations dictate that instruments will be reused. The best decontamination methods feasible should be applied to all disposable instruments and to materials slated for reuse. The practice of reusing instruments designed for single use only should not be allowed. All instruments should be protected as much as possible from contact with infected tissues.

According to the WHO guidelines:

> Procedures that employ heat and NaOH (either consecutively or simultaneously) appear to be sterilizing under worst-case conditions (e.g., infected brain tissue partly dried onto surfaces). Moreover, hot alkaline hydrolysis reduces biological macromolecules to their constituent sub-units, thereby cleaning as well as inactivating.

After removal of adherent material and decontamination of exterior surfaces, instruments that can be should be disassembled after decontamination of exterior surfaces. The disassembled parts may be subjected to various decontamination methods as determined by their material's resistance to damage during decontamination. Parts that can be autoclaved or treated with sodium hydroxide or bleach should be treated with the agent of choice. Manufacturer's information and consultation should be obtained if the decontamination methods are in doubt as to either efficacy or equipment degradation effect. Automated washers or other decontaminating vessels that consolidate instruments for cleaning must not be used for prion-contaminated equipment.

8.19.7 Sterilization — Medical Instruments

Since prions are not inactivated by the usual methods of sterilization, surgical procedures may account for some cases of CJD.[32] Incidences of iatrogenic transmission of TSEs via contaminated medical devices have been reported. (See review by Antloga et al.[1,24]) Surgical instruments sterilized by conventional methods have transmitted CJD. Zoebeley et al. (1999) studied the infectivity of stainless steel surfaces exposed to scrapie mouse brain homogenate. Their studies confirmed that the agent is not removed by extensive washing even with follow-up treatment with 10% formaldehyde solution.[18]

To reduce the potential for transmission of CJD and other TSEs via medical instruments, the effectiveness of cleaning and sterilization procedures must be firmly established. For decontamination of areas that are suspected to be contaminated, such as tables and Stryker saws in autopsy rooms, repeated wetting or wiping with $2 N$ sodium hydroxide is recommended. Autopsy instruments may also be soaked in a 5% sodium hypochlorite solution with at least a 20,000 ppm free chlorine residual for 2 hours, or 96% formic acid. These alternatives may be corrosive to stainless steel.[19]

Three methods have been used to reduce contamination of medical instruments. All have disadvantages. Exposure to household bleach (may corrode fine instruments, mechanical parts, stainless steel. Prolonged autoclaving is not feasible for large instruments or surfaces. Immersion in concentrated sodium hydroxide solutions is corrosive to instruments and may cause severe burns.[23] Guanidine thiocyanate solutions may be usable as an alternative reagent for surgical instruments.[23]

Recent results with a peracetic acid-based sterilant indicate that it may be a safe and effective means of prion inactivation on medical devices.[1]

8.19.8 Reuse Policies

Despite the cost savings afforded by reuse of medical devices, some facilities have discontinued reuse because of concern about transmission of TSEs. Policies to eliminate reuse of such devices should be developed considering the use of the device and what types of tissues contacted.

On the basis of the available scientific data, only critical (e.g., surgical instruments) and semi-critical devices contaminated with high-risk tissue such as brain, spinal cord, and eye tissue from high-risk patients those with known or suspected infection with CJD require special treatment.[7] A very low risk of transmission is associated with blood or blood products. Therefore, devices such as angioplasty catheters may potentially be reused, provided that stringent methods of cleaning and sterilization are used. Previous assumptions concerning the risk potential of various tissues may not be correct in animal systems. Mouse skeletal muscle can propagate prions and accumulate substantial titers of these pathogens.[7,20]

For certain devices that have the potential for iatrogenic transmission of CJD effective inactivation methods such as prolonged autoclaving at 121°C may be impossible or impractical, i.e., the tonometer tips used in optometry.

In these situations alternative procedures such as noncontact tonometry or use of disposable tip covers should be considered.[25]

Surgical instruments that are going to be re-used may be mechanically cleaned in advance of subjecting them to decontamination. Mechanical cleaning will reduce the bio-load and protect the instrument from damage caused by adherent tissues. If instruments are cleaned before decontamination, the cleaning materials must be treated as infectious waste, and the cleaning station must be decontaminated. The instruments are then treated by one of the decontamination methods before reintroduction into the general instrument sterilization processes. (*Note*: Some scientist believe that instruments should be decontaminated before mechanical cleaning, and then handled as per general instrument sterilization processes.)

Single-use instruments should be used in patients with known or suspected CJD. The United Kingdom Department of Health now requires use of such instruments in all tonsillectomies and other procedures presenting risk of iatrogenic transmission.[21]

Any endoscopic accessory used in a CJD patient should be treated as highly infectious and should be incinerated. Manufacturers have indicated that repair of endoscopes contaminated with CJD agents is not possible at present.[35]

A recent survey of dental practices in the United Kingdom found that most of the dental practitioners surveyed did not actively seek to identify patients at risk of prion diseases, and in many cases recommended practices for providing safe dental care for such patients were not in place.[36]

8.19.9 Decontamination — Work Surfaces

TSE infectivity can remain over long periods of time on surfaces, and the following precautions should be used whenever feasible:

- Use disposable cover sheets. (*Note*: According to WHO *Infection Control Guidelines for Transmittable Spongiform Encepalopathies* "transmission to humans has never been recognized to have occurred from environmental exposure.")
- Mechanically clean and disinfect equipment and surfaces to prevent build up of potentially contaminated materials.

According to WHO *Infection Control Guidelines for Transmittable Spongiform Encephalopathies,* by using specific decontamination routines, TSE contaminated by surfaces can be disinfected. The WHO recommendation is to flood the surface with NaOH or sodium hypochlorite, let the solution remain in place for 1 hour, and rinse the surface with water.

All work areas should be selected to contain contaminants prior to required decontamination efforts. Work surfaces should be protected from contamination as feasible by the use of enamel vessels, heat-stable plastic containers, or disposable trays. The enamel and heat-stable plastics should be purchased to withstand the required decontamination if reuse is deemed necessary.

8.19.10 PPE

The PPE requirements apply to patient care, instrument decontamination, and transfer of contaminated tissues. When TSE exposure potential exists, personnel should wear impermeable and disposable (single-use) protective outer clothing to include gloves and eye protection. Splash protection is also needed when splash of body fluids may occur, such as during spill events. Appropriate respiratory protection is also needed and should be determined on a case by case basis.

All disposable PPE should be secured in leak-proof containers for ultimate disposal by incineration. The sides, tops, and bottoms of all carriers should always be kept clean, with surface disinfection used should any 'to be disposed' material inadvertently touch the disposal container exteriors.

8.19.11 Waste Disposal

Infectious waste consists of disposable materials that have been in contact with blood and other potentially infectious materials (OPIM). Patient care items such as bandages, dressing, swabs, diapers, pads, and bedding are all potential repositories of these infectious materials. In addition, cultures and tissues used for diagnostic purposes and their containers may be infectious. Hard-surface materials such as bed pans, urine bottles, syringe needles, intravenous or catheter lines, scalpels, drainage tubes, and other routinely used medical equipment may contain infectious materials.

TSE infectious healthcare waste is the term used to describe infective tissues from persons with confirmed or suspected TSEs. These people may have active infections or may have received high infectivity tissue or tissue products (cornea, dura matter, or human growth hormone) from persons now known to be infected with TSE.

All TSE infectious waste contaminated by high infectivity tissues or cerebral spinal fluid (CSF) must be collected in impermeable, leak-resistant plastic bags or containers. These containers must be labeled *Biohazard* and incinerated.

8.19.12 Decontamination — Wastes and Waste-Contaminated Materials

Waste liquid and solid residues should be decontaminated with the same care and precautions recommended for any other exposure to TSE agents. All waste liquids and solids must be captured and treated as infectious waste.Liquids used for cleaning should be decontaminated *in situ* by the addition of NaOH or hypochlorite or equivalent procedures, and may then be disposed of as routine hospital waste. Absorbents, such as sawdust, may be used to stabilize liquids that will be transported to an incinerator; however, they should be added after decontamination. Cleaning tools and methods should be selected to minimize dispersal of the contaminant by splashing, splatters, or aerosols. Great care is required in the use of brushes and scouring tools. Where possible, cleaning tools such as brushes, toweling, and scouring pads, as well as tools used for disassembling contaminated apparatus, should either be disposable or selected for their ability to withstand the disinfection procedures.

Upon completion of the cleaning procedure, all solid wastes including disposable cleaning materials should be collected and decontaminated. Incineration is highly recommended. The cleaning station itself should then be decontaminated using one of the methods. Automated cleaning equipment must not be used for any instrument or material that has not previously been thoroughly decontaminated.

8.19.13 Recommended Practices — Waste Minimization

Tissues and fomites that are known or suspected to be contaminated with prions should be segregated from all other wastes to minimize the amount of waste that must be treated with the more rigorous procedures required for inactivation and disposal.

8.19.14 Recommended Practices — Disposal of Wastes

Contaminated solid wastes such as table covers, instrument pads, and disposable clothing should be double-bagged and incinerated.[19] As a routine procedure, hydroxide-treated wastes should be treated in a gravity-displacement autoclave in sealed, heat-resistant containers that can stand the pressures involved. This practice will make the procedure safer for the operator and reduce the potential for spills and autoclave damage from the hydroxide solution. Alternatively, hydroxide-treated wastes may be processed in a porous-load autoclave, provided that the containers used can withstand the higher temperatures (134–138°C), and the initial and final high vacuum stages that are features of this process.[13]

Some facilities now pre-treat wastes or the ash from incineration with sodium hydroxide to ensure inactivation of the agent.

8.19.15 Spills

Spills of potentially TSE infectious materials in a hospital ward should be removed using absorbent material and the surface disinfected. Absorbent material can be contained and incinerated or decontaminate. Secure leak-proof containers, e.g., double bagging, should be used for the safe handling of clinical waste and external contamination of the waste container should be avoided.

BIBLIOGRAPHY

Antloga, K., Meszaros, J., Malchesky, P.S., and McDonnell, G.E., Prion disease and medical devices, *ASAIO J.*, 46(6), S69–S72, 2000.

ASHRAE, *HVAC Applications Handbook 1999* (1-P) American Society of Heating, Refrigerating, and Air-Conditioning Engineers, Atlanta, 1999.

Bagg, J., Sweeney, C.P., Roy, K.M., Sharp, T., and Smith, A., Cross infection control measures and the treatment of patients at risk of Creutzfeldt–Jakob disease in U.K. general dental practice, *Br. Dent. J.*, 191(2), 87–90, 2001.

Baron, H. and Prusiner, S.B., Prion diseases, in *Biological Safety Principals and Practices*, 3rd ed., D.O. Fleming, D.O. and Hunt, D.L., Eds., ASM Press, Washington, D.C., 2000, pp. 187–208.

Bosque, P.J., Ryou, C., Telling, G., Peretz, D., Legname, G., DeArmond, S.J., and Stanley, B. Prions in skeletal muscle, *Proc. Natl. Acad. Sci. USA*, 99(6). 3812–3817, 2002.

Brown, P., Liberski, P.P., Wolff, A., and Gajdusek, D.C., Resistance of scrapie infectivity to steam autoclaving after formaldehyde fixation and limited survival after ashing at 360 degrees C: practical and theoretical implications, *J. Infect. Dis.*, 161(3), 467–472, 1990.

Brown, P., Wolff, A., and Gajdusek, D.C., A simple and effective method for inactivating virus infectivity in formalin-fixed tissue samples from patients with Creutzfeldt–Jacob disease, *Neurology*, 40(6), 887–890, 1990.

Brown, P., Rau, E.H., Johnson, B.K., Bacote, A.E., Gibbs, C.J., Jr., and Gajdusek, D.C., New studies on the heat resistance of hamster-adapted scrapie agent: threshold survival after ashing at 600 degrees C suggests an inorganic template of replication, *Proc. Natl. Acad. Sci. USA*, 97(7), 3418–3421, 2000.

Caspi, S., Halimi, M., Yanai, A., Sasson, S.B., Taraboulos, A., and Gabizon, R., The anti-prion activity of Congo red: putative mechanism, *J. Biol. Chem.*, 273(6), 3484–3489, 1998.

Code of Federal Regulations, Title 40, Subpart C, Part 261.22: Characteristic of corrosivity.

Code of Federal Regulations, Title 49, Subchapter C, Part 173.136: Class 8 definitions.

Collins, S., Law, M.G., Fletcher, A., Boyd, A., Kaldor, J., and Masters, C.L., Surgical treatment and risk of sporadic Creutzfeldt–Jacob disease: a case-control study, *Lancet*, 353, 693–697, 1999.

Darbord, J.C., Inactivation of prions in daily medical practice, *Biomed. Pharmacother.*, 53(01), 34–38, 1999.

DHHS, Biosafety in Microbiological and Biomedical Laboratories, 4th ed., HHS Publication No. (CDC) (99-xxxx), U.S. Department of Health and Human Services, Public Health Service, Centers for Disease Control and Prevention and National Institutes of Health, U.S. Government Printing Office, Washington, D.C., 1999.

Fagih, B. and Eisenberg, M.J., Reuse of angioplasty catheters and risk of Creutzfeldt–Jakob disease, *Am. Heart J.*, 137(6), 1173–1178, 1999.

Gale, P., Young, C., Stanfield, G., and Oakes, D., Development of a risk assessment for BSE in the aquatic environment, *J. Appl. Microbiol.*, 84(4), 467–477, 1998.

Laurenson, I.F., Whyte, A.S., and Fox, C., Iatrogenic prion infection, *N. Engl. J. Med.*, 345(11), 840–841, 2001.

Madec, J.Y., Vanier, A., Dorier, A., Bernillon, J., Belli, P., and Baron, T., Biochemical properties of protease resistant prion protein PrPsc in natural sheep scrapie, *Arch. Virol.*, 142(8), 1603–1612, 1997.

Manuelidis, L., Decontamination of Creutzfeldt–Jakob disease and other transmissible agents, *J. Neurovirol.*, 3(1), 62–65, 1997.

McKenzie, D., Bartz, J., Mirwald, J., Olander, D., Marsh, R., and Aiken, J., Reversibility of scrapie inactivation is enhanced by copper, *J. Biol. Chem.*, 273(40), 25545–25547, 1998.

MRC, Guidelines for the Handling of Recombinant DNA Molecules and Animal Viruses and Cells in 1977, The Medical Research Council of Canada (1977; rev. 1979, 1980)

Patterson, P., Incident may have exposed patients to CJD in surgery, *OR Manager*, 16(12), 1, 7–8, 2000.

Paul, J., Prion, from crazy cows to iatrogenic Creutzfeldt–Jakob disease: which risk in laboratory or in hospital?, *Pathol. Biol. (Paris)*, 43(2), 114–120, 1995.

Richmond, J.Y. and Mckinney, R.W., Eds., Biosafety in Microbiological and Biomedical Laboratories, 4th ed., Centers for Disease Control, National Institutes of Health, U.S. Government Printing Office, Washington, D.C.(1999)

Rey, J.F., Endoscopic disinfection: a worldwide problem, *J. Clin. Gastroenterol.*, 28(4), 291–297, 1999.

Rutala, W.A. and Weber, D.J., Creutzfeldt–Jakob disease: recommendations for disinfection and sterilization, *Clin. Infect. Dis.*, 32(9), 1348–1356, 2001.

Saborio, G.P., Permanne, B., and Soto, C., Sensitive detection of pathological prion protein by cyclic amplification of protein misfolding, *Nature*, 14(411), 810–813, 2001.

Schreuder, B.E., Geertsma, R.E., van Keulen, L.J., van Asten, J.A., Enthoven, P., Oberthur, R.C., de Koeijer, A.A., and Osterhaus, A.D., Studies on the efficacy of hyperbaric rendering procedures in inactivating bovine spongiform encephalopathy (BSE) and scrapie agents, *Vet. Rec.*, 142(18), 474–480, 1998.

Taylor, D.M., Brown, J.M., Fernie, K., and McConnell, I., The effect of formic acid on BSE and scrapie infectivity in fixed and unfixed brain tissue, *Vet. Microbiol.*, 58, 167–174, 1997.

Taylor, D.M., Impaired thermal inactivation of ME7 scrapie agent in the presence of carbon, *Vet. Microbiol.*, 27, 403–405, 1991.

Taylor, D.M., Inactivation of prions by physical and chemical means, *J. Hosp. Infect.*, 43(suppl.), S69–S76, 1999.

Taylor, D.M., Inactivation of transmissible degenerative encephalopathy agents: a review, *Vet. J.*, 159(1), 10–17, 2000.

Taylor, D.M., Fernie, K., and McConnell, I., Inactivation of the 22A strain of scrapie agent by autoclaving in sodium hydroxide, *Vet. Microbiol.*, 58(2–4), 87–91, 1997.

Taylor, D.M., Fernie, K., McConnell, I., and Steele, P.J., Observations on thermostable subpopulations of the unconventional agents that cause transmissible degenerative encephalopathies, *Vet. Microbiol.*, 64(1), 33–38, 1998.

Taylor, D.M., Fernie, K., McConnell, I., Ferguson, C.E., and Steele, P.J. Solvent extraction as an adjunct to rendering: the effect on BSE and scrapie agents of hot solvents followed by dry heat and steam, *Vet. Rec.*, 143(1), 6–9, 1998.

Timberlin, R.H., Walker, C.A., Millison, G.C. et al., Disinfect ion studies with two strains of mouse pass aged scrapie agent, *J. Neurol. Sci.*, 59, 355–369, 1983.

UTBSC, *Guidelines for the Handling of Recombinant DNA Molecules, Animal Viruses and Cells, Microorganisms, and Parasites*, The University of Toronto Biosafety Committee, Toronto, 1979.

UTBSC, *Laboratory Biosafety Guidelines*, 2nd ed., The University of Toronto Biosafety Committee, Toronto, 1979,

UTBSC, *University of Toronto, Biosafety Policies and Procedures Manual*, The University of Toronto Biosafety Committee, Toronto, 2000.

Walia, J.S. and Chronister, C.L., Possible iatrogenic transmission of Creutzfeldt–Jakob disease via tonometer tips: a review of the literature, *Optometry*, 72(10), 649–652, 2001.

Waste Reduction, Inc., *Waste Reduction*, http://www.wr2.net/, accessed March 11, 2002.

WHO, WHO Infection Control Guidelines for Transmissible Spongiform Encephalopathies, Report of a World Health Organization Consultation, March 23–26, Geneva, Switzerland (available at http://www.who.int/emc-documents/tse/whocdscsraph2003c.html).(1999)

Yamamoto, M., Horiuchi, M., Ishiguro, N., Shinagawa, M., Matsuo, T., and Kaneko, K., Glycidol degrades scrapie mouse prion protein, *J. Vet. Med. Sci.*, 63(9), 983–990, 2001.

Zobley, E., Flechsig, E., Cozzio, A., Enari, M., and Weissmann, C., Infectivity of scrapie prions bound to a stainless steel surface, *Mol. Med.*, 5(4), 240–3, 1999.

CHAPTER 9

Medical Setting Infection Control

Renee Dufault, Rita Smith, and Martha J. Boss

CONTENTS

1-56670-606-8/03/$0.00+$1.50
© 2003 by CRC Press LLC

Infection prevention in the hospital environment is one of the goals for all healthcare workers. The issues discussed in this chapter are in addition to those discussed in the preceding chapter and address the specifics of hospital, dental office, and medical clinic infection control practices.

9.1 NOSOCOMIAL INFECTIONS

Unfortunately, it is not uncommon these days for an individual to be admitted to a hospital for a simple procedure and then die of a hospital-acquired, or nosocomial, infection. Approximately 5 to 10% of all patients acquire an infection while in the hospital. Such nosocomial infections occur most frequently in patients whose immune systems are weak or weakened because of age, underlying diseases, or medical or surgical treatments. These patients are known as immunocompromised and are susceptible to infectious disease agents. Examples of immunocompromised patients include infants, the elderly, organ transplant recipients, patients infected with human immunodeficiency virus (HIV), and those patients undergoing cancer therapy. It is important to note that nosocomial infections can affect patients no matter where they stay in the hospital.

A nosocomial infection may prolong the length of stay for the patient and increase the costs of hospitalization if the patient survives. An article appearing in the 1998 *Emerging Infectious Diseases Journal*, published by the Centers for Disease Control and Prevention (CDC), estimated that, in 1995 alone, nosocomial infections cost $4.4 billion and contributed to more than 88,000 deaths — one death every 6 minutes (Weinstein, 1998; Cooper, 2002). This estimate was derived from data collected by the CDC during a 10-year study in the 1970s to determine the nationwide nosocomial infection rate (Haley, 2002). The data were collected from a random sample of patient medical records and used to estimate the nosocomial infection rate among 6449 U.S. hospitals from 1975 to 1976. At that time, the nosocomial infection rate was estimated to be 5.7 infections per 100 admissions, or 2 million infections per year (Haley et al., 1985a). In 2000, there were approximately 5810 hospitals in the United States (AHA, 2002). The nationwide nosocomial infection rate is not known today and will not be known, because no accurate or mandatory tracking system is in place in the United States to monitor the number of hospital-acquired infections or deaths caused by them.

9.1.1   National Nosocomial Infections Surveillance System

A voluntary reporting system is in place at the CDC to monitor certain types of infections acquired by patients in hospitals. This system is known as the National Nosocomial Infections

Surveillance System (NNIS). The NNIS Began in 1970 with 62 participating hospitals in 31 states (CDC, 2000). As of 2000, approximately 315 hospitals were participating in the NNIS (CDC, 2002a). The NNIS currently receives data on certain types of nosocomial infections from approximately 5.4% of the nation's hospitals and is not accepting new applications for membership at this time (CDC, 2002a). The NNIS requires each participating hospital, or member, to submit data on specific report forms (CDC, 2002b). Blank report forms can be downloaded from the CDC website and are used by members to collect data on site-specific infections acquired by patients in intensive-care hospital units or who have undergone certain surgical procedures. It bears repeating that although nosocomial infections can affect patients anywhere in the hospital, the NNIS only tracks data on selected infections, such as those acquired by patients in hospital intensive-care units or who have undergone certain surgical procedures (CDC, 2002b). The NNIS does not include long-term care facilities, such as rehabilitation, mental health, and nursing homes (CDC, 2000).

9.1.2 Hospital Management Changes

Meanwhile, the hospital industry is changing, particularly due to managed-care organizations and the aging population, both of which have grown explosively. With the shift of surgical care to outpatient surgical centers, the number of hospitals decreased from 7126 in 1975 to 5810 hospitals in 2000. Hospitals have become fewer and smaller, but the patient population has become more severely ill and immunocompromised and therefore more susceptible to nosocomial infections (Jarvis, 2001). With only the sickest patients being admitted, hospitals are becoming more like large intensive care units (Weinstein, 1998). It has long been recognized that stronger infection surveillance and prevention programs are needed in hospitals to curb nosocomial infection rates.

9.1.3 Study of the Efficacy of Nosocomial Infection Control

In 1985, the CDC published the results of the Study of the Efficacy of Nosocomial Infection Control (SENIC). During this landmark study, researchers evaluated many hospital infection control programs and found that hospitals with the lowest nosocomial infection rates had strong surveillance and prevention programs (Gaynes et al., 2001). Such programs included the following elements: organized surveillance and control activities, a trained and effective infection control physician, one infection control nurse per 250 beds, and a system for reporting infection rates to practicing surgeons (Haley et al., 1985b).

9.2 INFECTION PREVENTION

Transmission of infection within a hospital requires three elements:

1. Infectious microorganism or source of infectious agent
2. Susceptible person who can serve as a host
3. Means of transmission for the microorganism to the susceptible host.

Susceptibility to infection by microorganisms varies greatly from person to person. Some persons may be immune to infection or may be able to resist colonization by an infectious agent; others exposed to the same agent may become asymptomatic carriers of the microorganism, and still others may develop clinical disease (Garner, 1996). Sources of the infecting microorganisms in hospitals may include patients, visitors, and healthcare or ancillary personnel; these people may have symptoms of disease, may be in the incubation period of disease, may be colonized by an infectious agent but have no apparent disease, or may be chronic carriers of infectious agents (Garner, 1996).

Inanimate environmental objects can become contaminated by dirty hands. Examples of such objects include telephones, call lights, door knobs, tabletops, bed rails, toilet seats, sinks, and so on. Infectious microorganisms can be found on most surfaces. Researchers recently checked contamination levels in soft toys at six different doctors' waiting rooms and found that 90% of the toys had moderate to heavy bacterial contamination (Anon., 2002). The published findings in the *British Journal of General Practice* (cited in Anon., 2002) warned that such contamination may actually spread infections to already sick children. In yet a different study, researchers checked contamination levels on 36 pagers belonging to doctors at a large urban teaching hospital and found that 50% of them carried at least one disease-causing microorganism (Anon., 2001).

9.2.1 Hand Washing

The single most important means for preventing the transmission of harmful microorganisms to susceptible persons is hand washing. Washing hands as promptly and thoroughly as possible between patient contact and after contact with blood, body fluids, secretions, excretions, and equipment or articles contaminated by them reduces the rate of nosocomial infection (Garner, 1996). Proper hand washing is a proven strategy for preventing the spread of infections.

9.2.2 Isolation Precautions

A variety of infection control and prevention measures are used for reducing the risk of microorganism transmission in the hospital setting. These measures are known as *Isolation Precautions* and were developed by the CDC for use in hospitals. Isolation Precautions (Garner, 1996) include the use of:

- *Standard Precautions*, which are designed for the care of all patients in hospitals, regardless of their diagnosis or presumed infection status
- *Transmission-Based Precautions*, which are added precautions to be used with Standard Precautions for patients known or suspected to be infected with highly transmissible microorganisms spread by airborne or droplet transmission or by contact with dry skin or contaminated surfaces

The three types of Transmission-Based Precautions are:

- Airborne
- Droplet
- Contact

9.2.3 Standard Precautions

The use of Standard Precautions by healthcare and ancillary personnel is the primary method for preventing nosocomial infections. Standard Precautions (Garner, 1996) apply to:

- Blood
- All body fluids, secretions, and excretions, except sweat, regardless of whether or not these contain visible blood
- Non-intact skin
- Mucous membranes

The Association for Professionals in Infection Control and Epidemiology, Inc. (APIC) recommends the use of the following Standard Precautions by healthcare and ancillary personnel (Jennings and Manian, 1999):

- Wear gloves when your hands are likely to be in contact with blood or body fluids, mucous membranes, skin that has open cuts or sores, or contaminated items or surfaces.
- Wear a protective gown or apron when you are likely to soil your clothes with blood or body fluids.
- Wear gloves whenever you are handling laboratory specimens and tubes of blood; check to make sure the specimen is sealed.
- Use caution when handling contaminated sharps. Dispose of them immediately after use, in a puncture-resistant container; avoid recapping needles; and use a one-handed recapping technique or a mechanical device such as a forceps to remove needles.
- While performing procedures, use techniques that minimize the splashing or spraying of body fluids; use protective eyewear and mask if needed.
- Use a pocket mask or other ventilatory device when giving cardiopulmonary resuscitation (CPR).
- Clean up spills of blood or body fluids promptly using gloves, a towel, and a disinfectant.
- Place soiled linen in an impermeable bag and close it or tie it shut.
- Clean, disinfect, or sterilize contaminated equipment between uses and before sending equipment out for repairs.
- Do not eat, drink, apply lip balm, or handle contact lenses in an area where exposure is likely.
- If your job poses a reasonable potential for exposure to blood or body fluids, get the hepatitis B vaccine; be sure to be up to date on all your vaccinations.
- Wash your hands immediately if they become contaminated with blood or body fluids; wash your hands routinely before and after contact with a patient and after you take off your gloves.
- Report any blood or body fluid exposures promptly to your manager and occupational health services staff.
- Apply Standard Precautions to all patients, regardless of their diagnosis, and to all contaminated equipment and materials; use your best judgment in determining when protective barriers are necessary.

In addition to the Standard Precautions hospitals use for *all* patients and situations, patients with certain infections may need additional infection control measures for the protection of other patients and hospital staff. The appropriate isolation precautions for these patients must be individualized for each case. Refer to the facility's policies and procedures about isolation precautions for more information (e.g., tuberculosis or resistant organisms).

9.3 INFECTION SURVEILLANCE AND CONTROL

Five key activities constitute an infection surveillance and control program:

1. Identify infections.
2. Analyze infection data.
3. Implement guidelines for the prevention of infections.
4. Implement guidelines for control of infections.
5. Report infection data.

The above activities relate to infections present in patients on admission and those infections that are hospital acquired (nosocomial). The use of Standard Precautions on all individuals, regardless of their diagnosis, helps to reduce occupational exposures of healthcare personnel.

The healthcare environment includes invasive and noninvasive diagnostic areas, operative suites for invasive surgical procedures, and hospital rooms where patients spend much of their time. Patients are often the source of infectious diseases, which include bacterial and viral infections. Individuals in contact with these patients and their blood, body fluids, secretions, excretions, tissues, non-intact skin, and mucous membranes must be prepared to do their jobs while minimizing their risk of occupational exposure to the infectious organisms. Infection control activities must also be targeted at preventing the spread of an infectious disease to other patients. Standard precautions must always be followed by all healthcare personnel.

9.3.1   Transmission-Based Precautions

Individuals who are responsible for performing procedures in patient rooms must be aware of and meticulously follow the hospital's transmission-based precautions, which must be observed whenever a person enters or leaves an isolation room. Patients should have dedicated equipment whenever possible. Easy-to-read isolation signs should be posted to indicate what personal protective equipment (PPE) is required and to help boost compliance of staff and visitors. The signs must be posted outside the patient room doors, where they are easily seen by everyone.

9.3.2   Training and Education

Proper education and training of new personnel must include information on hospital policies for isolation, and healthcare staff are encouraged to let others know when transmission-based precautions are violated. The education must include information on the rationale for isolation procedures (e.g., how the infectious organism is spread) and strategies to interrupt nosocomial spread of the organisms. When multi-drug-resistant organisms are discussed, such as methicillin-resistant *Staphylococcus aureus* (MRSA) and vancomycin-resistant enterococci (VRE), scientific data on survival of these organisms on inanimate objects must be included.

Case Example

Hospital personnel reported to an infection control staff member that they observed a person who was drawing morning lab specimens go from one contact isolation room to another without a change of isolation gown or gloves. The infection control staff member communicated the observations to the appropriate supervisor and was invited to present an in-service training event to individuals assigned to draw blood. Guidelines for contact isolation were reviewed and included the following concerns:

- Multi-drug-resistant organisms, such as MRSA and VRE, have the ability to survive for many days on surfaces in a patient's room. These organisms can easily be transported to other patient rooms through contamination of inanimate objects, clothing, and hands.
- A person in contact isolation may not have the same infection as another person on the same precautions. Different strains of MRSA and VRE are known to exist and differ depending on their drug sensitivities.

9.4 NOSOCOMIAL TRANSMISSION — REDUCTION STRATEGIES

Mold and fungi thrive in a dusty environment, and *Aspergillus* infections can cause severe illness and death in high-risk populations, including the very young, the elderly, and individuals with compromised immune systems. Environmental staff must receive training on proper cleaning procedures that remove dust safely from patient care areas. Education should be provided about their role in preventing fungal infections by properly removing dust from these areas. Providing a visual picture of a cadaver lung with aspergillosis is very effective.

Case Example

An Infection Control staff member noticed several positive *Aspergillus* culture reports within a 2-week period; to see any positive *Aspergillus* reports was unusual in the current setting. The Infection Control staff member reviewed the medical records of each patient who had a positive result to identify common risk factors. A line listing of the patients' room assignments and dates of hospital admissions was completed. The one similarity for four of the five patients was their stay in one intensive-care unit (ICU) room. The Infection Control staff member performed a visual inspection of the room and found several potential dust sources. In the room under investigation, a window air conditioner was installed to help keep patients more comfortable, and visible dust was observed on the outside of the unit housing. Dust was also found on the tops of the monitors, which were hung over the head of each bed.

Environmental cultures were done for various areas of the room where dust was present, and all cultures were positive for *Aspergillus* species. While waiting for the culture results, the Infection Control staff met with the ICU nurse manager to discuss immediate actions that would reduce dust accumulation in the ICU room. One quick fix would be to remove the uncovered plastic bins that previously held medical supplies and replace them with an enclosed container that could be easily cleaned. The number of supplies kept in each room could be decreased to reduce clutter and assist in daily cleaning by the Environmental Services staff.

The Infection Control staff requested a meeting with the ICU nurse manager, the directors of Environmental Services and Facilities Engineering, and several members of the hospital administration to discuss the findings and develop appropriate interventions to correct the existing problem immediately and to prevent future occurrences. The Environmental Services director agreed to the following responsibilities:

- Have the staff perform a thorough cleaning of the ICU room.
- Instruct the staff in removing dust through their daily cleaning routine (including high dusting with a damp cloth).
- Direct staff to wipe visible dust off the air conditioner vents during daily cleaning.

The Facility Engineering director would:

- Direct staff to remove the air conditioner unit.
- Provide a separate room where the air conditioning unit could be disassembled, cleaned, and disinfected before being placed back in the window.
- Revise the existing cleaning procedure for window air conditioner units so that regular filter cleaning could be accomplished even if the room is occupied for an extended period (the final procedure was to be discussed and approved at the next hospital Infection Control committee meeting).
- Establish a monthly filter cleaning schedule.

The ICU nursing manager agreed to:

- Create and post a calendar in the ICU on which the Facility Engineering staff could document when filters are cleaned.
- Establish a procedure to monitor the documented cleaning dates and report any concerns to the Facility Engineering director.
- Remove open bins from all the ICU rooms and purchase closed supply containers that are easily cleaned.

The Infection Control staff responsibilities would include:

- Making daily rounds in the ICU
- Inspecting rooms with window air conditioners to look for signs of dust on equipment, shelves, and vents
- Reviewing the documented cleaning dates for gaps in maintenance and reporting any gaps to the Facility Engineering director

Since the institution of the above interventions, no further positive *Aspergillus* cultures have been reported.

9.5 FACILITY CLEANING: MICROORGANISMS AND INFECTIOUS AGENTS

The primary goal of a healthcare facility cleaning program is to prevent the spread of infectious agents among patients and healthcare workers. Environmental Services professionals play an important role in achieving this goal. The following considerations are important for ensuring infection control success:

- Daily cleaning reduces the amount of microorganisms in the patient care environment; always clean from least soiled to more soiled areas and from top to bottom in patient rooms.
- Change the disinfectant cleaning solution and mops every three to four rooms if a single bucket is used. Change the solution and mops every six to eight rooms if two buckets are used.
- Always change the solution and mops when the solution or mops appear dirty.

9.5.1   Isolation Rooms

Isolation rooms require:

- Daily cleaning, with spot washing of walls around light switches, doorknobs, and other visible soiled areas
- Use of clean mops, cleaning cloths, and clean mop water between rooms

9.5.2   Terminal Cleaning

Resistant organisms such as MRSA and VRE can survive on objects for 5 to 7 days. In order to prevent the spread of these organisms to other patients, the following should be done:

- Dispose of all disposable items.
- Change cubicle curtains.
- Carefully clean and disinfect all patient care items, including chairs, tables, ledges, call lights, telephones, sinks, showers, and toilets.
- Use the bucket method of cleaning (using a spray bottle for cleaning may not provide appropriate cleaning and disinfection).
- Dust.

9.5.3   Bucket Method of Cleaning

The bucket method for room cleaning includes the following steps:

1. Dip a cleaning instrument or tool into a bucket filled with approved disinfectant.
2. Clean items.
3. Allow the cleaned items to remain wet for 10 minutes.

9.5.4   Dusting

Dusting reduces potential pathogens for *Aspergillus* infection in hospitalized patients and allergies in employees. Dust can be removed without making patients sick by using chemically treated cloths and mops or cloths dampened with approved disinfectant. Do not shake the cloth or mop, as this releases fungal spores. While dusting ceilings and vents, report any stains and/or wet areas immediately for repair. Fungus will start to grow on ceiling tiles within 72 hours.

9.6 HOSPITALS AND SOURCE CONTROL

Source control (in this case, isolating patients) is already in practice in virtually every hospital and carries a substantial price. Individually housing patients in highly ventilated and filtered, segregated rooms is an expensive proposition. Hospital-wide, this type of isolation strategy is too costly to be practical.

9.6.1 Filtration

Because the vast majority of microbes are associated with particles, high-efficiency particulate air (HEPA) filtration may be used to control the spread of infectious agents. HEPA filters are, by industry definition, 99.97% effective against particles in the 0.3-μm size range. Thus, HEPA filters are effective against bacteria that tend to be 0.3 μm and larger. Tuberculosis bacteria tend to agglomerate in clumps averaging 1.5 μm and should be effectively controlled by HEPA filters.

9.6.2 Air Quality Issues

Patients with infectious diseases require good air quality. Infections that have triggered their hospitalization reduce the ability of these patients to fight additional secondary infections and may render them immune deficient. Patients with HIV or active acquired immune deficiency syndrome (AIDS) cases, hepatitis, and tuberculosis are especially susceptible to other infections. Air with low pathogen concentrations is essential for these patients.

Another factor dictating the need for filtered air in rooms housing these patients is that they, themselves, may also be infectious. For example, tuberculosis is very easily transmitted through air; one nurse was found to have contracted tuberculosis simply by walking past the room of a male patient with a fulminant case. To her knowledge, she had never entered the patient's room, yet that patient was thought to have been her only contact. For the protection of both employees and patients, areas containing such infectious patients must be highly filtered, and ventilation must be adequate to control the spread of tuberculosis bacteria. The medical community is concerned about the spread of multi-drug-resistant tuberculosis (MDRTB); persons with fulminant strains of this form of TB cannot be successfully treated with conventional antibiotics.

9.7 SENSITIVE POPULATIONS

Everyone else not included in the categories listed previously in this section can be considered susceptible to infection, depending on a variety of factors: age, nutritional state, stress levels, previous exposures, etc. Visitors, volunteers, and staff can be both infected and infectious. Highly filtered air is essential in preventing person-to-person infections.

9.7.1 Orthopedic Surgery

Orthopedic surgery patients are probably the most vulnerable. Such surgery often involves the complete replacement of diseased bone or tissue with synthetic materials. Studies have shown that inadequate sterilization of the replacement materials and surgical instruments contributes to the majority of infections that result from this type of surgery. However, organisms that can be airborne, such as *Staphylococcus aureus*, and are not usually a problem for healthy individuals can become lethal when permitted to colonize a deep-wound surgical infection. Even more disturbing is the fact that *S. aureus* has developed resistance to methicillin, the antibiotic of choice in managing this organism. In addition, an environmental pathogen that invades a deep wound is often protected from antibiotics, making control of such infections difficult.

To minimize the risk of airborne infections during orthopedic surgery, practitioners have demanded laminar-flow operating rooms (both horizontal and vertical airflows). Also, many surgeons operate in fully encapsulated, HEPA-filtered suits. In essence, these operating rooms are clean rooms with efficiencies as low as the class 10 range. Interestingly, surgeons are increasing their use of encapsulated suits out of concern for their own well-being, as well as for their patients' benefit. The high-speed cutting tools used in orthopedic surgery create significant aerosols from blood and fluids at the wound site. Therefore, any bloodborne pathogens present in these fluids could be inhaled, thus infecting the operating-room staff.

9.7.2   Transplant Patients

Transplant patients of all kinds are at particular risk from infection. The initial risk may be from deep-wound infections (similar to those that can occur following orthopedic surgery). In addition, transplant patients may have deliberately been placed in an immunocompromised state prior to surgery to prevent risk of organ rejection.

9.7.3   Chronic Immune-System Repression

Transplant patients are kept in a chronic state of immune-system repression through the use of such drugs as cyclosporin to prevent rejection. At any point, these patients are in jeopardy of developing life-threatening infections. Particularly in the early days of the procedure, transplant patients were as likely to succumb to postoperative infections, such as pneumonia, as to actual organ rejection.

9.7.4   Stem-Cell Replacement

Among the riskiest transplant procedures is stem-cell replacement. These operations are usually done for leukemia patients whose bone marrow has to be completely destroyed and replaced with healthy stem cells from bone marrow from a donor. Until the replacement becomes functional, the patient literally has completely lost his ability to fight infection. If the circulating antibodies become depleted, the patient is helpless to fight an infection. Obviously, clean air with a minimum of microbial contaminants is essential for managing these patients.

9.7.5   Elderly and Children

The elderly, many of whom are victims of chronic illnesses, are at risk from exposure to infectious diseases. The same can be said for all chronic-disease sufferers, regardless of age. Newborns are at risk because their immune systems are embryonic and are naive to many infections found in the general population. Children, while somewhat less susceptible to infection than newborns, are prone to a host of childhood diseases, as well as to the plethora of diseases that affect the adult population.

The inescapable conclusion is that air filtration coupled with proper air balancing is the best means of reducing and controlling hospital-acquired infections, both from an efficiency standpoint and from a cost perspective. The challenge is presented to all segments of the air quality industry to deal with these timely, contemporary problems. Given past history, this industry is expected to meet the challenge.

9.8 DENTISTRY AND HANDWASHING

The CDC's "Recommended Infection-Control Practices for Dentistry" (1993) lists the following handwashing requirements for dental healthcare workers (DHCWs).

Nonsterile gloves are appropriate for examinations and other nonsurgical procedures; sterile gloves should be used for surgical procedures. Before treatment of each patient, DHCWs should wash their hands and put on new gloves; after treatment of each patient or before leaving the dental operatory, DHCWs should remove and discard gloves, then wash their hands. DHCWs always should wash their hands and reglove between patients. Surgical or examination gloves should not be washed before use, nor should they be washed, disinfected, or sterilized for reuse. Washing of gloves may cause penetration of liquids through undetected holes in the gloves and is not recommended. Deterioration of gloves may be caused by disinfecting agents, oils, certain oil-based lotions, and heat treatments, such as autoclaving. DHCWs should wash their hands before and after treating each patient (i.e., before glove placement and after glove removal) and after barehanded touching of inanimate objects likely to be contaminated by blood, saliva, or respiratory secretions. Hands should be washed after removal of gloves because gloves may become perforated during use, and DHCWs' hands may become contaminated through contact with patient material. Soap and water will remove transient microorganisms acquired directly or indirectly from patient contact; therefore, for many routine dental procedures, such as examinations and nonsurgical techniques, hand washing with plain soap is adequate. For surgical procedures, an antimicrobial surgical hand scrub should be used.

Thus, the DHCWs should wash their hands:

- Before treating a patient (before putting their gloves on)
- After removing gloves
- After touching equipment potentially contaminated with body fluids

Note that these procedures do not address what DHCWs touch while wearing gloves and treating the patient.

9.8.1 Exposure Routes

Patients may be exposed to a variety of microorganisms via blood or oral or respiratory secretions. These microorganisms may include:

- Cytomegalovirus
- Hepatitis B virus (HBV)
- Hepatitis C virus (HCV)
- Herpes simplex virus types 1 and 2
- Human immunodeficiency virus
- *Mycobacterium tuberculosis*
- *Staphylococcus* spp.
- *Streptococcus* spp.
- Other viruses and bacteria — specifically, those that infect the upper respiratory tract

Infections may be transmitted in the dental operatory through several routes, including:

- Direct contact with blood, oral fluids, or other secretions
- Indirect contact with contaminated instruments, operatory equipment, or environmental surfaces
- Contact with airborne contaminants present in either droplet spatter or aerosols of oral and respiratory fluids

9.8.2 Dental Instrument Sterilization and Disinfection

Placing instruments into a container of water or disinfectant/detergent as soon as possible after use will prevent drying of patient material and make cleaning easier and more efficient. Before sterilization or high-level disinfection, instruments should be cleaned thoroughly to remove debris. Cleaning may be accomplished by thorough scrubbing with soap and water or a detergent solution or with a mechanical device (e.g., an ultrasonic cleaner). The use of covered ultrasonic cleaners, when possible, is recommended to increase efficiency of cleaning and to reduce handling of sharp instruments.

All critical and semicritical dental instruments that are heat stable should be sterilized routinely between uses. Critical and semicritical instruments that will not be used immediately should be packaged before sterilization. Proper functioning of the sterilization cycle should be verified by periodic use (at least weekly) of spore tests. Impervious-backed paper, aluminum foil, or plastic covers should be used to protect items and surfaces (e.g., light handles or x-ray unit heads) that may become contaminated by blood or saliva during use and that are difficult or impossible to clean and disinfect. Between patients, the coverings should be removed (while DHCWs are gloved), discarded, and replaced (after ungloving and washing of hands) with clean material.

9.8.3   Air and Water Lines and Intraoral Dental Devices

A closed water system/self-contained water system to the dental operatory unit is recommended to ensure consistent water quality. The self-contained water system allows for weekly cleaning of the dental unit waterlines. The daily addition of an approved waterline cleaner to distilled water is beneficial in reducing bacterial contamination in dental waterlines. Bacterial counts in dental waterlines can be reduced to < 200 CFU/ml with the addition of a nontoxic chlorine dioxide additive (i.e., MicroCLEAR) to the closed water supply.

When a self-contained water system is not present., the quality of water delivered through conventional or open water systems in the dental operatory may be improved with the addition of a comprehensive dental waterline system that filters and purifies incoming water.

Because retraction valves in dental unit water lines may cause patient material aspiration back into the handpiece and water lines, antiretraction valves (one-way flow-check valves) should be installed to prevent fluid aspiration and to reduce the risk of transfer of potentially infective material. Routine maintenance of the antiretraction valves is necessary to ensure effectiveness. The dental unit manufacturer should be consulted to establish an appropriate maintenance routine. The items to be maintained include the handpieces and the antiretraction valves themselves.

The following instruments have internal surfaces that may become contaminated with patient tissue during intraoral use:

- High-speed handpieces
- Low-speed handpiece components
- Prophylaxis angles
- Other reusable intraoral instruments attached to, but removable from, the dental unit air or water lines
- Ultrasonic scaler tips and component parts
- Air/water syringe tips

Restricted physical access to internal instrument surfaces limits the cleaning and disinfection or sterilization with liquid chemical germicides. Thus, surface disinfection by wiping or soaking in liquid chemical germicides is not an acceptable method for reprocessing these instruments. Heating processes capable of equipment sterilization are recommended by the CDC.

According to manufacturers, virtually all high-speed and low-speed handpieces in production today are heat tolerant, and most heat-sensitive models manufactured earlier can be retrofitted with heat-stable components. High-speed handpieces should be run to discharge water and air for a minimum of 20 to 30 seconds after use on each patient. This procedure aids in physically flushing out patient material that may have entered the turbine and air or water lines. Use of an enclosed container or high-velocity evacuation should be considered to minimize the spread of spray, spatter, and aerosols generated during discharge procedures. Overnight or weekend microbial accumulation in water lines can be reduced substantially by removing the handpiece and allowing water lines to run and to discharge water for several minutes at the beginning of each clinic day. Sterile saline or sterile water should be used as a coolant/irrigator when surgical procedures involving the cutting of bone are performed. Some dental instruments have components that are heat sensitive or are permanently attached to dental unit water lines.

9.8.4 Extraoral Dental Devices

Some items may not enter the patient's oral cavity but are likely to become contaminated with oral fluids during treatment procedures:

- Handles or dental unit attachments of saliva ejectors
- High-speed air evacuators
- Air/water syringes

These components should be covered with impervious barriers that are changed after each use or, if the surface permits, carefully cleaned and then treated with a chemical germicide having at least an intermediate activity level. If impervious barriers are punctured or if their integrity is compromised allowing the protected surface to be contaminated, this surface should be cleaned and treated with a chemical germicide having at least an intermediate activity level. As with high-speed dental handpieces, water lines to all instruments should be flushed thoroughly after the treatment of each patient and at the beginning of each clinic day.

9.8.5 Single-Use Instruments and Sharps Disposal

Single-use disposable instruments such as the following should be used for one patient only and discarded:

- Prophylaxis angles
- Prophylaxis cups and brushes
- Tips for high-speed air evacuators
- Saliva ejectors
- Air/water syringes

These items are neither designed nor intended to be cleaned, disinfected, or sterilized for reuse. Dental sharps such as injection needles, scalpels, and orthodontic wires should be disposed of in a puncture-resistant infectious waste sharps container.

9.8.6 Dental Laboratory Disinfection

Laboratory materials and other items that have been used in the mouth (e.g., impressions, bite registrations, fixed and removable prostheses, orthodontic appliances) should be cleaned and disinfected before being manipulated in the laboratory, whether on-site or in a remote location. These items also should be cleaned and disinfected after being manipulated in the dental laboratory and before placement in the patient's mouth. A chemical germicide having at least an intermediate activity level is appropriate for such disinfection.

9.9 PRIONS

Prions are small proteins that can alter brain function, leading to dementia, behavioral disorders, and death. Prions are infectious and present a unique infection control problem because prions exhibit unusual resistance to conventional chemical and physical decontamination methods. Brown et al. (2000) subjected an infectious prion agent to temperatures ranging from 600 to 1000°C in an effort to mimic the conditions found in the primary and secondary chambers of medical waste incinerators. The prion survived temperatures of at least 600°C, and the researchers expressed concern that the infectious agent may not be fully destroyed in the residual ash remaining after the incineration process. The findings of this particular study are worrisome because some prions, such as the one that causes Creutzfeldt–Jakob (CJD) disease, are often fatal. If these agents can remain infective even after incineration, then they will present a hazard in the hospital environment if infection control programs do not have effective prion specific disinfection and sterilization procedures.

9.9.1 Creutzfeldt–Jakob Disease

Recommendations to prevent cross-transmission of infection from medical devices contaminated by CJD are made for critical and semicritical devices contaminated with high-risk tissue (i.e., brain, spinal cord, and eye tissue) from high-risk patients — those with known or suspected infection with CJD. CJD is the most common of the rare prion diseases. The incubation period of CJD may be as short as 3 months but may occur as long as 30 years after exposure. Direct brain-to-brain inoculation appears to have the shortest incubation time. Presenting symptoms of CJD are dementia and behavioral problems. The disease rapidly progresses once symptoms begin, and death occurs within 7 to 12 months from the onset of symptoms in most cases. The expected incidence of CJD is one case per million population per year. Other human prion diseases are listed below:

- Kuru (shivering disease)
- New-variant CJD (nvCJD; identified as being caused by the same strain of agent that has caused bovine spongiform encephalopathy [BSE], feline spongiform encephalopathy, and transmissible spongiform encephalopathies and is also linked to eating contaminated beef)
- Germann–Straussler–Scheinker (GSS)
- Fatal familial insomnia (FFI)
- Atypical dementias requiring further investigation

9.9.2 Screening for CJD Prions

Persons scheduled for some neurosurgical procedures must be considered potentially high risk for transmission of CJD prions when certain tissues and/or fluids are manipulated. If a brain biopsy is performed and a specific lesion has not been identified, the procedure is treated as a CJD case. The patient should be screened with the following questions and treated as a suspected CJD case if the answers are yes:

- Is there a family history of CJD prion disease?
- Has there been a rapidly progressing dementia or cerebellar ataxia?

When invasive procedures involve contact with infectious materials, additional procedures must be followed to inactivate prions and prevent the spread of infection to other patients and healthcare workers. Infectious materials include cerebrospinal fluid, brain, spinal cord, cornea, lymph glands, kidney, lung, urine, and blood. Noninfectious materials include sweat, tears, saliva, phlegm, stool, and breast milk. Noninfectious materials require standard precautions to prevent transmission.

9.9.3 Controlling CJD Transmission in the Hospital

Areas of possible exposure in the hospital are mainly:

- Operating rooms, where neurosurgical (brain, cerebrospinal, spinal cord) and neuroophthalmology procedures are performed on a known or suspected case of CJD
- Pathology laboratory, where central nervous system tissues are handled
- Autopsy areas
- Central sterile processing departments, where instrument decontamination and sterilization occur

9.9.4   Methods of Control When CJD Is Known or Suspected

Try to avoid an invasive procedure, such as a brain biopsy. If one must be done, schedule the procedure at the end of the day. Care must be taken to avoid self-inoculation with sharp objects. If an invasive procedure involving infectious tissues will be done in the operating room, the physician should notify the operating room scheduling office to alert the operating room personnel, who should notify the Infection Control office of the procedure. If an invasive procedure involving infectious tissues will be done in the morgue, the physician should notify the Infection Control office.

9.9.5   Before the Procedure Begins

Pharmacy should prepare a solution of 1 gallon of 1-N sodium hydroxide (NaOH) for use in decontamination of contaminated environmental surfaces. When the Environmental Services director or designee has been notified about the procedure, that person will make arrangements for Environmental Services staff to:

- Clean/disinfect the room after the procedure.
- Contain and label trash that will have to be incinerated.
- Arrange for the contaminated trash/disposable sharps to be removed and sent off-site for incineration.
- Coordinate with the neuropathologist or the surgical pathologist handling frozen sections to ensure proper tissue handling for optimal diagnostic work-up.

9.9.6   During the Procedure

- Restrict traffic and access to high-risk patient care areas; limit the amount of equipment in the room during the procedure.
- Use single-use disposable products and equipment and incinerate them after use in the case of known CJD.
- Cover all patient and fluid or tissue contact surfaces with disposable drapes before the invasive procedure to minimize the need for surface decontamination after the case.
- Completely isolate or drape anesthetic and respiratory equipment near the patient's head to prevent accidental splatter or contamination of the equipment.
- Confine and contain surgical and autopsy procedures.
- For surgery, wear double gloves and disposable hats, masks, eyewear, gowns, and shoe covers; incinerate all disposable items following use.
- For autopsy, chain mail gloves are worn over a pair of latex gloves and under double additional latex gloves.
- When gloves come into contact with brain, spinal cord, or adjacent tissue, the gloves are considered contaminated and should be changed.
- Use a nonelectrical (manual) saw for a craniotomy to decrease splatter and aerosolization. If an electric saw must be used, cover the equipment with a plastic drape.

9.9.7   Management of Surgical Instruments

Preferably use disposable equipment or older equipment that can be thrown away after the procedure. During the procedure, instruments should be wiped off with a damp cloth or operating room sponge and kept moist to prevent infectious fluid or tissue from drying on the instruments. When reusable instruments are used and CJD is suspected, the following equipment decontamination (steam sterilization) steps are required:

1. Disassemble equipment
2. Prevacuum steam sterilize at 134°C for at least 18 minutes

When reusable instruments cannot be disassembled or steam sterilized, completely submerge items (or keep them continuously wet) in 1-N NaOH for 60 minutes. Rinse and dry the equipment. Once instruments are decontaminated, reassemble the instruments and wrap and sterilize them in the usual fashion. Instruments initially soaked in sodium hydroxide must also be sterilized with steam. These instruments must not be placed in a Sterad sterilizer; no information is currently available as to the safety of interactions between sodium hydroxide and the Sterad (plasma) sterilizer.

9.9.8   Management of Contaminated Environmental Surfaces

When contamination of exposed surfaces with potentially infectious materials is suspected, decontamination is performed by continually wetting the surfaces with 1-N NaOH for 60 minutes, collecting and retaining all contaminated liquids (including rinse water and suctioned materials) from surgery and autopsy, and placing all spent materials into a biohazard waste bag for incineration. Incinerate contaminated trash, needles, and other disposables.

9.9.9   Specimen Labeling

Coordinate with the neuropathologist or surgical pathologist responsible for frozen sections before tissue is obtained. If sufficient tissue is excised at biopsy, a portion should be frozen for eventual western blot analysis for prion protein. An additional portion of the tissue should be fixed in formalin for histopathologic study. Clearly label all specimens that are sent to the laboratory and which are potentially infected with CJD. Specimens should not sit in the operating room specimen refrigerator without formalin for any length of time (i.e., more than a few minutes).

9.10 OCCUPATIONAL EXPOSURE TO INFECTIOUS MATERIAL

- Contamination of unbroken skin with internal body fluids or tissue: Wash with detergent and abundant quantities of warm water (avoid scrubbing), rinse, and dry.
- Needlesticks or lacerations with internal body fluids or tissue: Gently encourage bleeding, wash (avoid scrubbing) with warm soapy water, rinse, dry, and cover with a waterproof dressing.
- If percutaneous exposure to cerebrospinal or brain tissue is suspected: Provide brief exposure (1 minute) to 1-M NaOH (or 1:10 diluted bleach can be considered for maximum safety), then wash the area with soap and water.
- Splashes into the eye or mouth with any tissue or fluid: Irrigate with either saline (eye) or tap water (mouth).

9.11 CONCLUSION

Hospital-acquired or nosocomial infections can be spread person to person or by touching contaminated surfaces. Some infectious microorganisms have the ability to survive for many days outside a susceptible host and can be easily transported from one area or person to another through contamination of inanimate objects, clothing, and hands. This is a serious problem today, with only the sickest patients being admitted to hospitals, which are fast becoming more like large intensive-care units. With new infectious diseases emerging, nosocomial infections are becoming one of the leading causes of death in the United States. New emphases must be placed on infection prevention programs in the hospital environment. If infections are prevented, they do not need to be controlled.

Healthcare and ancillary personnel must be provided with training that enables them to effectively use isolation precautions. In addition, personnel must receive information about how specific infectious organisms spread and their ability to survive outside the susceptible host. Healthcare and ancillary personnel must be vigilant in adhering to the recommended practice of hand washing, because hand washing done properly is a proven strategy for preventing the spread of infections.

Special attention must be paid to the healthcare facility cleaning program, the goal of which is to prevent the spread of infectious agents among patients and healthcare workers. Environmental Services professionals (e.g., hospital housekeepers) play an important role in achieving this goal, as daily cleaning reduces the number of microorganisms in the patient care environment. Environmental Services professionals must be properly trained to do their job effectively.

REFERENCES AND RESOURCES

AHA, *Fast Facts on U.S. Hospitals from Hospital Statistics*, American Hospital Association, www.aha.org, 2002.

Anon., Creutzfeldt–Jakob disease and other prions, *APIC News*, 1–10, 2001a.

Anon., Doctors' pagers crawling with germs, *Reuters Health*, Nov. 2, 2001b.

Anon., Teddy bears spread infection — Researchers, *Reuters*, Jan. 27, 2002.

Brown, P., Rau, E.H., Johnson, B.K., Bacote, A.E., Gibbs, Jr., C.J., and Gajdusek, D.C., New studies on the heat resistance of hamster-adapted scrapie agent: threshold survival after ashing at 600°C suggests an inorganic template of replication, *Proc. Natl. Acad. Sci.*, 97(7), 3418–3421, 2000.

CDC, Recommended Infection-Control Practices for Dentistry, MRW, 42(RR-8), 1993.

CDC, Monitoring Hospital-Acquired Infections to Promote Patient Safety: United States, 1990–1999, MRW, 49, 149–152, 2000.

CDC, About NNIS, Centers for Disease Control and Prevention, www.cdc.gov/ncidod/hip/NNIS, 2002a.

CDC, NNIS Forms, Centers for Disease Control and Prevention, www.cdc.gov/ncidod/hip/NNIS, 2002b.

Cooper, M., *Hospital Infections and Drug-Resistance Rise in U.S.*, www.chiro.org, 2002.

Fauerbach, L. and Hawkins, K., Three world health organization documents on control measures for transmissible spongiform encephalopathies, *APIC*, May/June, 7–9, 2001.

Garner, J.S., Hospital Infection Control Practices Advisory Committee (CDC): guidelines for isolation precautions in hospitals, *Infect. Control Hosp. Epidemiol.*, 17, 53–80; *Am. J. Infect. Control*, 24, 24–52, 1996.

Gaynes, R., Richards, C., Edwards, J., Emori, T.G., Horan, T., Alonso-Echanove, J., Fridkin, S., Lawton, R., Peavy, G., and Tolson, J. NNIS system hospitals: feeding back surveillance data to prevent hospital-acquired infections, *Emerg. Infect. Dis.*, 7(2), 295–298, 2001.

Haley, R.W., Culver, D.H., White, J.W., Morgan, W.M., and Emori, T.G.The nationwide nosocomial infection rate: a new need for vital statistics, *Am. J. Epidemiol.*, 121, 159–167. (1985a)

Haley, R.W., Culver, D.H., White, J.W., Morgan, W.M., Emori, T.G., Munn, V.P., and Hooton, T.M., The efficacy of infection surveillance and control programs in preventing nosocomial infections in U.S. hospitals, *Am. J. Epidemiol.*, 121(2), 182–205, 1985b.

Haley, R.W., The University of Texas Southwestern Medical Center, Division of Epidemiology, personal communication, 2002.

Jarvis, W., Infection control and changing health-care delivery systems, *Emerg. Infect. Dis.*, 7(2), 170–173, 2001.

Jennings, J. and Manian, F.A., *APIC Handbook of Infection Control*, Association for Professionals in Infection Control and Epidemiology, Inc., Washington, D.C., 1999, pp. 124–125.

Muscarella, L., Assessing the risk of Creutzfeldt–Jakob disease, *Infect. Control Today*, August, 28–30, 2001.

Rutala, W.A. and Weber, D.J., Healthcare epidemiology — Creutzfeldt–Jakob disease: recommendations for disinfection and sterilization, *Clin. Infect. Dis.*, 9(32), 1348–1356, 2001.

Steelman, V.M., Creutzfeldt–Jakob disease: decontamination issues, *Infect. Control Steriliz. Technol.*, 2(9), 32–38, 1996.

Weinstein, R.A., Nosocomial infection update, *Emerg. Infect. Dis.*, 4(3), 416–419, 1998.

CHAPTER 10

Decontamination and Assessment

Brian Wight and Martha J. Boss

CONTENTS

1-56670-606-8/03/$0.00+$1.50
© 2003 by CRC Press LLC

This section presents sample site-assessment specification language and an analysis of specification development to be used for Biosafety Level 1 or 2 decontamination. Site-specific information may, of course, substantially alter decontamination requirements. Before specifications can be issued, an investigation and site assessment must be conducted. The following text provides an assessment and decontamination hierarchy. Site-specific considerations may substantially alter this hierarchy, and the scope of work should reflect the needed alterations. Safety planning to provide protocols for physical safety and to ensure limited worker exposures must be developed prior to any initial entry.

10.1 INITIAL ON-SITE DETERMINATION OF CURRENT CONDITIONS

(See Figure 10.1.) *Note:* Mediation may be by abatement of the hazard or evacuation of the building.

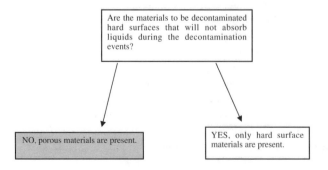

Figure 10.1 Decision tree based on surface type.

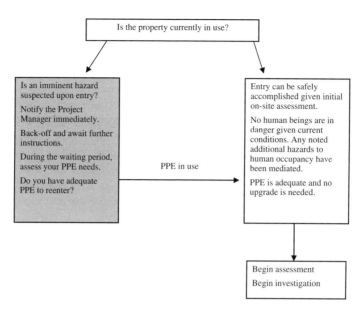

Figure 10.2 Decision tree based on property use.

10.2 ASSESSMENT PHASE

Fill in all rows of the assessment checklist with an answer, and leave *no* blanks. Take pictures, if that task is in the scope of your work. Do *not* take pictures if they are not required for your particular task or without approval by the project manager. Follow the personal protective equipment (PPE) protocols in the site planning documents (Figure 10.2). If questions arise as to safety or health, consult the project manager, who will refer any unresolved questions to safety and health personnel. Do not take any chances with your health.

10.3 DECONTAMINATION ASSESSMENT PHASE

For biological contamination and subsequent remediation, design documents may be required. Design actions that ultimately lead to specification generation and the assemblage of bids from approved remediation contractors may be required. This document presented herein is not a substitute for site-specific specifications.

10.3.1 Porous Materials

If possible, dispose of these as sanitary debris. Check with local and state codes. Federal mandates for class II biologicals are not currently codified. Local and state codes may be the defining criteria for disposal of class II biologicals. Most mold contamination remains a class II situation. If pretreatment prior to removal of these materials is needed, the spent biocide residual in the porous matrix may have to be considered during disposal.

For instances when the porous materials must be saved, a cost–benefit analysis should be conducted. Currently, guarantees of residual safety for most biocides cannot be substantiated by peer-reviewed research findings. This means that while the biological risk may be ameliorated, the risk associated with the residual biocide may become preeminent. Often the cost of saving these materials eliminates the ultimate benefit.

Mixed results have been documented for a combination of biocides and ozone treatments. Often ultraviolet (UV) lighting is used as an adjunct to ozone treatments. Essentially, the biocides and ozone are poisons. The UV lighting is used to lyse the spore cases or encysted biologicals. Repeated treatments are often required.

Layering, such as is common in installed wall-to-wall carpeting with underlying pad and wood base, must be considered. Stretched, jute-backed carpet often has a treated jute backing, and the residual stiffening agents and glue are not amenable to repeated biocide treatment. The carpet manufacturer should be consulted as to the predicted outcome of biocide treatments.

Treating underlying foam padding, whether in carpet or upholstered furniture, presents several problems. The foam is compressible and will load with biocide liquids. Transmission of the biocide throughout the layers is often not uniform, and drying is also often not uniform. Drying is an important consideration both for reuse and also in terms of the added biocide effect imparted by the drying event.

Hidden porous materials, such as insulation, may be present in the lining of appliances. Frost-free refrigerators have a defrost cycle, and the presence of insulation that may be contaminated during the intake of air in the defrost cycle and potential leakage into refrigerator layers must be considered. Refrigerators and freezers that leak a freon/mineral-oil mix from compressor and evaporative coil lines will contribute to the potential organic food base for these appliance surfaces and the surrounding areas.

Dishwashers insulated to run quietly will have insulation in the dishwasher doors, and any breach of the door may have allowed this insulation to be contaminated.

Items that can be washed and dried, preferably using bleach, often can be saved; however, a dryer that vents into the surrounding environs should not be used. Dryers with wet basin vents should also not be used; these vents should be removed and replaced with venting to an outdoor environment. Venting must not be to crawl spaces. Items that can be drycleaned can sometimes be saved; however, the drycleaning establishment must be notified that a class II biological risk is present. This notification should be in writing and must be secured to the items prior to delivery for cleaning. Items delivered to laundry or drycleaning facilities should be bagged. Alcohol bags that dissolve on contact with water are recommended for laundry, but these alcohol bags must not be used for drycleaning!

Family heirlooms, corporate historical files, and other items of personal and professional significance to the owners can be fogged with biocide. Fogging should occur in a confined, negative air pressure in a contained and filtered environment. The protocols used are similar to asbestos protocols, with the added considerations for biological contaminants and biocide usage. Copying and other means of reproducing paper copy should be considered, with appropriate safeguards for copy services. The filtration systems in photocopier exhausts can potentially become contaminated during photocopying events.

Fiberglass and foam liners, whether in ductwork or as other building components, lofts with time and is often a nest for biologicals. Decontamination may involve both biocide treatment and compression of the liners. Compression is accomplished mechanically either with the application of chemical products or by relining with hard-surface, inner compression cores.

10.3.2 Hard-Surface Materials

Painted hard-surface materials may provide a growth matrix. Decontamination of painted surfaces requires the application of biocide with additional chemical concentration and dwell time vs. decontamination of bare metal. The potential oxidation of all metals must be considered during the planning stages. Oxidation may occur many feet from a biocide application as the chemical moves through the ambient air stream. Semiporous surfaces such as gypsum board and oak tag may require sequential treatment to assure that adequate wetting and drying cycles have occurred during biocide application.

Wicking of the biocide up and through wood or other semiporous materials must also be considered. Most wood finishes will not withstand the application of concentrated biocides. Sequential application may be one answer to saving the wood finish. The concentration of biocide used in sequential applications will be one that is not sufficient to resuspend the wood finish. Repeated applications over time, however, will deposit sufficient biocide to guarantee the required residual effect.

Nonliquid means of biocide application may also be used on hard surfaces. Issues associated with dwell time, energy use, and repletion of treatment must all be considered. No treatment, whether chemical-liquid biocides, gas-delivered biocides, biocide powders, or the application of energy sources such as UV lighting, should be considered as a guarantee. The isolation of point sources of contamination may be easier with hard surfaces. These areas can then be treated with an extra biocide or ultraviolet energy application at the outset. Cabinetry and the hidden surfaces on furnishings may require additional treatment of dead air spaces associated with false bottoms. The juncture points of cabinets may also contain vertical plenum spaces where biologicals can amplify.

The interior of toilet bowls and the make-up water tank must also be decontaminated. Any failure of the toilet seal should also be considered as problematic, as leaking sewage and waters will contribute to the potential biological amplification under toilets. Hard-surface structures with numerous crevices or folds will present additional problems. An example is flexible ductwork; each of the folds in the ductwork can become a biological amplification site.

10.4 PROJECT MANUAL

A project manual is necessary to accurately define the work and to establish the contract terms and conditions. The project manual consists of bidding requirements, contract forms, conditions of the contract, and technical specifications. The bidding requirements, contract forms, and conditions of the contract are typically called the *upfront* specifications. A technical specification is the part of the contract documents or project manual that describes the technical work requirements necessary to complete the decontamination. Specifications are intended to complement the drawings and present the specific materials, equipment, biocides, and procedures required to complete the decontamination. Drawings are the part of the contract documents that describes the geometry of the work. This chapter describes the information required in a decontamination technical specification.

A technical specification is typically separated into three parts: Part 1, General; Part 2, Products; and Part 3, Execution. Each part generally contains several main items, or subparts.

- Part 1 (General) defines the summary, related work, references, definitions, submittals, quality assurance/quality control, project conditions, and certificates.
- Part 2 (Products) describes the technical requirements for the products and materials used in the execution of the work.
- Part 3 (Execution) defines the specific work practices and procedures to complete the work.

These parts and sample subparts are discussed here, with sample sections following the discussion. Keep in mind that the specifications are site specific, and the sample sections presented here may or may not be applicable to your project needs.

10.5 PART 1: GENERAL

10.5.1 Summary

This subpart describes the work covered by the technical specifications. In many cases, several decontamination technical specifications are needed to define the work accurately. For example, if a project has mold contamination on the drywall and in the air-handling unit (AHU), the project manual would likely have a drywall decontamination specification and an AHU decontamination specification. Summary statements usually include statements of work and/or scope of work items. Items listed in summary form may include:

- Removal techniques for the biological contaminants
- Procedures and equipment required to protect workers and occupants of the regulated area from contact with airborne biocides, biological contaminants, dust, and debris
- Containment, storage, transportation, and disposal of the generated biological contaminants/wastes

10.5.2 Related Work

This subpart references other specifications that may be related to the work covered by the technical specifications. Referencing is done to reduce duplication of methods that can be used by more than one technical specification. Text duplication from one specification sector to another is avoided to eliminate the potential for describing work methods differently from one specification sector to another. For example, if a large decontamination project has a site-specific health and safety specification, that specification is listed as related work instead of repeating the health and safety requirements throughout the separate specification text elements.

10.5.3 References

This subpart presents references that will be used in the execution of the work and in obtaining products and materials. These references typically include federal, state, and local regulations and codes. A project may include several technical specifications that reference the same federal, state, and local regulations. In these cases, the project manual may include a reference specification section in each technical specification, or a separate composite section may be developed either as a stand-alone document or in parallel to separate listings in each technical specification.

10.5.4 Definitions

This subpart defines the terms in the technical specification that will be used in the execution of the work and in obtaining products and materials. Defining the terms reduces the potential for conflict when two entities define the terms differently. For example, the contractor may define a *disposable bag* as a trashcan liner bought at the local grocery store, when the actual definition may be a 6-mil-thick, leaktight, plastic bag that is prelabeled and used for transporting waste to the disposal site. Another example is the term *contractor*, which, as used in this chapter, means the abatement contractor. Other contractors may also be present on-site and a clear definition of who, what, and where is consistently required in all project documents. In order to determine that everyone starts out with the same definitions for certain terms, some definitions are listed within the body of specifications. When regulatory definitions are used, the standards or regulations may also be listed.

Sample Definitions

- *Adequately wet*: To sufficiently mix or penetrate with liquid to prevent the release of particulate (defined in 40 CFR 61, Subpart M, and EPA 340/1-90-019).
- *Aggressive method*: Removal or disturbance of building material by sanding, abrading, grinding, or other method that breaks, crumbles, or disintegrates material.
- *Amended water*: Water containing a wetting agent or surfactant with a surface tension of at least 29 dynes per square centimeter when tested in accordance with ASTM D 1331.

- *Authorized person*: Any person authorized by the contractor and required by work duties to be present in the regulated areas.
- *Building inspector*: Individual who inspects buildings for biological decontamination adequacy.
- *Certified Industrial Hygienist (CIH)*: Industrial hygienist certified in the practice of industrial hygiene by the American Board of Industrial Hygiene (ABIH).
- *Clean room*: Uncontaminated room having facilities for the storage of employees' street clothing and uncontaminated materials and equipment.
- *Competent person*: In addition to the definition in 29 CFR 1926, Section.32(f), a person who is capable of identifying existing hazards and selecting the appropriate control strategy and has the authority to take prompt corrective measures to eliminate these hazards.
- *Contractor/supervisor*: Individual who supervises decontamination work and has the necessary training to be deemed competent.
- *Critical barrier*: One or more layers of plastic sealed over all openings into a regulated area or any other similarly placed physical barrier, sufficient to prevent airborne transmittal of biological contaminants and/or migration of biocide fogs in a regulated area to an adjacent area.
- *Decontamination area*: Enclosed area adjacent and connected to the regulated area and consisting of an equipment room, shower area, and clean room, which are used for the decontamination of workers, materials, and equipment contaminated with biological contamination or biocides.
- *Demolition*: The wrecking or taking out of any load-supporting structural member and any related razing, removing, or stripping of building materials.
- *Disposal bag*: A 6-mil-thick, leaktight, plastic bag that is prelabeled and used for transporting waste from containment to disposal site.
- *Disturbance*: Activities that disrupt the matrix of biological contaminants, crumble or pulverize biological contaminants, or generate visible debris from biological contaminants. Disturbance includes the cutting away of small amounts of materials containing biological contaminants, no greater than the amount that can be contained in one standard-sized glovebag or wastebag, not larger than 60 inches in length and width, in order to access a building component.
- *Equipment room or area*: An area adjacent to the regulated area used for the decontamination of employees and their equipment.
- *Employee exposure*: That exposure to airborne biological contamination and/or biocides that would occur if the employee were not using respiratory protective equipment.
- *High-efficiency particulate air (HEPA) filter*: Filter capable of trapping and retaining at least 99.97% of all monodispersed particles 0.3 μm in diameter.
- *Industrial hygienist*: Professional qualified by education, training, and experience to anticipate, recognize, evaluate, and develop controls for occupational health hazards.
- *Intact*: Biological contaminated material that has not crumbled, been pulverized, or otherwise deteriorated.
- *Modification*: A changed or altered procedure, material, or component of a control system that replaces a procedure, material, or component of a required system.
- *Negative exposure assessment*: A demonstration by the contractor to show that biological contamination is not in evidence, given sampling results.
- *Permissible exposure limits (PELs)*: Concentration of chemicals not in excess of the Occupational Safety and Health Administration (OSHA)-established limits for an 8-hour time-weighted average (TWA).
- *Regulated area*: Area established by the contractor to demarcate areas where biological decontamination work is conducted, debris and waste from such biological contamination work accumulate, and airborne concentrations of biocides or biological contaminants exceed (or there is a reasonable possibility they may exceed) the permissible exposure limit (defined in 29 CFR 1926, Section.1101).
- *Removal*: All operations where biological contaminants are taken out or stripped from structures or substrates, including demolition operations.
- *Repair*: Overhauling, rebuilding, reconstructing, or reconditioning of structures or substrates, including encapsulation or other repair of biological contaminants attached to structures or substrates.

- *Spills/emergency clean-ups*: Clean-up of sizable amounts of waste and debris which has occurred, for example, when water damage occurs in a building and sizable amounts of biological contaminants are dislodged. A competent person evaluates the site and biological contaminants to be handled and, based on the type, condition, and extent of the dislodged material, decides on clean-up methods to be used.
- *Worker*: Individual (not designated as the competent person or a supervisor) who performs biological decontamination work and has been trained to perform said work safely.

10.5.5 Submittals

This subpart is used to specify the information required in each submittal. Submittal items include, but are not limited to, manufacturers' data sheets, Material Safety Data Sheets, health and safety plans, training certificates, and physician's statements. Various submittals may be required before work begins. Because the generation of these documents can involve time and effort, these costs must be included in all project estimates. Examples of submittal items are provided below.

10.5.5.1 Materials and Equipment List

Materials and equipment lists include manufacturers' data for all materials and equipment to be used, including brand name, model, capacity, performance characteristics, and any other pertinent information. Following are examples of such materials and equipment:

- Vacuum cleaning equipment (must have manufacturer's certifications showing compliance with ANSI Z9.2)
- Water filtration equipment
- HEPA local exhaust equipment, other ventilation equipment, or pressure differential monitor for HEPA local exhaust equipment
- Air monitoring equipment
- Respirators and PPE
- Duct tape, disposal containers, disposal bags, sheet plastic, polyethylene sheeting
- Wetting agents, biocides, and coating materials (must have certificates from the manufacturers stating that materials meet the applicable specified performance requirements)
- Prefabricated or constructed decontamination units
- Material Safety Data Sheets (for all chemicals proposed)
- Encapsulants (must have test results and certificates from the manufacturers substantiating compliance with performance requirements of this specification)
- Miscellaneous necessary items, such as scrapers; brushes; brooms; staple guns; tarpaulins; shovels; rubber squeegees; dust pans; other tools; scaffolding; staging; enclosed chutes; wooden ladders; lumber necessary for the construction of containments; Underwriters Laboratories (UL)-approved temporary electrical equipment, material, and cords; groundfault circuit interrupters; water hoses of sufficient length; fire extinguishers; first aid kits; portable toilets; logbooks; log forms; markers with indelible ink; spray paint in bright colors to mark areas; project boundary fencing; etc.

10.5.5.2 Drawings

Descriptions, detailed project drawings, and site layouts may be required and should include worksite containment area techniques, local exhaust ventilation system locations, decontamination and load-out units, other temporary waste storage facilities, access tunnels, location of temporary utilities (electrical, water, sewer), and boundaries of each regulated area.

10.5.5.3 Qualifications and Organization Report

The Qualifications and Organization Report is a written report providing evidence of qualifications for personnel assigned to the work may be required. This report should include copies of all required certifications and training documentation. The contractor should furnish a written Qualifications and Organization Report providing evidence of:

- Qualifications of the contractor, the contractor's project supervisor, designated competent person, supervisors, and workers
- Designated independent Certified Industrial Hygienist (CIH)
- Independent testing laboratory (including name of firm, principal, and analysts who will perform analyses)
- All subcontractors to be used, including disposal transportation and disposal facility firms, subcontractor supervisors, subcontractor workers, and any others assigned to perform biological decontamination or abatement and support activities

The report should include an organization chart showing the contractor's staff organization for this project by name and title, chain of command, and reporting relationship with all subcontractors. The report should be signed by the contractor, the contractor's on-site project manager, designated competent person, independent CIH, designated testing laboratory, and the principals of all subcontractors to be used.

The contractor should provide in writing:

- The name, address, telephone number, and résumé of the contractor's designated competent person/site supervisor
- Evidence that the full-time designated competent person is qualified and experienced in the administration and supervision of biological decontamination or abatement projects, including:
  - Exposure assessment and monitoring
  - Work practices and abatement methods
  - Protective measures for personnel
  - Setting up and inspecting abatement work areas
  - Evaluating the integrity of containment barriers
  - Placement and operation of local exhaust systems
  - Biological-contaminant-generated waste containment and disposal procedures
  - Decontamination unit installation and maintenance requirements
  - Site safety and health requirements
  - Notification of other employees on-site

The duties of the competent person should include the following:

- Control entry to and exit from the regulated area
- Supervise employee exposure monitoring
- Ensure that all employees working within a regulated area wear the appropriate PPE, are trained in the use of appropriate methods of exposure control, and use the hygiene facilities and decontamination procedures specified
- Ensure that engineering controls in use are in proper operating condition and are functioning properly

The designated competent person should be responsible for compliance with applicable federal, state, and local requirements and the contractor's Site-Specific Safety and Health Plan (SSHP). The designated competent person should be on-site at all times this project is conducted.

The owner should provide the name, address, telephone number, résumé, and other information specified below for the designated CIH who has been selected to:

- Review the contractor's Biological Decontamination Plan
- Review the contractor's Training Program
- Direct air monitoring

The designated CIH should be a person who is board certified in the practice of industrial hygiene as determined and documented by the American Board of Industrial Hygiene and has a minimum of 2 years of comprehensive experience in planning and overseeing biological decontamination or abatement activities. A copy of the designated CIH's current, valid ABIH certification should be included. The designated CIH should be on-site at all times during the initial first week of biological decontamination/abatement and visit the site at least once per week for the duration of biological decontamination or abatement activities and should be available for emergencies. In addition, the designated CIH should submit the names, addresses, telephone numbers, and résumés of additional industrial hygienists (IHs) and industrial hygiene technicians (IHTs) who will be assisting the designated CIH in performing on-site tasks.

Training documentation will be required for all employees performing biological decontamination or abatement operations. Such documentation should be submitted on a contractor-generated form titled Certificate of Worker's Acknowledgment, to be completed for each employee in the same format and containing the same information as the example certificate at the end of this section. Training course completion certificates (initial and most recent update refresher) required by the information checked on the form should be attached.

The contractor should provide the name, medical qualifications, address, telephone number and résumé of the physician who will or has performed the medical examinations and evaluations of the persons who will conduct the biological decontamination or abatement work tasks. The physician should be currently licensed by the state where the workers will be or have been examined, have expertise in pneumoconiosis, and should be responsible for the determination of medical surveillance protocols and review of examination and test results performed in compliance with 29 CFR 1926, Section 1101, paragraph Medical Requirements. The physician should be familiar with the hazards of the site and the scope of the project.

The names of at least two persons who are currently trained in first aid and cardiopulmonary resuscitation (CPR) by the American Red Cross or other approved agency should be designated and should be on-site at all times during site operations. They should be trained in Universal Precautions and the use of PPE as described in the Bloodborne Pathogens Standard of 29 CFR 1910, Section.1030, and should be included in the contractor's Bloodborne Pathogen Program. These persons may perform other duties but should be immediately available to render first aid when needed. A copy of each designated person's current, valid First Aid and CPR certificate should be provided.

The CIH should provide the name, address, and telephone number of the independent testing laboratory selected to perform the sample analyses and report the results.

The contractor should provide written evidence that the landfill to be used is approved for biological decontamination/abatement disposal by the U.S. Environmental Protection Agency (EPA) and state and local regulatory agencies. Copies of signed agreements between the contractor (including subcontractors and transporters) and the biological waste disposal facility to accept and dispose of all sanitary and construction waste generated during the performance of this contract should be provided. Qualifications should be provided for each subcontractor or transporter to be used, indicating previous experience in transport and disposal of biological waste to include all required state and local waste hauler requirements for sanitary waste. The contractor and transporters should meet the Department of Transportation (DOT) requirements of 49 CFR 171, 49 CFR 172, and 49 CFR 173, as well as the registration requirements of 49 CFR 107 and other applicable state or local requirements. The disposal facility should also meet applicable state or local requirements.

10.5.5.4 Training Program

The training program is a copy of the written project site-specific training material that will be used to train on-site employees. The training document should be signed by the contractor's designated CIH and competent person. Prior to commencement of work, each worker should be instructed by the independent CIH and competent person in the following project-specific safety training:

- The hazards and health effects of the specific types of biological contaminants to be abated
- The content and requirements of the SSHP
- Hazard Communication Program
- Heat and/or cold stress monitoring specific to this project
- Air monitoring program and procedures
- Medical surveillance (including medical and exposure recordkeeping procedures)
- Security procedures
- Specific work practice controls and engineering controls required

10.5.5.5 Medical Requirements

Medical requirements are a physician's written opinion that should address worker fitness to wear all respirators required for site work.

10.5.5.6 Respiratory Protection Program

The Respiratory Protection Program includes records of the respirator program and a copy of the contractor's own Respiratory Protection Program. The independent CIH should establish minimum respiratory protection requirements based on measured or anticipated levels of mold and fungi and treatment biocides. The contractor's Respiratory Protection Program should include at least the following elements:

- Company policy, used for the assignment of individual responsibility, accountability, and implementation of the Respiratory Protection Program
- Standard operating procedures covering the selection and use of respirators (respirator selection should be determined by the hazard to which the worker is exposed)
- Medical evaluation of each user to verify that the worker may be assigned to an activity where respiratory protection is required
- Training in the proper use and limitations of respirators

Note that filters and cartridges used in biological decontamination cannot be reused after workers go through the decontamination sequence. The potential for growth of microorganisms on wetted filter material must always be considered when filters and cartridges have been used in biocontaminated areas.

- Respirator fit-testing (i.e., quantitative, qualitative, and individual functional fit checks)
- Regular cleaning and disinfection of respirators
- Routine inspection of respirators during cleaning and after each use when designated for emergency use
- Storage of respirators in convenient, clean, and sanitary locations
- Surveillance of regulated area conditions and degree of employee exposure (e.g., through air monitoring)
- Regular evaluation of the continued effectiveness of the respiratory protection program
- Recognition and procedures for the resolution of special problems as they affect respirator use (e.g., facial hair coming between the respirator facepiece and face or interfering with valve function, prescription eye wear or contact lens usage)
- Proper training in putting on and removing respirators

10.5.5.7 Hazard Communication Program

A Hazard Communication Program should be established and implemented in accordance with 29 CFR 1910.1200. Material Safety Data Sheets (MSDSs) should be provided for all hazardous materials brought onto the worksite.

10.5.5.8 Site-Specific Safety and Health Plan

The contractor should develop and submit a written comprehensive Site-Specific Safety and Health Plan (SSHP) at least 10 days prior to the preremediation conference. The Safety and Health Plan should address on-site work to be performed by the contractor and subcontractors. The plan should be prepared, signed, and dated by the contractor's competent person and the project supervisor. The SSHP should include a discussion of:

- PPE to be used
- Location and description of regulated areas, including clean and dirty areas, access tunnels, and decontamination units (clean room, shower room, equipment room, storage areas such as a load-out unit)
- Initial exposure assessment
- Level of supervision
- Method of notification of other employers at the worksite
- Abatement method to include containment and control procedures
- Interface of trades involved in the construction
- Sequencing of work
- Storage and disposal procedures and plan
- Type of biocide and coating agents to be used
- Location of local exhaust equipment
- Air monitoring methods (personal, environmental and clearance)
- Bulk sampling and analytical methods (if required)
- Detailed description of the method to be employed in order to control the spread of biological contaminants wastes and airborne concentrations
- Fire and medical emergency response procedures
- The security procedures to be used for all regulated areas

10.5.5.9 Reports

Reports include the results of exposure assessment and air monitoring, local exhaust ventilation records, and pressure differential recordings.

10.5.5.10 Licenses, Permits, and Notifications

Necessary licenses, permits, and notifications should be obtained in conjunction with the project's biological decontamination or abatement, transportation, and disposal actions and timely notification furnished of such actions as required by federal, state, regional, and local authorities.

10.5.5.11 Clean-Up and Disposal

The clean-up and disposal waste section includes the shipment record forms to be used.

10.5.6 Quality Assurance/Quality Control

This subpart is used to specify the quality assurance/quality control requirements. In general, a system of checks and balances is defined and established, including specific details of any testing programs (e.g., air monitoring or clearance sampling) and who is completing (e.g., a third party, owner, or engineer) and paying (e.g., owner and contractor) for the testing programs. Typically, the contractor and competent person are required to have a minimum number of years of experience in work similar to the work being completed.

Testing program elements are usually determined by a CIH, and the contractor maintains responsibility for the work force. Thus, decisions as to exposure monitoring for personnel will be made by the contractor's CIH. Quality assurance monitoring that duplicates the contractor's personnel monitoring, area monitoring, and final clearance monitoring is usually conducted by a third party in an oversight role for the project owner. The third party team includes CIHs and industrial hygiene technical staff.

10.5.6.1 *Sample Air-Monitoring Results and Documentation*

The daily air-monitoring log should contain the following information for each sample:

- Sampling and analytical method used
- Date sample collected
- Sample number
- Sample type (P, preabatement; E, environmental; C, abatement clearance)
- Location, activity, facility name where sample collected
- Sampling pump manufacturer and model and serial numbers, beginning flow rate, end flow rate, average flow rate (liters/minute)
- Calibration date, time, method, location, name of calibrator, signature
- Sample period (start time, stop time, elapsed time [minutes])
- Total air volume sampled (liters)
- Sample results
- Laboratory name, location, analytical method, analyst, and confidence level

Also included should be the printed name and a signature and date block for the industrial hygienist who conducted the sampling and for the industrial hygienist who reviewed the daily air monitoring log to verify the accuracy of the information.

10.5.7 Project Conditions

This subpart is used to describe the existing site conditions and conditions related to the work. These conditions affect the work practices of the abatement contractor. Existing site considerations may include:

- Determining the cause of the mold growth and whether or not the problem area has been repaired
- Describing the work environment

Note that descriptions of the work environment should specify whether the contaminated area is occupied, unoccupied, or partially occupied. Information related to location in a building should be provided (e.g., on a fourth floor with limited access for hauling material and waste out of the building; back area of an open office space; near children, a school, or a daycare center or in a retirement village or nursing home). The extent of the area should be estimated (e.g., contamination is limited to the drywall).

- Limiting the work to certain hours
- Specifying if utilities (e.g., sanitary, telephone, water, and electricity) are available
- Presenting analytical sample results
- Specifying that all, part, or none of the ventilation and electrical systems can be locked out/tagged out
- Describing the area of contamination

10.5.8 Clearance Criteria

The clearance criteria describe the requirements for determining when the mold has been abated by asking questions such as the following: How many samples should be taken? What type of samples should be taken (air, contact, bulk, or all)? How long will analytical results take to obtain? Are the results considered clean when the levels are less than those found outdoors? What happens if results indicate that the area is not clean? How long should the area be contained and under a negative pressure enclosure (NPE)?

The established clearance criteria are agreed to during specification writing with design input from the industrial hygiene staff, CIHs, and engineers. These criteria are then presented as elements of the specifications. Both the third-party oversight (independent) team and the contractor are required to provide documentation of testing conducted. This documentation is necessary for air and contact monitoring and for all other monitoring events required for a specific site.

When abatement is complete, sanitary waste has been removed from the regulated areas, and the final clean-up has been completed, the independent CIH usually conducts a visual inspection to document that the areas is safe. Barricade, warning signs, and boundary warning tape are not removed until this inspection has been conducted. For some sites, visual inspection alone is not sufficient, and removal of site security items awaits the receipt of clearance confirmation sampling results.

10.6 PART 2: PRODUCTS

The products section is used to specify the minimum technical requirements for the products and materials that will be used in the execution of the work. When selecting the products and materials technical requirements, care should be used to determine if the materials and products will react with the environment and nearby environment.

The use of heavy doses of chlorine bleach solutions will impact workers and may impact nearby building occupants. Additionally, certain biocides may react with elements of the building that may not be removed or decontaminated. For example, chlorine corrodes some metal components, including aluminum wiring, and alcohol can dry rubber gaskets, producing a spidering effect that degrades gasket integrity. The following questions, then, must be asked: Is the biocide selected the correct one for the mold growth? Has it been effectively used to remediate the mold that is present? What is the required contact time for the mold concentrations present? Is the manufacturer's recommended contact time for a lower concentration than the concentration that will be used?

Plastic sheeting is normally required to isolate the area. What thickness should the sheeting be? What color should the sheeting be? What is the minimum length and width of sheeting? Plastic bags are often used to dispose of the waste materials. What is the minimum size and thickness of the bags? Vacuums and negative pressure systems are often used. What is the minimum motor size? Vacuums and NPEs should be HEPA equipped. What is the minimum HEPA filter rating? What are the air monitoring equipment requirements? What type of duct tape, ladders, signs, extension cords, and other equipment will be used?

10.7 PART 3: EXECUTION

The execution section describes the procedures and work practices necessary to execute the work. Typically, execution of the work proceeds as defined by a specification and in the following manner:

- Prepare the Health and Safety Plan
- Obtain all permits

- Isolate the heating, ventilating, and air conditioning (HVAC) and electrical systems using lock-out/tagout procedures
- Set up personal and equipment decontamination areas
- Set up critical barriers and NPE system
- Decontaminate the building components starting with the highest concentrations and working to the lowest; describe the exact minimum requirements (destructive removal should not be used)
- Complete a final cleaning
- Perform clearance sampling

The following sample execution items should be made very site specific in the specification sections. Both textual elements and drawings will help to provide needed information, especially in regard to project siting issues (e.g., location of decontamination units, load-out units, regulated areas, negative pressure containments, utilities, staging and support areas, transportation routes).

10.7.1 Personal Protective Equipment

Biological contaminants, like chemical contaminants, may require the use of PPE. Engineering controls are, of course, the first choice. The main difference between biological and chemical contamination is that biological contaminants can reproduce. Thus, improper use and decontamination of PPE can lead to worker exposures in the future, particularly after the initial decontamination attempt.

10.7.1.1 Whole Body Protection

Personnel should be provided with whole body protection, and such protection should be worn properly. The contractor's competent person should select and approve the whole body protection to be used. The competent person should examine work suits worn by employees at least once per work shift for rips or tears that may occur during performance of work. When rips or tears are detected while an employee is working, those rips and tears should be immediately mended or the work suit should be immediately replaced. Disposable whole body protection should be disposed of as sanitary waste upon exiting from the regulated area. Reusable whole body protection can be either disposed of as sanitary waste upon exiting from the regulated area or properly laundered in accordance with a 10% bleach solution. Whole body protection should not be removed from the work site by a worker to be cleaned.

10.7.1.2 Coveralls

Disposable, breathable coveralls with a zipper front and hood should be provided. Sleeves should be secured at the wrists, and foot coverings should be secured at the ankles.

10.7.1.3 Underwear

Disposable underwear should be provided. If reusable underwear is used, the items should be disposed of as sanitary waste. Abatement workers should not remove contaminated reusable underwear worn during abatement of biological contaminants from the site to be laundered.

10.7.1.4 Gloves

Gloves should be provided to protect the hands. Where there is the potential for hand injuries (i.e., scrapes, punctures, cuts, etc.) a suitable glove should be provided and used. Neoprene or nitrile, 4-mil-thickness gloves should be provided.

10.7.1.5 Foot Coverings

Cloth socks should be provided and worn next to the skin. Footwear, as required by OSHA and appropriate for safety and health hazards in the area, should be worn. Rubber boots that can be decontaminated should be used. Disposable protective foot covering should be disposed of as sanitary waste. All disposable boot felt liners should be removed prior to use on-site.

10.7.1.6 Head Coverings

Hood-type disposable head coverings should be provided as components of disposable coveralls. In addition, protective headgear (hard hats) should be provided as required. Hard hats should only be removed from the regulated area after being thoroughly decontaminated with a 10% bleach solution.

10.7.1.7 Protective Eye Wear

Eye protection provided should be in accordance with ANSI Z87.1.

10.7.2 Hygiene Facilities and Practices

The contractor should establish a decontamination area for the decontamination of employees, material, and equipment. The contractor should ensure that employees enter and exit the regulated area through the decontamination area. A temporary negative pressure decontamination unit that is adjacent and attached in a leaktight manner to the regulated area should be provided. The decontamination unit should have an equipment room and a clean room separated by a shower. Equipment and surfaces of containers filled with biological contaminants should be cleaned prior to removing them from the equipment room or area. Surfaces of the equipment room should be wet wiped two times after each shift. Materials used for wet wiping should be disposed of as sanitary waste.

Hotwater service may be obtained from the building hotwater system, provided backflow protection is installed by the contractor at the point of connection. If sufficient hot water is not available, the contractor should provide a minimum 40-gal electric water heater with a minimum recovery rate of 20 gal/hr and a temperature controller for each showerhead. The contractor should provide a minimum of two showers. Flow and temperature controls should be located within the shower and should be adjustable by the user. The wastewater pump should be able to accommodate 1.25 times the showerhead flow rate at a pressure head sufficient to satisfy the filter head loss and discharge line losses. The pump should supply a minimum 25-gal/min flow with 35 ft of pressure head. Used shower water should be disposed of in the sanitary sewer system.

The floor of the clean room of the decontamination unit should be kept dry and clean at all times. Water from the shower should not be allowed to wet the floor in the clean room. Surfaces of the clean room and shower should be wet-wiped two times after each shift change with a disinfectant solution. Proper housekeeping and hygiene requirements should be maintained. Soap and towels should be provided for showering, washing, and drying. Any cloth towels provided should be disposed of as sanitary waste.

10.7.3 Load-Out Unit

A temporary load-out unit that is adjacent and connected to the regulated area should be provided. The load-out unit should be attached in a leaktight manner to each regulated area. Surfaces of the load-out unit and access tunnel should be adequately wet-wiped two times after each shift change. Materials used for wet wiping should be disposed of as sanitary waste.

10.7.4 Decontamination Area Entry Procedures

The contractor should ensure that employees entering the decontamination area through the clean room or clean area remove street clothing (including underwear) in the clean room and deposit this street clothing in lockers; put on protective clothing and respiratory protection before leaving the clean room or clean area; and pass through the equipment room to enter the regulated area.

10.7.5 Decontamination Area Exit Procedures

The contractor should ensure that these procedures are followed in sequence:

- Before employees leave the regulated area, they should continue to wear their respirators while removing all gross contamination and debris from their work clothing.
- Employees should remove their protective clothing in the equipment room and deposit the clothing in labeled impermeable bags or containers for disposal. (Employees should not remove their respirators in the equipment room or in the shower.)
- Employees should shower prior to entering the clean room.
- After showering, employees should enter the clean room before changing into street clothes.

10.7.6 Regulated Areas

All biological decontamination work should be conducted within regulated areas. The term *control areas* should be considered synonymous with regulated areas throughout the specification and work plan. The regulated area should be demarcated to minimize the number of persons within the area and to protect persons outside the area from exposure to airborne biological contaminants and/or biocides. Where critical barriers or negative pressure enclosures are used, they should demarcate the regulated area. Access to regulated areas should be limited to authorized persons. The contractor should control access to regulated areas, ensure that only authorized personnel enter, and verify that contractor-required medical surveillance, training, and respiratory protection program requirements are met prior to allowing entrance.

10.7.7 Warning Signs and Tape

Warning signs and tape printed in English should be provided at the regulated boundaries and entrances to regulated areas. The contractor should ensure that all personnel working in areas contiguous to regulated areas comprehend the warning signs. Signs should be positioned to allow personnel to read the signs and take the necessary protective steps required before entering the area. Warning signs should display the following legend in the lower panel:

DANGER:
BIOCIDE IN USE.
AUTHORIZED PERSONNEL ONLY.

Spacing between lines should be at least equal to the height of the combined uppercase of any two lines.

10.7.8 Local Exhaust Ventilation

Local exhaust ventilation units should conform to ANSI Z9.2. Filters on local exhaust system equipment should conform to ANSI Z9.2 and UL 586. Filters should be UL labeled.

10.7.9 General Requirements

Personnel should wear and utilize protective clothing and equipment as specified. The contractor should not permit eating, smoking, drinking, chewing, or applying cosmetics in the regulated area. Personnel of other trades not engaged in abatement activities should not be exposed at any time to the biological decontamination within regulated areas or the biocides being used therein. Power to the regulated area should be locked out and tagged in accordance with 29 CFR 1910, and temporary electrical service with groundfault circuit interrupters should be provided as needed. Temporary electrical service should be disconnected when necessary for wet removal.

10.7.10 Protection of Adjacent Work or Areas

Abatement should be performed without contamination of adjacent work or areas. Where such work or an area has been contaminated, the area should be decontaminated. This includes inadvertent spilling of dirt, dust, or debris for which it is reasonable to conclude that biological contamination from the regulated areas may exist. When these spills occur, work should stop in all effected areas immediately and the spill should be cleaned. When satisfactory visual inspection and air sampling analysis results are obtained and have been evaluated by the independent CIH, work can proceed.

10.7.11 Objects

Mobile objects, furniture, and equipment should be removed from the work area before abatement work begins.

10.7.12 Building Ventilation System and Critical Barriers

Building ventilating systems supplying air into or returning air out of a regulated area should be shut down and isolated by a lockable switch or other positive means in accordance with 29 CFR 1910, Section.147. Airtight critical barriers should be installed on building ventilation openings located inside the regulated area that supply or return air from the building ventilation system or serve to exhaust air from the building. The critical barriers should consist of two layers of polyethylene. Edges abutting wall, ceiling, and floor surfaces should be sealed with industrial grade duct tape.

10.7.13 Compliance Methods

The contractor should employ proper handling procedures in accordance with specified requirements. The specific abatement techniques and items identified should be detailed in the contractor's SSHP and should include, at the least, details of construction materials, equipment, and handling procedures. The contractor should use the following engineering controls and work practices in all operations, regardless of the levels of exposure:

- Vacuum cleaners equipped with HEPA filters to collect debris and dust containing biological contaminants
- Wet methods or wetting agents to control employee exposures during building material handling, mixing, removal, cutting, application, and clean-up (except where it can be demonstrated that the use of wet methods is unfeasible due to, for example, electrical hazards or equipment malfunction and in roofing)
- Prompt clean-up and disposal in leaktight containers of wastes and debris

- Inspection and repair of polyethylene in work and high-traffic areas
- Cleaning of equipment and surfaces of containers prior to removing them from the equipment room or area

10.7.14 Negative-Pressure Enclosure System

The NPE system should be as shown on the detail sheets. The system should provide at least four air changes per hour inside the containment. The local exhaust unit equipment should be operated 24 hr/day until the containment is removed and should be leakproof to the filter and equipped with HEPA filters. Air movement should be directed away from employees and toward a HEPA filtration device. The NPE should be smoke tested for leaks at the beginning of each shift. The building ventilation system should not be used as the local exhaust system for the regulated area. The local exhaust system should terminate outdoors unless an alternate arrangement is allowed by the owner's CIH. All filters used should be new at the beginning of the project and should be periodically changed as necessary and disposed of as sanitary waste.

10.7.15 Clean-Up and Disposal

All biological contaminated waste, including contaminated wastewater filters, scrap, debris, bags, containers, equipment, and contaminated clothing, should be collected and placed in leaktight containers such as double plastic bags, sealed double-wrapped polyethylene sheets, or other approved containers. For temporary storage, sealed impermeable containers should be stored in a waste load-out unit or in a storage/transportation conveyance (e.g., roll-off waste boxes) in a manner acceptable to and in an area assigned by the competent person. Procedures for hauling and disposal should comply with state, regional, and local standards. The contractor should complete, and give to the contracting officer, final completed copies of the waste shipment records for all shipments of waste material. Each waste shipment record should be signed and dated by the waste transporter and disposal facility operator.

10.8 CERTIFICATE OF WORKER'S ACKNOWLEDGMENT

Certificates of worker acknowledgment are often used to summarize the various individual requirements that have been met. Example items are provided here.

Project Name

Contract No. _____

Project Address _____

Contractor Firm Name _____

Employee's Name _____

Social Security Number _____-_____-_____

Working with biological contaminants and biocides can be dangerous. Your employer's contract for the above project requires that you be provided with:

- Formal training specific to the type of work you will perform and project-specific training, which you will complete
- Proper personal protective equipment, including a respirator, and that you are trained in its use
- A medical examination to evaluate your physical capacity to perform your assigned work tasks, under the environmental conditions expected, while wearing the required personal protective equipment

The above will be provided at no cost to you. By signing this certification, you are acknowledging that your employer has met these obligations to you. The contractor's competent person will check the block(s) for the type of formal training you have completed. Review the checked blocks prior to signing this certification.

Formal Training

- I have completed a 1-hour training class on the elements of 29 CFR 1910, Section 1200.

Project-Specific Training

- I have been provided and have completed the project-specific training required by this contract. My employer's designated competent person conducted the training.

Respiratory Protection

- I have been trained in accordance with the criteria in the contractor's respiratory protection program. I have been trained in the dangers of handling and breathing contaminants and in the proper work procedures and use and limitations of the respirator(s) I will wear. I have been trained in and will abide by the facial hair and contact lens use policy of my employer.

Respirator Fit-Test Training

- I have been trained in the proper selection, fit, use, care, cleaning, maintenance, and storage of the respirator(s) that I will wear. I have been fit-tested in accordance with the criteria in the contractor's Respiratory Program and have received a satisfactory fit. I have been assigned my individual respirator. I have been taught how to perform a proper positive and negative pressure fit-check upon donning negative pressure respirators each time.

Medical Examination

- I have had a medical examination within the last 12 months that was paid for by my employer. The examination included health history and pulmonary function tests and may have included an evaluation of a chest x-ray. A physician made a determination regarding my physical capacity to perform work tasks on the project while wearing personal protective equipment, including a respirator. I was personally provided a copy and informed of the results of that examination. My employer's industrial hygienist evaluated the medical certification provided by the physician and checked the appropriate blank below. The physician:
  - Found no limitations to performing the required work tasks
  - Identified physical limitations to performing the required work tasks

Date of the medical examination _____

Employee's signature _____Date _____

Contractor's industrial hygienist signature _____ Date _____

10.9 SPECIAL PROCEDURES FOR FLOODING

The procedures specified when flooding occurs exemplify procedures that must be used on real property when biological risk agents are suspected. Flooding presents special challenges. Any materials or furnishings that might have absorbed water (furniture, building materials, mattresses) could weigh over five times more wet than when dry. Water weighs over 60 lb/ft^3 and can wick farther up some materials than the level of the water so an item might absorb more water than expected. Safety procedures are included here as an adjunct to the biological safety procedures and protocols. The following procedures should be used to limit risks after a flood event.

10.9.1 Universal Precautions

Universal Precautions are used with the assumption that a hazard exists, whether that hazard actually does or not, unless proven otherwise. Therefore, protective measures are used until the hazard is proven not to exist. Post-flood buildings have a high probability of having health or safety hazards. Whenever you are unsure about how hazardous a situation is, always use caution until the situation is proven otherwise.

10.9.2 Health Precautions

Any tetanus shot that was received more than 5 years ago is assumed to be ineffective protection. Other immunizations might also be needed based on local health department recommendations. If you cannot remember when you last received a particular shot, assume it to be ineffective. Persons injured while working in post-flood buildings need to be up to date on their tetanus shots. Also, other shots may be required; pay attention to news bulletins from health organizations.

All persons should use the Universal Precautions when entering post-flood buildings. Some people, due to preexisting health conditions, should not enter until after the areas or buildings are completely cleaned up, decontaminated, and dried out. The following preexisting health conditions preclude entry into contaminated post-flood buildings:

- Severe asthma
- Mold allergies
- Chronic respiratory disease
- Hypersensitivity respiratory reactions to bacteria or mold
- Hypersensitivity pneumonitis or humidifier fever
- Immunocompromised status

Even after the buildings are cleaned, decontaminated, and dried, people with these conditions should leave the building if they develop symptoms, at least until the problem can be investigated. Anyone who develops unusual symptoms, such as the following, should seek *immediate* medical attention:

- Wheezing
- Difficulty breathing
- Chest tightness
- Chronic cough
- Fever
- Rashes or hives
- Extreme respiratory irritation

10.9.3 Washing

Wash hands and face frequently with antibacterial soap and drinking-quality water. When washing hands, scrub the areas under nails with a fingernail brush; dirt under the nails can harbor contaminated material. Wash hands and face before eating anything or smoking; contaminated material from dirt on the face and hands can be transferred to food or cigarettes and ingested or inhaled. Avoid touching the eyes, mouth, ears, or nose with dirty hands. Keep in mind that personal cleanliness is important in the prevention of illness or disease.

10.9.4 Cuts, Abrasions, Lacerations, and Puncture Wounds

Wash all cuts, abrasions, lacerations, and puncture wounds immediately with antibacterial soap and drinking-quality water for at least one minute, then apply an antibacterial salve and bandage. Have all deep cuts treated immediately by a medical professional. Infection can set in rapidly after injury. When in doubt about treating an injury, seek medical care.

10.9.5 General Biosafety Precautions

- Consider all pooled water inside and outside of the building to be biological or chemical exposure hazards, unless proven otherwise by qualified personnel.
- Do not permit children to play in water pools or mud.
- Attempt to drain and dry the pools as soon as possible.
- Empty flooded basements as soon as possible, but take care to ensure that the foundation will not collapse during draining (pressure from the water in the ground surrounding the foundation could cause the foundation to collapse).
- Do not use showers, toilets, or other facilities until certain that the sanitary lines from the building are clear; sewer water could back up into the building if the sewer or septic system is not working correctly.
- Move all debris immediately to disposal containers, such as dumpsters, or placed in plastic garbage bags and sealed. Do not accumulate piles of debris that could be microbiological breeding grounds or hiding places for wild animals.

10.9.6 General Chemical Safety

Unusual odors or irritation of the skin and mucous membranes should be considered to be signs of toxic chemical exposure, unless proven otherwise by qualified personnel. Be aware that some toxic chemicals do not have odors that warn of their presence. If irritation of the skin or mucous membranes occurs, leave the area immediately, wash the affected skin area with soap and water, and then be checked by medical personnel.

10.9.7 Cleaning and Decontamination Procedures

Surfaces should always be cleaned and decontaminated. The following procedure is one used for cleaning and decontaminating surfaces that have been under water inside post-flood buildings:

- Remove debris and materials that cannot be shoveled or scooped
- Shovel or scoop up dirt and mud and remove from the building
- Wash all surfaces with clean water
- Wash with a soap or detergent solution
- Rinse with clean water
- Apply a disinfectant solution
- After 15 to 20 minutes, rinse off the disinfecting solution

- Remove as much water as possible using a wet/dry vacuum or dry cloths
- Air-dry as rapidly as possible, without damaging the item

Water used in cleaning should be clean water, but it does not have to be of the same quality as drinking water.

10.9.8 Bleach

Disinfectant solutions can be made from household bleach that contains at least 5.25% sodium hypochlorite. For porous, dirty surfaces (e.g., wood, cloth, concrete), one cup of bleach should be used for every 10 cups (about 1/2 gallon) of water. For nonporous, dirt-free surfaces (e.g., metal, glass, plastic), one cup of bleach should be used for every 100 cups (about 6 gallons) of water. Note that the necessary concentration will vary based on the surface to be disinfected. The more porous or rough a surface is, the more concentrated the bleach solution should be because porous or rough surfaces cannot be cleaned as effectively as nonporous or smooth surfaces.

Bleach can corrode, etch, lighten, or otherwise negatively affect some materials, depending on the concentration. Small sections of a material should be tested with the bleach solution first to see if it affects the material. Be sure to leave the bleach solution on for as long as you would during the decontamination process. If the bleach solution harms the material, other disinfectants, such as Lysol® or Pine-Sol®, can be used. Bleach should never be mixed with any other products unless the product label states that it is okay, because the bleach could react with those products and produce hazardous gases. If in doubt, do not mix them.

10.9.9 Personal Protective Equipment

Required equipment should include:

- Long-sleeved shirts
- Long pants
- Goggles
- Head protection against bumps and falling debris
- Heavy-soled shoes or boots
- Work gloves

Quality respirators are necessary in areas where dust, mist, or fibers are being generated into the air from clean-up or demolition work and are recommended in areas that have a musty odor. Heavy-soled rubber boots or waders are necessary when walking through water pools or deep mud.

Gloves must be worn when using any cleaner, detergent, or disinfectant because the cleaner can cause skin problems. Furthermore, most cleaners and disinfectants contain respiratory irritants, whether or not masking fragrances have been added. Areas where cleaners and disinfectant solutions are mixed and used should be well ventilated. Read and follow all safety precautions on the labels of the cleaner and disinfectant products.

10.9.10 Food and Drinking Water

Drink only water you know is safe for drinking. Safe water is usually water in sealed bottles that have not been submerged under water, water that has been stated as being safe by health officials, or water that has been treated according to health department guidelines. Any containers used for water should be washed and decontaminated before refilling. Discard all food not in tin cans as unfit to eat, and discard all tin cans of food that are swollen, leaking, or corroded. For the remaining tin cans, the Centers for Disease Control and Prevention (CDC) recommends removing their labels and washing and disinfecting them. Be sure to mark them so their contents can be identified later. In all cases, when in doubt, throw it out.

10.9.11 Building Structure

Unless qualified personnel state otherwise, the following are signs of unsafe structural conditions:

- Buildings moved off of or shifted on their foundations
- Washed-out soil around foundations
- Large cracks or gaps in foundations or basement walls that did not exist before the flood
- Missing floor joist, main beam, or porch roof supports
- Sagging roofs, floors, or ceilings
- Floors that bounce or give when walked on
- Walls that move when pushed
- Gaps between steps and porches
- Leaning walls
- Loose ceiling or wall materials
- Doors or windows stuck for reasons other than swelling due to water or for which the frames are racked

For buildings with chimneys, fireplaces, or other interior brick or stone structures, consider unusual gaps, cracks, loose materials, sags, misalignments, or leaning in the structure to be signs of weak structure. Never enter a building that has an unsafe structural condition until a qualified person checks out the building and the structure is properly braced or repaired. If the condition is found after entering the building, everyone should leave the building immediately, an inspector should be called in, and unsafe materials should be removed or structures braced before work résumés inside the building.

10.9.12 Electrical

Consider all mud, debris, and water pools to be hiding potential electrical shock, laceration, or slip hazards; chemical or biological exposure hazards; or even wild animals. Verify that all power is out in the area before walking through mud or water or before clearing debris. Walk with a shuffle through mud and water pools when entering them for the first time. Lift debris in piles with poles or sticks to check for hazards or wild animals before moving the debris. Inspect the building using only flashlights; never use open flames of any kind. Consider any downed power lines within one block of the building to be potential shock hazards until proven otherwise. Electricity can travel for great distances through water, fence materials, or other conductors, and some wires might be hidden in the mud.

Consider all wiring in buildings to be shock hazards until it has been checked out by a building inspector or electrician. Until then, turn the power off at the service panel of the building. Only persons knowledgeable about electrical shock hazards should be asked to shut off the power. All electric circuit breakers, groundfault circuit interrupters (GFCIs), and fuses that were under water must be replaced. Switches and outlets that were under water can be cleaned and reused if still functional, but, again, when in doubt, throw it out. All electrical motors that were under water require cleaning, drying, and inspection by a qualified person before being put back in service. All light fixtures that were under water need to be opened, cleaned, dried, and checked before being put back in service.

Do not connect electrical generators to the electrical systems of the building. This could be a shock hazard to those in the building or those working on power lines. Use generators to power only devices connected to extension cords. Make sure that all extension cords are protected by GFCIs and overload protectors. Make sure that the extension cords have adequate capacity to handle the equipment they are being used for and that they are approved for use in wet areas. Do not use frayed or damaged extension cords. Follow all equipment safety precautions, and do not operate any equipment that you are not skilled in using without supervision. Use only wet/dry shop vacuums for vacuuming water and wet materials. If possible, pipe the vacuum exhaust out of the building using additional lengths of vacuum hose. The exhaust could contain water aerosol from the material being vacuumed. This aerosol might carry microbiological materials.

10.9.13 Liquid Propane (LP), Natural Gas, and Fuel Oil Lines

Combustion appliances and equipment can cause carbon monoxide poisoning when used in a building, unless proven safe for use under the circumstances by qualified personnel. Use all combustion equipment, such as gas-powered electrical generators and grills outside of the building. Make certain to locate them where their exhaust will not enter the building. Only heaters made to be used indoors should be used indoors; however, use them with caution and provide adequate ventilation. Follow manufacturers' precautions about using combustion equipment. Symptoms of dizziness, chronic headaches or nausea, excessive tiredness, or a cherry red skin color can indicate carbon monoxide poisoning, and medical care should be sought immediately.

Consider all gas lines to be leaking unless proven otherwise by leak checks. Gas lines should be cut off at the service supply until after clean-up is completed and gas appliances have been serviced. All gas control valves on gas-combustion appliances that were under water must be replaced. Leak checks must be performed on all lines when the appliances are returned to service. At any time, and even if the gas has been turned off, gas odors should be considered to be a sign of a leak, unless proven otherwise by fire or utility personnel (gas can travel underground from leaks in other locations). When odors are detected, the building should be evacuated immediately and fire or utility personnel called in to check for leaks.

Oil tanks are also considered to be leaking unless proven otherwise. Shut the line off at the tank until after clean-up is completed and the oil furnace has been serviced. When the furnace is put back in service, check for leaking lines.

10.9.14 Building Materials

Assume that any building materials (carpet, padding, wallboard, wallpaper, ceiling tiles) that are moist or wet 24 hours after the water recedes has mold growing on or in it, even if you cannot see or smell it. Replaceable building materials that cannot be thoroughly cleaned, decontaminated, and rapidly dried should be discarded. Irreplaceable building materials should be cleaned and decontaminated by professionals as soon as possible. It may be possible to clean, decontaminate and reinstall wall paneling made from wood laminates or vinyl. Low-cost paneling (particleboard, for example) should be discarded. Consider all wall and floor coverings (e.g., wallpaper, carpet, padding, and vinyl flooring) and insulation other than foam insulation to be contaminated with mold growth and discard them if they are replaceable. Foam insulation must be cleaned, decontaminated and dried thoroughly. Irreplaceable floor and wall coverings should be professionally cleaned and decontaminated as soon as possible. Remember, if in doubt, throw it out.

Consider all enclosed wall, ceiling, and floor cavities that were under water to be areas where toxic mold or bacteria are growing. These cavities must be opened, cleaned, decontaminated, and thoroughly dried. In general, walls that were under water should be stripped to the studs and outer skin of the building up to about one foot above the flood line. The remaining wall cavity above the flood line should be checked for mold growth, and areas where mold is found growing should also be opened. One side of floor and ceiling cavities usually can be exposed for work. Note that checking for mold growth in ceiling cavities above the flood line might also be prudent if these areas have gotten wet. Walls, ceilings, and floor cavities with irreplaceable sheeting materials or wall coverings will require access holes made in each stud or joist cavity to allow cleaning, disinfecting, and drying. These cavities should be professionally cleaned and decontaminated.

Building materials that are made of particle- and waferboard and were under water should be discarded. Some of these materials swell when wet and never return to their previous shape. Mold might also have grown within the material and be nearly impossible to remove. Buildings built before 1975 might have asbestos or lead paint. Asbestos was used primarily as insulation or as a tape on heating systems. Consider all white fibrous material used on heating system components to be asbestos and extremely hazardous. Loose or friable asbestos must be removed. Paint can be tested for lead using testing kits available at some building supply centers. If in doubt, have questionable materials checked by qualified personnel. Dry the interior of the building as rapidly as possible using dehumidifiers, heated air, and outdoor ventilation air. Using a wet/dry vacuum to pull water out of the materials will also help speed drying.

10.9.15 Personal Property

Assume that any material that is moist or wet 24 hours after the water recedes has mold growing on or in it, even if you cannot see or smell it. Most paper items and clothing and linens made from natural materials are highly susceptible to fungal growth. These items should be taken care of first. All personal items that are being kept and that were under water should be rinsed off. Clothing and linens should be laundered in hot water and dried in a dryer or sent to a dry cleaner. Nonessential paper items should be discarded. Other paper items should be air dried. Photographs can be wiped off and air dried. If possible, copy essential paper items after they have dried and discard the original. If you cannot tend to the paper items quickly, rinse and freeze them until you can. Discard all health and beauty supplies, cosmetics, bandages, and medicines that were under water. Children's toys that are being kept should be cleaned and decontaminated before the children play with them.

All other personal property will most likely be contaminated also. All replaceable property that cannot be cleaned, disinfected, and dried thoroughly, such as upholstered furniture and mattresses, should be discarded. Furnishings made from particle- or waferboard should also be discarded. Property that does not readily absorb water, such as metal or quality wood furniture, should be cleaned and decontaminated. Invaluable property that has absorbed water should be professionally cleaned and decontaminated. If possible, upholstering and fabric on irreplaceable furniture should be replaced. If these cannot be replaced, the fabric should be removed and decontaminated and the stuffing replaced. Consider all electric appliances that were under water to be shock hazards. All appliances will have to be cleaned, decontaminated, dried thoroughly, and checked before being used. Some appliances might have to be discarded. Qualified appliance service personnel should do the work on larger appliances and probably on the smaller ones, too.

10.9.16 Heating, Ventilation, and Air Conditioning Systems

The interior surfaces of HVAC equipment that has been under water are reservoirs for mold and bacteria growth. The interior components of the air-handling unit (i.e., furnace, air conditioner, or central air system) should be inspected, cleaned, and decontaminated by professionals. Insulation inside the air-handling unit might have to be replaced if it is damaged or if it has mold growing on it.

Fans will need to be removed, cleaned, decontaminated, and dried thoroughly before being placed back in the air-handling unit. Qualified service personnel should replace the gas control valves on gas-combustion units. They should also clean, check, and service the heating and air conditioning equipment and the control systems of all air-handling units that were under water. Registers or diffusers can be removed, washed, decontaminated, and reinstalled. Unlined ductwork can be disassembled, washed, decontaminated, dried, and reassembled by persons doing the clean-up if they have the necessary skills. Lined ductwork should be checked and cleaned by professionals. If the lining in the ductwork is damaged or has mold growing on it, the insulation should be replaced. Duct board ducts should be replaced. Exhaust fans should be removed, cleaned, decontaminated, and dried thoroughly before being reinstalled and put back in service.

BIBLIOGRAPHY

ANSI, ANSI Z9.2 – Fundamentals Governing the Design and Operation of Local Exhaust Systems, American National Standards Institute, Washington, D.C., 2001.

ANSI, ANSI Z87.1 – Practice for Occupational and Educational Eye and Face Protection, American National Standards Institutes, Washington, D.C., 1998.

ANSI, ANSI Z88.2 – Respiratory Protection, American National Standards Institutes, Washington, D.C., 1992.

ASTM, ASTM D 4397 – Standard Specification for Polyethylene Sheeting for Construction, Industrial, and Agricultural Applications, American Society for Testing and Materials. West Conshohocken, PA, 2000,

ASTM, ASTM E 84 – Standard Test Methods for Surface Burning Characteristics of Building Materials, American Society for Testing and Materials, West Conshohocken, PA, 2001.

ASTM, ASTM E 96 – Standard Test Methods for Water Vapor Transmission of Materials, American Society for Testing and Materials, West Conshohocken, PA, 2002.

ASTM, ASTM E 119 – Standard Test Methods for Fire Tests of Building Construction and Materials, American Society for Testing and Materials, West Conshohocken, PA, 2000.

Cassinelli, M.E. and O'Connor, P.F., Eds., NIOSH Manual of Analytical Methods (NMAM®), 4th ed., DHHS (NIOSH) Publication 94-113, Washington, D.C., August, 1994.

Compressed Gas Association, CGA G-7 – Compressed Air for Human Respiration, Compressed Gas Association, Chantilly, VA, 1990.

Compressed Gas Association, CGA G-7.1 – Commodity Specification for Air, Compressed Gas Association, Chantilly, VA, 1989.

DoT, Hazardous Materials Program Procedures, Code of Federal Regulations, Title 49 Part 107 (49 CFR 107), U.S. Department of Transportation, Washington, D.C.

DoT, General Information, Regulations, and Definitions, Code of Federal Regulations, Title 49 Part 171 (49 CFR 171), U.S. Department of Transportation, Washington, D.C.

DoT, Hazardous Materials Table, Special Provisions, Hazardous Materials Communications, Emergency Response Information, and Training Requirements, Code of Federal Regulations, Title 49 Part 172 (49 CFR 172), U.S. Department of Transportation, Washington, D.C.

DoT, Shippers – General Requirements for Shipments and Packagings, Code of Federal Regulations, Title 49 Part 173 (49 CFR 173), U.S. Department of Transportation, Washington, D.C.

Fisk, E.R., *Construction Project Administration,* 4th ed., Prentice Hall, Englewood Cliffs, NJ, 1992.

National Fire Protection Association, NFPA 701 – Standard Methods of Fire Tests for Flame Propagation of Textiles and Films, National Fire Protection Association, Quincy, MA, 1999.

USEPA, Hazardous Substances, Code of Federal Regulations, Title 29 Parts 171 and 172 (29 CFR 171 and 29 CFR 172), U.S. Environmental Protection Agency, Washington, D.C.

USEPA, Occupational Safety and Health Standards, Code of Federal Regulations, Title 29 Part 1910 (29 CFR 1910). U.S. Environmental Protection Agency, Washington, D.C.

OSHA, Safety and Health Regulations for Construction, Code of Federal Regulations, Title 29 Part 1926 (29 CFR 1926), U.S. Occupational Safety and Health Administration, Washington, D.C.

USEPA, National Emissions Standards for Hazardous Air Pollutants, Code of Federal Regulations, Title 40 Part 61 (40 CFR 61), U.S. Environmental Protection Agency, Washington, D.C.

CHAPTER 11

Legionella and Cooling Towers

Martha J. Boss and Dennis W. Day

CONTENTS

This case study illustrates the investigative and remediation principles for dealing with *Legionella* in cooling towers and associated systems. The OSHA technical manual describes the *Legionella* threat as follows:

Legionella pneumophila *is often present in hot water tanks, washing systems, and pools of stagnant water, but health effects are not observed until the contaminants become aerosolized within the building confinements.*

11.1 *LEGIONELLA PNEUMOPHILA*

Legionella pneumophila was first identified in 1977 by the CDC as the cause of an outbreak of pneumonia that caused 34 deaths at a 1976 American Legion Convention in Philadelphia. *L. pneumophila* had undoubtedly caused previous pneumonia outbreaks, but the slow growth and special growth requirements of the organism prevented earlier discovery. The diseases produced by *Legionella* are called legionellosis. More than 34 species of *Legionella* have been identified, and more than 20 are linked with human diseases.

Legionella pneumophila causes the pneumonia known as Legionnaires' disease and the flu-like Pontiac fever and has also been implicated in wound infections, pericarditis, and endocarditis without the presence of pneumonia. The factors that cause the same organism to produce two illnesses with major differences in attack rate and severity are not known. The *L. pneumophila* bacteria are Gram-negative rods that exist in a number of distinguishable serogroups. Each serogroup contains further subtypes that have different surface structures on the cell membrane and can be distinguished by special tests. Evidence indicates that some *Legionella* serogroups are more virulent than others. *L. pneumophila* serogroup 1 is the most frequently identified form of the bacterium isolated from patients with Legionnaires' disease. Other serogroups and subtypes of the bacterium are frequently isolated from water sources. Serogroups 4 and 6 are the next most frequently linked with disease.

11.2 LEGIONNAIRES' DISEASE

Legionnaires' disease has an incubation period of 2 to 10 days. Severity ranges from a mild cough and low fever to rapidly progressive pneumonia and coma. Early symptoms include malaise, muscle aches, and slight headache, while later symptoms include high fever (up to 105°F), a dry cough, and shortness of breath; gastrointestinal symptoms, including vomiting, diarrhea, nausea, and abdominal pain, are common.

The disease is treated with erythromycin or a combination of erythromycin and rifampin. Legionnaires' disease is frequently characterized as an opportunistic disease that most frequently attacks individuals who have an underlying illness or weakened immune system. The most susceptible include:

- The elderly
- Smokers
- Immunosuppressed patients
- Patients with chronic obstructive pulmonary disease (COPD)
- Organ transplant patients
- Persons taking corticosteroid therapy

11.3 PONTIAC FEVER

Pontiac fever is a nonpneumonia, flu-like disease associated with, and likely caused by, the *Legionella* bacterium. Pontiac fever has an attack rate of 90% or higher among those exposed and a short incubation period of 1 to 3 days. Complete recovery usually occurs in 2 to 5 days without medical intervention.

11.4 LEGIONELLOSIS: PROBABLE VS. CONFIRMED

A *probable* case of Legionnaires' disease is a person who has experienced an illness clinically compatible with Legionnaires', has a single antibody titer of 256 or higher, and can be associated with a population of individuals who have experienced confirmed cases of the disease (outbreak). A *confirmed* case of Legionnaires' disease requires a physician's diagnosis of pneumonia based on a chest x-ray and/or positive laboratory test results. A laboratory test is necessary for confirmation because the symptoms and x-ray evidence of Legionnaires' disease resemble those of other types of pneumonia. Various methods are used to confirm the presence of the disease.

11.4.1 Culture

The definitive laboratory methods of confirming the disease presence include culturing viable *Legionella* cells from sputum, bronchial washing, or autopsy on special media. Further cultured cell identification can be used to identify the species and serogroup. Special tests may determine isolate subtypes. Test sensitivity to detect the disease is reported to be about 70%.

11.4.2 Urine Antigen Test

The detection of antigen from *L. pneumophila* in the urine is considered a reliable measure of the disease. Antigenic materials may include *L. pneumophila* cells or portions of these cells in the urine during and after the disease. Presence of antigen in the urine is a strong legionellosis disease indicator. A patient may have a positive response for several months following the disease. Test sensitivity is limited because the only commercially available urinary antigen test detects only serogroup 1 forms of *L. pneumophila*. Fortunately, 80 to 90% of the clinically diagnosed cases are caused by serogroup 1. The Centers for Disease Control and Prevention (CDC) recommends only the radioimmunoassay (RIA) test because the latex antigen (LA) test has a high false-positive rate. The absence of a positive urinary test is not proof that a patient did not have Legionnaires' disease but merely indicates the absence of antigen in the urine at the time of the test.

11.4.3 Direct Fluorescent Antibody Staining

Direct fluorescent antibody (DFA) staining of lung aspirates can detect *L. pneumophila*. This test is frequently negative during the initial stages of the disease, as few organisms are present in the aspirate or sputum, and it requires an antigen-specific reagent. Due to the multitude of serogroups and subtypes of *L. pneumophila*, a test will be negative if the exact antigen-specific reagent is not included.

11.4.4 Serology (Antibody Titers)

An increase in the antibody level in the infected person's serum occurs several weeks after the onset of the disease. Pontiac fever also produces an elevated antibody titer, but the flu-like symptoms do not match those of Legionnaires' disease. A fourfold increase in the antibody titer coupled with a physician's diagnosis of pneumonia is considered a reliable disease indicator. The titer is measured by comparing the antibody level 4 to 8 weeks after onset (convalescent titer) to an initial (acute) titer at the beginning of the disease. Frequently, only convalescent titers have been measured from individuals who have had symptoms of the disease.

For situations in which these cases are associated with an outbreak of Legionnaires' disease, a single titer of 256 to 1 or higher is generally used as a presumptive indication of disease (probable case). Antibody strength is determined by the number of serum dilutions that elicit a positive antibody response and the reciprocal value of the number of dilutions is the antibody titer. For example, an antibody titer of 256 means a positive antibody test of the patient's serum following serial dilutions of 1:2, then 1:4, then 1:16, etc., until the 1:256 dilution point is reached. The indirect fluorescent antibody (IFA) test is the accepted diagnostic tool for demonstrating *L. pneumophila* exposure. Another widely used antibody response test is the enzyme-linked immunosorbent assay method (ELISA). The CDC believes that direct comparison of the results of IFA and ELISA is not reliable, as insufficient data are available to compare the two. The ELISA method has gained wide medical acceptance as a useful means of demonstrating exposure to *Legionella*.

11.5 TRANSMISSION

The relative likelihood of contracting Legionnaires' disease is dependent on:

- Water source contamination levels
- Susceptibility of the person exposed
- Intensity of exposure to the contaminated water

Disease transmission usually occurs via inhalation of a water aerosol contaminated with the organism. Aspiration of contaminated water into the lungs may also causes the disease. In the Philadelphia Legionnaires' disease outbreak, the cooling tower of the hotel was identified as the likely source of the disease, although domestic water sources were not evaluated. The disease has been associated with domestic hotwater systems in a number of outbreaks.

11.6 SOURCE IDENTIFICATION

L. pneumophila bacteria are widely distributed in water systems; tend to grow in biofilms or slime on the surfaces of lakes, rivers, and streams; and are not eradicated by the chlorination levels normally used to purify domestic water systems. Low and even nondetectable levels of the organism can colonize a water source and grow to high concentrations under the proper conditions. Conditions that promote growth of the organism include:

- Heat
- Sediment
- Scale
- Supporting (commensal) microflora in water
- Algae
- Amoebae
- Protozoa
- Other bacteria

Support occurs as these organisms provide nutrients (algae, flavobacteria, and *Pseudomonas*) or harbor the *L. pneumophila* bacteria (amebae and protozoa). Because of *L. pneumophila* bacteria's ability to remain viable in domestic water systems, this bacteria is capable of rapid multiplication under these conditions:

- Stagnation
- Temperatures between 20 and 50°C (68 to 122°F), with an optimal growth range of 35 to 46°C (95 to 115°F)
- pH between 5.0 and 8.5
- Sediment, which tends to promote growth of commensal microflora
- Microorganisms

11.7 CONTAMINATED WATER SOURCES

Water sources that frequently provide optimal conditions for growth include:

- Cooling towers
- Evaporative condensers
- Fluid coolers that use evaporation to reject heat
- Industrial processes that use water to remove excess heat

- Domestic hotwater systems with water heaters that operate below 60°C (140°F) and deliver water to taps below 50°C (122°F)
- Humidifiers and decorative fountains that create a water spray and use water at temperatures favorable to growth
- Spas and whirlpools
- Dental water lines, which are frequently maintained at temperature above 20°C (68°F) and sometimes as warm as 37°C (98.6°F) for patient comfort
- Stagnant water in fire sprinkler systems
- Warm water for eye washes and safety showers

Water stored below 20°C (68°F) is generally not a source for amplified *L. pneumophila* levels; however, high levels of bacteria have been measured in the water supplying ice machines. The amplification source was thought to be heat from the icemaker condenser. No cases of Legionnaires' disease have been linked to consumption of ice made from contaminated water.

11.8 MONITORING AIR

An air sample applied to special culture plates by a sampler sometimes demonstrates the presence of the organism in the air; however, negative results are frequent because of the difficulty in maintaining the viability of the organism on the culture plates. Special culture plate material and sample handling must occur in order to increase the air sampling reliability.

11.9 WATER

Analysis of water samples from a source suspected of being contaminated with *L. pneumophila* is a valuable means of identifying potential disease sources. A qualified microbiological laboratory experienced in *Legionella* detection can determine the number of organisms present in colony forming units (CFU) per volume of water and identify the different serogroups.

11.10 PHYSICAL SURVEY AND WATER SAMPLING PROTOCOL

- Obtain or prepare a simple schematic diagram of the water services.
- Record the following locations:
  - Incoming supply and/or private source
  - Storage tanks, water treatment systems, and pumps
  - Water heaters and boilers
  - All cooling towers, evaporative condensers, and fluid coolers
  - Any evaporative cooling systems or humidifiers
  - Ornamental fountains, whirlpools, eyewashes, safety showers, or other water sources within or near the facility
- Record the type and locations of:
  - Fittings used (e.g., taps, showers, valves)
  - Pipework materials
  - All systems served by the cooling tower water, including sump tanks, condensers, and indirect evaporative cooling coils in air handling units
- Trace the service route from the point of entry of the water supply.
- Assess the condition of:
  - Pipes
  - Jointing methods
  - Insulation
  - Heat sources

- Insulation in water storage tanks
- Disconnected fittings
- Dead legs
- Check for cross-connections with other services.

Once you have identified these features, take water samples from:

- The incoming water supply
- Each storage tank and water heater
- A representative number of faucets for each of the hot and cold water systems in the facility
- All cooling towers, evaporative condensers, humidifiers, spas, showers
- Water entering or leaving any other type of fitting or piece of equipment under particular suspicion

Do not overlook any potential water sources in the building. Water should be sampled from:

- Ice machines
- Hand spray bottles
- Decorative fountains
- Plastic injection-molding equipment

11.10.1 Water Sampling Procedure

Wear appropriate personal protective equipment (PPE), including respiratory protection. Do not flush the system to be sampled before collecting samples. Use sterile sampling containers (provided by the analytical laboratory) that have been autoclaved at 121°C for 15 minutes and are made of polypropylene.

11.10.1.1 Water

A 1-L sample is usually preferable. The minimum sample amount is 250 mL. Sampling bottles that contain sodium thiosulfate at a concentration of 0.5 cc of 0.1-N solution of sample water are preferred. Sodium thiosulfate inactivates any residual halogen biocide.

11.10.1.2 Temperature

Measure the temperature of the sampled water. Do not measure the temperature by placing the thermometer in the sample container. When measuring the temperature from faucets, showers, and water fountains, measure the water stream flowing from the water source. Record the initial water temperature, the amount of time necessary to run the water for the temperature to stabilize, and the final temperature. To avoid cross-contamination of the samples, sanitize the thermometer with isopropyl alcohol before measuring the temperature of each sample.

11.10.1.3 Transportation

As soon as possible after collection, water samples and swabs should be transported to and processed in a laboratory proficient at culturing water specimens for *Legionella* species. Samples may be transported at room temperature but must be protected from temperature extremes.

11.10.1.4 Analysis

Test samples for the presence of *Legionella* species by using semiselective culture media. Use standard laboratory procedures. Detection of *Legionella* species antigen by the DFA technique is not suitable for environmental samples. Use of the polymerase chain reaction (PCR) for identification of *Legionella* species is recommended as a screening tool.

11.10.2 Cooling System Sampling Sites

Collect samples of sludge, slime, or sediments, particularly where accumulations occur. Sampling sites include:

- Cooling towers
  - Make-up water (water added to system to replace water lost by evaporation, drift, and leakage)
  - Basin (area under tower for collection of cooled water)
  - Sump (section of basin from which cooled water returns to heat source)
  - Heat sources
  - Chillers
- Humidifiers
- Swamp coolers
- Building water services
- Evaporative condensers

11.10.3 Hospital Sampling Sites

Hospital sampling sites include:

- Potable water systems
- Incoming water mains
- Water softeners
- Holding tanks/cisterns
- Water heater tanks (inflow and outflow sites)
- Potable water outlets (faucets or taps, showers), especially outlets located in or near patients' rooms
- Humidifiers (nebulizers)
- Bubblers for oxygen
- Water used for respiratory therapy equipment
- Decorative fountains
- Irrigation equipment
- Fire sprinkler system (if recently used)
- Whirlpools/spas

11.10.4 Swabs

When obtaining swab samples always used prepackaged sterile swabs. Collect culture-swabs of the internal surfaces of faucets, aerators, and showerheads. Use sterile, screw-top container, such as a 50-cc plastic centrifuge tube, submerge each swab in 5–10 cc of sample water taken from the sampling location.

11.10.4.1 Swab Sampling Sites

Swab samples should be obtained from the following locations:

- Potable water systems
- Faucets (proximal to aerators)
- Faucet aerators
- Shower heads
- Internal components of cooling towers (e.g., splash bars and other fill surfaces)
- Areas with visible biofilm accumulation

11.10.4.2   Domestic Water Heaters

Take a sample of water from the bottom drain. Collect a sample of water from the outlet pipe if the plumbing provides for access.

11.10.4.3   Faucets and Showers

Collect a before-flush, initial-flow sample of water. This sample is intended to indicate the contamination level at the sample point or fitting. Collect an after-flush sample of water when the maximum temperature has been reached. The final sample should reveal the quality of the water being supplied to the sample point or fitting. Collect sterile swab samples from faucets or shower heads:

- Remove the fitting.
- Vigorously swab the interior.
- Swab samples may be positive for *Legionella* even when water samples from the source are negative.

11.10.4.4   Cooling Towers

Take a sample from the incoming supply to the tower and from any storage tanks or reservoirs in the system (e.g., chilled-water return tanks or header tanks). Take a sample from the basin of the cooling tower at a location distant from the incoming make-up water and another sample from the water returning from the circulation system at the point of entry to the tower. Take a sample of any standing water in the condensate trays or from the cooling coils.

11.10.4.5   Humidifiers, Swamp Coolers, and Spas

Take a sample from the water reservoirs. Sample the incoming water supply if it is accessible. Take swabs of showerheads, pipes, and faucets and rehydrate from water taken from the sampling site. Swab areas of scale build-up (e.g., remove showerheads, faucet screens, and aerators).

11.10.5   Sample Transportation

Prepare samples for shipment carefully:

- Wrap vinyl tape clockwise around the neck of each bottle to hold the screw cap firmly in place.
- Seal the interface between the cap and the bottle.
- Wrap absorbent paper around bottles.
- Place the bottles in resealable plastic bags
- Place the sealed plastic bags in an insulated container (styrofoam chest or box).

Samples should be stored at room temperature ($20 \pm 5°C$) and processed within 2 days. Samples should not be refrigerated or shipped at reduced temperature and should be protected from temperature extremes such as sunlight or other external heat or cold sources. Ship samples to the laboratory using overnight delivery. If shipping samples on a Friday, make arrangements for weekend receipt.

11.10.6   Water Sampling Guidelines

The contaminant levels requiring action vary depending upon the source of exposure, based on the assumption that some routes or exposure result in a greater dose to the lung. Humidifiers and

Table 11.1 Levels of *Legionella* (CFU/mL water)

Location	Action 1[a]	Action 2[b]
Cooling tower	100	1000
Domestic water	10	100
Humidifier	1	10

[a]Action 1 is prompt cleaning and/or biocide treatment of the system.

[b]Action 2 is immediate cleaning and/or biocide treatment and taking prompt steps to prevent employee exposure.

similar devices such as misters and evaporative condensers, which produce an aerosol mist that can be directly inhaled, should be controlled to lower levels of contaminant. The numbers provided in Table 11.1 are only guidelines, and the goal is zero detectable *Legionella* in a water source. Levels of *Legionella* equal to or greater than the values in the table constitute a need for the action described.

11.10.7 Microbiological Analysis

11.10.7.1 Cultured Samples

The process of growth and isolation can be time consuming, and results typically require 7 to 14 days from the time of submission. Water samples are cultured on special buffered charcoal yeast extract (BCYE) culture media. Selective isolation processes to eliminate other microbial overgrowth can determine the number of colony-forming units of *L. pneumophila* per milliliter of water. Cultured samples can also be analyzed to identify specific serogroups. Matching the serogroup and subtype of organism in the patient to that found in a water source is considered strong evidence of an associated link.

11.10.7.2 Direct Fluorescent Antibody

Direct fluorescent antibody (DFA) conjugate tests stain the organism with a fluorescent dye and can be useful in screening water samples. DFA tests, however, are unable to distinguish between live and dead bacteria. The DFA test may also have some cross-reactivity with other bacteria. Results can be available in one or two days. Use caution in interpreting the results, because the potential exists for both false-positive and false-negative results.

11.10.7.3 DNA Amplification

A relatively new method for rapid, specific organism detection employs a PCR process to amplify and then detect portions of DNA unique to *L. pneumophila*. Results can be produced in one day. Preliminary evidence indicates that sensitivity and specificity are comparable to those of cell culture.

11.11 INTERPRETING SAMPLE RESULTS

Because total eradication of *Legionella* may not be possible, an acceptable control strategy is to minimize the number of organisms present in a water source. A private consulting firm and microbiological laboratory (PathCon, Inc., Norcross, GA) has introduced suggested guidelines for control based on the number of colony-forming units of *L. pneumophila* per milliliter of water. These guidelines vary depending on the water source, a recognition by the authors of the PathCon guidelines that dose is related both to the potential for exposure and to concentration. For example, recommended contaminated water exposure limits for a humidifier, which would involve direct exposure to an aerosol, are lower than those for a cooling tower, where the opportunity for exposure is normally less. Work operations such as maintenance on cooling towers may involve direct exposure to cooling tower mist, and precautions to minimize exposure are always necessary.

11.12 COMMUNITY HEALTH CONCERNS

An outbreak of Legionnaires' disease among workers may have its origin in the community and may not be related to the work environment. A Legionnaires' outbreak is both an occupational and a public health concern, and the investigation may include local public health departments and the CDC. To minimize employee risk and maximize the effectiveness of effort, close coordination among the Occupational Safety and Health Administration (OSHA), other public agencies, and the employer is imperative.

11.13 INVESTIGATIONS

Investigation protocols are based on differing levels of suspected risk for exposure to *Legionella*. All cases require sound professional judgment in deciding the appropriate course of action. A level-one investigation may be initiated when workplace water sources are probably contaminated with *Legionella* or one case of Legionnaires' disease has been reported. A level-two investigation should be conducted when more then one case of Legionnaires' disease has been reported or a Legionnaires' disease outbreak has occurred for which two or more cases can be attributed to a work site. The outbreak is considered still in progress if at least one of the cases has occurred in the last 30 days. Prompt actions should be undertaken to provide maximum protection to employees and eliminate the hazard. Both types of investigations follow the same general pattern:

- Preliminary opening conference
- Walk-through of the facility to conduct a physical assessment of the water systems
- More detailed examination of the systems, including a review of maintenance records
- Assessment of findings
- Closing conference to present control actions based on the findings

11.13.1 Level One Investigation

Use the following procedure when Legionnaires' disease may be related to the work environment.

11.13.1.1 Step 1: Systems Overview

A facilities engineer or experienced member of the building maintenance staff should be available to explain system operation and assist in the walkthrough investigation. The overview of water systems should include:

- Plumbing systems
- Heating, ventilation, and air conditioning (HVAC) systems
- Water reservoirs
- Hot and cold domestic water systems
- Water heaters
- Distribution pipes
- Water coolers
- Water treatment equipment

- Connections to process water systems protected (or unprotected) by backflow preventers
- Storage tanks
- Decorative fountains
- Misters
- Whirlpools and spas
- Tepid-water eyewashes and safety showers
- Humidifiers
- Water for cooling industrial processes

The HVAC system review should include:

- Cooling towers
- Evaporative condensers
- Fluid coolers
- Humidifiers
- Direct evaporative air cooling equipment
- Indirect evaporative air cooling equipment
- Air washers for filtration
- Location of the fresh-air intakes relative to water sources

A review of maintenance records should include:

- Temperature checks of domestic water
- Visual and physical checks of cooling towers
- Reports of cooling tower water quality assessment and chemical treatment

Investigate recent major maintenance or changes in the system's operation. Determine if recent or frequent losses of water pressure from the incoming water supply have occurred due to line breakage or street repairs. The failure of a backflow prevention device under loss of pressure can contaminate the system. Identify the locations in the system where water is allowed to stagnate:

- Storage tanks
- Unused plumbing pipe sections/deadlegs
- Infrequently used faucets

Check for cross-connections between domestic and process water systems and note the condition and type of backflow prevention devices.

11.13.1.2 Step 2: Walkthrough Investigation

Equipment you will need includes:

- Thermometer for measuring water temperatures
- Flashlight
- Film or video camera

Measure and record the water temperature drawn from each storage-type water heater. This temperature may be significantly below the gauge temperature of the water heater because of heat stratification. Note the presence of rust and scale in this water. Record the maximum temperature of water at faucets connected to each water heater in the system. Record temperatures at locations near, intermediate, and distant from the heaters. (*Note:* In order to reach the maximum temperature, it may be necessary to run the water for several minutes.)

Determine the water temperature and the stagnation potential of coldwater storage tanks used for reserve capacity or to maintain hydrostatic pressure. These tanks should be protected from temperature extremes and covered to prevent contamination. Record the temperature of the domestic coldwater lines at various locations within the facility. Note both the initial temperature and the final equilibrium temperature on the coldwater line. Record the time required to reach equilibrium, as an indicator of the potential system stagnation. Evaluate cooling towers, evaporative condensers, and fluid coolers for:

- Biofilm growth
- Scale buildup
- Turbidity

Record the location of the tower relative to:

- Fresh-air intakes
- Kitchen exhausts
- Leaves
- Plant material
- Sources of organic material

Note the presence and condition of drift eliminators, the basin temperature of the water (if the cooling tower is currently being operated), and the location and condition of the sumps for the cooling towers, evaporative condensers, and fluid coolers. These sumps are sometimes located indoors to protect them from freezing. Record the locations of any cross-connections between the cooling tower water system and any domestic water system. These may supply a back-up source of cool water to refrigeration condenser units or serve to supply auxiliary cooling units. The lack of a regular maintenance schedule or water-treatment program for a cooling tower or evaporative condenser system strongly suggests a potential for *Legionella* contamination.

11.13.1.3 *Step 3: Assessment*

If no potential problems are identified, if the operating temperature measured at the water heaters is 60°C (140°F) or above, and if the delivery temperature at distant faucets is 50°C (122°F) or higher, no further action will be necessary. If the system is poorly maintained and operating temperatures are below recommended minimums, then recommendations for corrective action should be made.

11.13.1.4 *Step 4: Control Actions*

Disinfect the domestic water system by:

- Heat treatment
- Chlorination
- Cleaning and disinfecting the cooling tower system (according to the Wisconsin Division of Health's "Protocol for Control of Legionella in Cooling Towers" or a similar process for cleaning heat-rejection systems that follows sound practices to minimize potential for *Legionella* growth)
- Eliminating dead legs in the plumbing system
- Insulating plumbing lines
- Installing heat tracing to maintain proper temperatures
- Eliminating rubber gaskets
- Removing or frequently cleaning fixtures such as aerators and showerheads

The absence of proper operating conditions alone is sufficient for assuming that the water system can pose an unnecessary risk to the employees. Take water samples after completion of the control actions to confirm that the corrective measures were successful. The employer may want to obtain samples before starting corrective actions to assess the extent of the problem but still should take necessary corrective actions even if the results of presampling are negative. Water sampling can reduce false negatives in that a contaminated portion of the system may have been missed. The absence of *Legionella* organisms at the time of sampling does not ensure that the system will remain negative.

If, after control actions, the *Legionella* levels in a water source exceed the guidelines:

- Re-examine the water system to determine if potential contamination points within the system were overlooked
- Reassess control procedures to determine if they were performed properly
- Repeat the procedures as needed until contamination levels meet the guidelines

11.13.2 Level Two Investigation

A level two investigation is similar to a level one investigation with several additional steps. Supplemental actions include:

- Medical surveillance of all employees currently on sick leave to identify any new cases
- Employee awareness training on the disease to minimize employee concerns and aid in early recognition of new cases
- Assessment of past sick-leave absences for undetected cases of the disease
- Collection of water samples during the walk-through assessment

11.13.2.1 Step 1: System Overview and Assessment

Assess water systems as described for a level one investigation. Estimate the size of the building and the number of water services during the initial walkthrough and prearrange supply and shipping of the required number of sterile sample containers with the appropriate laboratory.

11.13.2.2 Step 2: Second Walkthrough Survey and from Step 1 Water Sampling

During this step, visual assessments are verified and sampling completed.

11.13.2.3 Step 3: Employee Awareness Program Development and Sick Leave Monitoring

Ensure that employees understand the early disease symptoms and seek medical assistance promptly, but do not alarm the workers. Stress the importance of the need to know the health status of all employees on sick leave.

11.13.2.4 Step 4: Review Worker Absences to Detect Other Cases

Identify all employees who have taken 3 or more consecutive days of sick leave from approximately 6 weeks before the case of Legionnaires' disease was identified and up to the present. Request that those employees who may have had pneumonia during this period undergo additional voluntary tests for evidence of Legionnaires' disease.

11.13.2.5 *Step 5: Assess Worker Absence Survey and Water Systems Analysis*

If evidence indicates more than one case of Legionnaires' disease at the workplace, then the site should be treated as having an outbreak. Take immediate control of all water sources to eliminate potential for exposure, and take measures to eliminate the hazard. No action is necessary if the results of the investigation are negative; that is,

- All water and HVAC systems are well maintained and in good operating condition.
- All water sample results are negative or acceptably low.
- No new cases of the disease have been identified at the work site.

Note: Under these circumstances, assume that the site is not the origin of the identified case.

11.13.2.6 *Step 6: Control Actions*

The control actions are the same as for a level-one investigation.

11.13.3 Ongoing Outbreak

If the evidence indicates that two or more Legionnaires' disease cases have occurred at a site, and at least one of the cases was within the last 30 days, assume that an outbreak is in progress and requires a high-priority investigation and prompt action. Conduct a level-two investigation as outlined above, and take the following precautions to protect building occupants:

- Immediately initiate control measures to prevent additional exposures to all water systems that have a reasonable potential for worker exposure, including:
  - Hot and cold domestic water
  - Cooling towers
  - Humidifiers
  - Other potential sources of water exposure
- Collect appropriate water samples to determine *Legionella* levels before shutting down the water systems.
- Have a member of the building maintenance or engineering staff explain how the water system operates and conduct a proper controlled shutdown; these control actions need not require facility shutdown.

Temporary provisions can allow work to continue:

- Bottled water can be supplied.
- Water heaters can be shut off to eliminate hot-water access.
- Temporary cooling towers can allow work to continue.

11.14 COOLING TOWERS, EVAPORATIVE CONDENSERS, AND FLUID COOLERS

The purpose of cooling towers, evaporative condensers, and fluid coolers is to reject heat from system fluids through evaporation. Cooling towers remove heat from condenser water via direct-contact evaporation in a wet airstream. This cooled water circulates through the condenser side of a mechanical refrigeration unit to absorb heat. As the fluid in the condenser returns to a liquid state, heat is given off. This heat is then absorbed by the cooling tower waters. Some of the cooling tower waters in the process of absorbing the heat absorb enough heat to change from liquid water to steam mists — the evaporative phenomenon.

Evaporative condensers are located directly inside the wet airstream, and water passing over the coils directly cools the refrigerant. Evaporative condensers take heat from their surroundings. The fluids within the coils take in this heat, and these interior fluids convert to a more gaseous state. The subsequent wet airstream exposure returns the fluid within these coils to a more liquid state by absorbing the fluid's heat. The coil area where the wet airstream surrounds the coils may be termed the condensate coil side.

Fluid coolers are employed for industrial processes and as computer-room air conditioners. Fluid coolers have heat-exchanger coils directly in the wet airstream and function similarly to evaporative condensers. All of these systems use a fan to move air through a recirculated water system. Thus, a considerable amount of water vapor is introduced into the surroundings despite the presence of drift eliminators designed to limit vapor release. In addition, this water may be in the ideal temperature range for *Legionella* growth, 20 to 50°C (68 to 122°F).

11.14.1 Inspection and Maintenance

Visual inspection and periodic maintenance are the best ways to control growth of *Legionella* and related organisms. Good maintenance is necessary both to control *Legionella* growth and for effective operation. The system should be properly monitored and maintained to prevent build-up of scale and sediment and biofouling, all of which support *Legionella* growth and reduce operating efficiency.

11.14.2 Biocide

Unfortunately, measurements of water quality such as total bacterial counts, total dissolved solids, and pH have not proven to be good indicators of *Legionella* levels in cooling towers. Periodic biocide use is needed to ensure control of *Legionella* growth. Traditional oxidizing agents such as chlorine and bromine have been proven effective in controlling *Legionella* in cooling towers. Little information exists on the demonstrated effectiveness of many commercial biocides for preventing *Legionella* growth in actual operations.

11.14.2.1 Commercial Biocide Treatments

According to the OSHA Technical Manual (OSHA, 1999):

Little information exists on the demonstrated effectiveness of many commercial biocides for preventing *Legionella* growth in actual operations. Recent Australian studies indicate that Fentichlor (2,2'-thiobis[4-chlorophenol]) used weekly for 4 hours at 200 ppm, or bromo-chloro-dimethyl-hydantoin (BCD) in a slow-release cartridge at an initial concentration of 300 pp.m are effective in controlling the growth of *Legionella*. There are no U.S. suppliers of Fentichlor, although the chemical licensed by the EPA for water treatment in cooling towers. Towerbrom 60M™, a chlorotriazine and sodium bromide salt mixture, has been reported to be effective when alternated with BCD for control of *Legionella* in U.S. studies of *Legionella* contamination of cooling towers. The Australian study also indicates that quaternary ammonium compounds, widely used for control of bio-fouling in cooling towers, are not effective in controlling *Legionella*.

Bromine is an effective oxidizing biocide. It is frequently added as a bromide salt and generated by reaction with chlorine. Bromine's effectiveness is less dependent than chlorine on the pH of the water; it is less corrosive; and it also produces less toxic environmental by-products.

The effectiveness of any water-treatment regimen depends on the use of clean water. High concentrations of organic matter and dissolved solids in the water will reduce the effectiveness of any biocidal agent. Each sump should be equipped with a "bleed," and make-up water should be supplied to reduce the concentration of dissolved solids.

11.14.2.2   Chlorination

Continuous chlorination at low free residual levels can be effective in controlling *Legionella* growth. The proper oxidant level must be established and maintained because free residual chlorine above 1 ppm may be corrosive to metals in the system and may damage wood used in cooling towers. Also, free residual levels below 1 ppm may not adequately control *Legionella* growth. Frequent monitoring and control of pH is essential for maintaining adequate levels of free residual chlorine. Above a pH of 8.0, chlorine effectiveness is greatly reduced. Proper control of pH will maintain the effectiveness of chlorination and minimize corrosion.

11.14.2.2.1 Chlorine and Organics

Chlorine also combines with organic substances in water to form toxic by-products that are of environmental concern. Do not rely on chlorine odor as an indicator of sufficient mixing. Instead, use real-time monitoring instruments or colorimetric papers/badges/sorbent tubes to gauge the residual chlorine amount. Chlorine odor is actually the odor of chloramines produced as the chlorine reacts with organic proteins. Thus, chlorine odor may not indicate that sufficient residual and unreacted chlorine molecules remain in the system.

11.14.2.2.2 Continuous Chlorination

To maintain concentrations of free residual chlorine at 1 to 2 mg/L at the tap requires the placement of flow-adjusted, continuous injectors of chlorine throughout the water distribution system. Adverse effects of continuous chlorination include accelerated corrosion of plumbing, resulting in system leaks and production of potentially carcinogenic trihalomethanes. However, when levels of free residual chlorine are below 3 mg/L, trihalomethane levels are kept below the maximum safety level recommended by the EPA.

11.14.2.3   Bromination

Bromine is an effective oxidizing biocide that is frequently added as a bromide salt and generated by reaction with chlorine. The effectiveness of bromine is less dependent than chlorine on the pH of the water, bromine is less corrosive, and it produces less toxic environmental by-products.

11.14.3 Sump Treatment

The effectiveness of any water-treatment regimen depends on the initial and continued use of clean water. High concentrations of organic matter and dissolved solids in the water will reduce the effectiveness of any biocidal agent. Each sump should be equipped with a bleed, and make-up water should be supplied to reduce the concentration of dissolved solids. One of the most effective means of controlling the growth of *Legionella* is to maintain sump water at a low temperature. System design should recognize the value of operating with low sump-water temperatures. Sump-water temperatures depend on:

* Tower design
* Heat load
* Flow rate
* Ambient dry-bulb and wet-bulb temperatures

Under ideal conditions, sump-water temperatures in evaporative devices approach the ambient wet-bulb temperature.

11.14.4 Drift Eliminators and Other Design Features

High-efficiency drift eliminators are essential for all cooling towers. Older systems can usually be retrofitted with high-efficiency models. A well-designed and well-fitted drift eliminator can greatly reduce water loss and potential for exposure. Other important design features include:

- Easy access or easily disassembled components to allow cleaning of internal components including the packing (fill)
- Enclosure of the system to prevent unnecessary drift of water vapor
- Features to minimize the spray generated by these systems

11.14.5 Cleaning Frequency

Cooling towers should be cleaned and disinfected at least twice a year. Normally this maintenance will be performed before initial start-up at the beginning of the cooling season and after shut-down in the fall. Systems with heavy biofouling or high levels of *Legionella* may require additional cleaning. Any system that has been out of service for an extended period should be cleaned and disinfected. New systems require cleaning and disinfecting because construction material residue can contribute to *Legionella* growth.

11.14.6 Wisconsin Cleaning Protocol

Acceptable cleaning procedures include those described in the Wisconsin Protocol. This procedure calls for:

- Initial shock treatment with 50 ppm free residual (total) chlorine
- Addition of detergent to disperse biofouling
- Maintenance of 10 ppm chlorine for 24 hours
- A repeat of the cycle until no visual evidence of biofilms remains

To prevent exposure during cleaning and maintenance, wear proper personal protective equipment:

- Coated Tyvek®-type suit with a hood
- Impermeable protective gloves
- Properly fitted respirator with a high-efficiency particulate air (HEPA) filter and cartridges as needed to preclude exposure to biocide chemicals

11.14.7 Recordkeeping

A description of the operating system (which includes all components cooled by the system) and details of the make-up water to the system should be readily available. Written procedures for proper operation and maintenance of the system should include standard operating procedures for using:

- Scale and corrosion inhibitors
- Antifoaming agents
- Biocides or chlorine

Logbooks should list dates of inspections and cleanings, water-quality test results, and maintenance.

11.15 DOMESTIC HOTWATER SYSTEMS

The term *domestic* applies to all nonprocessed water used for lavatories, showers, drinking fountains, and other personal supply applications in commercial, residential, and industrial settings. Cool zones within these systems are defined as areas where the water is below 60°C (140°F). Disease transmission from domestic hot water may be by inhalation or aspiration of *Legionella*-contaminated aerosolized water. Large water heaters like those used in hospitals or industrial settings frequently contain cool zones near the base where cold water enters and scale and sediment accumulate. The temperature and sediment in these zones can provide ideal conditions for *Legionella* amplification.

Water systems designed to recirculate water and minimize dead legs will reduce stagnation. Dead legs are defined as capped spurs or nonrecirculated plumbing lines that allow hot water to stagnate. These areas may facilitate stagnation and cooling to <50°C regardless of the circulating-water temperature. Increasing the flow rate from the hotwater-circulation system may help lessen the likelihood of water stagnation and cooling. Segments may have to be removed to prevent colonization.

If potential for scalding exists, appropriate, fail-safe, scald-protection equipment should be employed. Pressure-independent, thermostatic mixing valves at delivery points can reduce delivery temperatures. Installation of blending or mixing valves at or near taps to reduce the water temperature below 60°C (140°F) can cause *L. pneumophila* to multiply even in short segments of pipe containing water. Rubber fittings within plumbing systems have been associated with persistent colonization, and replacement of these fittings may be required for *Legionella* species eradication. Point-of-use water heaters can eliminate stagnation of hot water in infrequently used lines. Proper hotwater line insulation and heat tracing of specific lines can help maintain distribution and delivery temperatures.

11.15.1 Maintenance

To minimize the growth of *Legionella* in the system:

- Hot water should be stored at a minimum of 60°C (140°F) and delivered at a minimum of 50°C (122°F) to all outlets.
- The hotwater tank should be:
  - Drained periodically to remove scale and sediment
  - Cleaned with chlorine solution if possible
  - Thoroughly rinsed to remove excess chlorine before reuse
- Eliminate dead legs when possible, or install heat tracing to maintain 50°C (122°F) in the lines.
- Remove rubber or silicone gaskets. These gaskets provide nutrients for the bacteria, and removing them will help control organism growth. Frequent flushing of these lines should also reduce growth.
- Run domestic hotwater recirculation pumps continuously; these pumps should be excluded from energy conservation measures.

11.15.2 Control

Control measures include the following procedures:

- Raise the water-heater temperature to control or eliminate *Legionella* growth.
- Pasteurize the hotwater system by raising the water-heater temperature to a minimum of 70°C (158°F) for 24 hours and then flushing each outlet for 20 minutes.
- Flush all taps with the hot water because stagnant areas can reseed the system. Exercise caution to avoid serious burns from the high water temperatures used in pasteurization.

- Periodically chlorinate the system at the tank and lines to maintain a level of 10-ppm free residual chlorine.
- Flush all taps until chlorine is thoroughly mixed within the system.
- Install in-line chlorinators in the hotwater line; however, chlorine is quite corrosive and will shorten the service life of metal plumbing.
- Control the pH, which is extremely important to ensure that adequate residual chlorine remains.
- Use metal ions such as copper or silver (which have a biocidal effect) in solution.
- Provide an ozonization system that injects ozone into the water.
- Supply ultraviolet (UV) radiation by installing commercial, in-line UV systems on incoming water lines or on recirculating systems; eliminate stagnant zones to maximize the effectiveness of this treatment. Scale build-up on the UV lamp surface can rapidly reduce light intensity and requires frequent maintenance to ensure effective operation.

11.16 COLDWATER SYSTEMS

Domestic coldwater systems are not a major problem for *Legionella* growth. Maintaining coldwater lines below 20°C will limit the potential for amplification of the bacteria. Elevated *Legionella* levels have been measured in ice machines in hospitals. Coldwater lines near heat sources in the units are believed to have caused the amplification. Cross-contamination of the domestic coldwater system with other systems should always be suspected. If significant contamination of the domestic coldwater system occurs, the source of contamination must be determined. If the coldwater lines have significant contamination, hyperchlorination can eradicate *Legionella*. Run faucets until the chlorine is mixed throughout the system (the chlorinated water is allowed to remain in the system). Free chlorine levels of 20 and 50 ppm are allowed to remain for two hours and one hour, respectively.

11.16.1 Plumbing Lines

All connections to process water should be protected by a plumbing-code-approved device (e.g., backflow preventer or air gap). Inspect the system for dead legs and areas where water may stagnate. Elimination of these sections or frequent flushing of taps to drain the stagnant areas may be necessary to limit growth of the organism. Insulate coldwater lines that are close to hotwater lines to reduce the temperature in the line.

11.16.2 Dental Water Lines

Dental water lines are a common sources of water contaminated with high concentrations of microorganisms including *Legionella*; however, to date an increased risk of disease among dental staff or patients has not been demonstrated. Operating conditions for dental water lines are especially appropriate for *Legionella* proliferation because water is stagnant a majority of the time, narrow plastic tubing encourages biofilm formation, and the water temperature is usually 20°C (68°F) or higher. Some systems maintain water at 37°C (98.6°F). Food and Drug Administration (FDA)-approved methods to minimize risk include filtration of water at the point of use and using replaceable in-line, 1-μm filters

11.16.3 Water Tanks

Water tanks that allow water to remain uncirculated for long periods can promote growth of bacteria. These tanks should be designed to reduce storage time to a day or less. If this cannot be accomplished, the tanks should be eliminated. Water tanks should be covered to prevent contamination and protected from temperature extremes.

11.17 HEATING, VENTILATION, AND AIR CONDITIONING SYSTEMS

Heating, ventilation, and air conditioning systems can disseminate contaminated water aerosols. Water-aerosol sources are classified as either external or internal.

11.17.1 External Sources

External sources may emit contaminated aerosolized water drawn into a system's fresh-air intake. Mist discharged from cooling towers, evaporative condensers, and fluid coolers can be ingested by the HVAC fresh-air intake. Fresh-air intakes typically are concrete plenums located at grade level that supply fresh air to air handlers in the basement or lower levels of buildings. They can collect organic material (e.g., leaves and dirt) and water from rain or irrigation. When evaluating this path, you should consider:

- Prevailing wind direction and velocity
- Building effects (e.g., low-pressure zones on leeward sides of buildings and on roof)
- Architectural screen walls
- Distance from tower to intake
- Direct paths such as through an open window

When evaluating external sources, examine the potential for direct transmission. Indirect transmission paths through the HVAC system may be convoluted, and the bacteria may die from desiccation in the airstream and impaction on internal surfaces such as filters and duct lining.

On the other hand, HVAC systems that are contaminated with other biological growth may serve as amplification sites for *Legionella*, especially if condensate films or liquid collection sources are present. When draining properly, the water that passes through the condensate pans of cooling coils in an air handler is normally not a source of growth because of the low temperature of the water condensate.

11.17.2 Internal Sources

Internal sources may provide contaminated aerosolized water that is then disseminated by the air-distribution system. Internal sources include HVAC ducts. Contaminated water from domestic water systems, fire sprinklers, refrigeration condensers, or other systems can leak from pipes into HVAC ducts, where they are aerosolized and distributed by the system. HVAC system humidifiers can be hazards.

11.17.2.1 Heated-Pan Humidifiers

Heated-pan humidifiers use a heat source to evaporate water from a pan open to the airstream. Intermittent use of the device coupled with a warm pan of water may support *Legionella* growth. Contaminant-free water is essential.

11.17.2.2 Direct Steam-Type Humidifiers

Direct steam-type humidifiers inject boiler-generated steam directly into the airstream. They normally operate above 70°C (158°F), and *Legionella* cannot survive at that temperature.

11.17.2.3 Atomizing Humidifiers

Atomizing humidifiers use mechanical devices with pneumatic air to create a water mist that evaporates into the airstream. Contaminant-free water is essential.

11.17.2.4   Direct Evaporative Air Coolers

Direct evaporative air coolers mix water and air in direct contact to create a cool, wet airstream by evaporation. They include sumps, which may stagnate when not in use.

11.17.2.5   Indirect Evaporative Air Cooling in Dryer Climates

In dryer climates, one common design circulates cool water from a cooling tower sump through a water coil in the supply airstream. If the coil develops a leak, then pumped cooling tower water will be injected directly into the supply air. If the sump water is contaminated with *Legionella*, the supply airstream will become contaminated with *Legionella*.

Indirect evaporative air cooling is also found in air-to-air heat exchangers. One side of the heat exchanger is an evaporative cooled wet airstream, and the other side supplies air for the conditioned space. If the heat exchanger leaks, the wet airstream can mix with supply air. If the wet airstream is contaminated with *Legionella*, the supply air will become contaminated with *Legionella*. Many air-handling systems designed for dryer climates employ direct evaporative air cooling using:

- Wet evaporative coolers
- Slinger air coolers
- Rotary air coolers

The cooling devices mix water and air in direct contact to create a cool, wet airstream by evaporation. If these systems are using 100% outside air in a dry climate, the water sump temperature may be low and will not represent a significant risk. Improperly operated and maintained systems that use warm, stagnant sump water can present a significant risk.

11.17.2.6   Residential Humidifiers

Residential humidifiers are small, freestanding, portable units that use an internal fan and wet media to disseminate a wet airstream. These humidifiers have sumps that are frequently contaminated with *Legionella*.

Daily cleaning is necessary to maintain acceptable water quality, but these units seldom receive appropriate maintenance, and their use in the commercial or industrial workplace is strongly discouraged.

11.17.2.7   Computer-Room Air Conditioners

Computer-room air conditioners typically include humidifiers and frequently are not well maintained. They may contain a sump filled with contaminated water.

11.17.3   Design

The following are issues to consider when designing HVAC systems to minimize risk from *Legionella* contamination and most apply to all types of microbial contamination:

- Minimize the use of water reservoirs, sumps, and pans. Chemically untreated, stagnant, sources of warm water provide an ideal environment for *Legionella* growth. Provide a way to drain water sumps when not in use; an electric solenoid valve on the sump drain is one alternative. If an HVAC sump is used during the hours when a building is occupied, drain the sump during unoccupied hours. Provide a bleed for water sumps so that dissolved solids do not form sediments in the sump. Slope and drain sumps from the bottom so that all the water can drain out and allow the pan to dry.

- Locate HVAC fresh-air intakes so that mist from a cooling tower, evaporative condenser, or fluid cooler is not drawn into the system. Use the recommended minimum distances between cooling towers and fresh-air intakes. Various building codes and the *Guidelines for the Assessment of Bioaerosols in the Indoor Environment* (American Conference of Governmental Industrial Hygienists, ACGIH) provide these distances.
- Design indirect evaporative cooling systems to prevent the heat exchanger from mixing wet systems with the air-distribution systems. These designs should also include fail-safe measures and control or monitoring devices.
- Use steam or atomizing humidifiers instead of units that use recirculated water (atomizing humidifiers must have contaminant-free water).
- Do not use raw steam from the central heating boiler that contains corrosion inhibitors and antiscaling chemicals.

11.17.4 Operation and Maintenance

Operate all HVAC equipment in accordance with current design interpretations. The professional engineering team that designed the systems should provide these interpretations. Test all HVAC equipment periodically to ensure that performance is as designed. Water reservoirs must be properly drained and bled to prevent sediment accumulation. Inactive sumps must be included as maintenance items despite their current inactivity status. Maintenance failures can produce contaminated, stagnant water that can become an ideal environment for *Legionella* growth if heated, including heating by sunlight or conduction. Designers should always consider operation and maintenance during design; inadequate access ways, portals, clean-out devices, and water or chemical sources can defeat the best intentions.

11.18 EMPLOYEE AWARENESS PROGRAM

An employee awareness program informs employees of a potential outbreak, educates employees about the disease risk and consequences, and should be part of a level-two investigation or response to any Legionnaires' disease outbreak. This program is of critical importance to aid in early recognition of the disease. Program elements should supplement the case-identification program to detect previously undetected cases of the illness at the work site and should help alleviate employee concerns about the disease.

The employer should implement the following program elements immediately upon recognition of more than one probable or confirmed disease case in the work place:

- An initial employee training session to provide basic information about the disease and actions being taken to investigate the problem
- An ongoing general information service to
  - Provide updates
  - Answer questions
- Medical and psychological counseling services when an outbreak has occurred

11.18.1 Sample Letter to Employees (OSHA, 1999)

Below is a sample letter and supplemental information on the disease that the employer can use for informing employees of a potential or actual outbreak.

Date:

Memo to: All employees

From: [management official]

Subject: Legionnaires' disease

On _____, we were notified that one of the employees of our company had contracted legionellosis, commonly referred to as Legionnaires' disease. The employee is assigned to _____ on _____ shift. We want to share with you some general information concerning the disease. In addition, we want to tell you what we are currently doing here at _____ to ensure that necessary steps are being taken to address health concerns.

Legionellosis, or Legionnaires' disease, is a type of pneumonia caused by *Legionella* bacteria. Legionnaires' disease is not contagious, and you cannot catch it from another person. The bacteria are common and grow in water. People often receive low-level exposure in the environment without getting sick. Persons who are heavy smokers or are elderly or whose ability to resist infection is reduced are more likely to contract Legionnaires' disease than healthy individuals.

We are cooperating fully with local health officials who are investigating this matter. Most cases of legionellosis are isolated and are not associated with an outbreak. According to the Centers for Disease Control and Prevention in Atlanta, there are between 10,000 and 50,000 cases of Legionnaires' disease every year in the United States.

To date, _____ cases of the disease have occurred among employees in this facility. To identify any other cases, we will review sick-leave records for the period _____ to _____. Employees who took more than three consecutive days of sick leave will be identified, and we will attempt to determine if anyone in that group experienced pneumonia-like symptoms (fever, shortness of breath, cough). Those who have used three or more consecutive days of sick leave during this period can expect to be contacted by a representative of our company for an interview. If you experienced a pneumonia-like illness in the past two months but used fewer then three consecutive days of sick leave, contact _____ to arrange an interview.

To ensure that you are being protected during the interim, we are also instituting a medical surveillance program to identify any new or old cases. Part of this surveillance will be asking you to answer a few questions about your illness when you call in sick to your supervisor. In addition, we are offering counseling and employee information services. If you would like to take advantage of these services or want more information, contact your manager. For the present, please pay attention to the following information about what you should do now:

If you are not sick, there is no need for you to see a doctor.

If you are now sick with a cough and fever:

See your private doctor or contact _____ to arrange to see a physician.

Tell the physician that you work in a building that may be involved in a Legionnaires' disease outbreak.

If you see a physician, notify _____ so that your illness can be tracked.

If you have any concerns or questions concerning this issue, please contact your manager. Your health and safety are of great concern to us, and we will be grateful for your cooperation in this matter. As further information develops we will keep you informed.

11.18.2 Sample Interview with Employees Calling in on Sick Leave (OSHA, 1999)

Interviewer: _____

Date: _____

Supervisor Survey Form

We are screening employee illnesses as a result of our Legionnaires' disease incident. You are not obligated to participate in the survey, but your participation will help you and your fellow workers. We recommend that you see a physician if you currently have pneumonia-like symptoms such as severe chills, high fever, a cough, and difficult breathing. Are you currently experiencing these symptoms?

Yes _____ No _____ Prefer not to answer _____

If your answer to the question is "No," do not complete the rest of this form. If your answer is "Yes," please read the brief statement below and provide the information requested after the statement. If your answer is "Prefer not to answer," please provide only the information requested after the statement.

Statement

You will be contacted by_____ to obtain additional information necessary to complete

our survey.

Employee's name _____

Work telephone number _____

Home telephone number _____

Shift (day/swing/graveyard/rotating)_____

Branch _____

Organization _____

Employee's supervisor _____

Supervisor's work telephone number _____

Date _____

Please forward to _____ by 10:00 a.m. each day.*

* *Note:* For every day that this form is filled out, forward it to the appropriate company personnel by a set time (i.e., 10:00 A.M.).

11.18.3 Sample Information Sheets for Legionnaires' Disease (OSHA, 1999)

Legionnaires' disease is a common name for one of the several illnesses caused by *Legionella* bacteria. Legionnaires' disease is an infection of the lungs that is a form of pneumonia. A person can develop Legionnaires' disease by inhaling water mist contaminated with *Legionella*. *Legionella* bacteria are widely present at low levels in the environment in lakes, streams, and ponds. At low levels, the chance of getting Legionnaires' disease from a water source is very slight. The problem arises when high concentrations of the organism grow in water sources.

Water heaters, cooling towers, and warm, stagnant water can provide ideal conditions for the growth of the organism. Scientists have learned much about the disease and about the *Legionella* bacteria since it was first discovered in 1976. The following questions and answers will help you learn more of what is currently known about Legionnaires' disease.

Q. What are the symptoms of Legionnaires' disease?

A. Early symptoms of the illness are much like the flu. After a short time (in some cases a day or two), more severe pneumonia-like symptoms may appear. Not all individuals with Legionnaires' disease experience the same symptoms. Some may have only flu-like symptoms, but to others the disease can be fatal. Early flu-like symptoms include slight fever, headache, aching joints and muscles, lack of energy, tired feeling, and loss of appetite. Common pneumonia-like symptoms include high fever (102 to 105°F, or 39 to 41°C), cough (dry at first, later producing phlegm), difficulty in breathing or shortness of breath, chills, and chest pains.

Q. How common is Legionnaires' disease?

A. It is estimated that in the United States there are between 10,000 and 50,000 cases each year.

Q. How does a person get Legionnaires' disease?

A. A person must be exposed to water contaminated with *Legionella* bacterium. This exposure may happen by inhaling or drinking water contaminated with the *Legionella* bacteria. For example, inhaling contaminated water mist from a cooling tower, a humidifier, or even a shower or sink can cause the disease.

Q. How soon after being exposed will a person develop symptoms of the disease?

A. If infection occurs, disease symptoms usually appear within 2 to 10 days.

Q. Are some people at a higher risk of developing Legionnaires' disease?

A. Yes, some people have lower resistance to disease and are more likely to develop Legionnaires' disease. Some of the factors that can increase the risk of getting the disease include organ transplants (e.g., kidney, heart), age (older persons are more likely to get disease), heavy smoking, weakened immune system (cancer patients, HIV-infected individuals), underlying medical problems (e.g., respiratory disease, diabetes, cancer, renal dialysis), certain drug therapies (corticosteroids), or heavy consumption of alcoholic beverages.

Q. Is Legionnaires' disease spread from person to person?

A. No, Legionnaires' disease is not contagious and cannot be transmitted from one person to another.

Q. What causes Legionnaires disease?

A. Legionnaires' disease is caused by inhaling water contaminated with rod-shaped bacteria called *Legionella pneumophila*. Over 30 different species of *Legionella* exist, many of which can cause disease. *L. pneumophila* is the most common species that causes disease.

Q. Does everyone who inhales *Legionella* into the lungs develop Legionnaires' disease?

A. No, most people have resistance to the disease. It is thought that fewer than 5 out of 100 persons exposed to water contaminated with *Legionella* will develop Legionnaires' disease.

Q. Is Legionnaires' disease easy to diagnose?

A. No, the pneumonia caused by *Legionella* is not easy to distinguish from other forms of pneumonia. A number of diagnostic tests allow a physician to identify the disease. These tests can be performed on a sample of sputum, blood, or urine.

Q. How is Legionnaires' disease treated?

A. Erythromycin is currently the antibiotic of choice. Early treatment reduces the severity and improves chances for recovery. In many instances, this antibiotic may be prescribed without the physician knowing that the patient has Legionnaires' disease, because erythromycin is effective in treating a number of types of pneumonia.

Q. How did Legionnaires' disease get its name?

A. Legionnaires' disease got its name from the first outbreak in which the organism was identified as the cause. This outbreak occurred in 1976, in a Philadelphia hotel where the Pennsylvania American Legion was having a convention. Over 200 Legionnaires and visitors at this convention developed pneumonia, and some died. From lung tissue, a newly discovered bacterium was found to be the cause of the pneumonia and was named *Legionella pneumophila*.

Q. Is Legionnaires' disease a new disease?

A. No, Legionnaires' disease is not new, but it has only recently been identified. Unsolved pneumonia outbreaks that occurred before 1976 are now known to have been Legionnaires' disease. Scientists are still studying this disease to learn more about it.

Q. Are *Legionella* bacteria widespread in the environment?

A. Yes, studies have shown that these bacteria can be found in both natural and manmade water sources. Natural water sources including streams, rivers, freshwater ponds and lakes, and mud can contain the organism in low levels.

Q. Could I get the disease from natural water sources?

A. It is unlikely. In the natural environment, the very low levels of this organism in water sources probably cannot cause disease.

Q. What water conditions are best for growth of the organism?

A. Warm, stagnant water provides ideal conditions for growth. At temperatures between 68 and 122°F the organism can multiply. Temperatures of 90 to 105°F are ideal for growth. Rust (iron), scale, and other microorganisms can also promote the growth of *Legionella*.

Q. What common types of water are of greatest concern?

A. Water mist from cooling towers or evaporative condensers, evaporative coolers (swamp coolers), humidifiers, misters, showers, faucets, and whirlpool baths can be contaminated with the organism and if inhaled or swallowed can cause the disease.

Q. Can Legionnaires' disease be prevented?

A. Yes. Avoiding water conditions that allow the organism to grow to high levels is the best means of prevention. Specific preventive steps include regular maintenance and cleaning of cooling towers and evaporative condensers to prevent growth of *Legionella*. This should include twice-yearly cleaning and periodic use of chlorine or other effective biocide. Maintain domestic water heaters at 140°F (60°C). The water temperature should be 122°F or higher at the faucet. Avoid conditions that allow water to stagnate. Large water-storage tanks exposed to sunlight can produce warm conditions favorable to high levels of *Legionella*. Frequent flushing of unused water lines will help alleviate stagnation.

Q. Do you recommend that I operate my home water heater at 140°F?

A. Probably not if you have small children or infirm elderly persons who could be at serious risk of being scalded by the hot water. However, if you have persons living with you who are at high risk of contracting the disease, then operating the water heater at a minimum temperature of 140°F is probably a good idea.

Q. What can be done if a water system is already contaminated or is suspected of being contaminated?

A. Special cleaning procedures can eliminate *Legionella* from water sources. In many cases, these procedures involve the use of chlorine producing chemicals or high water temperatures. Professional assistance should be sought before attempting to clean a water system.

Q. Can my home water heater also be a source of *Legionella* contamination?

A. Yes, but evidence indicates that smaller water systems such as those used in homes are not as likely to be infected with *Legionella* as are larger systems in work places and public buildings.

Q. Can *Legionella* bacteria cause other diseases?

A. Yes. In addition to Legionnaires' disease, the same bacteria also causes a flu-like disease called Pontiac fever.

Q. How does Pontiac fever differ from Legionnaires' disease?

A. Unlike Legionnaires' disease, which can be a serious and deadly form of pneumonia, Pontiac fever produces flu-like symptoms that may include fever, headache, tiredness, loss of appetite, muscle and joint pain, chills, nausea, and a dry cough. Full recovery occurs in 2 to 5 days without antibiotics. No deaths have been reported from Pontiac fever.

Q. Are there other differences between Legionnaires' disease and Pontiac fever?

A. Yes. Unlike Legionnaires' disease, which occurs in only a small percentage of persons who are exposed, Pontiac fever will occur in approximately 90% of those exposed. In addition, the time between exposure to the organism and appearance of the disease (called the incubation period) is generally shorter for Pontiac fever than for Legionnaires' disease. Symptoms of Pontiac fever can appear within 1 to 3 days after exposure.

11.18.4 Legionnaires' Disease Case Identification

Examine sick leave records (from 6 weeks before the earliest known case to the present) to identify and then interview all employees who used 3 or more consecutive days of sick leave. If an employee experienced a pneumonia-like illness, the surveillance questionnaire should be completed. Employees who feel that they might have had symptoms of Legionnaires' disease but did not use three or more consecutive days of sick leave should also be interviewed. Employees who have experienced a pneumonia-like illness and have seen a physician should be requested to sign a medical release form to allow the company and/or OSHA to obtain additional information from their physicians.

Employees participating in medical surveys must be informed that the purpose of obtaining a proper diagnosis and sharing this information with the workers is to protect them and their fellow workers against the potential threat of legionellosis. Employees should be informed about the Privacy Act rights that protect their own medical information; physician-patient confidentiality must not be violated. Necessary medical information should be communicated only with the patient's written permission. All medical records should be handled in accordance with 29 CFR 1913.10. (*Note:* The company safety manager may need to obtain medical releases from the employees interviewed, so that amplifying information can be obtained from a company health unit or the employee's physician.) The physicians of all employees who have seen a physician and have signed a medical release will be interviewed using the physician survey questionnaire for legionellosis. Arrangements similar to those described above should be sought for permanent contract employees controlled by separate contractor organizations in the building (e.g., janitors, cafeteria workers, and security personnel).

Based on an interview with the employees' physicians, potential cases should be considered for clinical tests to detect additional cases. The test most likely to be used is a serological test to determine the antibody level of the individual. A single antibody titer of 1/256 or greater based on a physician's diagnosis of pneumonia should be interpreted as a probable case of Legionnaires' disease. If serum collected from patients during the early phase of the illness (acute phase) is available, then antibody titer levels should be determined from these samples to determine the convalescent-to-acute titer ratio. A fourfold increase in this titer will be sufficient to confirm a case of Legionnaires' disease.

Other diagnostic tests may also be appropriate. If a potential case occurred recently, then a urine antigen test may detect *Legionella pneumophila* serogroup-1 antigen. A positive urine antigen test for a diagnosed pneumonia case is also accepted as evidence of a confirmed case. This test is available only for *Legionella pneumophila* serogroup-1 infections. Samples taken from currently symptomatic individuals can be cultured for *Legionella*. A positive culture indicates confirmation. If this process detects one or more additional cases of disease, the facility should be considered to have experienced an outbreak. The immediacy of the action will depend on whether the outbreak is ongoing or occurred 30 days or more in the past. Prompt action to control exposure at the site should be taken if there is evidence that the outbreak is still occurring. Whatever the circumstances, control procedures should be initiated and medical surveillance of the workforce should continue to detect any new cases of disease and identify the water source responsible for the outbreak.

11.18.5 Sample Health Surveillance Questionnaire for Legionellosis (OSHA, 1999)

Because records show that you took sick leave for three consecutive days or more, we would like to ask you a few questions.

Name _____

Age _____

Sex _____

Work location _____

Home phone _____

Work phone _____

Dates of absence(s) _____

Stated reason for absence _____

Have you had any of the following symptoms?

Symptom	Yes	No	Comment
Fever			**Highest temperature**
Cough			
Headache			
Diarrhea			
Shortness of breath			
Chest pain			
Did you see a physician about these symptoms?			
Chest x-ray taken?			
Diagnosed as having pneumonia?			
Tested for legionellosis?			
Admitted to a hospital?			

If you were hospitalized, please provide the following information:

Physician's name _____

Physician's telephone _____

Physician's address _____

Admitting hospital _____

Admitting physician _____

Admission date _____

Date released _____

Interviewer_____

Date _____

11.18.6 Sample Physician Survey Questionnaire for Legionellosis (OSHA, 1999)

We are calling to inform you that _____, a patient of yours, is an employee at _____. He has signed a medical release giving us permission to contact you to obtain information about his recent illness. This questionnaire will be used to determine if your patient's recent illness could be classified as a pneumonia that may have been caused by exposure to *Legionella* at the workplace.

Name of physician_____

Address _____

Phone_____

Date of visits: (1st)_____ (2nd) _____ (3rd) _____

11.18.7 Sample Epidemiological Questionnaire (OSHA, 1999)

Employee's name _____

Age _____

Gender_____

Race_____

Home address _____

Date you first became ill: (—/—/—)

Symptom or Findings	Yes	No	Comment
Cough			
Shortness of breath			
History of fever			
Physical findings:			
Abnormal chest or lung findings			
Rales			
Dyspnea			
Cyanosis			
Temperature			
Other			
Chest x-ray done			Findings
Sputum culture			Results Laboratory
Sputum cultured for *Legionella*			Results Laboratory
Diagnostic testing:			
Urine antigen test			
Direct fluorescent antibody serology tests			
Indirect fluorescent antibody			
ELISA			
Laboratory			
Diagnosis or impression			

Number of days you were ill _____

Number of other individuals in your family who were ill_____

Symptoms experienced by other family members _____

Job description_____

Primary work area _____

Areas in _____building where you spend any time _____

Hours per week in each area _____

Since —/—/— how many friends have been diagnosed with pneumonia? _____

Query	Yes	No	Comment
Are you taking any oral steroid medications?			List
Have you been exposed to water mists?			Where
Were you exposed to water mist during the10 days prior to your illness?			Where
Do you shower at work?			Where
			How often
Do you eat lunch at work?			Where
Do you take breaks?			Where
Do you use the restrooms?			Where
Do you smoke in the restrooms?			Where
Do you spend extra time in the restrooms?			Where
Do you attend training courses outside of the building?			Where
Do you have a second job?			What
			Where
Where else do you spend time?			Where
Do you go to a health club?			Where
			Frequency
Do you use hot tubs or whirlpool spas?			Which
			Where
Do you attend church			Where
			Frequency
Have you had dental work performed?			Where
			Frequency
Do you go grocery shopping?			Where
			Frequency
Do you go to the movies or to theaters?			Where
			Frequency
Do you go to shopping malls?			Where
			Frequency
Do you go to other public places?			Where
			Frequency

11.19 WATER TREATMENT PROTOCOLS FOR FACILITIES WITH LEGIONNAIRES' OUTBREAK ISSUES

This section describes actions required to abate the threat of further infection in a building in which an outbreak of Legionnaires' disease has occurred. An outbreak is defined here as medically confirmed cases of Legionnaires' disease that are epidemiologically associated with a building or some portion of a building. This definition usually means that two or more confirmed cases of Legionnaires' disease have been identified within a 6-week period at the site. Under most circumstances, evacuation of the building is not recommended. Isolating individuals who are at high risk of contracting the disease from all potential sources of infection is required. Individuals at high risk include:

- Immunosuppressed
- Persons who have had organ transplants
- Individuals receiving chemotherapy including corticosteriods
- Other individuals in poor health

A medical monitoring program must be instituted to track all workers currently on sick leave. The building must be inspected to identify all potential *Legionella* sources including HVAC cooling systems (cooling towers, evaporative condensers), domestic water systems, humidifiers, and any sources of water maintained above 20°C (68°F) that have a potential for being aerosolized. Take water samples for analysis to determine serotypes and subtypes of *L. pneumophila*. This information will be helpful in identifying the source of the disease if the subtype of *L. pneumophila* has been identified in the afflicted worker population. Determine the number of colony-forming units per unit of water

Because of the 10-day to 2-week delay in obtaining sample results, corrective action should begin immediately. Because sampling for *Legionella* can be inconclusive, sampling results alone should not determine the appropriate course of action in a building where an outbreak has occurred. Flush and disinfect the water in these suspected sources. All potential sources of contamination will be assumed to be contaminated and treated accordingly in the event that an outbreak has occurred. After the treatment, collect and analyze water samples for the colony-forming units of *L. pneumophila* to determine the effectiveness of the treatment. Upon reuse of a water system following treatment, periodic maintenance and regular water sampling are essential to ensure that the maintenance continues to be effective. Make all monitoring results available to building occupants.

11.19.1 Cooling Towers and Evaporative Condensers

Both cooling towers and evaporative condensers use a fan system to move air through a recirculated water system and introduce a considerable amount of water vapor into the surroundings even with drift eliminators designed to limit vapor release. In addition, this water is typically in the 20 to 50°C (68 to 122°F) range, ideal for *L. pneumophila* growth.

11.19.1.1 Personal Protective Clothing

Provide personal protective equipment to workers who will be performing the disinfection to prevent their exposure to chemicals used for disinfection and aerosolized water-containing *Legionella*. Protective equipment may include full-length protective clothing, boots, gloves, goggles, and a full- or half-face mask that combines a HEPA filter and chemical cartridges to protect against airborne chlorine levels of up to 10 mg/L.

11.19.1.2   Water Sampling Protocol

Before starting decontamination, collect an adequate number of water samples in sterile containers. These samples should be cultured to determine the degree of contamination and the subtype of *L. pneumophila* before treatment. Collect at least three water samples (200-mL to 1-L volume). Include water from the incoming make-up water supply, the basin of the unit most distant from the make-up water source, and recirculated water from the HVAC system at its point of return to the unit.

11.19.1.3   Cleaning Mobilization

- Shut off cooling-tower.
- If possible, shut off the heat source.
- Shut off fans, if present, on the cooling tower and evaporative condenser (CT/EC).
- Shut off the system blowdown (purge) valve. Shut off the automated blowdown controller, if present, and set the system controller to manual.
- Keep make-up water valves open.
- Close building air-intake vents within at least 30 m of the CT/EC until after the cleaning procedure is complete.
- Continue operating pumps for water circulation through the CT/EC.

11.19.1.4   Initial Chemical Treatment

Clean and disinfect the entire cooling system, including attached chillers and storage tanks (sumps). Record the type and quality of all chemicals used for disinfection, the exact time the chemicals were added to the system, and the time and results of measurements of free residual chlorine (FRC) and pH. Shock-treat cooling tower water at 50-ppm FRC. Use fast-release, chlorine-containing disinfectant in pellet, granular, or liquid form. Examples of these disinfectants include sodium hypochlorite (NaOCl) or calcium hypochlorite ($Ca[OCl]_2$) calculated to achieve initial FRC of 50 mg/L:

- 3.0 lb (1.4 kg) industrial-grade NaOCl (12–15% available Cl) per 1000 gallons of cooling tower/evaporative condenser (CT/EC) water
- 10.5 lb (4.8 kg) domestic-grade NaOCl (3–5% available Cl) per 1000 gallons of CT/EC water
- 0.6 lb (0.3 kg) $Ca(OCl)_2$ per 1000 gallons of CT/EC water

Note: If significant biodeposits are present, additional chlorine may be required. If the volume of water in the CT/EC is not known, the volume may be estimated (in gallons) by multiplying the recirculation rate (in gal/min) by 10 or the refrigeration capacity (in tons) by 30.

Add dispersant simultaneously with the disinfectant (or within 15 minutes). The dispersant is best added by dissolving it in water and adding the solution to a turbulent zone in the water system. An example of a low or nonfoaming, silicate-based dispersant is automatic-dishwasher compounds. Dispersants are added at 10 to 25 lbs (4.5 to 11.25 kg) per 1000 gallons of CT/EC water.

11.19.1.5   Water Circulation

After adding disinfectant and dispersant, continue circulating the water through the system. Monitor FRC by using an FRC-measuring device, such as a swimming pool test kit, and measure the pH with a pH meter every 15 minutes for 2 hours. Add chlorine as needed to maintain FRC at ≥10 mg/L (10 ppm) for 24 hours. Adjust pH to 7.5 to 8.0 to maintain the biocidal effect of chlorine. The pH may be lowered by using any acid (e.g., muriatic acid or sulfuric acid) that is compatible with the treatment chemicals. After adding disinfectant and dispersant or after the FRC level is stable at 10 mg/L, monitor levels at 2-hour intervals and maintain the FRC at 10 mg/L for 24 hours.

11.19.1.6 Draining the System

After the FRC level has been maintained at 10 mg/L for 24 hours, drain the system. CT/EC water may be safely drained. If necessary, the drain-off may be dechlorinated by dissipation or chemical neutralization with sodium bisulfite.

11.19.1.7 Inspection Cycle

Refill and repeat initial cleaning system. Inspect the system for visual evidence of biofilm. If found, repeat initial cleaning system

11.19.1.8 Secondary Cleaning Sequence

After water from the second chemical disinfection has been drained, shut down the CT/EC. Perform mechanical cleaning (cooling tower design may require modified procedures). Inspect all water contact areas for sediment, sludge, and scale. Using brushes and/or a low-pressure water hose, thoroughly clean all CT/EC water contact areas including basin, sump, fill, spray nozzles, and fittings. Replace components as needed. If possible, clean CT/EC water contact areas within the chillers. Refill the system. Bring the chlorine level up to 10 ppm and circulate for one hour. Flush the system. Refill the system with clean water in accordance with an effective water treatment program

11.19.1.9 Testing To Return Unit to Service

Identify and eliminate all water leaks in the cooling water system. Sample the cooling water for *L. pneumophila* colony-forming units. If the sample culture results indicate detectable levels of *L. pneumophila*, repeat the chlorination and resample the water. If sampling indicates acceptable levels of *L. pneumophila*, return the unit to service.

11.19.1.10 Return Unit to Service

Once the nondetectable level for *L. pneumophila* has been achieved, institute maintenance as outlined in the Wisconsin Protocol to ensure continued safe and proper operation. Inspect the equipment monthly, and drain and clean quarterly. Treat circulating water to control microorganisms, scale, and corrosion.

Maintenance should include systematic use of biocides and rust inhibitors, preferably supplied by continuous feed. Monthly microbiologic analyses should be performed to ensure control of bacteria. Document operation and maintenance in a log or maintenance records book. Test the cooling-system water at the following intervals to verify that no significant growth of *Legionella* has occurred:

- Test weekly for the first month after return to operation.
- Test every 2 weeks for the next 2 months.
- Test monthly for the next 3 months.

The standard for *Legionella* concentration throughout the 6 months of monitoring is fewer than 10 CFU/mL (based on PathCon guidelines). If no water samples exceed this level, monitoring may be suspended.

11.19.1.11 Ongoing Maintenance

The maintenance program must continue indefinitely. If any sample contains 10 or more CFU/mL of *Legionella*, take immediate steps to reduce levels to acceptable limits. These steps may include increased frequency of application or concentration of biocides, pH adjustment, or additional shock treatments. Take new water samples and begin the testing schedule again.

11.19.2 Domestic Water Systems

Domestic water systems are designed to provide heated water for washing, cleaning, and consumption. A large building may have multiple independent systems that usually include a boiler or heater, a circulating piping system, and pipes terminating in taps and fixtures. Operating temperatures vary depending on system design, energy conservation programs, and intended use of the water. Water heaters should be kept at a minimum of 60°C (140°F), and all water should be delivered at each outlet at a minimum of 50°C (122°F). For treatment to be effective, the stagnant zones must be removed from the system. Identify all parts of the domestic water systems where water may stagnate, such as dead legs or laterals that have been capped off and storage tanks that have dead zones that are not frequently used. Rubber and plastic gaskets in the plumbing system may also serve as *Legionella* growth media. Eliminate or minimize use of these materials and substitute materials not conducive to *Legionella* growth. Identify and test the integrity of all backflow preventers to assure protection of domestic water from cross-contamination with process water through a building-code-approved method.

11.19.2.1 Water Sampling Protocol

Collect water samples before beginning treatment to determine potential contamination. Draw from 200 mL to 1 L of water from the draw-off valves of all water heaters into sterile containers. Check the temperature of the water in these units to determine if the sample temperature is significantly lower than the water heater setpoint temperature. Sample a representative number of domestic hotwater faucets or outlets. Do not flush the faucet before taking a sample, because the end section of the water system may be a source of contamination. Collect a 200-mL to 1-L preflush sample of the first hot water drawn from the outlet. Allow the water to run and measure the temperature. Collect a second, postflush sample when the water temperature is constant. Submit the water samples to a laboratory to measure the colony-forming units of *Legionella* per milliliter of water. Use the clean-up procedure below to treat all hotwater systems that have either been tested and found to contain detectable levels of *Legionella* or have been assumed to be contaminated.

11.19.2.2 Initial Cleaning

Disinfect the system using any effective chemical, thermal, or other treatment method. For example, pasteurize the hotwater system by heating the water to at least 70°C (158°F) and maintain this temperature for a minimum of 24 hours. Maintain the temperature at 70°C (158°F) and continuously flush each faucet on the system with hot water for 20 minutes. Use an accepted chemical disinfectant such as chlorine or an acceptable biocide treatment to clean the system. Thoroughly flush the system after treatment to remove all traces of the corrosive and possibly toxic chemicals. Maintain domestic water heaters at 60°C (140°F) and water delivered at the faucet at a minimum of 120°F (50°C). Where these temperatures cannot be maintained, control *Legionella* growth with a safe and effective alternative method.

11.19.2.3   Initial Sampling

After treatment, resample the hot water from each storage tank. If *Legionella* is detected, repeat the initial cleaning and resample the water system. If no measurable levels are found and all other potential sources have also been addressed, begin a regular maintenance and testing cycle.

11.19.2.4   Regular Maintenance and Testing

Use the PathCon criteria for *Legionella* in domestic water systems during the monitoring period. Test the domestic hot- or warm-water system for *Legionella* on the following schedule to assure that recontamination has not occurred:

- Weekly for the first month after resumption of operation
- Every 2 weeks for the next 2 months
- Monthly for the next 3 months

If a level of 10 or more colony-forming units per milliliter of water is present, treat the system again using the initial cleaning and sampling sequence. Resume the weekly testing aspect of regular maintenance and testing. If the levels are between 1 and 9 CFU/mL, continue monthly water sampling indefinitely. Continue efforts to determine the source of contamination. If levels remain below 1 CFU/mL, no further monitoring is necessary. Make test results available to building residents.

11.19.3   Tepid Water Systems

Warm or tepid water systems that dilute domestic hot water from a water heater with cold water upstream from the outlet source are not recommended. Warm water left in these lines is at ideal temperatures for amplification of *L. pneumophila*. Localized mixing at the source to temper very hot water is more acceptable. Another alternative is instantaneous point-of-delivery heating of water using individual steam heating systems at each outlet.

11.19.4   Domestic Coldwater Systems

Domestic coldwater systems are designed to provide water for drinking, washing, cleaning, and toilet flushing. *L. pneumophila* will not amplify at low temperatures such as those found in these systems. Coldwater storage and delivery should be at less than 20°C (68°F) to minimize growth potential. Coldwater lines near hotwater lines should be insulated. Try to eliminate stagnant places in the system, such as dead legs or storage tanks that are not routinely used. Detectable levels of *L. pneumophila* in the system may indicate contamination of the source water supply and should represent the maximum allowable level in the system. If sampling of the system indicates a level of contamination significantly greater than that of the incoming domestic water supply system, treat the system and identify the source of contamination or amplification.

11.19.4.1   Initial Cleaning

By definition, coldwater systems have no provision for heating water and therefore disinfection cannot be by heat treatment. Clean and disinfect all coldwater systems, including storage tanks, drinking fountains, water lines, and water outlets. Use an acceptable chemical disinfectant such as chlorine or other biocide. Ensure that coldwater systems are maintained so that conditions do not promote growth of *Legionella*. Maintain a temperature of 20°C (68°F). Keep residual chlorine in the range of 1 to 2 ppm. (*Note:* This level of chlorination may be excessively corrosive to metal pipes and containers; other alternatives may be needed.) Take samples according to sampling guidelines. If analysis shows no detectable *Legionella* and all other potential sources have been addressed, flush the system. Twelve hours before reentry, flush all coldwater outlets and fountains for 4 minutes.

11.19.4.2 Initial Sampling

Return the building to normal operation. Test the domestic coldwater system for *Legionella* according to the following schedule:

- Weekly for the first month after resumption of operation
- Every 2 weeks for the next 2 months
- Monthly for the next 3 months

The same criteria used for hotwater systems described above will also be used for the coldwater system during the monitoring period. A level of 10 or more colony-forming units per milliliter of water requires retreatment of the system. Following retreatment, resume weekly testing and repeat the initial sampling schedule. If levels are between 1 and 9 CFU/mL, continue monthly sampling of the water source indefinitely. Try to identify the source of contamination. If *Legionella* levels remain below 1 CFU/mL, additional monitoring is not necessary.

11.19.5 Heating, Ventilation, and Air-Conditioning Air-Distribution Systems

Under normal conditions, HVAC systems are not likely to be sources of *L. pneumophila* unless water contaminated with the bacteria enters the system. Condensate pans on coiling coils should not serve as a water source, as amplification of the bacteria can occur in water temperatures below 20°C (68°F). Improperly drained condenser pans may produce tepid conditions that can encourage microbial and fungal growth. Proper maintenance will lessen problems related to other diseases such as humidifier fever and asthmatic responses and will minimize the possibility of a Legionnaires' outbreak. For a Legionnaires' disease outbreak to be linked directly with the HVAC system, *Legionella*-contaminated water must enter the system and be aerosolized and delivered to building occupants.

11.19.5.1 Initial Inspection and Remedial Action

Examine the systems to rule out the possibility that the HVAC system is a source of *Legionella*. Inspect the entire air distribution system (including return and exhaust systems) for visual evidence of water accumulation. Eliminate all water leaks and remove any standing water found in the system. Replace or eliminate any water-damaged insulation in the system. Operate the HVAC system using 100% outside air for 8 hours before returning the building to normal operation.

11.19.5.2 Ventilation

Following a return to normal operation, keep outside air supply rates as high as possible for one month. At a minimum, the outdoor air requirements of ASHRAE Ventilation Standard 62-1989 must be met.

11.19.6 Humidifiers and Misters

Many HVAC systems supply humidified air to building occupants to maintain comfort. Improperly maintained humidifiers can be both amplifiers and disseminators of a variety of bioaerosols.

Generally, the cool temperatures in HVAC systems are not conducive to growth of *L. pneumophila*. Humidifier options include:

- Coldwater humidifiers in HVAC systems must be connected to a domestic water source and provided with a drain line to remove the water. They require rigorous maintenance to ensure that the water source does not contribute to potential problems.
- Stand-alone, console-type humidifiers that recirculate water for humidification are not recommended, because the water in these systems can quickly become contaminated with microorganisms.

Ideally, HVAC humidifiers should use steam-injection systems to eliminate potential microbe problems.

11.19.6.1   Initial Inspection

Because HVAC humidifiers discharge into the air distribution systems, inspect the unit for standing water and treat according to the HVAC air distribution system protocol provided earlier.

11.19.6.2   Initial Cleaning

Where water in humidifiers has been sampled and shown to contain measurable *Legionella* or where such water has been assumed to be contaminated with *Legionella*, use the following protocol:

- Disinfect water in piping or reservoirs, feeding the humidifier with chlorine or other effective biocides.
- Sample the humidifier water to assure that the *Legionella* has been eliminated. Samples must have no detectable colony-forming units of *Legionella*. If one or more colony-forming units are detected, repeat treatment and sampling.
- Before using the humidifier, flush the piping and reservoir thoroughly to remove biocides.
- Return the humidifier to operation.

11.19.6.3   Regular Maintenance and Testing

Ensure that an adequate maintenance program is in effect to reduce the *Legionella* growth. Water storage temperatures should be above or below 20 to 50°C (68 to 122°F), and the system must be kept clean. Use the schedule provided here to test the water system of the unit to detect recontamination with *Legionella*:

- Weekly for the first month
- Every 2 weeks for the next 2 months
- Monthly for the next 3 months

The criterion for *Legionella* in humidifier water systems during monitoring is fewer than 1 CFU/mL. If any sample shows 1 or more colony-forming units of *Legionella* per milliliter, retreat and retest the system. If no samples exceed the criterion, suspend monitoring and continue the maintenance program indefinitely.

11.19.6.4   After Mechanical Cleaning

Fill the system with water, and add chlorine to achieve a FRC level of 10 mg/L. Circulate water for 1 hour, then open the blowdown valve and flush the entire system until the water is free of turbidity. Drain the system, open any air intake vents that were closed prior to cleaning, and fill the system with water.

RESOURCES

AWWA, A Procedure for Disinfecting Water Mains, American Water Works Association, Denver, CO, 1981.

Best, M., Goetz, A., and Yu, V.L., Heat eradication measures for control of nosocomial Legionnaires' disease, *Am. J. Infection Control*, 12(1), 26–30, 1984.

Broadbent, C.R., *Legionella* in Cooling Towers: Practical Research, Design, Treatment and Control Guidelines, paper presented at 1992 Int. Symp. on *Legionella*, American Society for Microbiology, Jan. 26–29, Orlando FL, 1992.

CIBSE, *Minimizing the Risk of Legionnaires' Disease*, Chartered Institution of Building Services Engineers, London, 1987.

England, A.C. et al., Sporadic legionellosis in the U.S.: the first 1000 cases, *Ann. Inter. Med.*, 94, 164, 1981.

Gilpin, R.W., Kaplan, A.M., and Goldstein, E.F., Quantitation of *Legionella pneumophila* in one thousand commercial and industrial cooling towers, in *Proc. 48th Int. Water Conf.*, Oct. 24–26, Pittsburgh, PA, 1988, pp. 13–19.

Health Department Victoria, Guidelines for the control of Legionnaires' disease, in *Environmental Health Standards*, Melbourne, Australia, 1989.

HSE, The Prevention or Control of Legionellosis (including Legionnaires' Disease) Approved Code of Practice, Health and Safety Series Booklet HS (G)70, Library and Information Services, Health and Safety Executive, London, 1995.

Morris, G.K. and Shelton, B.G., *Legionella in Environmental Samples: Hazard Analysis and Suggested Remedial Actions*, Pathogen Control Association, Norcross, GA, 1991.

Muder, R.R., Yu, V.L., and Woo, A.H., Mode of transmission of *Legionella pneumophila*, *Arch. Intern. Med.*, 146, 1607–1612, 1986.

Muraca, P., Stout, J.E., and Yu, V.L., Comparative assessment of chlorine, heat, ozone, and UV light for killing *Legionella pneumophila* within a model plumbing system, *Appl. Environ. Microbiol.*, 53(2), 447–453, 1987.

Muraca, P.W., Stout, J.E., Yu, V.L., and Ying, Y.C., Legionnaires' disease in the work environment: implications for environmental health, *Am. Ind. Hyg. Assoc.*, 49(11), 584–590, 1988.

Muraca, P.W., Yu, V.L., and Goetz, R.N., Disinfection of water distribution systems for *Legionella*: a review of application procedures and methodologies, *Infect. Control Hosp. Epidemiol.*, 11(2), 79–88, 1990.

Nalco Chemical Company, Cooling Water Chlorination, Technifax, TF-132, Naperville, IL, 1986.

Nguyen, M.H., Stout, J.E., and Yu, V.L., Legionellosis, *Lower Respir. Tract Infect.*, 5(3), 561–584, 1991.

OSHA Technical Manual Section III: Chapter 7, Legionnaires' Disease, OSHA, Washington, D.C., 1999.

Stout, J.E., Yu, V.L., and Muraca, M.S., Isolation of *Legionella pneumophila* from the cold water of hospital ice machines: implications for origin and transmission of the organism. *Infection Control*, 7786(4), 141–146, 1985.

Stout, J.E., Yu, V.L., Muraca, M.S., Joly, J.J., Troup, N., and Tompkins, L.S., Potable water as a cause of sporadic cases of community-acquired Legionnaires' disease, *New Engl. J. Med.*, 326, 151–155, 1992.

Williams, J.F. et al., Microbial contamination of dental unit waterlines: prevalence, intensity and microbiological characteristics, *J. Am. Dental Assoc.*, 124, 59–65, 1993.

Wisconsin Division of Health, Control of *Legionella* in Cooling Towers: Summary Guidelines, Wisconsin Department of Health and Social Sciences, Madison, 1987.

CHAPTER 12

Biocides

Martha J. Boss and Dennis W. Day

CONTENTS

This chapter discusses the rationale for choosing or specifying various biocides, the known additional risks imposed through the use of biocides, the testing required to prove biocide claims, and the limitations of Material Safety Data Sheets (MSDS) in providing hazard communication information. Antimicrobial agents are substances or mixtures of substances used to destroy or suppress the growth of harmful microorganisms, whether bacteria, viruses, or fungi, on inanimate objects and surfaces. Antimicrobial products contain about 300 different active ingredients and are marketed in several formulations: sprays, liquids, concentrated powders, and gases. More than 8000 antimicrobial products are currently registered with the U.S. Environmental Protection Agency (EPA) and sold in the marketplace. Nearly 50% of antimicrobial products are registered to control infectious microorganisms in hospitals and other healthcare environments. However, public health antimicrobial products tend to be low-volume products and thus constitute less than 5% of the estimated total market for antimicrobial products. Antimicrobial products are divided into two categories based on the type of microbial pest against which the product works.

12.1 NON-PUBLIC HEALTH PRODUCTS

Non-public health products are used to control the growth of algae, odor-causing bacteria, bacteria that cause spoilage, deterioration or fouling of materials, and microorganisms infectious only to animals. This general category includes products used in cooling towers, jet fuel, paints, and treatments for textile and paper products.

12.2 PUBLIC HEALTH PRODUCTS

Public health products are intended to control microorganisms infectious to humans in any inanimate environment. The more commonly used public health antimicrobial products include the following:

- Bactericidals, to kill bacteria
- Bactericides, to kill bacteria
- Bacteriostats, to inhibit the growth of bacterial cells
- Cidal agents, to kill cells
- Fungicides, to kill fungi
- Static agents, to inhibit the growth of cells (without killing them)

12.2.1 Sterilizers (Sporicides)

Sterilization is complete destruction or elimination of all viable organisms (in or on an object being sterilized). An object is either completely sterile or not sterile, nothing in between. Sterilization is used to destroy or eliminate all forms of microbial life, including fungi, viruses, and all forms of bacteria and their spores. Spores are considered to be the most difficult form of microorganism to destroy; therefore, the EPA considers the term *sporicide* to be synonymous with *sterilizer*. Sterilization is critical to infection control and is widely used in hospitals on medical and surgical instruments and equipment. Gaseous and dry-heat sterilizers are used primarily for sterilization of medical instruments. Liquid sterilants are primarily used for delicate instruments that cannot withstand high temperatures and gases. Chemical sterilizers include low-temperature gas (ethylene oxide) and liquid chemical sterilants. Following are features of the heat sterilization methods.

12.2.1.1 Heat

For heat sterilization, consider the type of heat, the application interval, and the temperature. Endospores of bacteria are considered the most thermoduric of all cells. The destruction of test or indicator endospores guarantees sterility.

12.2.1.2 Incineration (> 500°F)

Incineration literally burns organisms from equipment or from the interior of vessels. Nonporous and nonflammable objects that can survive the heat levels needed to destroy contained organisms can be incinerated. Incineration can also be used to destroy organisms in wastes, in that the integrity of the remaining materials for future use is not an issue.

12.2.1.3 Water Boiling (100°C)

Boiling water at 100°C for 30 minutes can be effective in killing microbial pathogens and vegetative forms of bacteria. To kill endospores, and therefore sterilize the solution, very long or intermittent boiling is required. Intermittent boiling is defined as boiling for three 30-minute intervals, followed by periods of cooling.

12.2.1.4 Autoclaving (121°C)

Autoclaving is another name for pressure cooking. A temperature of 121°C for 15 minutes at a pressure of 15 lb/in.^{2} sterilizes. The effective temperature (121°C) must be maintained for the full 15 minutes. Some materials will be destroyed at these temperatures through melting.

12.2.1.5 Dry Heat/Hot-Air Oven (160 and 170°C)

These ovens maintain a temperature of 160°C for 2 hours or 170°C for 1 hour. The ovens can be used for objects that will not melt and must remain dry.

12.2.1.6 Pasteurization

Pasteurization is the use of mild heat to reduce the number of microorganisms in a product or food. In the case of pasteurization of milk, the time and temperature depend on killing potential pathogens that are transmitted in milk (e.g., *Staphylococcus*, *Streptococcus*, *Brucella abortus*, and *Mycobacterium tuberculosis*). For pasteurization of milk, the following methods can be used:

- Batch method: 63°C for 30 minutes kills most vegetative bacterial cells, including pathogens such as *Streptococcus*, *Staphylococcus*, and *Mycobacterium tuberculosis*.
- Flash method: 71°C for 15 seconds has an effect on bacterial cells similar to the batch method; for milk, this method is more conducive to industry and has fewer undesirable effects on quality or taste.

12.2.2 Disinfectants

Disinfectants kill microorganisms, but not necessarily their spores, and are not safe for application to living tissues. They are used on hard inanimate surfaces and objects to destroy or irreversibly inactivate infectious fungi and bacteria but not spores. Examples include chlorine, hypochlorites, chlorine compounds, lye, copper sulfate, and quaternary ammonium compounds. Disinfectant products are divided into two major types: hospital and general use.

12.2.2.1 Hospital Disinfectants

Hospital disinfectants are the most critical to infection control and are used on medical and dental instruments, floors, walls, bed linens, toilet seats, and other surfaces.

12.2.2.2 General Use Disinfectants

General use disinfectants are the major type of products used in households, swimming pools, and water purifiers. Disinfectants and antiseptics are distinguished on the basis of whether they are safe for application to mucous membranes, and safety often depends on the concentration of the compound. For example, sodium hypochlorite (chlorine), as added to water, is safe for drinking, but Clorox®, an excellent disinfectant, is not safe to drink.

12.3 ANTISEPTICS AND GERMICIDES

Antiseptics and germicides are used to prevent infection and decay by inhibiting the growth of microorganisms. Because these products are used in or on living humans or animals, they are considered drugs and are thus approved and regulated by the Food and Drug Administration (FDA). Antimicrobial pesticides, such as disinfectants and sanitizers, are pesticides that are intended to:

- Disinfect, sanitize, reduce, or mitigate growth or development of microbiological organisms
- Protect inanimate objects, such as floors and walls, cabinets, toilets, industrial processes or systems, paints, metalworking fluids, wood supports, surfaces water, or other chemical substances

These products protect users from contamination, fouling, or deterioration caused by bacteria, viruses, fungi, protozoa, algae, or slime. Antimicrobial pesticides, as a regulatory category, do not include certain pesticides intended for food use but do encompass pesticides with a wide array of other uses. Antimicrobials are especially important because many are public health pesticides. They help to control microorganisms (viruses, bacteria, and other microorganisms) that can cause human disease. Antimicrobial public health pesticides are used as disinfectants in medical settings. Proper use of these disinfectants is an important part of infection control activities employed by hospitals and other medical establishments.

Disinfectants and sanitizers contain toxic substances. The ability of chemicals in other household products used for cleaning to cause health effects varies greatly, from those with no known health effect to those that are highly toxic. Read and follow label instructions carefully, and provide fresh air by opening windows and doors. If it is safe for you to use electricity and the home is dry, use fans both during and after the use of disinfecting, cleaning, and sanitizing products. Be careful about mixing household cleaners and disinfectants together (check labels for cautions on this). Mixing certain types of products can produce toxic fumes and result in injury and even death.

Antiseptics are microbiocidal agents harmless enough to be applied to the skin and mucous membrane, but they should not be taken internally. Examples include mercurials, silver nitrate, iodine solution, alcohols, and detergents.

12.4 SANITIZERS

Used to reduce, but not necessarily eliminate, microorganisms from the inanimate environment to levels considered safe as determined by public health codes or regulations.

12.4.1 Food-Contact Sanitizers

Food-contact sanitizers include rinses for such surfaces as dishes and cooking utensils and for equipment and utensils found in dairies, food-processing plants, and eating and drinking establishments. These products are important because they are used at sites where food products are stored and consumed.

12.4.2 Non-Food-Contact Sanitizers

Non-food-contact surface sanitizers include carpet sanitizers, air sanitizers, laundry additives, and in-tank toilet bowl sanitizers.

12.5 DECONTAMINATION METHODS AND BIOCIDES

Decontamination methods may be very similar to those used to abate asbestos or lead. The main difference is the use of biocides, which are chemicals designed to kill life. Thus, at concentrated levels, biocides are a danger to workers. Manufacturers' instructions must be carefully followed. When researching which biocides to use and the concentrations required, the following should be considered:

- Type of biological contamination: Does the manufacturer have data showing that the product is effective against the biological contaminants at the current contamination level? Dwell time required after application of the biocide should be calculated in terms of biocide effectiveness.
- Presence of electrical wiring or ventilation equipment that could be corroded by biocides
- Flammability hazards posed by the biocides and their application methods
- General building ventilation, workplace zoning, and level of occupancy of the building
- Ability to isolate or shut down the heating, ventilation, and air conditioning (HVAC) system
- Usage in false plenums, ductwork, flexible ductwork, horizontal plenums, and ventilation hoods
- Usage in furnace, boiler, or other combustion chamber areas
- Usage on condensers, face and bypass systems, and evaporative coils
- Usage in cooling towers, sumps, liquid-filled plumbing lines and vessels, misters
- Usage in humidifiers, swamp coolers, and sprinkler systems
- Slip, trip, and fall potential when used on flooring or polyethylene sheeting
- Dermal hazard potential if personal protective equipment (PPE) is breached
- Respiratory hazards for the chosen application methods
- Waste disposal requirements

These and other questions regarding biocides and decontamination methods must be approached on a site-specific basis. Issues regarding potential adverse effects on carpets, porous materials, heirlooms, antique finishes, and other real property not slated for disposal must also be considered.

Biocides come in several forms:

- Acids or bases that are pH altering
- Chlorines, bromines, or iodines
- Chlorine dioxide
- Hydrogen peroxide
- Hypochlorite granules

- Alcohols
- Phenols such as the active ingredients in Lysol®
- Sulfur compounds
- Quaternary ammonia
- Stabilized chemical mixes
- Ozone gases

Soaps have limited biocide properties and function primarily to remove biological contamination from surfaces. Some liquid soaps have added biocides to enhance their bacterial biocide effect. Bar soaps should not be used on decontamination sites, as without sufficient drying the bar and surrounding liquids may harbor biological contaminants. Ozone gas treatments used alone or in combination with ultraviolet light treatment have shown success in eliminating airborne spores. These treatments, however, must be repeated several times to cover all spore-release cycles following initial decontamination events.

Material Safety Data Sheets for biocides may discuss hazards based on the assumption that the biocides will be used for surface applications only. Manufacturers should be consulted whenever fogging, concentrated soaking applications, high-pressure delivery systems, or usage within confined areas is anticipated. Steam cleaning without the use of biocide washes or rinses is usually not effective and may make the situation worse. The core of the stream is very hot and perhaps produces steam, but the peripheries of the core often do not retain sufficient heat to kill biological contaminants.

Dry and wet vacuuming, if used in a cleaning cycle, may aerosolize additional biological fragments, spores, and particulates to which biological contaminant are attached. Thus, vacuuming and sweeping may require additional personal and area protection for workers and building inhabitants. In general, dry removal without prior treatment of areas with biocides should be carefully evaluated in terms of increased hazards.

The choice of biological decontamination methods must be determined by an Occupational Safety and Health Administration (OSHA)-competent person and should be reflected in contract documents and resulting plans.

Due to their spent biocide content, biocides and materials treated with biocides may be considered hazardous wastes. During the development of MSDS, manufacturers are required to determine disposal options for spent chemical, but materials laced with the chemical are not included in MSDS development. In determining appropriate disposal and labeling requirements, the following general regulatory requirements should be considered: 29 CFR 1910.1200 (Hazard Communication) and equivalent requirements in 29 CFR 1926; the Resource Conservation and Recovery Act (RCRA), in regard to storage, treatment, and disposal of wastes; and Department of Transportation (DOT) requirements for labeling, marking, placarding, and transportation protocols on public byways.

12.6 ACIDS AND ALKALIZERS

Acids and alkalizers cause changes in the microenvironment of the microbes.

12.6.1 Acids

If the microenvironment is maintained at about pH 3, organisms begin to die off. The longer this lower pH is maintained, the greater the die off. Acids are used in food preservation techniques.

12.6.2 Alkalizers

Alkalizers work against Gram-positive cocci, rods, spore-formers, and some viruses. *Mycobacterium* species are resistant to alkali

12.7 ALCOHOLS

Alcohols are effective killers of vegetative bacteria and fungi but are not effective against endospores and most viruses. They are used to enhance the effectiveness of other chemical agents and work by denaturing proteins and dissolving lipids. The effectiveness of various alcohols increases with increasing molecular weight; unfortunately, their negative impact on skin also increases. Ethanol (50–70%) and isopropanol (50–70%) denature proteins and solubilize lipids and are used as antiseptics on skin.

12.8 CHLORAMINES

Chloramines are produced by, and ultimately are a combination of, chlorine and ammonia. Chloramines are slow to volatilize, release the chlorine over long periods of time, are effective in contact with organic matter, and are used in root canal surgery and for general wound disinfection. Halazone is an example of a chloramine used for emergency disinfection of water. Chloramine is used in the treatment of public water supplies to reduce tastes and odors, the by-products of disinfection such as trihalomethanes (THMs), and the level of THMs in the water. The principal disadvantages of chloramines are that they are far weaker and slower acting disinfectants than chlorine and are especially weak for inactivating certain viruses.

When chloramine is used as the principal disinfectant, ammonia is added at a point downstream from the initial chlorine application so that microorganisms, including viruses, will be exposed to the free chlorine for a short period before the chloramine is formed. Hospitals and kidney dialysis centers must be alerted when chloramines are used for water supply disinfection. Cases of chloramine-induced hemolytic anemia in patients have been reported when their dialysis water was not appropriately treated. Otherwise, no ill effects associated with the ingestion of chloraminated drinking water are documented. Chloramines can be removed from water with very low flow rates (5 to 10 minutes contact time) through shell-base activated carbon, followed by mineral zeolite media for residual ammonia adsorption.

12.9 AMMONIA AND QUATERNARY AMMONIA

Ammonia and quaternary ammonia are detergents (quaternary ammonium compounds) that disrupt cell membranes and are used as skin antiseptics and disinfectants.

12.10 DYES

Dyes are used primarily in selective and differential media and can be used intravenously and as pills or applied to the skin in liquid form. Some dyes may be strong mutagenic agents, and the actions of some are unclear. When used as gaseous chemosterilizers, these disinfectant aerosol particles should be between 1 and 5 μm in size to be most effective:

12.10.1 Formaldehyde

Formaldehyde (8%), or formalin (40%), reacts with NH_2, SH, and COOH groups to disinfect by killing endospores. Formaldehyde is toxic to humans, works best in dry environment (better penetration), and crystallizes at room temperature.

12.10.2 Ethylene Oxide

Ethylene oxide is volatile, flammable, and offers good penetration. Ethylene oxide gas is an alkylating agent used to sterilize heat-sensitive objects such as rubber and plastics.

12.10.3 Beta-Propriolactone

Beta-propriolactone is nonflammable, and is more antimicrobial and less penetrating than ethylene oxide.

12.10.4 Glutaraldehyde

Glutaraldehyde is effective at room temperature, and its microbial activity increases with heat. It is effective against certain viruses, endospores, and *Mycobacterium* species. It may irritate skin or eyes. Examples of glutaraldehyde include Sonacide®, Cidex®, and Metracide®.

12.11 HALOGENS

Halogens include iodine, chlorine, bromine, and fluorine. The disinfectant usually recommended for mold removal is a solution of one part bleach to two parts water. Commercial disinfectants are also available through janitorial supply stores. Use a household or garden sprayer and spray all surfaces that have been touched by flood water or have been soaked by water from some other source. Use a brush or broom to force the solution into crevices.

12.11.1 Iodine

Tincture of iodine (2% I_2 in 70% alcohol) inactivates proteins and is used as an antiseptic on skin. Iodine is one of the oldest (300 to 400 years) and most effective germicidal agents. It is a broad-spectrum bactericide and a good fungicide with some viricidal action. It will kill spores and is an excellent disinfectant that is effective against protozoa (amebas). It is only slightly soluble in water; iodine is available as a tincture dissolved in alcohol. Problems arise when the alcohol evaporates and the concentration of iodine increases, which can cause burning of skin.

12.11.2 Iodophors

Iodophors are combinations of iodine and organic molecules (hydrocarbons). Iodophors work by inhibiting enzyme action and are more effective than iodine. They are nonirritating, good surfactants, and nonstaining.

12.11.3 Chlorine

Chlorine (Cl_2) gas forms hypochlorous acid (HClO), a strong oxidizing agent, and is used to disinfect drinking water and as a general disinfectant. Chlorine is used as a gas dissolved in water or in combination with other chemicals. The chlorine mode of operation is not completely understood but appears to be a strong oxidizing agent as result of the following reaction:

$$Cl_2 + H_2O \rightarrow HCl + HClO \rightarrow HCl + [O]$$

Hypochlorites are used domestically and industrially for disinfection. Hypochlorites were first advocated by Semmelweiss (1846–1848) to reduce incidence of childbed fever in hospitals, and they have a broad spectrum of kill. NaOCl (sodium hypochlorite) is the active agent in Clorox®. Chlorine is a universal disinfectant that is active against all microorganisms, including bacterial spores. Potential applications for chlorine as a disinfectant include:

- Work surfaces
- Glassware
- Fixed or portable equipment and cages
- Liquids treated for discard
- Before and after vivarium entry, as a footbath

Many active chlorine compounds are available at various strengths; however, the most widely used for chemical disinfection is sodium hypochlorite. Household or laundry bleach is a solution of 5.25% (or 52,500 ppm) sodium hypochlorite. Note that a 10% or 1:10 dilution of bleach will result in a 0.525% or 5250-ppm solution of chlorine. The Centers for Disease Control and Prevention (CDC) recommends 500 ppm (1:100 dilution of household bleach) to 5000 ppm (1:10 dilution of bleach), depending on the amount of organic material present, to inactivate the human immunodeficiency virus (HIV). The strength of chlorine to be used for disinfection must be clearly indicated when described in standard operating procedures.

Chlorine solutions will gradually lose strength, so fresh solutions must be prepared frequently. Diluted solutions should be replaced after 24 hours. The stability of chlorine in solution is greatly affected by the following factors:

- Chlorine concentration
- Presence and concentration of catalysts such as copper or nickel
- pH of the solution
- Temperature of the solution
- Presence of organic material
- Ultraviolet irradiation

The chlorine solution should have the following characteristics for maximum stability:

- Low chlorine concentration
- Absence or low content of catalysts such as nickel or copper
- High alkalinity
- Low temperature
- Absence of organic materials

Chlorine should be shielded from ultraviolet light by storage in the dark in closed containers. The following factors may or may not affect chlorine biocidal activity:

- pH — Chlorine is more effective at a lower pH.
- Temperature — An increase in temperature produces an increase in bactericidal activity.
- Concentration — A fourfold increase of chlorine will result in a 50% reduction in killing time, and a twofold increase results in a 30% reduction in killing time.
- Organic material — Organic material will consume available chlorine. If the organic material contains proteins, the reaction with chlorine will form chloramines that will have some antibacterial activity. Loss due to organic materials is more significant if minute amounts of chlorine are used. Footbaths are frequently contaminated with organic material and may require more frequent changing than the 24 hours previously stated.
- Hardness — Hardness of the water does not have a slowing effect on the antibacterial action of sodium hypochlorite.
- Addition of ammonia or amino compounds — Addition of ammonia and nitrogen compounds will slow the bactericidal action of chlorine.

Other available active chlorine sources include liquid chlorine, chlorine dioxide, inorganic chloramines, organic chloramines, and halazone.

Chlorine combines with protein and rapidly decreases in concentration when protein is present. This property gives rise to swimming pool odor which is often mistaken for the odor of chlorine. In actuality, that characteristic swimming pool odor indicates that the chlorine in the water has combined with organic contaminants and is off-gassing from the pool water. The organic source may be contamination in the pool (e.g., perspiration, urine, feces). Other natural non-protein materials and plastics and cationic detergents may also inactivate chlorine.

Chlorine is a strong oxidizing agent that is corrosive to metals and should not be used on the metal parts of machines that are subject to stress when in use. Do not autoclave chlorine solutions or materials treated with them, as the residual chlorine can vaporize resulting in an inhalation hazard. Do not use chlorine in combination with ammonia, acetylene, butadiene, butane, methane, propane or other petroleum gases, hydrogen, sodium carbide, benzene, finely divided metals, or turpentine. Chlorine may cause irritation to the eyes, skin, and lungs. Wear safety goggles, rubber gloves, aprons, or other protective clothing when handling undiluted solutions.

12.12 HEAVY METALS

Heavy metals are the most ancient of antiseptics and disinfectants. Heavy metals were used by Egyptians, in the form of gold ointments and dust, and were often buried with the corpse or mummies to provide salves and ointments in the afterlife. Heavy metals have an oligodynamic (all encompassing) action and are extremely effective. They work because of the strong affinity of the metals to proteins. Metallic ions bind and adhere to the sulfhydryl groups in proteins, and enzymatic bindings are created. Stronger concentrations act as protein precipitants. Low concentrations have a subtle interference on the metabolism of the cell. Examples of heavy metal usage as disinfectants include the use of copper for ionizing water and to control algae. DaVinci and others added gold dust to ointments for wounds.

Mercuric chloride inactivates proteins by reacting with sulfide groups and is used as a disinfectant, although it occasionally is also used as an antiseptic on skin. Mercurials (inorganic mercury compounds) have a long history, with their heyday occurring during World War I. Mercurials were replaced by organic mercury compounds such as mercurochrome, methiolate, and metaphen. These compounds were used as skin antiseptics but their effects are reversed when they are washed off. Due to the toxic effects of mercury, these compounds are no longer recommended for first aid or skin disinfection.

Silver nitrate ($AgNO_3$) precipitates proteins and is used as a general antiseptic and in the eyes of newborns. Silver, as a 1% silver nitrate solution (Argyrol™), has been used as an antiseptic and in the eyes of newborn, although this practice has been largely replaced by the use of antibiotics.

Zinc is used in combination with chlorine compounds as a mouthwash and in other combinations is an effective fungicide. Organometallics (organically activated metals such as heavy metals or organic radicals such as alcohol) are effective against Gram-positive cocci, diphtheroids, spore-forming rods, tuberculosis, and similar organisms and may be effective against viruses. They are extremely effective against mycoses and have virtually no effectiveness against Gram-negative rods. Tributyltin is an example of an organometallic that also has deodorizing qualities.

12.13 OXIDIZERS

Oxidizers supply boundless oxygen. In combination with mercurials, oxidizers have been used in wound cleaning. Examples include H_2O_2 (hydrogen peroxide), $KMnO_4$ (potassium permanganate), and zinc peroxide.

12.13.1 Ozone

Ozone generators sold as air cleaners intentionally produce the gas ozone. Ozone is a molecule composed of three atoms of oxygen. Two atoms of oxygen form the basic oxygen molecule — the oxygen we breathe that is essential to life. The third oxygen atom can detach from the ozone molecule and reattach to molecules of other substances, thereby altering their chemical composition. Ozone is a toxic gas with vastly different chemical and toxicological properties from oxygen (see Table 12.1). The same chemical properties that allow high concentrations of ozone to react with organic material outside the body give it the ability to react with similar organic materials that make up the body, with potentially harmful health consequences. Relatively low amounts can cause chest pain, coughing, shortness of breath, and, throat irritation. Ozone may also worsen chronic respiratory diseases such as asthma and compromise the ability of the body to fight respiratory infections. Whether in its pure form or mixed with other chemicals, ozone can be harmful to health. When inhaled, ozone can damage the lungs.

People vary widely in their susceptibility to ozone. Healthy people, as well as those with respiratory difficulty, can experience breathing problems when exposed to ozone. Exercise during exposure to ozone causes a greater amount of ozone to be inhaled and increases the risk of harmful respiratory effects. Recovery from the harmful effects can occur following short-term exposure to low levels of ozone, but health effects may become more damaging and recovery less certain at higher levels or from longer exposures. Several federal agencies have established health standards or recommendations to limit human exposure to ozone. No agency of the federal government has approved ozone generators for use in occupied spaces.

Table 12.1 Ozone Health Effects and Standards

Health Effects	Risk Factors	Health Standards
Potential risk of experiencing:	Factors expected to increase risk and severity of health effects are:	The Food and Drug Administration (FDA) requires ozone output of indoor medical devices to be no more than 0.05 ppm.
Decreases in lung function	Increase in ozone air concentration	The Occupational Safety and Health Administration (OSHA) requires that workers not be exposed to an average concentration of more than 0.10 ppm for 8 hours.
Aggravation of asthma	Greater duration of exposure for some health effects	The National Institute of Occupational Safety and Health (NIOSH) recommends an upper limit of 0.10 ppm, not to be exceeded at any time.
Throat irritation and cough		
Chest pain and shortness of breath	Activities that raise the breathing rate (e.g., exercise)	The Environmental Protection Agency's (EPA) National Ambient Air Quality Standard for ozone is a maximum 8-hour average outdoor concentration of 0.08 ppm.
Inflammation of lung tissue	Certain preexisting lung diseases (e.g., asthma)	
Higher susceptibility to respiratory infection		

12.13.1.1 Stratospheric vs. Atmospheric and Ambient Ozone

The phrase *good up high, bad nearby* has been used by the EPA to make the distinction between ozone in the upper and lower atmosphere. Ozone in the upper atmosphere (referred to as stratospheric ozone) helps filter out damaging ultraviolet radiation from the sun. Though ozone in the stratosphere is protective, ozone in the atmosphere, which is the air we breathe, can be harmful to the respiratory system. Harmful levels of ozone can be produced by the interaction of sunlight with certain chemicals emitted to the environment (e.g., automobile emissions and chemical emissions of industrial plants). These harmful concentrations of ozone in the atmosphere are often accompanied by high concentrations of other pollutants, including nitrogen dioxide, fine particles, and hydrocarbons. Whether pure or mixed with these or other chemicals, ozone can be harmful to health.

12.13.1.2 Concentrations

At concentrations that do not exceed public health standards, ozone has little potential to remove indoor air contaminants. For many of the chemicals commonly found in indoor environments, the reaction process with ozone may take months or years (Boeniger, 1995). For all practical purposes, ozone does not react at all with such chemicals. Ozone generators are not effective in removing carbon monoxide (Salls, 1927; Shaughnessy et al., 1994) or formaldehyde (Esswein and Boeniger, 1994). In an experiment designed to produce formaldehyde concentrations representative of an embalming studio, where formaldehyde is the main odor producer, ozone showed no effect in reducing formaldehyde concentration (Esswein and Boeniger, 1994). Other experiments suggest that body odor may be masked by the smell of ozone but is not removed by ozone (Witheridge and Yaglou, 1939). Ozone is not considered useful for odor removal in building ventilation systems (ASHRAE, 1989).

Some odorous chemicals will react with ozone. For example, in some experiments, ozone appeared to react readily with certain chemicals, including some chemicals that contribute to the smell of new carpet (Weschler et al., 1992b; Zhang and Lioy, 1994). Ozone is also believed to react with acrolein, one of the many odorous and irritating chemicals found in secondhand tobacco smoke.

For many of the chemicals with which ozone does readily react, the reaction can form a variety of harmful or irritating by-products (Weschler et al., 1992a,b, 1996; Zhang and Lioy, 1994). For example, in a laboratory experiment that mixed ozone with chemicals from new carpet, ozone reduced many of these chemicals, including those that can produce new carpet odor. However, in the process, the reaction produced a variety of aldehydes, and the total concentration of organic chemicals in the air increased rather than decreased after the introduction of ozone (Weschler et al., 1992b). In addition to aldehydes, ozone may also increase indoor concentrations of formic acid (Zhang and Lioy, 1994), both of which can irritate the lungs if produced in sufficient amounts. Some of the potential by-products produced by the reactions of ozone with other chemicals are themselves very reactive and capable of producing irritating and corrosive by-products (Weschler and Shields, 1997a,b; Weschler et al., 1996). Given the complexity of the chemical reactions that occur, additional research is needed to more completely understand the complex interactions of indoor chemicals in the presence of ozone.

Some studies show that ozone concentrations produced by ozone generators can exceed health standards even when one follows manufacturer's instructions. Many factors affect ozone concentrations, including the amount of ozone produced by the machines, the size of the indoor space, the amount of material in the room with which ozone reacts, the outdoor ozone concentration, and the amount of ventilation. These factors make it difficult to control the ozone concentration in all circumstances.

Results of some controlled studies show that concentrations of ozone considerably higher than these standards are possible even when a user follows the manufacturer's operating instructions.

The many brands and models of ozone generators on the market vary in the amount of ozone produced. In many circumstances, the use of an ozone generator may not result in ozone concentrations that exceed public health standards. But, many factors affect the indoor concentration of ozone so that under some conditions, ozone concentrations may exceed public health standards.

In one study (Shaughnessy and Oatman, 1991), a large ozone generator recommended by the manufacturer for spaces up to 3000 square feet was placed in a 350-ft^2 room and run at a high setting. The ozone in the room quickly reached concentrations that were exceptionally high — 0.50 to 0.80 ppm, which is 5 to 10 times higher than public health limits.

In an EPA study, several different devices were placed in a home environment in various rooms, with doors alternately opened and closed and with the central ventilation system fan alternately turned on and off. The results showed that some ozone generators, when run at a high setting with interior doors closed, would frequently produce concentrations of 0.20 to 0.30 ppm. A powerful unit set on high with the interior doors opened achieved values of 0.12 to 0.20 ppm in adjacent rooms. When units were not run on high and interior doors were open, concentrations generally did not exceed public health standards.

The concentrations reported above were adjusted to exclude that portion of the ozone concentration brought in from the outdoors. Indoor concentrations of ozone brought in from outside are typically 0.01 to 0.02 ppm, but could be as high as 0.03 to 0.05 ppm (Hayes, 1991; USEPA, 1996b; Weschler et al., 1989, 1996; Zhang and Lioy, 1994). If the outdoor portion of ozone were included in the indoor concentrations reported above, the concentrations inside would have been correspondingly higher, increasing the risk of excessive ozone exposure. None of the studies reported here involved the simultaneous use of more than one device. The simultaneous use of multiple devices increases the total ozone output and therefore greatly increases the risk of excessive ozone exposure.

The actual concentration of ozone produced by an ozone generator depends on many factors:

- Concentrations will be higher if a more powerful device or more than one device is used, if a device is placed in a small space rather than a large space, if interior doors are closed rather than open, if the room has fewer rather than more materials and furnishings that adsorb or react with ozone, and (provided that outdoor concentrations of ozone are low) if there is less rather than more outdoor air ventilation.
- The proximity of a person to the ozone-generating device can also affect one's exposure. The concentration is highest at the point where the ozone exits from the device and generally decreases as one moves further away.

Manufacturers and vendors advise users to size the device properly to the space or spaces in which it is used. Unfortunately, some manufacturers' recommendations about appropriate sizes for particular spaces have not been sufficiently precise to guarantee that ozone concentrations will not exceed public health limits.

Ozone generators typically provide a control setting by which the ozone output can be adjusted. The ozone output of these devices is usually not proportional to the control setting. The relationship between the control setting and the output varies considerably among devices. In experiments to date, the high setting in some devices generated 10 times the level obtained at the medium setting (USEPA, 1995). Manufacturers' instructions on some devices link the control setting to room size and thus indicate what setting is appropriate for different room sizes. However, room size is only one factor affecting ozone levels in the room. In addition to adjusting the control setting to the size of the room, users have sometimes been advised to lower the ozone setting if they can smell the ozone. Unfortunately, the ability to detect ozone by smell varies considerably from person to person, and one's ability to smell ozone rapidly deteriorates in the presence of ozone. While the smell of ozone may indicate that the concentration is too high, lack of odor does not guarantee that levels are safe.

At least one manufacturer is offering units with an ozone sensor that turns the ozone generator on and off with the intent of maintaining ozone concentrations in the space below health standards.

The EPA is currently evaluating the effectiveness and reliability of these sensors and plans to conduct further research to improve society's understanding of indoor ozone chemistry. The EPA will report its findings as the results of this research become available.

12.13.1.3 Particulates

Ozone does not remove particles (e.g., dust and pollen) from the air, including the particles that cause most allergies; however, some ozone generators are manufactured with an ion generator or ionizer in the same unit. An ionizer is a device that disperses negatively (and/or positively) charged ions into the air. These ions attach to particles in the air, giving them a negative (or positive) charge so that the particles may attach to nearby surfaces such as walls or furniture, or attach to one another and settle out of the air. The effectiveness of particle air cleaners, including electrostatic precipitators, ion generators, or pleated filters, varies widely.

12.13.1.4 Biological Effect

If used at concentrations that do not exceed public health standards, ozone applied to indoor air does not effectively remove viruses, bacteria, mold, or other biological pollutants. Some data suggest that low levels of ozone may reduce airborne concentrations and inhibit the growth of some biological organisms while ozone is present, but ozone concentrations would have to be 5 to 10 times higher than public health standards allow before the ozone could decontaminate the air sufficiently to prevent survival and regeneration of the organisms once the ozone is removed (Dyas et al., 1983; Foarde et al., 1997). Even at high concentrations, ozone may have no effect on biological contaminants embedded in porous material such as duct lining or ceiling tiles (Foarde et al., 1997). In other words, ozone produced by ozone generators may inhibit the growth of some biological agents while it is present, but it is unlikely to fully decontaminate the air unless concentrations are high enough to be a health concern if people are present. Even with high levels of ozone, contaminants embedded in porous material may not be affected at all.

12.13.1.5 Water

Ozone has been extensively used for water purification, but ozone chemistry in water is not the same as ozone chemistry in air. High concentrations of ozone in air, when people are not present, are sometimes used to help decontaminate an unoccupied space from certain chemical or biological contaminants or odors (e.g., fire restoration). However, little is known about the chemical by-products left behind by these processes (Dunston and Spivak, 1997). While high concentrations of ozone in air may sometimes be appropriate in these circumstances, conditions should be sufficiently controlled to ensure that no person or pet becomes exposed. Ozone can adversely affect indoor plants and damage materials such as rubber, electrical wire coatings and fabrics and artwork containing susceptible dyes and pigments.

12.14 PHENOLS

Phenolic compounds (e.g., carbolic acid, Lysol®, hexylresorcinol, hexachlorophene) denature proteins and disrupt cell membranes. They are used as antiseptics at low concentrations and as disinfectants at high concentrations. Carbolic acid was first used by Lister in the 1860s and became the standard against which all other disinfectants were compared. Phenol is a colorless-to-white solid when pure; however, the commercial product, which contains some water, is a liquid. It has a distinct odor that is sickeningly sweet and tarry. Most people begin to smell phenol in the air at about 40 ppb, and in water at about 1 to 8 ppm. Note that these levels are lower than the levels at which adverse health effects have been observed in animals that have breathed air containing phenol or have drunk water containing phenol. Phenol evaporates more slowly than water, and a moderate amount can form a solution with water. Phenol can catch on fire. It occurs naturally or can be manmade.

Phenol is used as a slimicide (a chemical toxic to bacteria and fungi characteristic of aqueous slimes), as a disinfectant, and in medicinal preparations such as over-the-counter treatments for sore throats. Phenol is ineffective against spores and is a poor viricide; however, it is a good fungicide in terms of destroying vegetative structures. It works in combination with soap and is good in a saline solution or at warm temperature.

After World War II, a variety of phenol derivatives, phenolics, were developed, including *ortho*-phenyl phenol, which is a cresol and is found in Lysol. Other cresols are used in creosote as environmental disinfectants and fungicides. Hexylresorcinol reduces surface tension and is used as an antiseptic, in mouthwashes, and in throat lozenges. Hexachlorophene (G-11) was widely used as a topical scrub and skin antisepsis. In 1972, hexachlorophene was shown to be easily absorbed through the skin. After absorption, hexachlorophene enters the blood stream and ultimately causes neurological damage. Newborns and the elderly, with less subcutaneous fat, are at greater risk for brain damage. These compounds work by denaturing cell proteins, inactivating enzymes, and damaging cell membranes.

12.15 SOAPS AND SYNTHETIC DETERGENTS

Soaps and synthetic detergents act by mechanical removal of contaminants by decreasing surface tension on the contaminated substrate. These compounds are mildly germicidal wetting agents. A detergent is any surface-tension depressant (keeps organisms spread out).

12.15.1 Anionic Detergents

For anionic detergents, the negatively charged portion of the molecule is the active part. These detergents are not considered broadbased germicides but may work against Gram-positive bacteria. $C_{12}H_{25}OSO_3$ attached to Na^+ is a common formulation. Examples are sodium laurel sulfate and Dreft.

12.15.2 Cationic Detergents

For cationic detergents, the positively charged portion of the molecule is the active part. Cationic detergents are very germicidal; CPC (cetyl pyridinium chloride) is used in mouthwash and tooth-paste (e.g., the mouthwash Cepacol®). Cationic detergents cause inactivation of enzymes and destruction of cell membranes. Quaternary ammonium compounds can be in concentrations as low as 1:30,000 and still be cidal. Zephiran, Phemerol, Diaparene, and Ceepryn are all examples of cationic detergents.

12.15.3 Nonionic Detergents

Nonionic detergents are not germicidal, are good surfactants, and are primarily used as laundry detergents.

12.16 IRRADIATION OR ULTRAVIOLET LIGHT

Irradiation usually destroys or distorts nucleic acids. Ultraviolet light is commonly used to sterilize the surfaces of objects. X-rays and microwaves are possibly useful, especially for endospores. Many spoilage organisms are easily killed by irradiation.

12.17 DESICCANTS

For drying (removal of H_2O), the reduced water activity ($\alpha w < 0.90$) produced by dessication will destroy cellular microorganisms in their vegetative state. Desiccation is not reliable for endospores or viruses. Desiccation occurs through application of heat, evaporation, freeze-drying, addition of salt or sugar, or application of desiccant chemicals.

12.18 LOW TEMPERATURES

Most organisms grow very little or not at all at $0°C$. Low temperatures, however, are not bactericidal.

12.19 EQUIPMENT DECONTAMINATION OR DISPOSAL

Personal protective equipment and other equipment used on site must be either decontaminated or properly disposed. Because biocides can destroy some equipment, more disposal may be needed than that estimated for asbestos or lead abatement projects. Respirators can be decontaminated using chlorine solution and sequential rinses. Respirator filters and cartridges cannot be decontaminated and should not be used for more than one work day. Storage of biologically contaminated respirator filters and cartridges may cause residual biological contaminants to amplify through reproduction. Thus, if respirator filters and cartridges are worn on successive day, workers will be exposed over and over again to the growing biological contamination in their filters and cartridges.

Amplification is particularly troublesome in HEPA filters, whether used on respirators or to filter interior air streams. When used in respirators, these filters, if contaminated, may continually expose workers within the very small breathing portal space provided by the interior of respirators. If filtration units are contaminated, air containing unacceptable levels of biological contaminants may be vented from the decontamination area.

All filtration units used for area air filtration must be checked to ensure that filtration continues to be adequate for occupied spaces. When at all possible, venting should be outdoors where human receptors are not present. Filtration may be checked using the same air monitoring techniques used to gauge biological contamination in workplace air. If air is vented when biocides are in use, the level of biocides present in vented air may also need to be checked. Vessels used to store biocides may not be usable again. The manufacturers' recommendations should be consulted and documented as to reuse of pails, buckets, mops, handheld tools, polyethylene sheeting, and other equipment exposed to biocides. Porous materials, such as fibrous booms and spill mats, cannot be decontaminated and should be disposed. Vacuums, whether dry or wet, and negative air machines should be decontaminated in accordance with manufacturers' recommendations.

REFERENCES AND RESOURCES

ALA, Residential Air Cleaning Devices: Types, Effectiveness, and Health Impact, American Lung Association, Washington, D.C., 1997.

Al-Ahmady, K.K., Indoor ozone, *Fl. J. Environmental Health*, June, 8–12, 1997.

ASHRAE, *ASHRAE Handbook of Fundamentals*, American Society of Heating, Refrigerating, and Air Conditioning Engineers, Atlanta, GA, 1989, p. 12.5.

Boeniger, M.F., Use of ozone generating devices to improve indoor air quality, *Am. Indust. Hygiene Assoc. J.*, 56, 590–598.(1995)

Dunston, N.C. and Spivak, S.M., A preliminary investigation of the effects of ozone on post-fire volatile organic compounds, *J. Appl. Fire Sci.*, 6(3), 231–242, 1997.

Dyas, A., Boughton, B.J., and Das, B.C., Ozone killing action against bacterial and fungal species: microbiological testing of a domestic ozone generator, *J. Clin. Pathol.*, 36, 1102–1104, 1983.

Esswein, E.J. and Boeniger, M.F., Effects of an ozone-generating air-purifying device on reducing concentrations of formaldehyde in air, *Appl. Occup. Environ. Hygiene*, 9(2), 139–146, 1994.

Favero, M.S. and Bond, W.W., Sterilization, disinfection, and antisepsis in the hospital, in *Manual of Clinical Microbiology*, American Society for Microbiology, Washington, D.C., 1991, pp. 183–200.

Foarde, K., van Osdell, D., and Steiber, R,. Investigation of gas-phase ozone as a potential biocide, *Appl. Occup. Environ. Hygiene*, 12(8), 535–542, 1997.

Hayes, S.R., Use of an indoor air quality model (IAQM) to estimate indoor ozone levels, *J. Air Waste Manage. Assoc.*, 41, 161–170, 1991.

National Antimicrobial Information Network, http://www.epa.gov, 2002.

OSHA Technical Manual, Section III, Chapter 7, Legionnaires' disease, OSHA, Washington, D.C., 1999.

Pierce, M.W., Janczewski, J.N., Roethlisbergber, B., Pelton, M., and Kunstel, K., Effectiveness of auxiliary air cleaners in reducing its components in offices, *ASHRAE J.*, November, 1996.

Rutala, W.A., APIC guideline for selection and use of disinfectants, *Am. J. Infect. Control*, 24, 313–342, 1996.

Salls, C.M., The ozone fallacy in garage ventilation, *J. Indust. Hygiene*, 9, 12, 1927.

Sawyer, W.A., Beckwith, H.I., and Skolfield, E.M., The alleged purification of air by the ozone machine, *J. Am. Med. Assoc.*, November, 13, 1913.

Shaughnessy, R.J. and Oatman, L., The use of ozone generators for the control of indoor air contaminants in an occupied environment, in *Proc. of the ASHRAE Conf. IAQ Õ91: Healthy Buildings*, ASHRAE, Atlanta, GA, 1991.

Shaughnessy, R.J., Levetin, E., Blocker, J., and Sublette, K.L., Effectiveness of portable indoor air cleaners: sensory testing results for indoor air, *J. Int. Soc. Indoor Air Quality Climate*, 4, 179–188, 1994.

USEPA, Ozone Generators in Indoor Air Settings, EPA-600/R-95–154, report prepared for the Office of Research and Development by R. Steiber, National Risk Management Research Laboratory, U.S. Environmental Protection Agency, Research Triangle Park, NC, 1995.

USEPA, Air Quality Criteria for Ozone and Related Photochemical Oxidants, EPA/600/P-93/004aF-cF, 3v, National Center for Environmental Assessment–RTP Office, U.S. Environmental Protection Agency, Research Triangle Park, NC, (PB-185582, PB96–185590 and PB96–185608), NTIS, Springfield, VA, 1996a.

USEPA, Review of National Ambient Air Quality Standards for Ozone: Assessment of Scientific and Technical Information, EPA-452/R-96–007, staff paper, Office of Air Quality Planning and Standards, U.S. Environmental Protection Agency, Research Triangle Park, NC, 1996b.

Weschler, C.J. and Shields, H.C., Production of the hydroxyl radical in indoor air, *Environ. Sci. Technol.*, 30(11), 3250–3268, 1996.

Weschler, C.J. and Shields, H.C., Measurements of the hydroxyl radical in a manipulated but realistic indoor environment, *Environ. Sci. Technol.*, 31(12), 3719–3722, 1997a.

Weschler, C.J. and Shields, H.C., Potential reactions among indoor pollutants, *Atmos. Environ.*, 31(21), 3487–3495, 1997b.

Weschler, C.J., Shields, H.C., and Naik, D.V., Indoor ozone exposures, *J. Air Pollut. Contr. Assoc.*, 39(12), 1562–1568, 1989.

Weschler, C.J., Brauer, M., and Koutrakis, P., Indoor ozone and nitrogen dioxide: a potential pathway to the generation of nitrate radicals, dinitrogen pentaoxide, and nitric acid indoors, *Environ. Sci. Technol.*, 26(1), 179–184, 1992a.

Weschler, C.J., Hodgson A.T., and Wooley, J.D., Indoor chemistry: ozone, volatile organic compounds, and carpets, *Environ. Sci. Technol.*, 26(12), 2371–2377, 1992b.

Weschler, C.J., Shields, H.C., and Naik, D.V., The Factors Influencing Indoor Ozone Levels at a Commercial Building in Southern California: More than a Year of Continuous Observations of Tropospheric Ozone, Air and Waste Management Association, Pittsburgh, PA, 1996.

Witheridge, W.N. and Yaglou, C.P., Ozone in ventilation: its possibilities and limitations, *ASHRAE Trans.*, 45, 509–522, 1996.

Zhang, J. and Lioy, P.J., Ozone in residential air: concentrations, I/O ratios, indoor chemistry, and exposures for indoor air, *J. Int. Soc. Indoor Air Quality Climate*, 4, 95–102, 1994.

CHAPTER 13

Laws and Regulations

James D. Hollingshead and Martha J. Boss

CONTENTS

1-56670-606-8/03/$0.00+$1.50
© 2003 by CRC Press LLC

Laws and regulations are currently available to address some aspects of biological risk. Industry guidelines, insurance provider decisions, and government guidelines supplement the regulations. Regulations are not codified as yet for many biological risks associated with buildings and indoor air quality. Industry standards are the primary source of information as to recommendations.

13.1 VARYING DEFINITIONS FOR CONTAMINANTS OF CONCERN

Workplace air sampling may be used to determine either ambient (general indoor environs) or point-source-generated air contaminants. Point-source-generated contaminants are those produced by specific industrial, agricultural, commercial, or other defined work efforts; however, the definitions used for Clean Air Act Amendment (CAAA) issues may be different from those associated with workplace air sampling. For example, particulate matter emissions in air are defined in 40 CFR 60.2 as any airborne, finely divided solid or liquid material, except uncombined water, emitted to the ambient air. This definition is much broader than the Occupational Safety and Health Administration (OSHA) definition. Essentially, OSHA defines particulates in terms of respirator usage; for example, a particulate-filter respirator is an air-purifying respirator, commonly referred to as a dust or a fume respirator, which removes most of the dust or fume from the air passing through the device. So, for OSHA, the term *particulate* refers to dust or fume and not to liquid material.

Even when definitions are consistent, some contaminants have not been quantified as to risk. For instance, biological contaminants (bacteria and molds) are currently not addressed by any U.S. Environmental Protection Agency (EPA) standards. The EPA-funded research into these issues was published in 1994 as *Review of Quantitative Standards and Guidelines for Fungi in Indoor Air.* Further work toward EPA regulations addressing biological risk issues has not resulted, and regulatory definitions of quantitative biological risks, according to the EPA, have not been established for most biological contaminants.

13.2 INDOOR AIR QUALITY STANDARDS

Maintenance and operations to ensure that designed systems continue to deliver quality air are not regulated. Industry standards and good practice doctrines provide some guidance. Still, real-world maintenance and operations are largely determined by the interpretation and implementation of specified requirements at the facility. Thus, a system originally designed to maintain a certain make-up air inflow may or may not be operated and maintained so as to continue to provide that airflow. Operating procedure changes, altered maintenance priorities, and retrofitting of air-handling systems may contribute to air-handling problems.

13.3 HAZARDOUS WASTE OPERATIONS

The OSHA General Duty Clause and its requirement that an employer guarantee a safe and healthful workplace is applicable and in force. Section 5(a)(1) of the Occupational Safety and Health Act (General Duty Clause) requires the following:

A. Each employer will furnish to each of his employees employment and a place of employment which are free from recognized hazards that are causing or likely to cause death or serious physical harm to his employees and will comply with occupational safety and health standards promulgated under this Act.
B. Each employee will comply with occupational safety and health standards and all rules, regulations, and orders issued pursuant to this Act which are applicable to his own actions and conduct.

For biological decontamination work, both 29 CFR 1910 (Industry Standard) and 29 CFR 1926 (Construction Standard) are applicable. The original Superfund regulations and the resultant OSHA regulations (29 CFR 1910.1v20 and 29 CFR 1920.65) list biological contaminants as hazards that may be associated with uncontrolled wastes. While Superfund provided funding for sites vacated before 1984, the Resource Conservation and Recovery Act (RCRA) requirements under 29 CFR 1910.120 and 29 CFR 1920.65 are applicable for active sites.

The emphasis in the development of site protocols under these regulations has been on dealing with chemical and radioactive uncontrolled wastes, not uncontrolled biological wastes. Because the original intent of these regulations was to address uncontrolled biological wastes, sites where negative health effects and consequent risks to the general public can be anticipated from these biological contaminants are covered by these regulations. Thus, hazardous waste operations and emergency response (HAZWOPER) training, site-specific health and safety plans, and safety program documents are required in order for a contractor to perform biological decontamination as a remediation method at these sites.

Judgements as to the applicability of these regulations, local health codes, insurance risk management, and civil or criminal liability should be made during the development of contract documents. Proper training, medical surveillance, and delineation of the on-site hierarchy of responsibility should be outlined in the contract documents. Specifications and plans similar to those required for asbestos work, and in some cases hazardous waste work, should be fully developed as contract documents to delineate the job requirements, regulatory requirements, and anticipated hazards.

13.4 OCCUPATIONAL SAFETY AND HEALTH ADMINISTRATION

Ventilation criteria or standards are included in OSHA regulatory codes for job- or task-specific worker protection. In addition, many OSHA health standards include ventilation requirements. OSHA's construction standards (29 CFR 1926) contain ventilation standards for welding. OSHA deals with local exhaust systems in 29 CFR 1910.94 (ventilation). OSHA's compliance policy regarding violation of ventilation standards is set forth in their *Field Inspection Reference Manual*.

The lack of indoor air quality standards for general building usage was the impetus for the OSHA indoor air quality rule making. In *Talking Points for Assistant Secretary Dear* (rev. 6/17; Communications Workers of America International, Occupational Safety and Health Conference, 9:00 a.m., Thursday, June 9, 1994), the following indoor air quality (IAQ) initiatives were proposed:

> We have stepped up to a big health problem and published a proposed rule that would regulate indoor air quality and environmental tobacco smoke to protect more than 20 million exposed workers. We have taken the action to prevent thousands of heart disease deaths, hundreds of lung disease deaths, and respiratory diseases and other ailments linked to these hazards. The environmental tobacco smoke provisions would apply to more than 6 million enclosed and indoor workplaces under OSHA jurisdiction, while the indoor air provisions apply to more than 4.5 million non-industrial worksites. Hearings are to begin in the fall.

This abandoned indoor air quality regulation developed by OSHA focused on:

- Maintenance and operation of heating, ventilation, and air conditioning (HVAC) systems to reduce health effects related to indoor air pollution
- Provisions for the control of specific contaminant sources:
  - Environmental tobacco smoke (ETS)
  - Bacteria, molds
  - Volatile organic compounds (VOCs)
- Training and recordkeeping requirements

Various proponents and challengers continued to question the need for this regulation, and the regulation was ultimately abandoned.

The quality of air breathed by the workforce is addressed through OSHA's permissible exposure limits (PELs), which establish limits on certain chemicals in the workplace. These limits do not apply to sensitized individuals, the immunocompromised, or the very young and very old, and in some instances they may be gender specific. Gender specificity implies that PELs and the research on which PELs were based may not be protective of men or women in certain stages of life, especially regarding reproductive potential.

13.4.1 Personal Protective Equipment

The OSHA Bloodborne Pathogen Standard (29 CFR 1910.1030) covers workplace exposures to pathogenic biologicals that may be bloodborne or are carried in blood-derived tissue fluids. This standard requires the use of barrier methods and medical consultation for workers. Barrier methods for all mold, fungi, and yeasts follow similar conventions as those contained in 29 CFR 1910.1030. If airborne levels exceed or may exceed those judged to be healthy, respirators must be worn.

Usually, individuals involved in a biological decontamination event should assume that respirators are required during some part of the decontamination work.

Other barrier methods such as splash shields, gloves, protective coveralls, boot covers, and hoods may also be needed. These barrier methods are intended to prevent skin exposure through broken skin or mucous membrane (nose, mouth, and genitals). The added benefit is that these barrier methods, when used correctly, eliminate most of the potential for workers to carry contamination home on their own clothing. Workers must always keep in mind that personal protective equipment (PPE), when used for any contamination whether chemical or biological, is usually not totally protective. Exposures are reduced but not eliminated. Thus, the choice of initial protective equipment should be the responsibility of a competent person, as defined by OSHA.

13.4.2 OSHA General Duty Clause

Regulations controlling training, worker protection, acceptable work practices, transfer of materials, and disposal of wastes have not been developed for most biological contaminations associated with molds, fungi, and yeasts. The OSHA General Duty Clause and its requirement that an employer guarantee a safe and healthful workplace is applicable and in force. For biological decontamination work, both 29 CFR 1910 (Industry Standard) and 29 CFR 1926 (Construction Standard) are usually applicable.

13.5 INSURANCE COVERAGE

Insurance adjusters make decisions as to policy coverage. Because mold intrusion is often associated with water intrusion, decisions as to the extent of water intrusion coverage may be paramount. Both the cause and ultimate outcomes associated with water intrusion are often extremely variable. Perhaps the most difficult decision is whether the mold intrusion is the direct and only cause of biological growth. In this regard, prior maintenance and building usage that have caused mold situations must be separated from mold problems caused by covered events.

If the insurance adjuster gives incorrect advice as to water intrusion remediation, subsequent mold amplification may be in part due to this bad advice. Section 7.20h of the *Loss Recovery Guide with Standards* (LRGS; William Yobe & Associates) recommends:

When in doubt about the water or moisture source responsible for mold formation, a competent person should be consulted.

Competent person is defined in Section 0.15 as "a person who is capable through training, education and/or experience to instruct on the matter or matters at hand." Although it is the opinion of William Yobe & Associates that only professionals (e.g., engineers, industrial hygienist) should be involved with cause and origin, forensic, or mold formation evaluations, the LRGS uses the terminology of competent person as a minimum guideline.

13.6 CLEAN AIR ACT AMENDMENTS

The Clean Air Act Amendments of 1990 (CAAA90), Public Law (PL) 101–549 (42 U.S. Code [USC] 7401–7671q), is the current federal legislation regulating the prevention and control of air pollution in our environment (outdoor air). This regulation describes air pollution control requirements for geographic areas in the United States with respect to the National Ambient Air Quality Standards (NAAQS). The following air pollution concerns are regulated in the context of the CAAA:

- Motor vehicles as sources of pollutants
- Routine industrial emissions of hazardous air pollutants
- Accidental releases of highly hazardous chemicals (risk management program/plan development)
- Commercial facilities that produce energy for sale, which are addressed in terms of acid deposition control (acid rain, acid particulate potential from stack emissions)
- Emissions of chlorofluorocarbons (CFCs), halons, and other halogenated chemicals from various sources (air conditioning systems, aerosol can propellant usage, fire suppression systems)

Remember that the CAAA do not guarantee clean air; rather, the intent is to provide a benchmark for the attainment of air quality standards for a region and to determine zones of chemical influence during an accident. These air quality standards do not require clean air in all areas within a region; rather, these standards require collective attainment for the region at large. Federal and state (delegated authority) compliance initiatives may focus on limiting individual sources of air pollution in order to attain these regional goals. Biologicals are not covered by these CAAA standards.

13.7 NATIONAL EMISSION STANDARDS FOR HAZARDOUS AIR POLLUTANTS (NESHAP)

The National Emission Standards for Hazardous Air Pollutants (NESHAPs) have been established in accordance with the Clean Air Act (CAA) as amended in the CAAA. The NESHAPs definition of a hazardous air pollutant is a pollutant listed in or pursuant to section 112(b) of the Act (meaning the CAA). The NESHAPs regulate asbestos as well as various volatile organic compounds (VOCs), semivolatile organics (SVOCs), and heavy metals (e.g., lead, cadmium, mercury). These EPA regulations cover the generally available air pollutants that could compromise both outdoor and indoor air quality. Other EPA regulations deal with specific source reduction, emergency releases of hazardous substances, and toxic chemical releases associated with construction/demolition activities. The NESHAPs do not cover biological hazards.

13.8 INDUSTRY STANDARDS: AMCA, ACGIH, ANSI/ASHRAE, NFPA, AND SMACNA

To date, ambient indoor air quality has not been regulated by a federal mandate. Industry standards such as those produced by the American Society of Heating, Refrigerating, and Air-Conditioning Engineers (ASHRAE) and the American Society for Testing and Materials (ASTM) provide engineers with criteria guidance for air handling and treatment systems. These standards are primarily focused on initial design efforts whether for new building construction or retrofitting of building components.

The Air Movement and Control Association (AMCA) is a trade association that has developed standards and testing procedures for fans. The American Conference of Governmental Industrial Hygienists (ACGIH) has published widely used guidelines for industrial ventilation. The American National Standards Institute (ANSI) has produced several important standards on ventilation, including ventilation for paint spray booths, grinding exhaust hoods, and open-surface tank exhausts. Four ANSI standards were adopted by OSHA in 1971 and are codified in 29 CFR1910.94; these standards continue to be important as guides to design. ANSI has recently published a new standard for laboratory ventilation (ANSI Z9.5). The American Society of Heating, Refrigerating Air-Conditioning Engineers (ASHRAE) is a society of heating and air conditioning engineers that has produced, through consensus, a number of standards related to indoor air quality, filter performance and testing, and HVAC systems. The National Fire Protection Association (NFPA) has produced a number of recommendations that become requirements when adopted by local fire agencies. NFPA 45 lists a number of ventilation requirements for laboratory fume hood use. The Sheet Metal and Air Conditioning Contractors National Association (SMACNA) is an association representing sheetmetal contractors and suppliers. SMACNA sets standards for ducts and duct installation.

Acceptable indoor air quality is the goal of these standards. Even if all the requirements are met, air quality goals may not be achieved due to diversity of sources and contaminants in indoor air; the range of susceptibility of the population; unacceptable ambient air brought into the building without first being cleaned (cleaning of ambient outdoor air is not required by this standard); improper system operation and maintenance; or occupant perception and acceptance of indoor air quality as affected by air temperature, humidity, noise, lighting, and psychological stress.

13.8.1   ASHRAE Guideline 1-1996: The HVAC Commissioning Process

ASHRAE Guideline 1 describes a commissioning process that will ensure that HVAC systems perform in conformity with design intent. It defines the commissioning process for each phase and describes all types and sizes of HVAC systems, from pre-design through final acceptance to postoccupancy and changes in building and occupancy requirements after initial occupancy. Guideline 1-1996 addresses system adjustments required to meet actual occupancy needs within the system capacity, including when building use changes and recommissioning are warranted. It provides formats for documenting occupancy requirements, design assumptions, and the resultant design intent for the HVAC system, including the owner's assumptions and requirements, sample specifications, design intent, basis of design, and expected performance. It provides for verification and functional performance testing (testing the system for acceptance by the owner). Operation and maintenance criteria are covered, as are guidelines for periodic maintenance and recommissioning, as needed. Guideline 1-1996 includes the procedures for conducting verification and functional performance testing and maintaining system performance after initial occupancy so as to meet design intent. It also includes recommendations for corrective measures implementation and provides guidelines and a program for operator and maintenance personnel training.

13.8.2   ASHRAE Guideline 4-1993: Preparation of Operating and Maintenance Documentation for Building Systems

ASHRAE Guideline 4 is an operations and maintenance (O&M) guidance document that addresses preparing and delivering documentation that is easy to use, is simple to prepare and update, provides accurate and adequate information, and is delivered on time. It covers the format, contents, delivery, and maintenance of HVAC building systems O&M documentation normally provided by the building design and construction team members.

13.8.3   ASHRAE Guideline 12-2000: Minimizing the Risk of Legionellosis Associated with Building Water Systems

ASHRAE Guideline 12 provides information and guidance to minimize *Legionella* contamination in building water systems, as well as specific environmental and operational guidelines that contribute to the safe operation of building water systems and minimize the risk of occurrence of legionellosis.

13.8.4   ANSI/ASHRAE Standard 41.2-1987 (RA-92): Methods for Laboratory Airflow Measurement

ANSI/ASHRAE Standard 41.2 provides procedures for laboratory testing of heating, ventilating, air conditioning, and refrigerating components and equipment and does not necessarily apply to field testing of installed equipment and systems. ANSI/ASHRAE 41.2 recommends airflow measurement practices necessary to provide adequate and consistent measurement procedures used in preparing other ASHRAE standards. The testing procedures are for testing air-moving, air-handling, and air-distribution equipment and components. The particular method(s) used must include appropriate operating tolerances, instrument accuracies, and instrument precision in order to achieve the objectives of the product test. The recommendations include consideration of density effects on accurate measurement of flow rates. The procedures are for application only to flow measurements of air at pressures to the equipment not exceeding 100 in.H_2O (25-kPa) gauge. This standard does not include procedures for testing fans, blowers, exhausters, compressors, and other air-moving devices, the principal function of which is to produce a stream of moving air and which fall within the scope of ANSI/ASHRAE Standard 51 (ANSI/AMCA Standard 210).

13.8.5   ANSI/ASHRAE Standard 41.3-1989: Methods for Pressure Measurement

ANSI/ASHRAE Standard 41.3 presents recommended practices and procedures for accurately measuring steady-state, nonpulsating pressures. This standard describes methods for measurement of pressures appropriate for use in other ASHRAE standards, limited to a range of 1 psia (6.9 kPa) to 500 psia (3450 kPa). The descriptions include type of pressure, range of suitable application, expected accuracy, proper installation and operation techniques for attaining the desired accuracy, and pressure devices, such as differential pressure (head) meters, elastic element (bellows, Bourdon tube, and diaphragm sensor) gauges, manometric gauges, and pressure-spring gauges. Reference to suitable ANSI/ASME and ANSI/ISA standards is made where appropriate.

13.8.6   ANSI/ASHRAE Standard 41.6-1994 (RA-01): Methods for Measurement of Moist Air Properties

ANSI/ASHRAE Standard 41.6 recommends practices and procedures for the measurement and calculation of moist air properties in order to promote accurate measurement methods for specific use in the preparation of other ASHRAE standards. This standard recommends procedures for measurement of moist air properties in connection with establishment of the desired moist air environment for tests of heating, refrigerating, humidifying, dehumidifying, and other air conditioning equipment and determination of moisture quality in airstreams moving through or within such equipment or spaces. This standard covers methods appropriate for use in ASHRAE standard methods of test for rating and for determining compliance with ASHRAE environmental standards. The method descriptions include condition ranges over which method use is practicable and the associated attainable accuracies, as well as proper method use techniques to achieve desired accuracy. Calibration, reference standards, and traceability to National Institutes of Standards and Technology (NIST) standards help ensure accurate measurements. Specific attention is given to the wet- and dry-bulb psychrometer and the dewpoint hygrometer, while other methods are also discussed.

13.8.7   ANSI/ASHRAE Standard 52.1-1992: Gravimetric and Dust Spot Procedures for Testing Air-Cleaning Devices Used in General Ventilation for Removing Particulate Matter

ANSI/ASHRAE Standard 52.1 establishes test procedures for evaluating the performance of air-cleaning devices for removing particulate matter to establish specifications for equipment required to conduct the test, to define methods of calculation from test data, and to establish formats for reporting the results obtained. This standard establishes measurement procedures to load the air-cleaning system with a standard synthetic dust and determine the ability of the air cleaning device to remove dust as loading proceeds. A uniform performance reporting methodology for evaluating resistance to airflow and dust-holding capacity is thus established.

13.8.7.1   Atmospheric Dust Spot/Dust Spot Efficiency Test

The atmospheric dust spot test determines the efficiency of a medium efficiency air cleaner and uses ambient atmospheric dust to compare the blackening of targets both upstream and downstream of the air-cleaning device. The removal rate is based on the cleaner's ability to reduce the soiling of a downstream clean paper target. Removal is dependent on the ability of the cleaner to remove very fine particles from the air. (*Note:* This test addresses the overall efficiency of removal of a complex mixture of dust; removal efficiencies for different size particles may vary widely.) Recent studies by the EPA compare ASHRAE ratings to filter efficiencies for particles by size and have shown that efficiencies for particles in the size range of 0.1 to 1 µg are much lower than the ASHRAE rating.

13.8.7.2   Weight Arrestance Test

The weight arrestance test evaluates low-efficiency filters designed to remove the largest and heaviest particles. These filters are commonly used in residential furnaces and/or air conditioning systems or as upstream filters for other air-cleaning devices. The test measures the percentage of the synthetic dust weight that is captured by the filter and includes:

- Feeding a standard synthetic dust into the air cleaner
- Determining the dust proportion (by weight) trapped on the filter

The test uses larger standard dust particles and is of limited value in assessing the removal of smaller, respirable- size particles from indoor air. The standard discusses differences in results from the weight arrestance and the atmospheric dust spot test. No comparable guidelines or standards are currently available for use in assessing the ability of air cleaners to remove gaseous pollutants or radon and its progeny. The standard does not measure the ability of the air cleaner to remove particles of specific diameters. The standard is not intended to test air cleaners exhibiting ASHRAE dust spot efficiencies of greater than 98%.

13.8.8   ANSI/ASHRAE Standard 52.2-1999: Method of Testing General Ventilation Air-Cleaning Devices for Removal Efficiency by Particle Size

ANSI/ASHRAE Standard 52.2 establishes a laboratory method and test procedure for evaluating the general performance of the ventilation air cleaning device as a function of particle size. This standard establishes testing equipment performance specifications and defines procedures for generating the aerosols required for conducting the test, including:

- Feeding a standard synthetic dust into the air cleaner (dust is fed at intervals to simulate accumulation of particles during service life)
- Determining filter performance in removing particles of specific diameters

It provides a method for counting airborne particles from 0.30 to 10 µm in diameter upstream and downstream of the air cleaning device and to calculate removal efficiency by particle size. The standard defines methods of calculating and reporting the results obtained from the test data and establishes a minimum efficiency reporting system that can be applied to the covered air cleaning devices.

13.8.9 ANSI/ASHRAE Standard 55-1992: Thermal Environmental Conditions for Human Occupancy, including Addendum 55a-1995

ANSI/ASHRAE Standard 55 describes the combinations of indoor space environment and personal factors that will produce thermal environmental conditions acceptable to 80% or more of the occupants within a space. The environmental factors addressed are temperature, thermal radiation, humidity, and air speed; the personal factors are those of activity and clothing. Comfort in the space environment is a complex subject, and the interaction of all of the factors must be addressed. This standard specifies thermal environmental conditions acceptable for healthy people at atmospheric pressure equivalent to altitudes up to 3000 m (10,000 ft) in indoor spaces designed for human occupancy for periods not less than 15 minutes. This standard does not address such nonthermal environmental factors as air quality, acoustics, and illumination or other physical, chemical, or biological space contaminants that may affect comfort or health.

13.8.10 ANSI/ASHRAE Standard 62-2001: Ventilation for Acceptable Indoor Air Quality

ANSI/ASHRAE Standard 62 specifies minimum ventilation rates and indoor air quality that will be acceptable to human occupants and minimize the potential for adverse health effects. This standard is intended to assist professionals in the proper design of ventilation systems for buildings. It specifies minimum ventilation rates and indoor air quality that will be acceptable to human occupants, that are intended to minimize the potential for adverse health effects, and that apply to all indoor or enclosed spaces that people may occupy, except where other applicable standards and requirements dictate larger ventilation amounts. The standard includes discussion of the release of moisture in residential kitchens and bathrooms and in locker rooms and from swimming pools. It considers chemical, physical, and biological contaminants that can affect air quality. Thermal comfort requirements are not included in this standard.

13.8.10.1 Features of Standard 62-1999

Important features of this ASHRAE standard are:

- Definitions of arrestance and efficiency
- Discussion of the additional environmental parameters that must be considered
- Recommendations for summer and winter comfort zones for both temperature and relative humidity
- A guideline for adjusting for activity levels
- Guidelines for making measurements

13.8.10.2 Comparative Testing

Uniform comparative testing procedures for evaluating the performance of air cleaning devices used in ventilation systems include:

- Ventilation rate procedure — Acceptable air quality is achieved by specifying a given quantity and quality of outdoor air based upon occupant density and space usage.
- Air quality procedure — This performance specification allows acceptable air quality to be achieved within a space by controlling for known and specifiable contaminants. The procedure uses the atmospheric dust spot test and the weight arrestance test. The values obtained with these two tests are not comparable.

Generally, a range of 15 to 60 ft³/min of outdoor air for each person in the area served by the HVAC system is recommended.

13.8.10.3 Carbon Dioxide as an Indicator of Ventilation Effectiveness

Carbon dioxide (CO_2) can be used as a rough indicator of the effectiveness of ventilation. CO_2 levels above 1000 parts per million (ppm) indicate inadequate ventilation with outdoor air. Formulas are given for calculating outdoor air quantities using thermal or CO_2 information.

Ventilation (outdoor air) requirements are on an occupancy basis; however, for a few types of spaces requirements are given on a floor area basis. Tables provide a process for calculating ventilation (outdoor air) on either an occupancy or floor area basis.

13.8.11 ANSI/ASHRAE Standard 110-1995: Method of Testing Performance of Laboratory Fume Hoods

ANSI/ASHRAE Standard 110 specifies quantitative and qualitative test methods for evaluating fume containment of laboratory fume hoods. This method of testing applies to conventional, bypass, auxiliary-air, and VAV laboratory fume hoods. It is intended primarily for laboratory and factory testing but may also be used as an aid in evaluating installed performance.

13.8.12 ANSI/ASHRAE Standard 111-1988: Practices for Measurement, Testing, Adjusting, and Balancing of Building Heating, Ventilation, Air Conditioning, and Refrigeration Systems

ANSI/ASHRAE Standard 111 describes methods for evaluating building heating, ventilation, air conditioning, and refrigeration systems. This standard applies to air moving and hydronic systems, including associated air moving, circulating heat transfer fluid systems, refrigeration, electrical power, and control systems. The purposes of this standard are to:

- Provide uniform and systematic procedures for making measurements in testing, adjusting, balancing, and reporting the performance of building heating, ventilation, air conditioning, and refrigeration systems in the field
- Provide the means for evaluating the validity of collected data considering system effects
- Establish methods, procedures, and recommendations for providing field-collected data to designers, users, manufacturers, and installers of system

This standard includes methods for determining temperature, enthalpy, velocity flow rate, pressure, pressure differential, voltage, amperage, wattage, and power factor. It establishes minimum system configuration requirements to ensure that the system can be field tested and balanced, minimum instrumentation required for field measurements, procedures for field measurements used in testing and in balancing, and a format of recording and reporting test results for use in evaluating conformance with design requirements.

13.8.13 ANSI/ASHRAE Standard 113-1990: Method of Testing for Room Air Diffusion

ANSI/ASHRAE Standard 113 specifies measurement techniques for determining air speed, air temperatures, and air temperature differences in occupied spaces such as offices or similar building spaces. This standard defines a repeatable method of testing the steady-state air diffusion performance of a system of air supply outlets in spaces such as offices, provides a means of determining the ability of an air distribution system to produce an acceptable thermal environment based on air motion or air speed and air temperature distribution at specified zone heating or cooling loads, and provides both building designers and owners with a tool to quantify the air diffusion performance in a building. This test method is applicable to both prototype and field installations and all types of supply outlets, but it is not applicable to the rating of individual air supply outlets.

13.8.14 ANSI/ASHRAE Standard 120-1999: Methods of Testing to Determine Flow Resistance of HVAC Ducts and Fittings

ANSI/ASHRAE Standard 120 establishes uniform methods of laboratory testing of HVAC ducts and fittings to determine their resistance to air flow. This standard may be used to determine the change in total pressure resulting from airflow in HVAC ducts and fittings. The test results can be used to determine ductflow losses in pressure loss per unit length. Fitting losses are reported as local loss coefficients. This standard does not cover interpretation of the test data.

13.8.15 ANSI/ASHRAE/SMACNA Standard 126-2000: Methods of Testing HVAC Air Ducts

ANSI/ASHRAE/SMACNA Standard 126 provides laboratory test procedures for the evaluation of HVAC air ducts. This standard may be used to determine HVAC airduct structural strength, dimensional stability, durability, and leakage characteristics. This standard does not cover:

- Effects of aerosols, solid particulates, corrosive environments, or combustibility
- Long-term effects
- Seismic qualifications
- Underground ducts
- Plenums and equipment casings
- Ductwork hangers

13.8.16 ANSI/ASHRAE Standard 129-1997: Measuring Air Change Effectiveness

ANSI/ASHRAE Standard 129 prescribes a method for measuring air change effectiveness in mechanically ventilated spaces and buildings that meet specified criteria. The air change effectiveness is a measure of the effectiveness of outdoor air distribution to the breathing level within the ventilated space. The method of measuring air change effectiveness compares the age of air where occupants are breathing to the age of air that would occur throughout the test space if the indoor air were perfectly mixed. The standard includes measurement procedures and criteria for assessing the suitability of the test space for measurements of air exchange effectiveness.

13.9 ASTM STANDARD E-1527-00 AND REVISIONS

The fourth edition of the *Standard Practice for Environmental Site Assessments: Phase I Environmental Site Assessment Process* (ASTM E-1527-00) was published in 2000 by the ASTM Committee E-50 on Environmental Assessment. This standard incorporates three important new terms:

- 3.3.7 — business environmental risk
- 3.3.16 — historical recognized environmental condition
- 3.3.23 — material threat

Standard E-1527-00 defines good commercial and customary practice for conducting an environmental site assessment of a parcel of commercial real estate with respect to the range of contaminants within the scope of the Comprehensive Environmental Response, Compensation, and Liability Act (CERCLA; also known as Superfund) and petroleum products.

The innocent landowner defense (ILD) to CERCLA liability concept is provided in 42 U.S. Code (USC) §9601(35) and §9607(b)(3) and was included as part of the Superfund Amendments and Reauthorization Act (SARA) of 1986. The E-1527 Standard codifies tasks that, when considered in concert with each other, may satisfy the ILD to CERCLA liability. This defense is that all appropriate inquiry into the previous ownership and uses of a property consistent with good commercial or customary practice in an effort to minimize liability (§9601(35)(B) has occurred.

The concept of a material threat has always been included within the standard and, more specifically, within the definition of a recognized environmental condition (REC). The Standard defines a material threat as a physically observable or obvious threat that is reasonably likely to lead to a release that, in the opinion of the environmental professional (EP), is threatening and might result in impact to public health or the environment. Biological contamination should be considered as a potential material threat. If this contamination has or will cause an uncontrolled biological waste to be generated, certain CERCLA requirements may be applicable.

13.10 CALIFORNIA TOXIC MOLD PROTECTION ACT OF 2001

Increasingly, states are stepping up to the challenge of issuing indoor air quality (IAQ) rule-making. Senate Bill 732 was introduced to the California legislature in 2001 in response to the growing concern surrounding indoor mold and the many lawsuits caused by it. After several revisions, the bill was signed into law and became effective on January 1, 2002. Even though the new law was technically effective in 2002, it will not be possible to implement any of the changes required by this law until mid-2003 at the earliest. By July 1, 2003, the department is required to report to the legislature on the progress made in determining the feasibility of establishing permissible exposure limits (PELs) and guidelines development.

13.10.1 Mold and Permissible Exposure Limits

The first provision of the Toxic Mold Protection Act involves determining whether establishment of PELs for mold in the indoor environment is feasible. These PELs are intended to establish levels, with an adequate margin of safety, which will avoid any significant risk to public health. While the law does include a definition of indoor environment, it does not specifically define *adequate margin of safety* or *significant risk*.

The California State Department of Health Services has been assigned the task of meeting the requirements of the law. The department has been instructed to convene a task force that will advise the department on the development of the PELs and include public health and environmental health officers, medical experts, certified industrial hygienists, mold abatement experts, representative of school districts and county offices of education, employees groups and employers groups, and other affected consumers. In preparing the PELs, the department and the task force have been instructed to consider and include the latest scientific data from authoritative bodies, such as the World Health Organization, the American Industrial Hygiene Association, the American Conference of Governmental Industrial Hygienists, the New York City Department of Health, the U.S. Environmental Protection Agency, and the Centers for Disease Control and Prevention.

The criteria for adoption of PELs limits include:

- Protection of susceptible populations
- Adoption of existing standards by other authoritative bodies
- Technical and economic feasibility for compliance
- Performance of toxicological studies that relate to mold

In consideration of susceptible populations such as people in hospitals, nursing homes, and other healthcare facilities, the department has the authority to develop an alternative set of PELs. Others considered to be at greater risk from exposure to mold include pregnant women, children under six years of age, the elderly, asthmatics, people allergic to mold, and immunocompromised individuals. Other requirements of the law include the electronic notification of the public that the preparation of PELs is about to begin. Notices are to be posted on the department Internet website to inform interested persons that work on PEL development has begun. The notices will include a list of the technical documents and other information to be used in the process and announce a public comment period for those with mold-related information. All information submitted will be made available to the public. After the PELs are established, the department has the authority to amend the limits if, in the department's opinion, the PELs are too stringent. The data will be reviewed and the PELs updated every 5 years as new technology or scientific evidence becomes available.

13.10.2 Assessment Standards

The department, with the aid of the task force, is required to

- Establish standards to assess the threat to human health by the presence of both visible and invisible or hidden mold in the indoor environment
- Develop standards for determining if the presence of mold constitutes a health threat, without requiring the use of air or surface testing

The resultant mold identification guidelines are intended to assist in the identification of mold, water damage, or microbial volatile organic compounds in indoor environments. To assist in the PEL determination, the department has been instructed to develop a building inspection form that is to be used to document the presence of mold. Alternative standards for determining the threat to health in healthcare or childcare facilities and other similar facilities may be adopted. Again, the department must provide a public notice, provide for public comment, and evaluate the established procedures at least every 5 years.

13.10.3 Method Development and Validation

The department has been required to develop and validate methods for detection and identification of mold using elements for collection of air, surface, and bulk samples; visual identification; olfactory identification; laboratory analysis; measurements of amounts of moisture; and presence of mold and other recognized analytical methods. In developing these standards the department is to consider the PELs they are establishing, the existing mold identification techniques, professional judgment, and toxicological reports.

13.10.4 Criteria for Personal Protective Equipment and Sampling

The criteria for personal protective equipment (PPE) to be used during remediation activities are to be evaluated by the department; however, the use of air or surface sampling by any commercial, industrial, or residential landlord to determine whether the PELs have been exceeded is not to be required.

13.10.5 Remaining Requirements

The remaining requirements of the Toxic Mold Protection Act are not applicable until the first January 1st or July 1st that occurs at least 6 months after the department adopts the PELs, threat assessment tools are in place, and guidelines for remediation of molds from the indoor environment have been developed and disseminated by the department.

13.10.5.1 Disclosure Statements

The remaining requirements include provisions for written mold presence disclosure statements to prospective buyers or tenants of any commercial or industrial real property before the transfer of title or prior to entering a rental agreement and to current tenants of residential properties. This disclosure is not required by any of the above purchasers or tenants if the presence of mold was remediated according to the mold remediation developed by the department.

13.10.5.2 Tenant or Resident Responsibilities

The tenants of leased facilities or residences who know that mold is present in the building or that a condition of chronic water intrusion or flood exists are required to inform the owner in writing as soon as is reasonably practicable. The tenant must then make the property available to the owner to provide responsible maintenance. The law does not require landlords of commercial, industrial, or residential properties to conduct air or surface tests to determine whether the presence of mold exceeds the PELs established by the department. This section of the law does not relieve any tenant who is contractually responsible for maintenance of the property from any aspect of that responsibility, including remediation.

13.10.5.3 Realtors

The law does not provide for the assignment of liability to any listing or selling agent if the error, inaccuracy, or omission was not within the personal knowledge of the transferrer or the listing or selling agent or was based on information provided in a timely manner by public agencies or by other persons providing relevant information by delivery of a report or opinion prepared by an expert dealing with matters within the relevant scope of the professional's license or expertise, and ordinary care was exercised in obtaining and transmitting it.

13.10.5.4 Professional Services and Education

Interestingly, the law specifically states that the department is not to require a landlord, owner, seller, or transferrer to be specially trained or certified or to utilize the services of a specially qualified professional to conduct the mold remediation. Instead, the department is to make available on its website information about contracting for the removal of mold in a building. This information is to be reviewed at least every five years and should provide the recommended steps to take to hire a remediation contractor and include basic health information available in existing publications. The department is to develop resources for the education of the public in matters related to health effects of mold, methods of prevention, and identification and remediation of mold growth. These materials are to be made available and produced in other languages to accommodate the diverse multicultural population of California.

13.11 BIOCIDE PATENT PROCESS

The following information is provided per the 2107 Guidelines for Examination of Applications for Compliance with the Utility Requirement. The utility requirement is used in any patent application to evaluate compliance with the utility requirements of 35 USC 101 and 112. The utility evaluation guidelines do not:

- Alter the substantive requirements of 35 USC 101 and 112
- Obviate the examiner's review of applications for compliance with all other statutory requirements for patentability
- Constitute substantive rulemaking
- Have the force and effect of law

U.S. Patent and Trademark Office (Patent Office) personnel are to adhere to the following procedures when reviewing patent applications for compliance with the useful invention (utility) requirement of 35 USC 101 and 112, first paragraph:

(A)   Read the claims and the supporting written description.
   (1) Determine what the applicant has claimed, noting any specific embodiments of the invention.
   (2) Ensure that the claims define statutory subject matter (i.e., a process, machine, manufacture, composition of matter, or improvement thereof).
   (3) If at any time during the examination it becomes readily apparent that the claimed invention has a well-established utility, do not impose a rejection based on lack of utility. An invention has a well-established utility if:
      (i) A person of ordinary skill in the art would immediately appreciate why the invention is useful based on the characteristics of the invention (e.g., properties or applications of a product or process), and
      (ii) The utility is specific, substantial, and credible.

(B)   Review the claims and the supporting written description to determine if the applicant has asserted for the claimed invention any specific and substantial utility that is credible:
   (1) If the applicant has asserted that the claimed invention is useful for any particular practical purpose (i.e., it has a "specific and substantial utility") and the assertion would be considered credible by a person of ordinary skill in the art, do not impose a rejection based on lack of utility.
      (i) A claimed invention must have a specific and substantial utility. This requirement excludes "throw-away," "insubstantial," or "nonspecific" utilities, such as the use of a complex invention as landfill, as a way of satisfying the utility requirement of 35 USC 101.
      (ii) Credibility is assessed from the perspective of one of ordinary skill in the art in view of the disclosure and any other evidence of record (e.g., test data, affidavits or declarations from experts in the art, patents, or printed publications) that is probative of the applicant's assertions. An applicant need only provide one credible assertion of specific and substantial utility for each claimed invention to satisfy the utility requirement.
   (2) If no assertion of specific and substantial utility for the claimed invention made by the applicant is credible and the claimed invention does not have a readily apparent well-established utility, reject the claim(s) under 35 USC 101 on the grounds that the invention as claimed lacks utility. Also reject the claims under 35 USC 112, first paragraph, on the basis that the disclosure fails to teach how to use the invention as claimed. The 35 USC 112, first paragraph, rejection imposed in conjunction with a 35 USC 101 rejection should incorporate by reference the grounds of the corresponding 35 USC 101 rejection.
   (3) If the applicant has not asserted any specific and substantial utility for the claimed invention and it does not have a readily apparent well-established utility, impose a rejection under 35 USC 101, emphasizing that the applicant has not disclosed a specific and substantial utility for the invention. Also impose a separate rejection under 35 USC 112, first paragraph, on the basis that the applicant has not disclosed how to use the invention due to the lack of a specific and substantial utility. The 35 USC 101 and 112 rejections shift the burden of coming forward with evidence to the applicant to:

(i) Explicitly identify a specific and substantial utility for the claimed invention; and

(ii) Provide evidence that one of ordinary skill in the art would have recognized that the identified specific and substantial utility was well-established at the time of filing. The examiner should review any subsequently submitted evidence of utility using the criteria outlined above. The examiner should also ensure that there is an adequate nexus between the evidence and the properties of the now claimed subject matter as disclosed in the application as filed. That is, the applicant has the burden to establish a probative relation between the submitted evidence and the originally disclosed properties of the claimed invention.

(C) Any rejection based on lack of utility should include a detailed explanation why the claimed invention has no specific and substantial credible utility. Whenever possible, the examiner should provide documentary evidence regardless of publication date (e.g., scientific or technical journals, excerpts from treatises or books, or U.S. or foreign patents) to support the factual basis for the *prima facie* showing of no specific and substantial credible utility. If documentary evidence is not available, the examiner should specifically explain the scientific basis for his or her factual conclusions.

(1) Where the asserted utility is not specific or substantial, a *prima facie* showing must establish that it is more likely than not that a person of ordinary skill in the art would not consider that any utility asserted by the applicant would be specific and substantial. The *prima facie* showing must contain the following elements:

(i) An explanation that clearly sets forth the reasoning used in concluding that the asserted utility for the claimed invention is neither specific and substantial nor well established;

(ii) Support for factual findings relied upon in reaching this conclusion; and

(iii) An evaluation of all relevant evidence of record, including utilities taught in the closest prior art.

(2) Where the asserted specific and substantial utility is not credible, a *prima facie* showing of no specific and substantial credible utility must establish that it is more likely than not that a person skilled in the art would not consider credible any specific and substantial utility asserted by the applicant for the claimed invention. The *prima facie* showing must contain the following elements:

(i) An explanation that clearly sets forth the reasoning used in concluding that the asserted specific and substantial utility is not credible;

(ii) Support for factual findings relied upon in reaching this conclusion; and

(iii) An evaluation of all relevant evidence of record, including utilities taught in the closest prior art

(3) Where no specific and substantial utility is disclosed or is well established, a *prima facie* showing of no specific and substantial utility need only establish that applicant has not asserted a utility and that, on the record before the examiner, there is no known well-established utility.

Patent Office personnel:

• Treat as true an applicant's statement of fact made in relation to an asserted utility, unless countervailing evidence can be provided that shows that one of ordinary skill in the art would have a legitimate basis to doubt the credibility of such a statement.

• Must accept an opinion from a qualified expert that is based upon relevant facts whose accuracy is not being questioned.

• Do not disregard the opinion solely because of a disagreement over the significance or meaning of the facts offered.

Once a *prima facie* showing of no specific and substantial credible utility has been properly established, the applicant bears the burden of rebutting it. The applicant can do this by amending the claims, by providing reasoning or arguments, or by providing evidence in the form of a declaration under 37 CFR 1.132 or a patent or a printed publication that rebuts the basis or logic of the *prima facie* showing.

The Patent Office must examine each application to ensure compliance with the useful invention or utility requirement of 35 USC 101. In discharging this obligation, however, Patent Office personnel must keep in mind several general principles that control application of the utility requirement. As interpreted by the federal courts, 35 USC 101 has two purposes:

1. 35 USC 101 defines which categories of inventions are eligible for patent protection. An invention that is not a machine, an article of manufacture, a composition or a process cannot be patented. See *Diamond v. Chakrabarty*, 447 U.S. 303, 206 USPQ 193 (1980); *Diamond v. Diehr*, 450 U.S. 175, 209 USPQ 1 (1981).
2. 35 USC 101 serves to ensure that patents are granted on only those inventions that are useful. This second purpose has a Constitutional footing — Article I, Section 8 of the Constitution authorizes Congress to provide exclusive rights to inventors to promote the useful arts. See *Carl Zeiss Stiftung v. Renishaw PLC*, 945 F.2d 1173, 20 USPQ2d 1094 (Fed. Cir. 1991).

Thus, to satisfy the requirements of 35 USC 101, an applicant must claim an invention that is statutory subject matter and must show that the claimed invention is *useful* for some purpose either explicitly or implicitly. Application of this latter element of 35 USC 101 is the focus of these guidelines. Deficiencies under the useful invention requirement of 35 USC 101 will arise in one of two forms:

1. It is not apparent why the invention is useful. This can occur when an applicant fails to identify any specific and substantial utility for the invention or fails to disclose enough information about the invention to make its usefulness immediately apparent to those familiar with the technological field of the invention. See *Brenner v. Manson*, 383 U.S. 519, 148 USPQ 689 (1966); *In re Ziegler*, 992 F.2d 1197, 26 USPQ2d 1600 (Fed. Cir. 1993).
2. An assertion of specific and substantial utility for the invention made by an applicant is not credible (a rare occurrence).

13.11.1 Specific and Substantial Requirements

To satisfy 35 USC 101, an invention must be useful. The Patent Office relies on the inventor's understanding of the invention in determining whether and in what regard an invention is believed to be useful. Patent Office personnel focus on and are receptive to assertions made by the applicant that an invention is useful for a particular reason. If an invention is only partially successful in achieving a useful result, a rejection of the claimed invention as a whole based on a lack of utility is not appropriate. See *In re Brana*, 51 F.3d 1560, 34 USPQ2d 1436 (Fed. Cir. 1995); *In re Gardner*, 475 F.2d 1389, 177 USPQ 396 (CCPA), reh'g denied, 480 F.2d 879 (CCPA 1973); *In re Marzocchi*, 439 F.2d 220, 169 USPQ 367 (CCPA 1971). See also *E.I. du Pont De Nemours and Co. v. Berkley and Co.*, 620 F.2d 1247, 1260 n.17, 205 USPQ 1, 10 n.17 (8th Cir. 1980).

The invention does not need to accomplish all of its intended functions or operate under all conditions. Partial success is sufficient to demonstrate patentable utility. In short, the defense of nonutility cannot be sustained without proof of total incapacity. Biocides or their application equipment could be only partially successful in achieving their utility and still could be patentable.

13.11.1.1 *Practical Utility*

The Court of Customs and Patent Appeals has stated: "'Practical utility' is a shorthand way of attributing real-world value to claimed subject matter." In other words, one skilled in the art can use a claimed discovery in a manner that provides some immediate benefit to the public. See *Nelson v. Bowler*, 626 F.2d 853, 856, 206 USPQ 881, 883 (CCPA 1980). Biocides would need to be of value in the real world and, thus, possess a practical utility.

13.11.1.2 Specific Utility

A *specific utility* is specific to the subject matter claimed. This contrasts with a general utility that would be applicable to the broad class of the invention. Patent Office personnel distinguish between situations where an applicant either:

- Discloses an invention's specific use or application
- Indicates that the invention may prove useful without specifically identifying an invention's specific use or application

Indicating that a biocide has useful biological properties would not be sufficient to define the specific utility of a biocide. The biocide would also have to be described as useful. The useful description may be made with or without specifically disclosing the ultimate use or application of the biocide.

13.11.1.3 Substantial Utility

A *substantial utility* defines a real world use. Utilities that require or constitute carrying out further research to identify or reasonably confirm a real world context of use are not substantial utilities. The following examples are situations that require or constitute carrying out further research to identify or reasonably confirm a real-world context of use and, therefore, do not define substantial utilities:

- Basic research (e.g., studying the claimed product properties or the mechanisms in which the material is involved)
- Methods of treating an unspecified condition, assaying for or identifying a material that has no specific and/or substantial utility, and making a material that has no specific, substantial, and credible utility
- A claim to an intermediate product used to make a final product that has no specific, substantial, and credible utility

Patent Office personnel are careful not to interpret the phrase *immediate benefit to the public* or similar wording to mean that products or services based on the claimed invention must be currently available to the public in order to satisfy the utility requirement. See, for example, *Brenner v. Manson*, 383 U.S. 519, 534–35, 148 USPQ 689, 695 (1966).

Any reasonable use that an applicant has identified for the invention that can be viewed as providing a public benefit is usually accepted as sufficient, at least with regard to defining a *substantial* utility. Biocides with reasonable use, even if not currently available to the public, may be patented. If further research is required to prove reasonable use, the biocide would not be immediately patentable. Similarly, the biocide research used to prove patentability would not in and of itself be patentable.

13.11.2 Therapeutic or Pharmacological Utility

The Federal Circuit has reiterated that therapeutic utility sufficient under the patent laws is not to be confused with the requirements of the Food and Drug Administration (FDA) with regard to safety and efficacy of drugs marketed in the United States. See *Scott v. Finney*, 34 F.3d 1058, 1063, 32 USPQ2d 1115, 1120 (Fed. Cir. 1994). FDA approval is not a prerequisite for finding a compound useful within the meaning of the patent laws. Therapeutic or pharmacological utility is not applicable to biocides used on inanimate objects. Only biocides used on a life form — humans — would be subject to the requirements listed as therapeutic or pharmacological.

13.11.3 Claimed Invention Focus

The claimed invention is the assessment focus as to whether an applicant has satisfied the utility requirement. Each claim (i.e., each invention) must be evaluated on its own merits for compliance with all statutory requirements. A dependent claim will define an invention that has utility — if the independent claim upon which it depends has defined an invention having utility. An exception is where the invention's utility defined in a dependent claim differs from the utility indicated for independent claim upon which it depends. Where an applicant has established utility for a species that falls within an identified genus of compounds and presents a generic claim covering the genus, as a general matter, that claim will be treated as being sufficient under 35 USC 101. Only where it can be established that other species clearly encompassed by the claim do not have utility will a rejection be imposed on the generic claim. In such cases, the applicant should be encouraged to amend the generic claim so as to exclude the species that lack utility.

It is common and sensible for an applicant to identify several specific utilities for an invention, particularly when the invention is a product (e.g., a machine, an article of manufacture or a composition of matter). Regardless of the category of invention that is claimed (e.g., product or process), an applicant need only make one credible assertion of specific utility for the claimed invention to satisfy 35 USC 101 and 35 USC 112

Additional statements of utility, even if not credible, do not render the claimed invention lacking in utility. See, for example, *Raytheon v. Roper*, 724 F.2d 951, 958, 220 USPQ 592, 598 (Fed. Cir. 1983), cert. denied, 469 U.S. 835 (1984). (When a properly claimed invention meets at least one stated objective, utility under 35 USC 101 is clearly shown.) If an applicant makes one credible assertion of utility, utility for the claimed invention as a whole is established.

Statements made by the applicant in the specification or incident to prosecution of the application before the Patent Office cannot, standing alone, be the basis for a lack of utility rejection under 35 USC 101 or 35 USC 112. See *Tol-O-Matic, Inc. v. Proma Produkt-Und Mktg. Gesellschaft m.b.h.*, 945 F.2d 1546, 1553, 20 USPQ2d 1332, 1338 (Fed. Cir. 1991). (It is not required that a particular characteristic set forth in the prosecution history be achieved in order to satisfy 35 USC 101.)

An applicant may include statements in the specification whose technical accuracy cannot be easily confirmed, if those statements are not necessary to support the invention's patentability with regard to any statutory basis. The Patent Office will not require an applicant to strike nonessential statements relating to utility from a patent disclosure, regardless of the technical accuracy of the statement or assertion it presents. Patent Office personnel are especially careful not to read into a claim unclaimed results or an invention's limitations or embodiments. Doing so can inappropriately change the relationship of an asserted utility to the claimed invention and raise issues not relevant to examination of that claim. See *Carl Zeiss Stiftung v. Renishaw PLC*, 945 F.2d 1173, 20 USPQ2d 1094 (Fed. Cir. 1991); *In re Krimmel*, 292 F.2d 948, 130 USPQ 215 (CCPA 1961).

The key element in determining whether a biocide can be patented is the acceptance of one specific utility. Other extraneous information provided during the patent process may or may not be of concern.

13.11.4 Asserted or Well-Established Utility

Upon initial examination, the patent examiner will review the specification to determine if any statements are included asserting that the claimed invention is useful for any particular purpose. A complete disclosure should include a statement that identifies the specific and substantial utility of an invention.

13.11.5 Specific and Substantial Utility

Applicants are required, when their applications are filed, to disclose the best mode known to them of practicing the invention. A statement of specific and substantial utility should fully and clearly explain why the applicant believes the invention is useful. Such statements will usually explain the purpose of or how the invention may be used. Regardless of form, the statement of utility must enable one who is ordinarily skilled in the art to understand why the applicant believes the claimed invention is useful.

Except where an invention has a well-established utility, the failure of an applicant to specifically identify why an invention is believed to be useful renders the claimed invention deficient under 35 USC 101 and 35 USC 112, first paragraph. In such cases, the applicant has failed to identify a specific and substantial utility for the claimed invention.

A statement that a composition has an unspecified biological activity or that does not explain why a composition with that activity is believed to be useful fails to set forth a specific and substantial utility. A disclosure that identifies a particular biological activity of a compound and explains how that activity can be utilized in a particular application of the compound does contain an assertion of specific and substantial utility for the invention.

13.11.6 Evaluating the Credibility of Asserted Utility

In most cases, an applicant's assertion of utility creates a presumption of utility that will be sufficient to satisfy the utility requirement of 35 USC 101. See *In re Langer,* 503 F.2d at 1391, 183 USPQ at 297. Langer and subsequent cases direct the Patent Office to presume that a statement of utility made by an applicant is true. To overcome the presumption of truth that an assertion of utility by the applicant enjoys, Patent Office personnel must establish that it is more likely than not that one of ordinary skill in the art would doubt (i.e., question) the truth of the statement of utility. The evidentiary standard to be used throughout *ex parte* examination in setting forth a rejection is a preponderance of the totality of the evidence under consideration. Patent Office personnel must determine if the assertion of utility is credible (i.e., whether the assertion of utility is believable to a person of ordinary skill in the art based on the totality of evidence and reasoning provided). An assertion is credible unless

- The logic underlying the assertion is seriously flawed.
- The facts upon which the assertion is based are inconsistent with the logic underlying the assertion.

Credibility as used in this context refers to the reliability of the statement based on the logic and facts that are offered by the applicant to support the assertion of utility.

Patent Office personnel should be careful, however, not to label certain types of inventions as incredible or speculative as such labels do not provide the correct focus for the evaluation of an assertion of utility. Incredible utility is a conclusion, not a starting point, for analysis under 35 USC 101. A conclusion that an asserted utility is incredible can be reached only after the Patent Office has evaluated both the assertion of the applicant regarding utility and any evidentiary basis of that assertion. The Patent Office will be particularly careful not to start with a presumption that an asserted utility is, per se, incredible and then proceed to base a rejection under 35 USC 101 on that presumption.

Rejections under 35 USC 101 have rarely been sustained by federal courts. The 35 USC 101 rejection has been sustained because the applicant either failed to disclose any utility for the invention or asserted a utility that could only be true if the utility violated a scientific principle, such as the second law of thermodynamics or a law of nature or was wholly inconsistent with contemporary knowledge in the art. See *In re Gazave*, 379 F.2d 973, 978, 154 USPQ 92, 96 (CCPA 1967). Reasonable assertions of utility for a biocide would generally be accepted.

In appropriate situations, the Patent Office may require an applicant to substantiate an asserted utility for a claimed invention. See *In re Pottier*, 376 F.2d 328, 330, 153 USPQ 407, 408 (CCPA 1967). (When the operativeness of any process would be deemed unlikely by one of ordinary skill in the art, it is not improper for the examiner to call for evidence of operativeness.) See also *In re Jolles*, 628 F.2d 1322, 1327, 206 USPQ 885, 890 (CCPA 1980); *In re Citron*, 325 F.2d 248, 139 USPQ 516 (CCPA 1963); *In re Novak*, 306 F.2d 924, 928, 134 USPQ 335, 337 (CCPA1962).

In *re Citron*, the court held that when an alleged utility appears to be incredible in the light of the knowledge of the art, or factually misleading, applicant must establish the asserted utility by acceptable proof; 325 F.2d at 253, 139 USPQ at 520. The court thus established a higher burden on the applicant where the statement of use is incredible or misleading. In such a case, the examiner should challenge the use and require sufficient evidence of operativeness.

The purpose of this authority is to enable an applicant to cure an otherwise defective factual basis for the operability of an invention. This is a curative authority (e.g., evidence is requested to enable an applicant to support an assertion that is inconsistent with the facts of record in the application). Patent Office personnel will indicate why the factual record is defective in relation to the assertions of the applicant and, where appropriate, what type of evidentiary showing can be provided by the applicant to remedy the problem.

Requests for additional evidence will be imposed rarely and only if necessary to support the scientific credibility of the asserted utility (e.g., if the asserted utility is not consistent with the evidence of record and current scientific knowledge). As the Federal Circuit recently noted, only after the Patent and Trademark Office (PTO) provides evidence showing that one of ordinary skill in the art would reasonably doubt the asserted utility does the burden shift to the applicant to provide rebuttal evidence sufficient to convince such a person of the invention's asserted utility.

13.11.6.1 *Evidence*

The character and amount of evidence needed to support an asserted utility will vary depending on what is claimed (*ex parte* Ferguson, 117 USPQ 229 [Bd. App. 1957]) and whether the asserted utility appears to contravene established scientific principles and beliefs. See *In re Gazave*, 379 F. 2d 973, 978, 154 USPQ 92, 96 (CCPA 1967); *In re Chilowsky*, 229 F.2d 457, 462, 108 USPQ 321, 325 (CCPA 1956).

The applicant does not have to provide evidence sufficient to establish that an asserted utility is true beyond a reasonable doubt. See *In re Irons*, 340 F.2d 974, 978, 144 USPQ 351, 354 (CCPA 1965). Nor must the applicant establish an asserted utility as a matter of statistical certainty. See *Nelson v. Bowler*, 626 F.2d 853, 856–57, 206 USPQ 881, 883–84 (CCPA 1980) (reversing the Board and rejecting Bowler's arguments that the evidence of utility was statistically insignificant; the court pointed out that a rigorous correlation is not necessary when the test is reasonably predictive of the response).

Evidence will be sufficient if, considered as a whole, this evidence leads a person of ordinary skill in the art to conclude that the asserted utility is more likely than not true. No predetermined amount or character of evidence must be provided by an applicant to support an asserted utility — therapeutic or otherwise. Predetermined evidence sets are not required to evaluate biocides.

As a general matter, evidence of pharmacological or other biological activity of a compound will be relevant to an asserted therapeutic use if a reasonable correlation between the activity in question and the asserted utility exists. See *Cross v. Iizuka*, 753 F.2d 1040, 224 USPQ 739 (Fed. Cir. 1985); *In re Jolles*, 628 F.2d 1322, 206 USPQ 885 (CCPA 1980); *Nelson v. Bowler*, 626 F.2d 853, 206 USPQ 881 (CCPA 1980). An applicant can establish this reasonable correlation by relying on statistically relevant data documenting the activity of a compound or composition, arguments or reasoning, documentary evidence (e.g., articles in scientific journals), or any combination thereof.

The applicant does not have to prove that a correlation exists between a particular activity and an asserted therapeutic use of a compound as a matter of statistical certainty, nor does he or she have to provide actual evidence of success in treating humans where such a utility is asserted. As the courts have repeatedly held, all that is required is a reasonable correlation between the activity and the asserted use. See *Nelson v. Bowler*, 626 F.2d 853, 857, 206 USPQ 881, 884 (CCPA 1980). A reasonable correlation between biocide biological activity and the biocide's asserted use must be established.

13.11.6.2 Structural Similarity

Patent Office personnel should evaluate the structural relationship existence and the applicant or a declarant reasoning. This reasoning is used to explain why structural similarity is believed to be relevant to the applicant's assertion of utility. See *In re Wooddy*, 331 F.2d 636, 639, 141 USPQ 518, 520 (CCPA 1964). (It appears that no one on Earth is certain as of the present whether the process claimed will operate in the manner claimed. Yet absolute certainty is not required by the law. The mere fact that something has not previously been done clearly is not, in itself, a sufficient basis for rejecting all applications purporting to disclose how to do it.) Structural similarity (and this includes chemical structure) may be used to establish utility.

13.11.7 Safety and Efficacy Considerations

The Patent Office must confine review of patent applications to the statutory requirements of the patent law. Other agencies of the government have been assigned the responsibility of ensuring conformance to standards established by statute for the advertisement, use, sale or distribution of drugs. The FDA pursues a two-prong test to provide approval for testing. Under that test, a sponsor must show that the investigation does not pose an unreasonable and significant risk of illness or injury and that an acceptable rationale for the study exists.

As a review matter, a rationale for believing that the compound could be effective must be established. If the use reviewed by the FDA is not set forth in the specification, the FDA review may not satisfy 35 USC 101. If the review is set forth in the specification, Patent Office personnel must be extremely hesitant to challenge utility. In such a situation, experts at the FDA have assessed the rationale for the drug or research study upon which an asserted utility is based and found it satisfactory. Thus, in challenging utility, Patent Office personnel must be able to carry their burden that there is no sound rationale for the asserted utility even though experts designated by Congress to decide the issue have come to an opposite conclusion. FDA approval, however, is not a prerequisite for finding a compound useful within the meaning of the patent laws.

The FDA does not approve biocides used on inanimate objects, unless these inanimate objects are used in food preparation. If the FDA approves a substance (biocide), this does not mean that the Patent Office will approve it. The PTO process is not similar to the FDA process. The Patent Office will examine a claimed invention or process to determine whether it is novel and not obvious in light of the prior art, and the FDA is checking for product efficacy and safety, so these agencies are concerned about two different things.

13.12 FIFRA AND EPA REGULATION OF ANTIMICROBIALS

Under the Federal Insecticide, Fungicide and Rodenticide Act (FIFRA), the EPA requires companies that register public health antimicrobial pesticide products to ensure the safety and effectiveness of their products before they are sold or distributed. Any registrant, dealer, retailer, or distributor who violates the Act may be assessed a civil penalty of not more than $5000 for each offense. (The previous EPA regulation may be changed due to the passage of the Food Quality Protection Act [FQPA]).

An inanimate object or surface does not have the qualities of a living organism. Antimicrobial agents used on inanimate objects and surfaces are regulated as pesticides by the EPA. Products intended for the control of fungi, bacteria, viruses, or other microorganisms in or on living humans or animals are considered drugs, not pesticides, and are therefore regulated by the FDA. See 40 CFR 152.5(d) and 152.8(a).

Manufacturers are required to submit to EPA detailed and specific information concerning the chemical composition of their product, effectiveness data to document their claims against specific microorganisms and to support the directions for use provided in labeling, labeling that reflects the required elements for safe and effective use, and toxicology data to document any hazards associated with use of the product.

13.12.1 Pesticide Data Submitters List

When applying for registration of a pesticide product, a registrant may develop and submit the required data, cite all previously submitted data, or cite selected data. When an applicant cites data previously submitted by another pesticide registrant, the applicant must make a valid offer to pay compensation to the owner of that data. The Pesticide Data Submitters List (available at http://www.epa.gov/opppmsd1/DataSubmittersList) is a compilation of names and addresses of registrants who wish to be notified and offered compensation for use of their data. It was developed to assist pesticide applicants in fulfilling their obligation as required by sections 3(c)(1)(f) and 3(c)(2)(D) of FIFRA and 40 CFR Part 152 subpart E regarding ownership of data used to support registration. The Pesticide Data Submitters List contains the names and addresses of companies who submitted data relating to certain pesticide chemicals who wish to receive such offers.

13.12.2 Testing and Outreach

Because public health products are crucial for infection control and because of the increased controversy regarding product effectiveness, the USEPA is conducting preregistration confirmatory and postregistration enforcement testing of certain public health products. More specifically, the EPA has entered into an interagency agreement with the FDA, and they are jointly testing all sterilants except gases (registered and those seeking registration) and registered products that make unsubstantiated claims of controlling the bacterium that causes tuberculosis (including sterilants and hospital disinfectants). These two types of public health products are the most crucial to infection control, and their failure could pose grave danger to the public and the medical community. The EPA has committed funds to ensure that the tests used to demonstrate the efficacy of antimicrobial products are reliable and reproducible and that amplified internal controls are in place to ensure the integrity of data submitted by registrants. (*Note:* At the time of this publication, the testing protocols were still being developed.) The EPA is in the process of developing a complaint system to handle concerns regarding ineffective products.

13.13 FOOD QUALITY PROTECTION ACT

The Food Quality Protection Act of 1996 amended FIFRA and the Federal Food Drug, and Cosmetic Act (FFDCA). These amendments fundamentally changed the way the EPA regulates pesticides. The requirements included a new safety standard — reasonable certainty of no harm — that must be applied to all pesticides used on foods. The FQPA made a number of changes to the way in which EPA regulates the universe of antimicrobial pesticides, including such diverse products as hospital disinfectants, cooling tower products, swimming pool products, and toilet bowl sanitizers. All of these changes are designed to streamline the registration process so that the EPA's premarket regulatory reviews are completed within statutory imposed deadlines. The key provisions are:

- Statutory goals for making decisions on different categories of applications for registration of antimicrobial pesticides and a requirement to report to Congress annually on the EPA's success in meeting the goals
- Statutory deadlines for issuance of regulations to streamline the registration process in order to meet the goals for decisions on applications
- An exemption from FIFRA of certain liquid chemical sterilants, thereby leaving the FDA with sole responsibility for their regulation
- Statutory changes to the EPA's procedures for reviewing minor changes to product labeling involving nonpesticidal claims
- A directive to work with states to eliminate duplicative, burdensome regulations

13.14 FIFRA AND EXEMPTIONS

On March 6, 2000, this version made corrections to the February 3, 2000, PR Notice 2000–1:

Pesticide Registration (PR) Notice 2000-1: Notice to Manufacturers, Formulators, Producers, and Registrants of Pesticide Products

Attention: Persons Responsible for Registration of Pesticide Products
Subject: Applicability of the Treated Articles Exemption to Antimicrobial Pesticides

This notice clarifies current EPA policy with respect to the scope of the treated articles exemption in 40 CFR 152.25(a). This exemption covers qualifying treated articles and substances bearing claims to protect the article or substance itself. The EPA does not regard this exemption as including articles or substances bearing implied or explicit public health claims against human pathogens. This notice addresses the types of claims which are not permitted for antimicrobial pesticide products exempt from registration under this provision and gathers together in one place guidance the Agency has offered in recent years on labeling statements which it believes would or would not be covered under this provision. This notice also explains the requirement that the pesticide in a treated article be registered for such use. This notice provides guidance to producers and distributors of pesticide treated articles and substances, and to producers and distributors of pesticides used as preservatives to protect treated articles from microbial deterioration.

The EPA regulations in 40 CFR 152.25(a) exempt certain treated articles and substances from regulation under the Federal Insecticide, Fungicide, and Rodenticide Act (FIFRA) if specific conditions are met. The specific regulatory language follows.

Section 152.25 Exemptions for Pesticides of a Character Not Requiring FIFRA Regulation

(a) Treated articles or substances. An article or substance treated with, or containing, a pesticide to protect the treated article or substance itself (for example, paint treated with a pesticide to protect the paint coating or wood products treated to protect the wood against insect or fungus infestation), if the pesticide is registered for such use.

Known as the Treated Articles Exemption, section 152.25(a) provides an exemption from all requirements of FIFRA for qualifying articles or substances treated with, or containing a pesticide, if:

- The incorporated pesticide is registered for use in or on the article or substance.
- The sole purpose of the treatment is to protect the article or substance itself.

The exemption gives two examples of treatments that are intended to protect only the treated article or substance itself. In the first case, paint is being protected from deterioration of the paint film or coating. In the second case, wood is being protected from fungus or insect infestations that may originate on the surface of the wood. Pesticides used in this manner are generally classified as preservatives. Other pesticides are incorporated into treated articles because of their ability to inhibit the growth of microorganisms, which may cause odors, or to inhibit the growth of mold and mildew. Because of this treatment, it is claimed that a fresher and more pleasing surface can be maintained. To qualify for the treated articles exemption, both conditions stated above must be met. If both are not met, the article or substance does not qualify for the exemption and is subject to regulation under FIFRA

In recent years, the marketplace has experienced a proliferation of products that are treated with pesticides and bear implied or explicit public health claims for protection against bacteria, fungi, and viruses, as well as specific claims against pathogenic organisms which may cause food poisoning, infectious diseases or respiratory disorders. Examples of such articles include toothbrushes, denture cleansers, children's toys, kitchen accessories such as cutting boards, sponges, mops, shower curtains, cat litter, vacuum cleaner bags, pillows, mattresses, and various types of finished consumer textiles. In many cases, these products have made public health claims that extend beyond the protection of the article itself and thus do not qualify for the treated articles exemption.

13.15 PUBLIC HEALTH ANTIMICROBIAL CLAIMS

Consumers have long associated the following widely used claims and references to microorganisms harmful to humans with products providing public health protection. Thus, the EPA considers an article or substance to make a public health claim if either explicitly or implicitly the manufacturer claims that the product:

- Controls specific microorganisms or classes of microorganisms that are directly or indirectly infectious or pathogenic to man (or both humans and animals) (examples of specific microorganisms include *Mycobacterium tuberculosis*, *Pseudomonas aeruginosa*, *Escherichia coli*, human immunodeficiency virus (HIV), *Streptococcus*, and *Staphylococcus aureus)*
- Is a sterilant, disinfectant, viricide, or sanitizer, regardless of the site of use of the product, and regardless of whether specific microorganisms are identified
- Has antibacterial, bactericidal, or germicidal activity or makes reference in any context to activity against germs or human pathogenic organisms implying public-health-related protection is made.
- Is a fungicide against fungi infections or fungi pathogenic to humans, or the product does not clearly indicate it is intended for use against non-public-health fungi
- Controls the spread of allergens through the inhibition or removal of microorganisms such as mold or mildew
- Will beneficially impact or affect public health by pesticidal means at the site of use or in the environment in which applied
- Has unqualified antimicrobial activity

13.16 NON-PUBLIC-HEALTH ANTIMICROBIAL CLAIMS

The EPA considers a product to make a non-public-health claim if any of the following claims are made:

- The product inhibits the growth of mildew on the surface of a dried paint film or paint coating.
- The product inhibits microorganisms which may cause spoilage or fouling of the treated article or substance.
- The product inhibits offensive odors in the treated article or substance.

The EPA considers terms such as *antimicrobial, fungistatic, mildew-resistant*, and *preservative* as being acceptable for exempted treated articles or substances, provided that these terms are properly, and very clearly, qualified as to their intended non-public-health use. Use of these terms in product names or elsewhere in the labeling in bolder text than accompanying information may render such qualifications inadequate.

13.17 INTERPRETATIONS OF PAST EPA LABELING CLAIMS

13.17.1   Odor- and Mildew-Resistant Properties May Be Claimed

Over the past 25 years, the EPA has issued several interpretations concerning the exemption from FIFRA regulations of certain types of antimicrobial treated article claims associated with mildew-resistant paint, films, and coatings. In the same period, the EPA has also issued other interpretations concerning certain types of odor-resistant antimicrobial-treated article claims. During this period, there has been widespread dissemination and adoption by the antimicrobial pesticide product community of these EPA interpretations regarding mildew-resistant and odor-resistant claims under the treated articles exemption.

Furthermore, the EPA continues to treat these general types of claims as covered by the term *to protect the treated article or substance itself* because mitigation of these non-public-health-related organisms can contribute to the protection of the appearance and maintenance of the intended useful life of the treated article or substance. Because, during this period, there has also been widespread misinterpretation of EPA's guidance, the agency has developed a representative set of statements designed to clarify its position in this area. Consequently, if they otherwise qualify for the exemption, properly labeled treated articles and substances bearing claims such as those described under Unit IV.B continue to be eligible for the treated articles exemption.

13.17.2   Product Names May Not Contain Public Health Claims

The EPA regards trademarked product names of treated articles or substances (or references to trademarked names of registered pesticides) as potential sources of public health claims that could render a product ineligible for the treated articles exemption just as could other direct or indirect public health claims on or in the packaging of a product or in its labeling or advertising literature. The Agency has maintained this position in enforcement actions against pesticide-treated articles, such as pesticide-treated cutting boards and other items, which bore names suggesting health or other benefits beyond mere preservation of the treated article itself. In determining the eligibility of a treated article or substance for the exemption, the agency will examine the product name, its context, labeling claims, and other related elements on a case-by-case basis.

13.17.3   Treated Article Labeling Claims

Products treated with antimicrobial pesticides with claims such as those described in Section 13.17.4 below are not likely to be acceptable under the treated articles exemption because they imply or express protection that extends beyond the treated article or substance itself. Products treated with antimicrobial pesticides registered for such use and which only bear claims for protection of the article or substance itself such as those described in Section 13.17.6 are likely to be acceptable and eligible for the treated articles exemption, assuming all other conditions have been met. Section 13.17.7 contains examples of appropriate qualifying and prominence statements that have been extracted from multiple enforcement proceedings dealing with claims that can be made for treated articles without obtaining registration.

13.17.4 Examples of Labeling Claims the EPA Is Likely to Consider Unacceptable under the Exemption

The following examples are not intended to be an all-inclusive listing of unacceptable treated article-labeling claims. If persons are not sure whether their antimicrobial pesticides are covered by the provisions of this section, the EPA encourages them to request a written opinion from the Antimicrobials Division at one of the addresses listed under Unit VII. The examples provided here represent claims or types of claims for a treated article that would lead to a requirement to register the article as a pesticide product:

- Antibacterial
- Bactericidal
- Germicidal
- Kills pathogenic bacteria
- Effective against *Escherichia coli* and *Staphylococcus*
- Reduces the risk of foodborne illness from bacteria
- Provides a germ-resistant surface
- Provides a bacteria-resistant surface
- Surface-kills common Gram-positive and -negative bacteria
- Surface-controls both Gram-positive and -negative bacteria
- Surface minimizes the growth of both Gram-positive and -negative bacteria
- Reduces risk of cross-contamination from bacteria
- Controls allergy-causing microorganisms
- Improves indoor air quality through the reduction of microorganisms

13.17.5 Examples of Labeling Claims the Agency Is Likely to Consider Acceptable under the Exemption

The following examples are not intended to be an all-inclusive listing of acceptable treated article labeling claims. If persons are not sure whether their antimicrobial pesticides are covered by the provisions of this section, the EPA encourages them to request a written opinion from the Antimicrobials Division.

13.17.5.1 *Mold- and Mildew-Resistant Claims*

The following are examples of claims that can be made for treated articles:

- Mildew-resistant paints have been treated with a fungistatic agent to protect the paint itself from the growth of mildew.
- Mildew-resistant paint contains a preservative that inhibits the growth of mildew on the surface of the paint film.
- Mildew-resistant products extend the useful life of articles by controlling deterioration caused by mildew.
- Algae-resistant products contain a preservative to prevent discoloration by algae (a fungistatic agent has been incorporated into the article to make it resistant to stain caused by mildew).
- These articles are treated to resist deterioration by mold fungus.
- These articles are treated to resist deterioration from mildew (fungistatic agent in this type of article makes it especially useful for resisting deterioration caused by mildew).

- Dry coating of this paint is mildew resistant.
- Dried paint film resists mold fungus.
- Dry enamel coating resists discoloration from mildew.
- Cured sealant is mildew resistant.
- Dried film resists stains by mold.
- A mold- or mildew-resisting component has been incorporated in this article to make its dry film mildew resistant.
- This type of paint is specially formulated to resist mildew growth on the paint film.
- These articles give a mildew-resistant coating.
- The mildew resistance of this type of outside house paint film makes it especially useful in high-humidity areas.
- These articles retard paint film spoilage.
- These articles resist film attack by mildew.

13.17.5.2   Odor-Resistant Claims

This product:

- Contains an antimicrobial agent to control odors and to prevent microorganisms from degrading the product
- Has been treated to resist bacterial odors
- Inhibits the growth of bacterial odors
- Resists microbial odor development
- Retards the growth and action of bacterial odors
- Guards against the growth of odors from microbial causes
- Guards against degradation from microorganisms
- Reduces odors from microorganisms
- Is odor resistant
- Acts to mitigate the development of odors

13.17.6   Antimicrobial Qualifying and Prominence Considerations

The EPA does not believe that claims such as *antimicrobial*, *fungistatic*, *mildew-resistant*, and *preservative* or related terms are consistent with the intent of 40 CFR 152.25(a) if they are (1) part of the name of the product; or (2) not properly qualified as to their intended non-public-health use. Examples of permissible statements would include, but not be limited to: *antimicrobial properties built in to protect the product* and *provides a mildew-resistant dried paint coating*. All references to the pesticidal properties and the required qualifying statements should be located together; should be printed in type of the same size, style, and color; and should be given equal prominence. Moreover, such references should not be given any greater prominence than any other described product feature. In addition, treated articles or substances intended for microbial odor control or article preservation in areas where foodborne or disease-causing organisms may be present have the potential to create the impression that the article provides protection against foodborne and disease-causing bacteria. This potential should be addressed through very careful narrowing and qualification of the non-public-health claims. A complete assurance that there is no misleading impression could be achieved through use of language such as:

- This product does not protect user or others against foodborne (or disease-causing) bacteria. Always clean this product thoroughly after each use.
- This product does not protect users or others against bacteria, viruses, germs, or other disease organisms. Always clean this product thoroughly after each use.

13.18 ADDITIONAL INFORMATION

13.18.1   Registration of Treated Articles Making Public Health Claims

Treated articles or substances with implied or explicit public health claims or which otherwise fail to qualify for exemption are pesticide products subject to all requirements of FIFRA. They may not be legally sold or distributed unless they are registered with the EPA or unless such claims have been removed and the article otherwise qualifies for exemption. To obtain a registration, an applicant must submit acceptable data supporting all the proposed claims under which the product will be marketed and meet all other applicable registration requirements. Refer to 40 CFR Parts 152, 156, and 158.

The EPA currently has no established protocols for the development of data to support public health claims on treated articles for which registration is sought. Acceptable protocols for product testing reflecting actual use conditions must be submitted and approved by the EPA prior to the development of these data. As part of this review process, the agency will require that these protocols be independently validated for accuracy and reproducibility. Antimicrobial-treated articles requiring registration must meet the same efficacy performance standards that are required for corresponding antimicrobial public health products.

13.18.2   The Term *Registered for Such Use*

In order to qualify for the treated articles exemption, 40 CFR 152.25(a) specifies that an article or substance must be treated with, or contain, a pesticide to protect the treated article or substance itself. The terms *treated with, or containing, a pesticide* and *if the pesticide is registered for such use* in 40 CFR 152.25(a) refer to actual incorporation or adding of an antimicrobial pesticide specifically registered for that use. To qualify under the treated articles exemption (assuming the article or substance otherwise qualifies), it is not sufficient that the antimicrobial pesticidal substance in the treated article merely resembles or has activity like a registered pesticide.

The antimicrobial pesticide in the treated article or substance must be present in the article or substance solely as the result of incorporating an antimicrobial pesticide which is registered for treating the specific article or substance. Because of the wide range of exposure scenarios associated with the use of treated articles such as cutting boards and conveyor belts used in the food processing industry and the wide range of household consumer uses, the agency has interpreted 40 CFR 152.25(a) to mean that the registration and the labeling of the antimicrobial pesticide intended for incorporation into the treated article or substance must include specific listings of the articles or substances that may be treated.

Accordingly, in registration actions over the past several years, the EPA has not permitted broad general use patterns, such as the preservation of hard surfaces, plastics, adhesives, or coatings for the registered pesticide. Instead, it has required that specific listings such as toys, kitchen accessories, and clothing articles be reflected in the product registration and labeling as a prerequisite for incorporation of the pesticide into an article or substance under 40 CFR 152.25(a).

13.18.3   Effective Date and Procedures

In order to remain in compliance with FIFRA and avoid regulatory or enforcement consequences as described here and below, it is the EPA's position that producers, distributors, and any other person selling or distributing pesticide treated articles and substances not in compliance with the agency's interpretation of 40 CFR 152.25(a) must bring their products, labeling, and packaging and any collateral literature, advertisements, or statements made or distributed in association with the marketing (sale or distribution) of the treated article or substance into full compliance with the regulation as clarified by this notice as soon as possible. Because some of the elements of this interpretation may not have been well understood by the regulated community, the agency expects that some companies may need up to a year in order to comply with those elements that have been clarified by this notice.

For the present, therefore, the agency is following the approach set forth in the April 17, 1998 Federal Register (63 FR 19256). Although non-public-health claims for microbial odor control and mold and mildew claims associated with deterioration, discoloration, and staining were not specifically mentioned in the April 17, 1998, Federal Register, such claims are also consistent with the enforcement approach set forth in that notice, as well as with this guidance, provided that they are properly, and very clearly, qualified as to their non-public-health use. The agency began to rely on the guidance provided in the notice on February 11, 2001. Products in commerce after that date that make statements, etc. that do not reflect the clarification offered in the notice risk being considered out of compliance with 40 CFR 152.25(a).

13.19 EPA TESTING AND ENFORCEMENT ACTIONS

In the early 1990s, the EPA implemented an Antimicrobial Product Testing Program to conduct postregistration testing of public health antimicrobial products. Under the testing program, the EPA conducts tests on registered public health antimicrobial products to verify their chemical potency and effectiveness based on their label claims. These tests also serve to validate results of efficacy data submitted by various testing laboratories in support of the product's registration. The testing program is divided into three phases: the first phase involves testing of liquid sterilants and the second phase involves testing of disinfectants that are also tuberculocides (products that kill the microorganism that causes tuberculosis). The third phase includes testing of hospital disinfectants that do not make tuberculocide claims. If any of the tested samples prove to be ineffective, enforcement action is taken to ensure appropriate remediation.

The first phase of the testing program, the testing of liquid sterilants, was completed in September 1993. Liquid sterilants were given the highest priority for testing because these products are most critical to public health (regulatory authority for most liquid sterilants has since been transferred by statute to the FDA). The agency's emphasis has now shifted to validating the efficacy of tuberculocides and other hospital disinfectants. Approximately 800 registered products will be tested during the next two phases. The EPA is first testing the 150 hospital disinfectants with tuberculocidal claims, and then test the 650 disinfectant products that are solely hospital disinfectants (i.e., make no tuberculocidal claims). However, if and when the EPA has a concern about the antimicrobial effectiveness of a product, that product is collected and tested as soon as possible.

A 1990 report issued from the Government Accounting Office (GAO) revealed that many of the public health antimicrobial products on the market do not work as claimed. In addition, concerns from the public health community about the effectiveness of these products prompted the agency to reassess its policy on efficacy testing of public health products.

Since 1996, when the second phase of the testing began, the EPA has collected and evaluated the effectiveness of 14 hospital disinfectants with tuberculocidal claims and three other hospital disinfectants. The data resulting from testing of these products are currently under review by the agency. Progress of the testing program is dependent upon timely collection of products from the manufacturers, adequate resources, and test method limitations. The EPA has made an effort to expedite testing and to increase the laboratory's testing capability through implementation of a screening process for hospital disinfectants. In addition, the agency is considering other options for validation of efficacy claims. The EPA has prioritized these products for testing on the basis of those for which the efficacy is most in doubt — that is, if the agency has received complaints of failing data on the product or products have failed previous agency testing.

The EPA has several active programs to assure that data submitted to the agency in support of product registrations are reliable. First, the EPA establishes detailed guidelines describing how studies must be performed. In addition, each scientific study submitted for agency review must have been conducted at a laboratory facility that follows the good laboratory practices (GLPs) regulations (40 CFR 160). The GLP standards are a management tool to ensure that studies are conducted according to certain scientific standards. Each laboratory conforms to GLP requirements by implementing standard operating procedures (SOPs) and maintaining quality assurance oversight through a quality assurance/quality control unit that conducts internal audits of raw data and laboratory practices.

The mission of EPA's GLP program is to assure the quality and integrity of studies submitted to the agency in support of pesticide product registration. The EPA accomplishes this mission by conducting data audits to assure compliance with the GLP regulations. The EPA conducts more than 300 study audits a year. These studies vary from chemical analysis of pesticides to long-term toxicity and carcinogenicity studies in mammals. Other audited studies may look at the effect of pesticides on the environment, residues of pesticides on commodities, and the efficacy of public health antimicrobial products.

Once the agency receives registration data (e.g., efficacy, product chemistry, and, if applicable, toxicology and environmental fate/effects), scientists from appropriate scientific disciplines thoroughly review the data. These reviews look not only at the substantive results but also for signs that the data may not be trustworthy (e.g., internal inconsistencies, discrepancies with tests run on similar products, or missing information on GLP compliance). If the EPA has concerns regarding the submitted data, additional data may need to be requested, or the agency may require that a laboratory audit be conducted.

The EPA routinely conducts inspections and study audits of the laboratories that generate efficacy data to determine compliance with GLPs. In 1989, efficacy studies became subject to compliance with GLPs. The laboratory inspection and data audit program allows for validation of studies submitted to EPA, identifies labs that conduct effectiveness studies and labs that do not adhere to GLP standards, and educates lab management on GLP regulations.

Ineffective public health antimicrobials have been an enforcement priority for EPA for several years. As part of that effort, EPA has paid special attention to compliance with the GLP regulations for public health antimicrobial studies. In 1996, the EPA inspected about 10 labs that conduct antimicrobial efficacy tests for compliance with GLPs. These 10 inspections reached 20% of EPA efficacy testing labs and accounted for 10% of labs the EPA inspected in 1996. Since the beginning of 1997, EPA has conducted GLP inspections at an additional 7 antimicrobial efficacy testing laboratories and by the end of 1997, the EPA will have completed 11 GLP inspections of these laboratories. In 1997, the EPA issued two Civil Administrative complaints against two registrants seeking a total of $16,400 in penalties for violation of GLPs. The EPA also issued Notices of Warning to one additional public health antimicrobial registrant and one contract/independent laboratory for GLP violations. The agency will take necessary regulatory action against potential submissions of fraudulent data and GLP violators. The EPA plans to continue its compliance monitoring activities of all laboratories that conduct studies to be submitted to the agency in support of pesticide product registration, including antimicrobial efficacy laboratories. In addition, the agency will conduct a laboratory inspection and data audit if, under its testing program, it obtains efficacy test results data that do not agree with those submitted by the registrant.

Since 1992, the EPA has issued civil administrative complaints against ten registrants of ineffective hospital sterilants and one registrant of two ineffective disinfectant products. The ineffective sterilants are Clidox-S™, Perfecto Germ-X/Ucarcide 602™; Cetylcide-G™; Wavicide-01 Concentrate™; Alcide Exspor™; Alcide ABQ™; Wipe Out™; Bionox™; Coldspor/Colcide 10™; Sporicidin™; Metricide Activated Dialdehyde Solution™; Metricide Plus 28™; Metricide Plus 14;™ and Metricide Plus 30™. The two ineffective disinfectants are Broadspec 128™

and Broadspec 256™. Production of Broadspec 128 has been discontinued. After retesting, Broadspec 256 was allowed to reenter the channels of commerce.

In 1995, the agency issued civil administrative complaints against six distributors for the sale and distribution of unregistered Wipe Out sterilants and disinfectants. Total proposed penalties were over $1.2 million dollars. The unregistered products are Wipe Out Disinfectant Towelettes™; QuickKit Biological Fluid Emergency Spill Kit™; Wipe Out Household or Office Disinfectant Spray™; Wipe Out Medi Disinfectant Wand™; Wipe Out Infection Control Travel Kit™; and Wipe Out Spray™. The agency took cancellation action against Wipe Out Cold Sterilizing Disinfecting Solution™, and a settlement was reached whereby the product will be cancelled. The product is currently subject to a Stop Sale, Use and Removal Order. Additionally, notices of warning were issued to three distributors for selling registered inefficacious Wipe Out products.

Recently, the EPA issued orders to stop the sale of various unregistered antimicrobial products making claims of effectiveness against germs and specific bacteria, such as *Salmonella*, *Escherichia coli*, and *Staphylococcus*. The products include cutting boards, brushes, sponges, and other products. In addition, the EPA has settled civil administrative actions for $220,000 against two manufacturers of unregistered antimicrobial products. In settlement of these cases, the respondents have agreed to take corrective actions to bring their products into compliance.

To detect ineffective pesticides quickly, the EPA is expanding its testing program to increase the number of public health pesticides it can test each year.

The agency also is considering how to use its registration authority to further complement its existing enforcement powers to regulate ineffective or unregistered public health pesticides. The EPA is developing regulations under the Food Quality Protection Act of 1996 that could impose new conditions for registering public health pesticides. (The majority of registered public health pesticides are antimicrobials.)

The EPA has taken several notable actions to stop the sale of products that were making public health claims but had not been registered by the EPA as public health pesticides. The agency expects to continue to vigorously pursue similar enforcement actions against unregistered products that can mislead the public and may pose a health threat.

The EPA receives information regarding unregistered pesticides through tips and complaints sent to the agency from outside sources or through field inspections. The agency investigates these allegations and, if it determines that a violation of FIFRA has occurred, pursues the appropriate enforcement action. As an aid in deterring other noncompliers, the EPA has repeatedly issued press releases and other public documents to widely communicate most antimicrobial enforcement actions. The EPA also developed and disseminated a brochure to registrants, distributors, and sellers of pesticides to assist them in identifying and avoiding unregistered, misbranded, and defective pesticides.

13.20 IMPORTATION AND SHIPMENT

The importation of etiologic agents and vectors of human diseases is subject to the Public Health Service Foreign Quarantine regulations. Companion regulations of the Public Health Service and the Department of Transportation specify packaging, labeling, and shipping requirements for etiologic agents and diagnostic specimens shipped in interstate commerce. The U.S. Department of Agriculture (USDA) regulates the importation and interstate shipment of animal pathogens and prohibits the importation, possession, or use of certain exotic animal disease agents that pose a serious disease threat to domestic livestock and poultry. Biological agents include infectious agents of humans, plants, and animals, as well as the toxins that may be produced by microbes and by genetic material potentially hazardous by itself or when introduced into a suitable vector. Etiologic agents and infectious substances are closely related terms in most transportation and transfer regulations. Biological agents may exist as purified and concentrated cultures but may also be present in other materials.

13.21 TRANSPORTATION VS. TRANSFER

The transportation and transfer of biological agents and hazardous materials are regulated. Transportation refers to the packaging and shipping of these materials by air, land, or sea, generally by a commercial conveyance. Transfer refers to the process of exchanging these materials between facilities.

13.22 CURRENT REGULATIONS FOR TRANSPORTATION

Transportation regulations regarding biological agents are aimed at ensuring that the public and the workers in the transportation chain are protected from exposure to any agent that might be in the package. Protection is achieved through requirements for rigorous packaging that will withstand rough handling and contain all liquid material within the package without leakage to the outside and through appropriate labeling, documentation of the hazardous contents, and training.

13.22.1 Interstate Shipment of Etiologic Agents (42 CFR Part 72)

See http://www.cdc.gov/od/ohs.

13.22.2 Hazardous Materials Regulations (49 CFR Parts 17 1–178)

These regulations apply to the shipment of both biological agents and clinical specimens. Information may be obtained from the Internet at: http://www.dot.gov.rules.html.

13.22.3 United States Postal Service: Mailability of Etiologic Agents (39 CFR Part 111)

These regulations are codified in the *Domestic Mail Manual 124.38: Etiologic Agent Preparations*. A copy of the manual can be obtained from the Government Printing Office by calling 1–202–512–1800 or on the Internet at http://www.access.gpo.gov.

13.22.4 Occupational Health and Safety Administration: Occupational Exposure to Bloodborne Pathogens (29 CFR Part 1910.1030)

These regulations provide minimal packaging and labeling requirements for transport of blood and body fluids within the laboratory and outside of it. Information may be obtained from local OSHA offices or on the Internet at http://osha.gov.

13.22.5 Dangerous Goods Regulations

The Dangerous Goods Regulations (DGRs) provide packaging and labeling requirements for infectious substances and materials and clinical specimens that have a low probability of containing an infectious substance. These regulations are followed by airlines and are derived from:

- Committee of Experts on the Transport of Dangerous Goods, United Nations Secretariat
- Technical Instructions for the Transport of Dangerous Goods by Air, provided by the International Civil Aviation Organization (ICAO)

A copy of the DGRs may be obtained by calling 1-800-716-6326 or on the Internet at http://www.iata.org or http://www.who.org.

13.23 GENERAL PACKAGING OF BIOLOGICAL AGENTS AND CLINICAL SPECIMENS

For biological agents or materials associated with human disease pathogens the packaging must be three layers and consist of:

- Primary receptacle
- Watertight secondary packaging
- Durable outer packaging

This packaging requires:

- *Infectious Substance* label on the outside of the package
- Certification that rigorous performance tests as outlined in regulations of the Department of Transportation (DOT), U.S. Postal Service (USPS), U.S. Public Health Service (PHS), and International Air Transport Association (IATA) have been met

Clinical specimens with a low probability of containing an infectious agent are also required to be triple packaged, but performance tests require only that the package shall not leak after a 4-foot drop test. The DOT, PHS, and IATA require a clinical specimen label on the outside of the package.

13.24 TRANSFER REGULATIONS

Biological agent transfer regulations are aimed at ensuring that the change in possession of biological materials is within the best interests of the public and the nation. They require documentation of personnel and facilities, justification of the biological agent need, and federal authority approval of the transfer process.

13.24.1 Importation of Etiologic Agents of Human Disease: Foreign Quarantine and Etiologic Agents, Hosts, and Vectors (42 CFR Parts 71 and 71.54)

These regulations require an import permit from the Centers for Disease Control and Prevention for importing etiologic agents of human disease and any materials, including live animals or insects, that may contain them. An application and information on importation permits can be obtained by calling 1-888-CDC-FAXX and entering document number 101000 or on the Internet at http://www.cdc.gov/od/ohs/biosfty/imprtper.htm.

13.24.2 Importation of Etiologic Agents of Livestock, Poultry, and Other Animal Diseases (9 CFR Parts 92, 94, 95, 96, 122, and 130)

These regulations require an import permit from the USDA, Animal and Plant Health Inspection Service (APHIS), Veterinary Services, to import or domestically transfer etiologic agents of livestock, poultry, other animals, and any materials that might contain these etiologic agents. Information can be obtained by calling 301-734-3277 or on the Internet at http://aphisweb.aphis.usda.gov/ncie.

13.24.3 Importation of Plant Pests: Federal Plant Pest Regulations; General; Plant Pests; Soil; Stone and Quarry Products; Garbage (7 CFR Part 330)

These regulations require a permit to import or domestically transfer a plant pest, plant biological agent, or any material that might contain them. Information can be obtained by calling 301-734-3277 or on the Internet at http://www.aphis.usda.gov/ppq/ppqpermits.htm.

13.24.4 Transfer of Select Biological Agents of Human Disease: Additional Requirements for Facilities Transferring or Receiving Select Agents (42 CFR Part 72.6)

Facilities transferring or receiving select agents must be registered with the CDC and each transfer of a select agent must be documented. Information can be obtained on the Internet at http://www.cdc.gov/od/ohs/lrsat.htm.

13.24.5 Export of Etiologic Agents of Humans, Animals, Plants and Related Materials: Department of Commerce (15 CFR Parts 730–799)

This regulation requires that exporters of a wide variety of etiologic agents of human, plant, and animal diseases, including genetic material, and products that might be used for culture of large amounts of agents will require an export license. Information may be obtained by calling the Department of Commerce (DOC) Bureau of Export Administration at 202-482-4811 or on the Internet at: http://bxa.fedworld.gov or http://www.bxa.doc.gov.

13.24.6 Interstate Shipment of Etiologic Agents (42 CFR Part 72)

A revision is pending that may result in additional package labeling requirements, but this has not been issued in final form as of the publication of the fourth edition of Biosafety in Microbiological and Biomedical Laboratories (BMBL) (DHHS, 1999). For further information on any provision of this regulation contact:

Centers for Disease Control and Prevention
Attn: External Activities Program Mail Stop F-05
1600 Clifton Road N.E.
Atlanta, GA 30333
Telephone: 404-639-4418; FAX: 404-639-2294

Note that the shipper's name, address, and telephone number must be on the outer and inner containers. Refer to additional provisions of the Department of Transportation (49 CFR Parts 171–180) Hazardous Materials Regulations, Figure 1; Packing and Labeling of Infectious Substances, Figure 2; and Packing and Labeling of Clinical Specimens.

13.25 RESTRICTED ANIMAL PATHOGENS

Nonindigenous pathogens of domestic livestock and poultry may require special laboratory design, operation, and containment features. The importation, possession, or use of the following agents is prohibited or restricted by law or by U.S. Department of Agriculture regulations or administrative policies. The following animal pathogens are so restricted:

- African horse sickness
- African swine fever virus

- Akabane virus
- Avian influenza virus
- *Besnoitia besnoiti*
- Bluetongue virus
- Borna disease virus
- Bovine infectious petechial fever agent
- Bovine spongiform encephalopathy (BSE)
- *Brucella abortus*
- *Brucellosis melitensis*
- *Burkholderia mallei* (*Pseudomonas mallei* Glanders)
- Camelpox virus
- Classical swine fever
- *Cochliomyia hominivorax* (screwworm)
- *Cowdria ruminantium* (heartwater)
- Creutzfeldt–Jakob disease variant
- Ephemeral fever virus
- Foot and mouth disease virus
- *Histoplasma* (*Zymonema*) *farciminosum*
- Louping ill virus
- Lumpy skin disease virus
- *Mycobacterium bovis*
- *Mycoplasma agalactiae, M. mycoides* (*mycoides*)
- Nairobi sheep disease virus (Ganjam virus)
- Newcastle disease virus (velogenic strains)
- Peste des petits ruminants (plague of small ruminants)
- Rift Valley fever virus
- Rinderspest virus
- Sheep and goat pox
- Swine vesicular disease virus
- Teschen disease virus
- *Theileria annulata, T. lawrencei, T. bovis, T. hirci*
- *Trypanosoma brucei, T. congolense, T. equiperdum* (dourine), *T. evansi, T. vivax*
- Venezuelan equine encephalomyelitis
- Vesicular exanthema virus
- Vesicular stomatitis
- Viral hemorrhagic disease of rabbits
- Wesselsbron disease virus

13.26 EXPORT LICENSE REQUIRED BY DEPARTMENT OF COMMERCE

The importation, possession, use, or interstate shipment of animal pathogens other than those listed above may also be subject to regulations of the USDA. A USDA/APHIS import permit is required to import any infectious agent of animals listed by USDA/APHIS as a restricted animal pathogen. Such a permit may be required to import any other infectious agent of livestock or poultry. An import permit is also required to import any livestock or poultry animal product such as blood, serum, or other tissues. Additional information can be obtained by writing to:

U.S. Department of Agriculture Animal and Plant Health Inspection Service Veterinary Services
National Center for Import and Export
4700 River Road, Unit #40
Riverdale, MD 20737–1231
Telephone: 301-734-3277; FAX: 301-734-8226
Internet: http://www.aphis.usda.gov/ncei

RESOURCES

40 CFR.
  Eligibility of Pesticide Products for Exemption from Registration as Treated Articles Pursuant to 40 CFR 152.25(a).
  40 CFR Parts 152, 156, and 158.
  40 CFR 152.5(d) and 152.8(a).
  40 CFR Part 152 subpart E, Sections 3(c)(1)(f) and 3(c)(2)(D) of the Federal Insecticide, Fungicide, and Rodenticide Act (FIFRA).
American Society of Heating, Refrigerating, and Air Conditioning Engineers (ASHRAE).
  Method of Testing General Ventilation Air Cleaning Devices for Removal Efficiency by Particle Size, ASHRAE Standard 52.2, 2000.
  Ventilation for Acceptable Air Quality, ASHRAE Standard 62, 1999.
  Method of Testing Air Cleaning Devices Used in General Ventilation for Removing Particulate Matter, ASHRAE Standard 55, 1992.
  ASHRAE Guideline 1–1989: Guideline for the Commissioning of HVAC Systems.
  Particulate Matter Emissions in Air (40 CFR 60.2).
California Senate Bill 732, Toxic Mold Protection Act.
Clean Air Act Amendments of 1990 (CAAA90), Public Law (PL) 101–549 (42 U.S. Code [USC] 7401–7671q).
DHHS (1999) Biosafety in Microbiological and Biomedical Laboratories, 4th ed., HHS Publication No. (CDC) (99-xxxx), U.S. Department of Health and Human Services, Public Health Service, Centers for Disease Control and Prevention and National Institutes of Health, U.S. Government Printing Office, Washington, D.C.
Draft Subpart W, CFR 158, Subpart W, Antimicrobials Data Requirements, 158.1100, General Requirements.
Environmental Protection Agency, 40 CFR Parts 152 and 156, RIN No. 2070-AD14m, Antimicrobial Pesticide Products; Other Pesticide Regulatory Changes, April 30, 1997.
Export of Etiologic Agents of Humans, Animals, Plants and Related Materials, Department of Commerce (15 CFR Parts 730 to 799).
Federal Insecticide, Fungicide, and Rodenticide Act (FIFRA).
Food Quality Protection Act (FQPA) of 1996, which amended the Federal Insecticide, Fungicide, and Rodenticide Act (FIFRA) and the Federal Food Drug, and Cosmetic Act (FFDCA), April 17, 1998, Fed. Registr. (63 FR 19256).
Hazardous Materials Regulations (49 CFR Parts 171–178).
Importation of Etiologic Agents of Human Disease (42 CFR Part 71): Foreign Quarantine; Etiologic Agents, Hosts, and Vectors (42 CFR Part 71.54).
Importation of Etiologic Agents of Livestock, Poultry, and Other Animal Diseases (9 CFR Parts 92, 94, 95, 96, 122, and 130).
Importation of Plant Pests (7 CFR Part 330): Federal Plant Pest Regulations; General; Plant Pests; Soil; Stone and Quarry Products; Garbage.
Indoor Air Quality, #59:15968–16039, CFR Title 29, *Federal Register*, April 5, 1994.
Interstate Shipment of Etiologic Agents (42 CFR Part 72).
Loss Recovery Journal, P.O. Box 71, Sharpsville, PA 16150–0071, e-mail: plm@surf724.com. The *Loss Recovery Journal* is the official publication of the Loss Recovery Guide with Standards (LRGS; William Yobe & Associates); see Vol. II, Issue I, July/August 2001.
Occupational Safety and Health Industry Standard (29 CFR 1910).
Pesticide Registration (PR) Notice 98-X, Notice to Manufacturers, Formulators, Producers and Registrants of Pesticide Products.
Standard Practice for Environmental Site Assessments: Phase I Environmental Site Assessment Process, 4th ed., ASTM E 1527–00), American Society for Testing and Materials (ASTM), Committee E-50 on Environmental Assessment, Atlanta, GA, 2000.
Transfer of Select Biological Agents of Human Disease (42 CFR Part 72.6): Additional Requirements for Facilities Transferring or Receiving Select Agents.
U.S. Postal Service (39 CFR Part 111): Mailability of Etiologic Agents.

CHAPTER 14

Proposed OSHA Tuberculosis Standard and CDC Guidance Comparison

Martha J. Boss and Dennis W. Day

CONTENTS

1 56670-606-8/03/$0.00+$1.50
© 2003 by CRC Press LLC

This chapter presents information detailed in Guidelines for Preventing the Transmission of *Mycobacterium tuberculosis* in Healthcare Facilities, published by the Centers for Disease Control and Prevention (CDC) in the Federal Register of October 28, 1994. The CDC, an agency of the U.S. Public Health Service (USPHS), follows the epidemiology of *M. tuberculosis* and periodically revises and updates its guidelines and recommendations to reflect changes in the diagnosis and treatment of tuberculosis. During development of 29 CFR 1910.1035, Occupational Exposure to Tuberculosis: Proposed Rule, the Occupational Safety and Health Administration (OSHA) reviewed this and other CDC guidance documents. The proposed 29 CFR 1910.0135 standard was originally presented in the Federal Register of October 17, 1997. At that time, OSHA proposed that the health standard be promulgated under Section 6(b) of the Occupational Safety and Health Act of 1970, 29 USC 655, to control occupational exposure to tuberculosis (TB). The preamble discusses differing OSHA vs. CDC decision logic.

The Occupational Safety and Health Administration has set forth what an employer must do to prevent or minimize occupational exposure in the employer's workplace. In the following text, the proposed OSHA 1910.1035 regulatory text and preamble statements are explored. Whenever possible, this discussion immediately precedes the CDC text. By comparing the OSHA vs. CDC decision logic, the OSHA worker protection directive can be compared to the CDC infection control directive. It is hoped that, as these issues are resolved, a final standard will be prepared that is consistent with both worker and patient needs, even though the OSHA emphasis will remain on workers only. (*Note:* Consistency of thought on biological issues is a need that has been addressed by the Government Accounting Office and Homeland Security initiatives.)

Tuberculosis is a well-recognized occupational hazard. Numerous epidemiological studies, case reports, and outbreak investigations provide evidence that employees who are exposed to aerosolized *M. tuberculosis* have become infected with TB and, in some cases, have developed active TB disease. Strains of multidrug-resistant TB (MDR-TB) are of particular concern. Individuals with MDR-TB may be infectious for weeks or months while awaiting effective drug treatment to render them noninfectious, and employees providing healthcare or other services to these individuals will be exposed. The risk of death from infections with MDR-TB is markedly increased vs. the risk of death with TB infections that respond to drug therapy. Outbreaks involving strains of MDR-TB have had mortality rates as high as 75%, with death occurring 4 to 16 weeks after the diagnosis of disease. OSHA's quantitative risk assessments show that the potential is high for TB infection for employees who work in close proximity to individuals with infectious TB; TB cases are not distributed evenly throughout the entire population; a relatively high prevalence of tuberculosis infection and disease exists in certain populations, such as residents of nursing homes and inmates of correctional institutions; and MDR-TB is on the rise (OSHA, 1997).

14.1 TRANSMISSION

Occupational exposure occurs through contact with air that may contain aerosolized *M. tuberculosis*. Occupational transmission occurs when an employee is working in the same environment with an individual with infectious TB or repairing air systems that may be carrying aerosolized *M. tuberculosis*. (OSHA, 1997)

According to the CDC, transmission is most likely to occur from patients who:

- Have unrecognized pulmonary or laryngeal TB
- Are not on effective anti-TB therapy
- Have not been placed in TB isolation

Patients who have MDR-TB can remain infectious for prolonged periods, which increases the risk for nosocomial and/or occupational *M. tuberculosis* transmission. Increases are related partially to the high risk for TB among immunosuppressed persons, particularly those infected with human immunodeficiency virus (HIV). Groups with a higher risk for progression from latent TB infection to active disease include:

- Persons who have been infected recently within the previous 2 years
- Children less than 4 years of age
- Persons with fibrotic lesions on chest radiographs
- Persons with certain medical conditions (e.g., HIV infection, silicosis, gastrectomy or jejuno-ileal bypass, a body weight ≥ 10% below ideal, chronic renal failure with renal dialysis, diabetes mellitus, immunosuppression resulting from receipt of high-dose corticosteroid or other immunosuppressive therapy, some malignancies). (CDC, 1994)

14.1.1 Droplet Nuclei

Individuals with infectious tuberculosis expel airborne particles called droplet nuclei when they cough, sneeze, or speak. These droplet nuclei contain the organism that causes tuberculosis. Normal air currents can keep these droplet nuclei airborne for long time periods, thus spreading them throughout a building. Employees who breathe *M. tuberculosis* droplet nuclei are at risk for TB infection, and employees may be exposed when laboratory procedures produce *M. tuberculosis* aerosols.

Because the CDC does not consider fomites (e.g., objects such as clothing or silverware) to present a hazard for *M. tuberculosis* transmission, the proposed OSHA standard is designed to eliminate or reduce airborne exposures only. Exposure to TB-contaminated air is a well-established exposure route; however, air contamination levels that cause the disease have not been quantified. Unlike toxic chemicals, an OSHA permissible exposure limit (PEL) for *M. tuberculosis* air concentration has not been determined. (OSHA, 1997)

In summary, *M. tuberculosis* is carried in airborne particles, or droplet nuclei, which can be generated when persons who have pulmonary or laryngeal TB sneeze, cough, speak, or sing. Particles are an estimated 1 to 5 μm in size. Normal air currents can keep these particles airborne for prolonged time periods and spread them throughout a room or building. (CDC, 1994)

14.1.2 Infection

Infection occurs when a susceptible person inhales droplet nuclei containing *M. tuberculosis*, and these droplet nuclei traverse the mouth or nasal passages, upper respiratory tract, and bronchi to reach the alveoli of the lungs. Once in the alveoli, the organisms are taken up by alveolar macrophages and spread throughout the body. Usually within 2 to 10 weeks after initial infection with *M. tuberculosis*, the immune response limits further multiplication and spread of the tubercle bacilli; however, some of these bacilli remain dormant and viable for many years. This condition is referred to as latent TB infection.

The probability that a person who is exposed to *M. tuberculosis* will become infected depends primarily on the infectious droplet nuclei concentration in the air and the exposure duration. Characteristics of the TB patient that enhance transmission include:

- Disease in the lungs, airways, or larynx
- Cough or other forceful expiratory measures
- Acid-fast bacilli (AFB) in the sputum
- Failure of the patient to cover the mouth and nose when coughing or sneezing
- Cavitation on chest radiograph
- Inappropriate or short chemotherapy duration
- Procedures such as sputum induction that can induce coughing or cause *M. tuberculosis* aerosolization (CDC, 1994)

14.1.3 Latent

Persons with latent TB infection usually have positive purified protein derivative (PPD) tuberculin skin-test results, do not have active TB symptoms, and are not infectious. (CDC, 1994)

14.1.4 Aged Populations

Persons ages 65 and older constitute a large repository of *M. tuberculosis* infection in the United States. Many were infected many decades ago when TB was a much more common disease. TB occurring in this age group may arise from preexisting infection of long duration, while other cases may be the result of recent infections. Elderly persons residing in nursing homes are at greater risk than elderly persons living in the community. In the CDC 1990 guidelines, *Prevention and Control of Tuberculosis in Facilities Providing Long-Term Care to the Elderly, 1984–1985*, data were cited indicating a TB case rate of 39.2 per 100,000 population, a rate that was twice that of elderly persons living in the community. The same document stated that CDC had found that the increased risk for nursing home employees was three times higher than the rate expected for employed adults of similar age, race, and sex. (OSHA, 1997)

14.1.5 Environmental Factors

Environmental factors that enhance the transmission probability include:

- Exposure in relatively small, enclosed spaces
- Inadequate local or general ventilation that results in insufficient dilution and/or removal of infectious droplet nuclei
- Recirculation of air containing infectious droplet nuclei (OSHA, 1997)

14.2 GENERAL PROGRAMS AND PLANS

Paragraph (c)(2)(vii)(A) of OSHA regulation 29CFR 1910.1035:

- Requires that a copy of the Exposure Control Plan is accessible to employees
- Assures that an employee can obtain and consult the Exposure Control Plan within a reasonable time, place, and manner
- Encourages employees to develop a complete understanding of the plan and its application, so that the program can be carried out by both employer and employees
- Serves as an on-site adjunct to the overall infection control program and may reinforce the training programs

To ensure access, the plan should be in a central location where employees may see it whenever they wish. OSHA does not specify where the plan must be kept. The employer is permitted to determine where the plan is kept provided that employees can access a plan copy at the workplace, within the workshift. Access to the computer or hard copy must be available to the employee. Policy document copies must be accessible in addition to any general policy statement or guiding document that may exist. For fixed work sites and primary workplace facilities, the plan must be maintained on-site at all times. When employees travel between work sites or when the employee's work is carried out at more than one geographical location, the plan may be maintained at the primary workplace facility.

Paragraph (c)(2)(vii)(B) requires that the Exposure Control Plan be reviewed at least annually and updated whenever necessary to reflect new or modified tasks, procedures, or engineering controls that affect occupational exposure and to include new or revised employee positions with occupational exposure. New and revised job classifications must be added to the lists of job classifications and tasks and procedures to assure full coverage of occupationally exposed employees. The updating must occur as soon as feasible and may not be postponed until the annual review. Paragraph (c)(2)(vii)(C) requires that the Exposure Control Plan be made available to the Assistant Secretary and the Director upon request for examination and copying.

Healthcare facilities in which respiratory protection is used to prevent inhalation of *M. tuberculosis* are required by OSHA to develop, implement, and maintain a respiratory protection program. All healthcare workers who use respiratory protection should be included in this program. Facilities that do not have isolation rooms and do not perform cough-inducing procedures on patients who may have TB may not need to have a respiratory protection program for TB. Such facilities should have:

- Written protocols for the early identification of patients who have signs or symptoms of TB
- Procedures for referring these patients to a facility where they can be evaluated and managed appropriately (OSHA, 1997)

According to the CDC, an effective TB infection-control program requires:

- Early identification
- Isolation
- Treatment

Based on the risk assessment results, a written TB infection-control plan should be developed and implemented for each:

- Facility area
- Healthcare worker occupational group that is not assigned to a specific facility area

The occurrence of drug-resistant TB in the facility or the community or a relatively high prevalence of HIV infection among patients or healthcare workers in the community may increase concern about *M. tuberculosis* transmission and may influence any infection-control decisions made. Healthcare facilities are likely to have a combination of low-, intermediate-, and high-risk areas or occupational groups during the same time period.

The primary emphasis of TB infection-control plans in healthcare facilities should be applying control measures that include:

- Administrative measures to reduce the risk for exposure to persons who have infectious TB
- Engineering controls to prevent the spread and reduce the concentration of infectious droplet nuclei
- Personal respiratory protective equipment in areas where there is still a risk for exposure to *M. tuberculosis* (e.g., TB isolation rooms)

In summary, tuberculosis infection-control program and plan implementation requires:

- Risk assessment
- TB infection-control plan development
- Early identification, treatment, and isolation of infectious TB patients
- Engineering controls
- Respiratory protection program
- Healthcare worker TB training, education, counseling, and screening
- Program effectiveness evaluation (CDC, 1994)

14.3 ADMINISTRATIVE CONTROLS

One of the most important steps in preventing TB transmission is the early detection of individuals who may have infectious TB. Employees who must have contact with such patients should be warned early and be able to use appropriate infection control practices to protect themselves from exposure. Exposure Control Plans should allow for the timely transfer and initiation of effective treatment of those individuals for whom the TB diagnosis is likely. By promptly administering effective treatment, these individuals can be rendered noninfectious, thus decreasing the time they are infectious and their potential for exposing employees and other people. As part of the Exposure Control Plan, employers must develop a procedure for the prompt identification of individuals with suspected or confirmed infectious TB. (OSHA, 1997)

The CDC has recommended that identification procedures be based on the prevalence and TB characteristics in the population served by the specific facility. The procedure requirements are discussed in the following subheadings.

14.3.1 Methodology

OSHA regulation 29CFR 1910.1035 states that the employer must define in its Exposure Control Plan how the determination as to suspected or confirmed infectious TB will be made. The following information sources should be considered:

- Information provided by a healthcare provider in advance of an individual's admission to the employer's facility that a suspected or confirmed infectious TB diagnosis has been made
- Determination as to whether an individual should be considered as having suspected infectious TB

OSHA defines suspected infectious TB as a potential disease state in which an individual is known, or with reasonable diligence should be known, by the employer to have one or more of the following conditions, unless the individual's condition has been medically determined to result from a cause other than TB:

- To be infected with *M. tuberculosis* and to have the signs or symptoms of TB
- To have a positive acid-fast bacilli smear
- To have a persistent cough lasting 3 or more weeks and two or more TB symptoms (e.g., bloody sputum, night sweats, anorexia, weight loss, and fever)

In situations where a medical diagnosis is not available either before or at the time of admission, employers must collect the information they need to make the determination. This can be accomplished in two ways. The employer can:

- Administer medical history questionnaires to individuals seeking services from the facility
- Observe individuals to ascertain their health status, looking for the signs and asking about the symptoms included in OSHA's definition that may indicate infectious TB

The employee collecting the information will have to be trained on how to conduct the investigation effectively and with respect for the privacy of the individual. (OSHA, 1997)

14.3.2 Responsibilities

The employer, per OSHA, must designate responsibilities for determining whether an individual should be considered as having suspected or confirmed infectious TB. Employees must be given clear instructions regarding their roles in the prompt identification of suspected or confirmed infectious TB cases. The Exposure Control Plan must designate those employees who make the determination as to whether an individual has suspected or confirmed infectious TB. (*Note:* The employer is ultimately responsible for ensuring that these employees know and use the proper criteria.) (OSHA, 1997)

14.3.3 Promptness

Prompt identification of an individual with suspected or confirmed infectious TB is important and allows isolation before the disease is spread through the facility. The CDC recommends that procedures be in place for prompt identification. OSHA expects that the determination will be made as soon as reasonably practical. Information regarding the signs or symptoms of suspected infectious TB must be reported and processed as soon as possible. (OSHA, 1997)

14.3.4 Effectiveness

An effective procedure, when implemented, will identify individuals as having suspected or confirmed infectious TB. OSHA assumes that many employers currently use effective procedures and has found them to be practical while recognizing that some identification failure will occur. If the employer finds that individuals with suspected and confirmed infectious TB are not being identified, the employer must investigate in order to determine what procedures should be modified. During an inspection, an OSHA compliance officer will review the adequacy of the procedures. A citation would not be issued solely on the basis of failure to identify an individual with suspected infectious TB because no identification system is foolproof. Failure to identify a number of individuals with undetected suspected or confirmed infectious TB would, however, be good evidence that the procedures or their implementation must be investigated and improved, and such a failure could result in a citation.

Paragraph (c)(2)(iii)(C) requires the employer to list all high-hazard procedures performed in the workplace. High-hazard procedures are performed on an individual with suspected or confirmed infectious tuberculosis and include those for which the potential for being *M. tuberculosis* exposure increases due to aerosolized *M. tuberculosis* generation. (OSHA, 1997)

Effective written policies and protocols must be developed and implemented to ensure the rapid identification, isolation, diagnostic evaluation, and treatment of persons likely to have TB through such measures as implementing effective work practices; educating, training, and counseling; and screening healthcare workers for TB infection and disease. Specific measures to reduce the risk for *M. tuberculosis* transmission include the following:

- Assign supervisory responsibility for designing, implementing, evaluating, and maintaining the TB infection-control program.
- Conduct a risk assessment to evaluate the risk for *M. tuberculosis* transmission.
- Develop a written TB infection-control program based on the risk assessment.
- Periodically repeat the risk assessment to evaluate the TB infection-control program.
- Develop, implement, and enforce policies and protocols to ensure early identification, diagnostic evaluation, and effective treatment.
- Provide prompt triage and appropriate patient management in the outpatient setting.
- Promptly initiate and maintain TB isolation for persons who may have infectious TB and who are admitted to the inpatient setting.
- Effectively plan discharge arrangements.
- Develop, install, maintain, and evaluate ventilation and other engineering controls to reduce the potential for airborne exposure to *M. tuberculosis*.
- Develop, implement, maintain, and evaluate a respiratory protection program.
- Use precautions while performing cough-inducing procedures.
- Educate and train healthcare workers about TB, effective methods for preventing *M. tuberculosis* transmission, and the benefits of medical screening programs.
- Develop and implement a program for routine periodic counseling and screening of healthcare workers for active TB and latent TB infection.
- Promptly evaluate possible episodes of *M. tuberculosis* transmission in healthcare facilities, such as PPD skin-test conversions among healthcare workers and epidemiologically associated cases among healthcare workers or patients
- Contact patients or healthcare workers who have TB and who were not promptly identified and isolated.
- Coordinate activities with the local public health department, emphasizing reporting and ensuring adequate discharge follow-up, including continuation and completion of therapy.

Supervisory responsibility for the TB infection-control program should be assigned to a designated person or group of persons with expertise in infection control, occupational health, and engineering. These persons should be given the authority to implement and enforce TB infection-control policies. If supervisory responsibility is assigned to a committee, one person should be designated as the TB contact person. Questions and problems can then be addressed to this person. (CDC, 1994)

14.4 BASELINE RISK ASSESSMENT

Rather than conducting baseline risk assessments, OSHA proposes that medical records reviews and community demographics be used to determine employee risks (Figure 14.1). Paragraph (c)(2)(vi) of 29CFR 1910.1035 requires that the employer claiming reduced responsibilities (paragraph (b), Application, and paragraph (g)(3)(iii)(D), Medical Surveillance) document confirmed infectious tuberculosis cases encountered in the work setting over the past 12 months. This documentation is to be included in the Exposure Control Plan.

Under paragraph (b), these employers are relieved from implementing certain OSHA standard provisions if all of the following are true:

- Individuals with suspected or confirmed infectious TB were not admitted or provided with care.
- In the past 2 years, no cases of confirmed infectious TB have been reported in the local county in one or both years.
- If any cases have occurred in one of the past 2 years, fewer than 6 confirmed infectious cases were reported in that year.
- No such cases have been encountered in the employees' work setting in the past 12 months.

Under paragraph (g)(3)(iii)(D), Medical Surveillance, employees with negative TB skin tests are to be provided with a TB skin test every 6 months if the employee works in:

- An intake area where early identification procedures are performed
- Facilities where 6 or more individuals with confirmed infectious TB have been encountered in the past 12 months

If the employer can document that fewer than 6 individuals with confirmed infectious TB have been encountered in the facility, employees in the intake area would only have to be provided with TB skin tests annually. The count of the number of confirmed infectious TB cases in the Exposure Control Plan is used to document if fewer than 6 individuals with confirmed infectious TB had been encountered in the past 12 months. This proof relieves the employer of the requirement to provide skin tests every 6 months for those affected employees.

The Occupational Safety and Health Administration proposes requiring employers to establish and maintain a medical record in accordance with 29 CFR 1910.1020 for each employee with occupational exposure to TB. The record must include:

- Name, Social Security number, and job classification of each employee
- A copy of all results of examinations and medical testing, including the employees' tuberculin skin test status, and follow-up procedures required by paragraph (g)
- The employer's copy of the physician's or other licensed healthcare professional's written opinion as required by paragraph (g)(7)
- A copy of the information provided to the physician or other healthcare professional required by paragraph (g)(6)

Similar provisions for collection and retention of such information have been included in other OSHA health standards, including, most recently, those for bloodborne pathogens (29 CFR 1910.1030) and cadmium (29 CFR 1910.1027). (OSHA, 1997)

CDC believes that a baseline risk assessment to evaluate the risk for *M. tuberculosis* transmission should be conducted in each area and occupational group. Appropriate infection-control interventions are then developed on the basis of actual risk. Regardless of risk level, patient management should not vary. Risk levels will determine:

- Index of suspicion for infectious TB among patients
- Frequency of healthcare worker PPD skin testing
- Number of TB isolation rooms

The risk assessment should be conducted by a qualified person or group of persons for the entire facility as well as specific areas within the facility, such as all inpatient and outpatient settings; the assessment should include healthcare worker groups throughout the facility rather than in a specific area.

Classification of risk for a facility, for a specific area, and for a specific occupational group should be based on the:

- Community TB profile
- Number of infectious TB patients admitted to the area or ward
- Estimated number of infectious TB patients
- Healthcare worker PPD test conversion analyses results
- Possible person-to-person *M. tuberculosis* transmission

All TB infection-control programs should include periodic reassessments of risk. Frequency should be based on the results of the most recent risk assessment. (CDC, 1994)

14.4.1 Minimal-Risk Category

OSHA limited program elements closely track the recommendations for facilities designated as having minimal risk under the CDC's guidelines. The CDC considers facilities to have minimal risk if TB is not present in the community or facility. The CDC's recommendations for such facilities include a written TB control plan, procedures for early identification and prompt transfer of individuals with suspected or confirmed infectious TB, and employee training. The CDC does not specifically recommend baseline skin testing; however, the CDC's guidelines do say that baseline testing would be advisable in these facilities so that, if an unexpected exposure does occur, conversions can be distinguished from positive skin test results caused by previous exposures. The CDC also recommends that a risk assessment should be conducted by such facilities each year. In the case of a minimal-risk facility, as defined by the CDC, this would essentially involve checking on the number of reported cases of TB in the community and within the facility, which is essentially what OSHA requires under the Exposure Control Plan as documentation to qualify for the limited program. (OSHA, 1997)

The CDC recommendation applies only to an entire facility that:

- Does not admit TB patients to inpatient or outpatient areas
- Is not located in a community with TB
- Has essentially no risk for exposure to TB patients in the facility

This category of facility may apply to many outpatient settings such as medical and dentist offices. (CDC, 1994)

14.4.2 Very-Low-Risk Category

This category generally applies only to an entire facility where patients with active TB are not admitted to inpatient areas but may receive initial assessment and diagnostic evaluation or outpatient management in outpatient areas (e.g., ambulatory-care and emergency departments); those patients

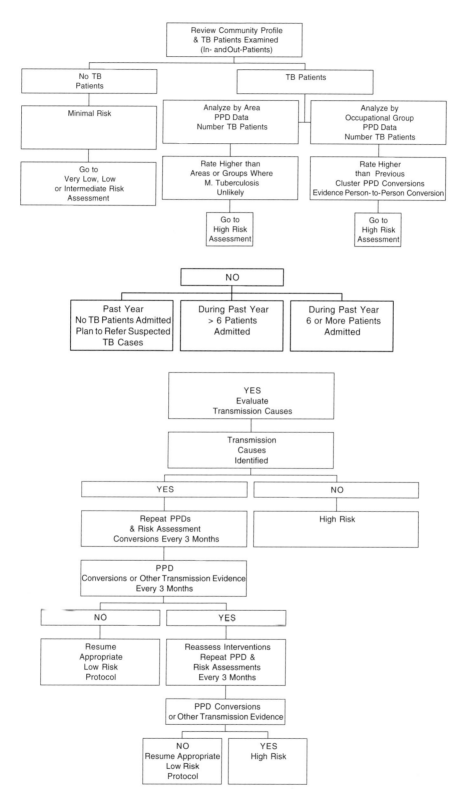

FIGURE 14.1   These charts, arranged sequentially from initial assessment to either the lower or higher risk assessment, provide the sequential decision logic path recommended by the CDC. (From CDC, Federal Register, October 28, 1994.)

who may have active TB and require inpatient care are promptly referred to a collaborating facility. In such facilities, the outpatient areas in which exposure to patients with active TB could occur should be assessed and assigned to the appropriate low-, intermediate-, or high-risk category. Categorical assignment will depend on:

- The number of TB patients examined in the area during the preceding year
- Evidence of nosocomial *M. tuberculosis* transmission in the area
- No patients with active TB examined in the outpatient area during the preceding year but TB cases reported in the community

The referring and receiving facilities should establish a referral agreement to prevent inappropriate management and potential loss to follow-up of patients suspected of having TB during evaluation in the triage system of a very low-risk facility. In some facilities in which TB patients are admitted to inpatient areas, a very low-risk protocol may be appropriate for areas or occupational groups that have only a very remote possibility of exposure to *M. tuberculosis* or outpatient facilities that do not provide initial assessment of persons who may have TB but do screen patients for active TB as part of a limited medical screening before undertaking specialty care. (CDC, 1994)

14.4.3 Low-Risk Areas

Low-risk areas or occupational groups are those in which:

- The PPD test conversion rate is not greater than that for areas or groups in which occupational exposure to *M. tuberculosis* is unlikely or than previous conversion rates for the same area or group.
- No clusters of PPD test conversions have occurred.
- Person-to-person *M. tuberculosis* transmission has not been detected.
- Fewer than six TB patients are examined or treated per year. (CDC, 1994)

14.4.4 Intermediate-Risk Areas

Intermediate-risk areas or occupational groups are those in which:

- The PPD test conversion rate is not greater than that for areas or groups in which occupational exposure to *M. tuberculosis* is unlikely or than previous conversion rates for the same area or group.
- No clusters of PPD test conversions have occurred.
- Person-to-person *M. tuberculosis* transmission has not been detected.
- Six or more patients with active TB are examined or treated each year. (CDC, 1994)

Areas in which cough-inducing procedures are performed on patients who may have active TB should, at the minimum, implement the intermediate-risk protocol. (CDC, 1994)

14.4.5 High-Risk Areas

High-risk areas or occupational groups are those in which:

- The PPD test conversion rate is significantly greater than for areas or groups in which occupational exposure to *M. tuberculosis* is unlikely or than previous conversion rates for the same area or group, and epidemiologic evaluation suggests nosocomial transmission.
- A cluster of PPD test conversions has occurred, and epidemiologic evaluation suggests nosocomial transmission of *M. tuberculosis*.
- Possible person-to-person *M. tuberculosis* transmission has been detected. If no data or insufficient data for adequate determination of risk have been collected, such data should be compiled, analyzed, and reviewed expeditiously. (CDC, 1994)

14.5 COMMUNITY TUBERCULOSIS PROFILE

A profile should be obtained from the public health department. This profile should include, at a minimum:

- Incidence (and prevalence, if available) of active TB in the community
- Drug-susceptibility patterns of *M. tuberculosis* isolates (i.e., the antituberculosis agents to which each isolate is susceptible or alternately resistant) from patients in the community (CDC, 1994)

14.6 CASE SURVEILLANCE

Data concerning the number of suspected and confirmed active TB cases among patients and healthcare workers in the facility should be systematically collected, reviewed, and used to:

- Estimate the number of TB isolation rooms needed using laboratory surveillance data on specimens positive for AFB smears or *M. tuberculosis* cultures, infection-control records, or hospital discharge diagnoses databases
- Recognize possible nosocomial transmission clusters
- Assess the potential occupational risk level

Drug-susceptibility patterns of *M. tuberculosis* isolates from TB patients treated in the facility should be reviewed to identify the frequency and patterns of drug resistance. This information may indicate a need to modify the initial treatment regimen or may suggest possible nosocomial transmission or increased occupational risk. (CDC, 1994)

14.7 HEALTHCARE WORKER PPD TEST SCREENING

Results of healthcare worker PPD testing should be recorded in the individual healthcare worker's employee health record and in a retrievable aggregate database of all healthcare worker PPD test results. Personal identifying information should be handled confidentially. PPD test conversion rates should be calculated at appropriate intervals to estimate the risk for PPD test conversions for each area of the facility and each specific occupational group not assigned to a specific area. (CDC, 1994)

14.7.1 Conversion Rate

To calculate PPD test conversion rates determine the number of:

- PPD-negative healthcare workers tested in each area or group (i.e., the denominator)
- PPD test conversions among healthcare workers in each area or group (i.e., the numerator)

PPD test conversion rates for each area or occupational group should be compared with:

- Rates for areas or groups in which occupational exposure to *M. tuberculosis* is unlikely
- Previous conversion rates in the same area or group to identify areas or groups where the risk for occupational PPD test conversions may be increased (CDC, 1994)

14.7.2 Significance

A low number of healthcare workers in a specific area may result in a greatly increased rate of conversion for that area or overestimated area risk analysis. Testing for statistical significance may assist interpretation. Lack of statistical significance may not rule out a problem (if the number of healthcare workers tested is low, there may not be adequate statistical power to detect a significant difference). Interpretation of individual situations is necessary. An epidemiologic investigation to evaluate the likelihood of nosocomial transmission should be conducted if PPD test conversions are noted.

The frequency and comprehensiveness of the healthcare worker PPD testing program should be evaluated periodically to ensure that all healthcare workers who should be included in the program are being tested at appropriate intervals. For surveillance purposes, earlier detection of transmission may be enhanced if healthcare workers in a given area or occupational group are tested on different scheduled dates rather than all being tested on the same date. (CDC, 1994)

14.8 MEDICAL RECORDS REVIEW

Review of medical records of patients who were examined in the facility and diagnosed as having TB may serve as a guide for developing or revising protocols. To evaluate the effectiveness of infection control parameters, a sample of medical records of TB patients should be reviewed periodically. Time intervals, treatment regimes, infection-control practices, and periodic reassessments should all be reviewed. (CDC, 1994)

14.8.1 Time Interval

Because time is of the essence in preventing transmission, the following intervals from the date of admission should be reviewed as to their initiation and resolution phases:

- TB was suspected.
- Specimens for AFB smears were ordered and collected.
- Tests were performed.
- Results were reported. (CDC, 1994)

14.8.2 Treatment Regimens

Medical record reviews of treatment regimes should include:

- Previous hospital admissions of TB patients before the onset of TB symptoms (patient-to-patient transmission may be suspected if active TB occurs)
- In-patients who have had prior hospitalizations during which exposure to another TB patient occurred
- If isolates from two or more TB patients have identical characteristic drug susceptibility or DNA fingerprint patterns (CDC, 1994)

Data from the case review should be used to determine modifications to:

- Protocols for identifying and isolating patients who may have infectious TB
- Laboratory procedures
- Administrative policies and practices
- Patient management protocols (CDC, 1994)

14.8.3 Infection-Control Practice Observations

Assessing adherence to the policies of the TB infection-control program should be part of the evaluation process and performed on a regular basis or whenever an increase occurs in the number of TB patients or healthcare worker PPD test conversions. Areas at high risk for *M. tuberculosis* transmission should be monitored more frequently than other areas: Reviews of patient medical records provide information on healthcare worker adherence to the TB infection-control program, and work practices related to TB isolation should be observed to determine if employers are enforcing, and healthcare workers are adhering to, these policies and if patient adherence is being enforced. If these policies are not being enforced or adhered to, appropriate education and other corrective action should be implemented. (CDC, 1994)

14.8.4 Periodic Reassessment

Follow-up risk assessment should be performed at the interval indicated by the most recent risk assessment. Based on the results of the follow-up assessment, problem evaluation may need to be conducted or the protocol may need to be modified to a higher or lower risk level. After each risk assessment, the staff responsible for TB control, in conjunction with other appropriate healthcare workers, should review all TB control policies to ensure that they are effective and meet current needs. (CDC, 1994)

14.9 ACTIVE (CONFIRMED) OR SUSPECTED TUBERCULOSIS

Paragraph (c)(2)(iii) of 29CFR 1910.1035 outlines the additional elements required of employers who have work settings where individuals with suspected or confirmed infectious TB are admitted or provided with medical services. The most important factors in preventing *M. tuberculosis* transmission are:

- Early identification of patients who may have infectious TB
- Prompt implementation of TB precautions for such patients
- Prompt initiation of effective treatment for those who are likely to have TB

The criteria used should be:

- Based on the prevalence and characteristics of TB in the population served by the specific facility
- Evaluated periodically and revised according to the results of the evaluation (CDC, 1994)

14.9.1 Diagnosis

A diagnosis of TB may be considered for any patient who has a persistent cough (a cough lasting for more than 3 weeks) or other signs or symptoms compatible with active TB (e.g., bloody sputum, night sweats, weight loss, anorexia, or fever). Diagnostic measures for identifying TB should be conducted for patients in whom active TB is being considered. (CDC, 1994)

14.9.1.1 Laboratories

Prompt laboratory results are crucial to the proper treatment of the TB patient and to early infection control initiation. Laboratories should use the most rapid methods available (e.g., fluorescent microscopy for AFB smears; radiometric culture methods for isolation of mycobacteria; r-nitro-a-acetylamino-b-hydroxy-proprophenone (NAP) test, nucleic acid probes, or high-performance liquid chromatography [HPLC] for species identification; and radiometric methods for drug-susceptibility testing). Laboratories that rarely receive specimens for mycobacteriologic analysis should refer the specimens to a laboratory that more frequently performs these tests. AFB sputum smear results should be available within 24 hours of specimen collection. (CDC, 1994)

14.9.1.2 *Probability*

The TB probability is greater among patients who have:

- Positive PPD test results or a history of positive PPD test results
- Previously had TB or have been exposed to *M. tuberculosis*
- Membership in a group at high risk for TB

Immunosuppressed patients who have pulmonary signs or symptoms that are ascribed initially to infections or conditions other than TB should be evaluated initially for coexisting TB. The evaluation for TB should be repeated if the patient does not respond to appropriate therapy for the presumed causes of the pulmonary abnormalities. (CDC, 1994)

14.9.2 Treatment

Patients with suspected or confirmed TB should be reported immediately to the appropriate public health department so that standard procedures for identifying and evaluating TB contacts can be initiated. Treatment initiation for suspected or confirmed TB patients who have confirmed active TB or who are considered highly likely to have active TB should be started promptly on appropriate treatment in accordance with current guidelines. In geographic areas or facilities that have a high prevalence of MDR-TB, the initial regimen used may need to be enhanced while the results of drug-susceptibility tests are pending. The decision should be based on analysis of surveillance data. While the patient is in the healthcare facility, anti-TB drugs should be administered by directly observed therapy (DOT), the process by which a healthcare worker observes the patient swallowing the medications. Continuing DOT after the patient is discharged should be strongly considered. This decision and the arrangements for providing outpatient DOT should be made in collaboration with the public health department. (CDC, 1994)

14.9.3 Ambulatory Care and Emergency Transfer

According to OSHA, transferring an individual with suspected or confirmed infectious TB:

- Protects employees within the facility by making sure the occupational exposure source is removed
- Benefits the patients with TB in that they receive help in locating and getting to a receiving facility with the capability for appropriately managing their care

OSHA Standard 29CFR 1910.1035 does not require any employer to transfer individuals with suspected or confirmed infectious TB. Transfer is an option that employers have that relieves them of many OSHA standard provisions, such as AFB isolation rooms. If an employer chooses to use the transfer option, the employer must include the procedure for implementing the transfer in the Exposure Control Plan.

An employer's duties regarding transfer of an individual with suspected or confirmed infectious TB will vary with the type of facility the employer operates and the work performed by employees. Transfer responsibilities of hospitals, long-term care for the elderly, correctional facilities, and hospices may include contacting the receiving facility, providing transport, and taking other steps to ensure the individual can get to the receiving facility. These types of facilities often exercise custodial care over such individuals and, hence, have more responsibility for assuring transfer completion. A homeless shelter or a facility that offers drug treatment for drug abuse but does not have custody over individuals may only include providing information about the receiving facility, contacting the facility, and providing directions to the facility. An employer who provides home healthcare or home-based hospice care has no obligation to transfer an individual from his or her home to a receiving facility.

Not all facilities will have the capabilities to admit or provide medical services to individuals with suspected or confirmed infectious tuberculosis. Consequently, these facilities will have to transfer such individuals to another facility where isolation rooms or areas are available. Paragraph (c)(2)(iii)(B) also requires that the employer develop policies and procedures for managing the care of individuals with suspected or confirmed infectious TB once they have been placed in isolation. The Exposure Control Plan must include procedures and policies addressing:

- Minimization of the time an individual with suspected or confirmed infectious TB remains outside an AFB isolation room or area
- Minimization of employee exposure in AFB isolation rooms or areas
- Delay of elective transport or relocation of individuals with infectious TB within the facility and, to the extent feasible, performance of services or procedures for such individuals in an AFB isolation room or area
- During transport outside of AFB isolation areas, masking of individuals with infectious TB and use of portable containment engineering controls
- Return of the individual to an AFB isolation room or area as soon as is practical

As is the case with any AFB isolation room or area, the means used to isolate an individual awaiting placement or transfer must achieve negative pressure and safely discharge air to the outside. Masking or segregation of individuals with suspected or confirmed infectious TB while those individuals are awaiting placement in isolation or transfer to another facility is done to assure that employee exposure is minimized to the extent feasible. (*Note:* Although the CDC recommends masking such individuals, OSHA presents a choice of masking or segregation because the agency believes that this practice is directly involved with patient medical management. OSHA's mission is to protect employees from occupational exposure to tuberculosis, not to dictate medical practice.

Where employers choose not to mask individuals with suspected or confirmed infectious TB when they are not in isolation rooms or areas or when such individuals cannot be masked (e.g., because they are combative), employers must segregate these individuals in a manner such that contact with employees who are not wearing respiratory protection is eliminated or minimized.

The time that a facility can permit an individual to await placement or transfer is limited to 5 hours. After that the individual must be placed in isolation. OSHA is concerned about the amount of time an individual who has been identified as having suspected or confirmed infectious TB should be permitted to stay in non-isolation areas. Individuals who must wait for extended periods of time before placement in AFB isolation or transfer may present an exposure risk to employees working in these areas even though these individuals may be masked. Note that the 5-hour cutoff is the amount of time allotted per facility to accomplish AFB isolation or transfer. More specifically, if an individual spent 4 hours awaiting transfer at an identifying facility, the receiving facility would still be allowed 5 hours to accomplish isolation, not just the 1 hour remaining since initial identification of the individual. The intent of the proposed facility-based 5-hour period is to allow the receiving facility adequate time to accomplish isolation and to recognize that the receiving facility should not be held responsible for circumstances beyond the facility's control (e.g., the time the individual waited before arrival at the receiving facility). If placement or transfer cannot be completed within 5 hours, it must be done as soon as possible thereafter. The employer must ensure that the facility has AFB isolation rooms or areas for the isolation of the individual until placement or transfer can be accomplished. (OSHA, 1997)

Triage of patients in ambulatory-care settings and emergency departments should include vigorous efforts to promptly identify active TB patients. Healthcare workers who are the first points of contact in facilities that serve populations at risk for TB should be trained to ask questions that will facilitate identification of patients with signs and symptoms suggestive of TB. Patients with signs or symptoms suggestive of TB should be evaluated promptly to minimize the amount of time they are in ambulatory-care areas. TB precautions should be followed while the diagnostic evaluation is being conducted for these patients. TB precautions in the ambulatory-care setting should include placing these patients in a separate area apart from other patients, ideally, in a room or enclosure meeting TB isolation requirements, and not in open waiting areas. These patients should be given surgical masks to wear and should be instructed to keep the masks on, as well as tissues and instructions to cover their mouths and noses with the tissues when coughing or sneezing.

Tuberculosis precautions should be followed for patients who are known to have active TB and who have not completed therapy until a determination has been made that they are noninfectious. Patients with active TB who need to attend a healthcare clinic should have appointments scheduled to avoid exposing HIV-infected or otherwise severely immunocompromised persons to *M. tuberculosis*. Designate certain times of the day for appointments for these patients and treat them in areas where immunocompromised persons are not treated.

Ventilation in ambulatory-care areas where patients at high risk for TB are treated should be designed and maintained to reduce the risk for *M. tuberculosis* transmission. General-use areas (e.g., waiting rooms) and special areas (e.g., treatment or TB isolation rooms in ambulatory areas) should be ventilated in the same manner as described for similar inpatient areas, with enhanced general ventilation or the use of air-disinfection techniques (e.g., ultraviolet germicidal irradiation [UVGI] or recirculation of air within the room through high-efficiency particulate air [HEPA] filters). Ambulatory-care settings in which patients with TB are frequently examined or treated should have TB isolation rooms available. Such rooms are not necessary in ambulatory-care settings in which patients who have confirmed or suspected TB are seen infrequently. A written protocol should be in place for early identification of patients with TB symptoms and referral to an area or a collaborating facility where the patient can be evaluated and managed appropriately. These protocols should be reviewed on a regular basis and revised as necessary. (CDC, 1994)

14.9.4 Pediatric Isolation

Pediatric patients with suspected or confirmed TB should be evaluated for potential infectiousness according to the same criteria as are adults (i.e., on the basis of symptoms, sputum AFB smears, radiological findings, and other criteria). Children who may be infectious should be placed in isolation until they are determined to be noninfectious. Such children include those who have laryngeal or extensive pulmonary involvement, pronounced cough, positive sputum AFB smears, or cavitary TB or those for whom cough-inducing procedures are performed. The infection source for a child with TB is often a member of the child's family. Visitors should be evaluated for TB as soon as possible. Until such evaluation or source identification is completed, visitors should wear surgical masks when in areas of the facility outside of the child's room and refrain from visiting common areas in the facility. (CDC, 1994)

14.9.5 Intensive-Care Unit Isolation

Tuberculosis patients in intensive-care units (ICUs) should be treated the same as TB patients in noncritical-care settings and should be placed in TB isolation. They should have respiratory secretions submitted for AFB smear and culture if they have undiagnosed pulmonary symptoms suggestive of TB. If readmitted to a healthcare facility, patients who are known to have active TB and who have not completed therapy should have TB precautions applied until a determination has been made that they are noninfectious. (CDC, 1994)

14.9.6 Isolation Rooms

An AFB isolation room or area is a room, area, booth, tent, or other enclosure maintained at negative pressure to adjacent areas — for example, a rigid enclosure on casters with a ventilation unit to achieve negative pressure, a window kit to safely exhaust the enclosure's air to the outside, and a digital pressure monitor to assure maintenance of negative pressure within the enclosure.

Placement of individuals with suspected or confirmed infectious TB in an AFB isolation room is the most effective way to prevent or lessen transmission. The primary purposes of AFB isolation rooms or areas are to:

- Isolate patients who are likely to have infectious TB from unprotected employees
- Prevent droplet nuclei escape from the room, thus preventing *M. tuberculosis* entry into the corridor and other facility areas where unprotected employees may be exposed
- Provide an environment that will promote reduction of the droplet nuclei concentration through various engineering controls (OSHA, 1997)

Paragraph (d)(4) of Work Practice and Engineering Controls of 29CFR 1910.1035 requires that all employers ensure that high-hazard procedures are conducted in an AFB isolation room or area. Thus, listing the high-hazard procedures will serve to identify those procedures that require special ventilation considerations. This will assist employees in determining which procedures must be performed using such engineering controls and, consequently, will help minimize employee exposure.

The primary purposes of the TB isolation rooms are to:

- Separate patients who are likely to have infectious TB from other persons
- Provide an environment that will allow reduction of the droplet nuclei concentration through various engineering methods
- Prevent the droplet nuclei escape from the TB isolation room and treatment room, thus preventing entry into the corridor and other areas of the facility

In hospitals and other inpatient facilities, any patient suspected of having or known to have infectious TB should be placed in a TB isolation room that has currently recommended ventilation characteristics. The number of persons entering an isolation room should be minimal. All acute-care inpatient facilities should have at least one TB isolation room. Grouping isolation rooms together in one area of the facility may reduce the possibility of transmitting *M. tuberculosis* to other patients and facilitate care of TB patients and the installation and maintenance of optimal engineering (particularly ventilation) controls. (CDC, 1994)

14.9.7 Programs and Plans

Paragraph (c)(2)(iii)(A) of 29CFR 1910.1035 requires that Exposure Control Plans include procedures for the prompt identification of individuals with suspected or confirmed infectious TB. The Exposure Control Plan must contain procedures for minimizing employee exposure in AFB isolation rooms or areas, such as minimizing the number of employees given access to isolation rooms and the time that employees spend in isolation rooms. Paragraph (c)(2)(ii) requires employers who transfer individuals with suspected or confirmed infectious TB to develop Exposure Control Plan procedures that address the following:

- Prompt identification of individuals with suspected or confirmed infectious TB
- Masking or segregation of individuals with suspected or confirmed infectious TB
- Transfer of such individuals to a facility with AFB isolation capabilities

The employer must also include in the Exposure Control Plan procedures for transferring individuals with suspected or confirmed infectious TB to facilities with AFB isolation capabilities. The procedures must address:

- Methods to ensure prompt transfer with minimal exposure to employees
- Transfer locations
- Transfer precautions

The Exposure Control Plan must contain policies for the delay of elective transport or relocation within the facility of individuals with suspected or confirmed infectious TB — for example, delaying the transfer of an inmate with suspected or confirmed infectious TB from one prison to another, where possible, until the inmate has been determined to be noninfectious. This delay should reduce the number of employees exposed and minimize the exposure of other inmates, thereby decreasing the disease transmission risk.

Paragraph (c)(2)(iii)(B) requires that the employer develop policies and procedures for managing the suspected or confirmed infectious TB patients placed in isolation. The Exposure Control Plan must include procedures and policies addressing:

- Minimization of the time an individual with suspected or confirmed infectious TB remains outside of an AFB isolation room or area
- Minimization of employee exposure in AFB isolation rooms or areas
- Delay of elective transport or relocation of individuals with infectious TB within the facility and, to the extent feasible, performance of services or procedures for such individuals in an AFB isolation room or area
- Masking of individuals with infectious TB or use of portable containment engineering controls during transport outside of AFB isolation rooms and return of the individual to an AFB isolation room or area as soon as is practical after completion of the service or procedure
- Delay of elective high-hazard procedures and elective surgery until an individual with suspected or confirmed infectious TB is determined to be noninfectious (elective high-hazard procedures such as pulmonary function testing or elective surgery such as noncritical dental procedures might easily be delayed, without compromising care) (OSHA, 1997)

Written policies for initiating isolation should specify the:

- Isolation indications
- Persons authorized to initiate and discontinue isolation
- Isolation practices to follow
- Isolation monitoring
- Management of patients who do not adhere to isolation practices
- Criteria for discontinuing isolation (CDC, 1994)

14.9.8 Cohorting

In rare circumstances, placing more than one TB patient together in the same room may be acceptable. This practice is sometimes referred to as *cohorting*. Because of the risk for patients becoming super-infected with drug-resistant organisms, patients with TB should be placed in the same room only if all patients involved:

- Have culture-confirmed TB
- Have drug-susceptibility test results available on a current specimen obtained during the current hospitalization
- Have identical drug-susceptibility patterns on these specimens
- Are on effective therapy

Having isolates with identical DNA fingerprint patterns is not adequate evidence for placing two TB patients together in the same room, because isolates with the same DNA fingerprint pattern can have different drug-susceptibility patterns. (CDC, 1994)

14.9.9 Isolation Measures

The Occupational Safety and Health Administration allows employers to determine what criteria should be included in the procedures for isolation. Paragraph (c)(2)(iii)(B) of 29CFR 1910.1035 requires that employers develop procedures for isolating and managing the care of individuals with suspected or confirmed infectious TB. Procedures must address:

- The indications for isolation
- Who is authorized to initiate and discontinue isolation
- Isolation practices
- Isolation monitoring
- Management of patients who will not comply with isolation practices
- Criteria for discontinuing isolation (OSHA, 1997)

Patients who are placed in TB isolation should be educated about the mechanisms of *M. tuberculosis* transmission and the reasons why they are being placed in isolation. They should be instructed to cover their mouths and noses with tissues when coughing or sneezing, even while in the isolation room, and to contain liquid drops and droplets before they are expelled. Efforts should be made to facilitate patient adherence to isolation measures (e.g., staying in the TB isolation room) and to address other problems that could interfere with adherence to isolation (e.g., management of the patient's withdrawal from addictive substances, including tobacco). Patients placed in isolation should remain in the isolation rooms with the door closed. (CDC, 1994)

14.9.10 Procedures

Policies addressing minimizing both the number of employees and time that such employees spend in isolation rooms can reduce exposure. Examples of Exposure Control Plan procedures are:

- In order to minimize the number of employees entering an isolation room, certain tasks or procedures that might normally be done by several different employees could be done by one person. A nurse coming into the room to administer daily TB treatment could also bring in the patient's breakfast at the same time rather than have a hospital dietitian deliver the meal. In addition, the employer must address minimization of time that employees spend in an isolation room or area.
- Rather than conducting an entire discharge-planning interview with an individual in person, an employee may be able to collect and convey a large part of the information over the phone with the individual. Personal contact could be limited to just the time necessary to obtain items requiring direct interaction, such as the individual's signature. (OSHA, 1997)

14.9.11 High-Hazard Procedures

High-hazard procedures, as defined in paragraph (b) of 29CFR 1910.1035, are procedures that are performed on an individual with suspected or confirmed infectious TB in which the probability of *M. tuberculosis* being expelled into the air is increased. Such procedures include, but are not limited to, endotracheal intubation and suctioning, diagnostic sputum induction, aerosol treatments (including pentamidine therapy), pulmonary function testing, bronchoscopy, irrigation of tuberculous abscesses, and homogenizing or lyophilizing infectious tissue. Included are autopsy, clinical, surgical, and laboratory procedures that may aerosolize *M. tuberculosis*.

In view of the increased probability of droplet nuclei generation associated with these procedures, all high-hazard procedures are required to be performed in rooms, areas, or booths that meet AFB isolation criteria (e.g., negative pressure) in order to contain the droplet nuclei and eliminate or minimize employee exposure. Paragraph (d)(4) stipulates that high-hazard procedures must be conducted in AFB isolation rooms or areas. (*Note:* If a procedure is to be performed outside of the isolation room, a time should be chosen when the procedure area is not being used by others.) (OSHA, 1997)

If possible, diagnostic and treatment procedures should be performed in the isolation rooms to avoid transporting patients through other facility areas. If patients who may have infectious TB must be transported outside their isolation rooms for medically essential procedures that cannot be performed in the isolation rooms, they should wear surgical masks that cover their mouths and noses during transport. Persons transporting the patients do not need to wear respiratory protection outside the TB isolation rooms. Procedures should be scheduled to be performed rapidly and when waiting areas are less crowded. (CDC, 1994)

14.9.12 Respiratory Protection

Because masking of an individual with suspected or confirmed infectious TB will reduce the number of droplet nuclei expelled into the air, OSHA requires the employer to develop policies addressing the masking of such individuals during transport outside of an AFB isolation room. (OSHA, 1997)

14.9.12.1 Surgical Masks

A barrier such as a surgical mask, when placed over the mouth of an individual who is coughing, will reduce the droplet nuclei formation. The mask will collect and contain the droplets as discharged before they have time to evaporate and form droplet nuclei; however, these masks do not prevent exposure to droplet nuclei and do not represent adequate physical barriers to the droplet nuclei aerosol transmission. A formed droplet nucleus can penetrate the fiber of a tissue or a surgical mask. So, although a simple surgical mask applied to a tuberculosis patient who must be transported outside the isolation room will prevent the dispersal of organisms as droplet nuclei, such a mask does not provide adequate protection to the individual who must breathe air containing droplet nuclei. (OSHA, 1997)

14.9.12.2 Respirators

The appropriate barrier is a well-fitted respirator that does not allow air leakage around the edges and blocks passage of microorganisms in the filter media (fibers or pores) through which air is inspired. According to OSHA, a respirator that does not have an exhalation valve can also be used to capture droplets being discharged. An exhalation valve would permit droplets to pass through and discharge into the air, where the droplets could evaporate and form droplet nuclei.

This usage statement by OSHA implies that the respirator is to be used on the patient; questions as to respiratory fitness to wear such a respirator may preclude such usage. Air-purifying respirator usage requires sufficient pulmonary function to adequately bring air into the respirator through a filter. Plugging the exhalation valve would create a more positive pressure within the respirator and make breathing more difficult. Gapping could occur around the respirator seal, allowing release of the droplets, including both droplet nuclei and other tissue fluids (perspiration, saliva). Many questions remain as to the efficacy of this type of respirator use, especially for untrained respirator users who are already sick. (OSHA, 1997)

All persons who enter an isolation room should wear respiratory protection. A patient's visitors should be provided with respirators to wear while in the isolation room and given general instructions on how to use the respirators. (CDC, 1994)

14.9.13 Waste

Paragraph (h)(1)(ii) of 29CFR 1910.1035 requires that clinical and research laboratory wastes that are contaminated with *M. tuberculosis* and are to be decontaminated outside of the immediate laboratory must be labeled with the biohazard symbol or placed in a red containers. This provision is intended to assure that employees are adequately warned that these containers require special handling and to serve as notice that certain precautions may be necessary should materials in the container be released (e.g., a spill). This provision closely follows the recommendations outlined in the CDC–NIH publication *Biosafety in Microbiological and Biomedical Laboratories* and is in accordance with the labeling requirements of paragraph (e)(2)(i)(D), Clinical and Research Laboratories, of the proposed standard. (OSHA, 1997)

Disposable items contaminated with respiratory secretions are not associated with *M. tuberculosis* transmission. These items should be handled and transported in a manner that reduces the risk for transmitting other microorganisms and should be disposed of in accordance with hospital policy and applicable regulations. (CDC, 1994)

14.9.14 Isolation Discontinuation

If a TB diagnosis cannot be ruled out; patients should remain in isolation until a determination has been made that the patient is noninfectious. Patients can, however, be discharged from the healthcare facility while still potentially infectious if appropriate post discharge arrangements can be ensured. The length of time required for a TB patient to become noninfectious after starting anti-TB therapy varies considerably. Isolation should be discontinued only when the patient:

- Is on effective therapy
- Is improving clinically
- Has had three consecutive negative sputum AFB smears collected on different days

Hospitalized active TB patients should be monitored for relapse by having sputum AFB smears examined regularly (e.g., every 2 weeks).

The two most common reasons why patients remain infectious despite treatment are (1) nonadherence to therapy (e.g., failure to take medications as prescribed), and (2) the presence of a drug-resistant organism. These reasons should be considered if a patient does not respond clinically to therapy within 2 to 3 weeks. Continued isolation throughout the hospitalization should be strongly considered for patients who have MDR-TB because of the tendency for treatment failure or relapse. (CDC, 1994)

14.9.15 Discharge Planning

Before a TB patient is discharged from the healthcare facility, the facility's staff and public health authorities should collaborate to ensure continuation of therapy. These discharge plans should be initiated and in place before the patient's discharge and should include, at a minimum:

- A confirmed outpatient appointment with the provider who will manage the patient until the patient is cured
- Sufficient medication to take until the outpatient appointment
- Placement into case management (e.g., DOT) or the public health department outreach programs

Patients who may be infectious at the time of discharge should only be discharged to facilities that have isolation capability or to their homes. Plans for discharging a patient who will return home must consider whether or not any of the household members were infected previously and if any uninfected household members are at very high risk for active TB if infected. Arrangements should be made to prevent these individuals from being exposed to the TB patient until a determination has been made that the patient is noninfectious. (CDC, 1994)

14.9.16 Operative Procedures

When operative procedures (or other procedures requiring a sterile field) are performed on patients who may have infectious TB, respiratory protection worn by the healthcare worker should protect:

1. Surgical fields from the respiratory secretions of the healthcare workers
2. Healthcare workers from infectious droplet nuclei that may be expelled by the patient or generated by the procedure

Respirators with exhalation valves and most positive-pressure respirators do not protect the sterile field. Surgical masks are designed to:

- Prevent the respiratory secretions of the person wearing the mask from entering the air
- Reduce the droplet nuclei expulsion into the air

Patients suspected of having TB should wear surgical masks when not in TB isolation rooms. These patients do not need to wear particulate respirators, which are designed to filter the air prior to inhalation by the person wearing the respirator. Patients suspected of having or known to have TB should never wear a respirator that has an exhalation valve, because this respirator type does not prevent droplet nuclei expulsion into the air. (CDC, 1994)

14.9.17   Cough-Inducing and Aerosol-Generating Procedures

Procedures that involve instrumentation of the lower respiratory tract or induce coughing can increase the likelihood of droplet nuclei being expelled into the air. These cough-inducing procedures include endotracheal intubation and suctioning, diagnostic sputum induction, aerosol treatments (e.g., pentamidine therapy), and bronchoscopy. Other procedures that can generate aerosols (e.g., irrigation of tuberculous abscesses, homogenizing or lyophilizing tissue, or other processing of tissue that may contain tubercle bacilli) are also covered by these recommendations.

Cough-inducing procedures should not be performed on patients who may have infectious TB unless the procedures are absolutely necessary and can be performed with appropriate precautions. All cough-inducing procedures performed on patients who may have infectious TB should be performed using local exhaust ventilation devices (e.g., booths or special enclosures) or in a room that meets the ventilation requirements for TB isolation.

Healthcare workers should wear respiratory protection when present in rooms or enclosures in which cough-inducing procedures are being performed. After completion of cough-inducing procedures, patients who may have infectious TB should remain in their isolation rooms or enclosures and not return to common waiting areas until coughing subsides. Patients should be given tissues and instructed to cover their mouths and noses with the tissues when coughing. TB patients who must recover from sedatives or anesthesia after a procedure (e.g., after a bronchoscopy) should be monitored in separate isolation rooms and should not be placed in recovery rooms with other patients. Before the booth, enclosure, or room is used for another patient, enough time should be allowed to pass for at least 99% of airborne contaminants to be removed. This time will vary according to the efficiency of the ventilation or filtration used. (CDC, 1994)

14.9.18   Bronchoscopy

If performing bronchoscopy in positive-pressure rooms (e.g., operating rooms) is unavoidable, TB should be ruled out as a diagnosis before the procedure is performed. If bronchoscopy must be performed to diagnose pulmonary disease and that diagnosis could include TB, the procedure should be performed in a room that meets TB isolation ventilation requirements and should include special considerations for the administration of aerosolized pentamidine. Patients should be screened for active TB before prophylactic therapy with aerosolized pentamidine is initiated. Screening should include obtaining a medical history and performing skin testing and chest radiography. Before each subsequent treatment with aerosolized pentamidine, patients should be screened for symptoms suggestive of TB (e.g., development of a productive cough). If such symptoms are elicited, a diagnostic evaluation for TB should be initiated. Patients who have suspected or confirmed active TB should take, if clinically practical, oral prophylaxis for *Pneumocystis carinii* pneumonia. (CDC, 1994)

14.10 EDUCATION AND TRAINING

OSHA requires that training records be maintained for 3 years beyond the date the training occurred. These records are not required to be kept confidential and so may become part of an employee's personnel file or part of a larger file, at the discretion of the employer. (OSHA, 1997)

All healthcare workers, including physicians, should receive education regarding TB that is relevant to persons in their particular occupational group. Training should be conducted before initial assignment, and the need for additional training should be reevaluated periodically (e.g., once a year). The level and detail of this education will vary according to the healthcare worker's job responsibilities and the level of risk in the facility (or area of the facility) in which the healthcare worker is employed. However, the program may include the following elements:

- Basic concepts of *M. tuberculosis*
- Transmission
- Pathogenesis
- Diagnosis
- Latent TB infection vs. active TB disease
- TB signs and symptoms
- Reinfection possibility
- Occupational exposure potential to the infectious TB patients
- TB prevalence in the community and facility
- Active TB patient isolation potential
- Situations with increased risk for exposure to *M. tuberculosis*
- Infection-control protocols that reduce transmission risk
- TB infection-control measure hierarchy
- Written policies and procedures
- Site-specific control measures (provided to healthcare workers in areas that require control measures in addition to basic TB infection-control program measures)
- Purpose of PPD skin testing
- Significance of positive PPD test results
- Skin-test program participation
- Latent TB infection prevention therapy
- Drugs (e.g., indications, use, effectiveness, potential adverse effects)
- Healthcare worker's responsibility to seek prompt medical evaluation; if PPD test conversion occurs or symptoms develop that could be caused by TB

(*Note:* Medical evaluation will enable healthcare workers who have TB to receive appropriate therapy and will help to prevent *M. tuberculosis* transmission to patients and other healthcare workers.)

- Active TB drug therapy principles
- Facility notification if the healthcare worker is diagnosed with active TB and procedures for initiation of contact investigation
- Providing currently infected healthcare workers with confidential and appropriate therapy and establishing noninfectious status before returning to duty
- Higher risks associated with TB infection concurrent with HIV infection or other causes of severely impaired cell-mediated immunity, including
  - More frequent and rapid development of clinical TB after infection with *M. tuberculosis*
  - Clinical disease presentation differences

- High mortality rate associated with MDR-TB
- Potential cutaneous anergy as immune function (as measured by CD4+ T-lymphocyte counts) declines
- Efficacy and safety of BCG (Bacille Calmette–Guérin) vaccination
- PPD screening principles among BCG recipients
- Voluntary work reassignment options for immunocompromised healthcare workers (CDC, 1994)

14.11 COUNSELING, SCREENING, AND EVALUATION

Employers must comply with paragraph (g) of 29CFR 1910.1035 and with the most current CDC recommendations in providing medical surveillance. OSHA explains that medical management and follow-up include diagnosis, and, where appropriate, prophylaxis and treatment related to TB infection and disease. The employer must provide medical management and follow-up for occupationally exposed employees with skin test conversions (see paragraph (g)(3)(i)(D)) or those who undergo an exposure incident whether or not they are categorized as occupationally exposed (see paragraphs (g)(1)(ii) and (g)(3)(i)(C)). In addition, any time an occupationally exposed employee develops signs and symptoms of infectious tuberculosis, medical management and follow-up are required (see paragraph (g)(3)(i)(B)). Paragraph (g)(2)(vi) explains that other related tests and procedures:

- Are defined as any TB-related tests and procedures determined to be necessary by the physician or other licensed healthcare professional, as appropriate
- Could include chest radiographs, sputum smears, or other testing determined to be necessary to make an assessment, a diagnosis, or medically manage the employee (OSHA, 1997)

A TB counseling, screening, and prevention program for healthcare workers should be established to protect both healthcare workers and patients. Healthcare workers with any of the following should be identified and evaluated as to active TB status:

- Positive PPD test results
- PPD test conversions
- Symptoms suggestive of TB

As necessary, these healthcare workers should be started on therapy or preventive therapy. The healthcare worker PPD screening program results will contribute to current infection control evaluations. (CDC, 1994)

14.11.1 Counseling

Providing services (e.g., social welfare, social work, teaching, law enforcement, legal aid) in facilities or in residences to individuals having suspected or confirmed infectious tuberculosis may put the employees providing these services at risk. For example, certain social workers may need to enter AFB isolation rooms or areas or visit homes of people who have suspected or confirmed infectious tuberculosis for the purposes of collecting information or providing discharge planning. While OSHA believes that collecting such information over the telephone in order to prevent occupational exposure is preferable, the agency realizes that situations may occur where direct contact with these isolated or confined individuals may be necessary. In these limited situations, these employees would be covered under the scope of the standard. There may also be situations where teachers may be providing tutoring to individuals isolated with suspected or confirmed infectious tuberculosis. Again, OSHA believes that such situations would be limited and that most educational instruction could be delayed until an individual was determined to be noninfectious.

However, where teachers must provide instruction to individuals identified as having suspected or confirmed infectious TB, those teachers would be covered under the scope of the standard. In addition, certain law enforcement officers might have to be in contact with individuals who have been identified as having suspected or confirmed infectious tuberculosis. For example, they may have to transfer such an individual from a correctional or detainment facility to a hospital for diagnosis or treatment. Because these workers must be in direct contact with the individual during transport, perhaps for long periods of time and probably in an enclosed vehicle, such employees could incur significant occupational exposure.

Paragraph (a)(9) would assure that such employees would be covered under the standard. Similarly, occasions may arise where attorneys must consult with clients or inmates who have been isolated or segregated because they have been identified as having suspected or confirmed infectious tuberculosis. Such attorneys would be covered under the standard in the limited situations where these consultations cannot be done by phone or delayed until the individual has been determined to be noninfectious. Under paragraph (a)(9), OSHA has specified certain employee groups that it believes would have to enter AFB isolation rooms or areas or homes where individuals are confined due to suspected or confirmed infectious TB, in order to provide services that may result in occupational exposure. OSHA requests comments and data as to whether there are other employee groups that may incur occupational exposure and thus need protection under this paragraph. (OSHA, 1997)

Because of the increased risk for rapid progression from latent TB infection to active TB in HIV-infected or otherwise severely immunocompromised persons, all healthcare workers should know if they have a medical condition or are receiving a medical treatment that may lead to severely impaired cell-mediated immunity. Healthcare workers should:

- Know their HIV status (existing guidelines for counseling and testing should be followed routinely).
- Seek the appropriate preventive measures.
- Consider voluntary work reassignments.

Employers should offer (but not compel a healthcare worker to accept) a work setting in which the healthcare worker would have the lowest possible risk for occupational exposure to *M. tuberculosis*. Evaluation of these situations should also include consideration of the provisions of the Americans with Disabilities Act of 1990 and other laws. (*Note:* Healthcare workers should know their HIV status if they work with or near patients who have drug-resistant TB.)

Limiting exposure to TB patients is the most protective measure that severely immunosuppressed healthcare workers can take to avoid becoming infected with *M. tuberculosis*. Healthcare workers who are known to be HIV-infected or otherwise severely immunosuppressed should be tested for cutaneous anergy at the time of PPD testing. Consideration should be given to retesting, at least every 6 months, those immunocompromised healthcare workers who are potentially exposed to *M. tuberculosis* because of the high risk for rapid progression to active TB if they become infected.

Information provided by healthcare workers regarding their immune status should be treated confidentially. If the healthcare worker requests voluntary job reassignment, the confidentiality of the healthcare worker should be maintained. Facilities should have written procedures on confidential handling of such information. (CDC, 1994)

14.11.2 Screening for Active Tuberculosis

Any healthcare worker who has a persistent cough (i.e., a cough lasting more than 3 weeks), especially in the presence of other signs or symptoms compatible with active TB (e.g., weight loss, night sweats, bloody sputum, anorexia, or fever), should be evaluated promptly for TB and should not return to the workplace until a diagnosis of TB has been excluded or the healthcare worker is on therapy and a determination has been made that the healthcare worker is noninfectious. (CDC, 1994)

14.11.3 Screening for Latent Tuberculosis

Administrators of healthcare facilities should ensure that physicians and other personnel not paid by, but working in, the facility receive skin testing at appropriate intervals for their occupational group and work location. Risk assessments should identify healthcare workers that have potential *M. tuberculosis* exposure and exposure frequency. This information should be used to determine which healthcare workers to include in the skin-testing program and the skin-testing frequency. Healthcare workers from risk groups with increased TB prevalence should be in the skin-testing program, so that converters can be identified and preventive therapy offered.

During the pre-employment physical or when applying for hospital privileges, healthcare workers who have *M. tuberculosis* exposure potential, including those with BCG vaccination, should have baseline purified protein derivative (PPD) skin testing. Healthcare workers with a positive PPD test and who are receiving adequate disease treatment or adequate infection preventive therapy may be exempt from PPD screening unless TB signs or symptoms develop. For healthcare workers without documented negative PPD test results during the preceding 12 months, the baseline PPD testing should employ the two-step method, which will detect boosting phenomena that might be misinterpreted as a skin-test conversion. Decisions concerning the two-step procedure use for baseline testing should be based on the facility's boosting frequency. For a negative PPD test, PPD testing should be repeated at regular intervals as determined by the risk assessment. Healthcare workers should be tested whenever they are exposed to TB patients without appropriate precautions.

Performing PPD testing of healthcare workers who work in the same area or occupational group on different scheduled dates, rather than testing all healthcare workers in the area or group on the same day, may lead to earlier detection of *M. tuberculosis* transmission. All PPD tests should be administered, read, and interpreted in accordance with current guidelines by specified and trained personnel.

At the time their test results are interpreted, healthcare workers should be informed as to both positive and negative PPD test results and that interpretation of an induration 5 to 9 mm in diameter depends on the healthcare worker's immune status and history of exposure to persons who have infectious TB. Healthcare workers who have indurations 5 to 9 mm in diameter should be advised that such results may be considered positive for healthcare workers who have been in contact with persons with infectious TB and who have HIV infection or other causes of severe immunosuppression (e.g., immunosuppressive therapy for organ transplantation).

When a healthcare worker who is not assigned regularly to a single work area has a PPD test conversion, appropriate personnel should identify the areas worked during the time when infection was likely to have occurred. Information derived can be considered in analyzing these area's transmission risk.

In any area of the facility where *M. tuberculosis* transmission occurred, a problem evaluation should be conducted. The skin-testing frequency should be determined according to the applicable risk category. PPD test results should be recorded confidentially in the individual healthcare worker's employee health record and in an aggregate database. The database can be analyzed periodically to estimate the risk for acquiring new infection in specific areas or occupational groups in the facility. (CDC, 1994)

14.11.4 Positive PPD or Active Tuberculosis

All healthcare workers with newly recognized positive PPD test results or PPD test conversions should be evaluated promptly for active TB through clinical examination and chest radiograph (Figure 14.2). If the history, clinical examination, or chest radiograph is compatible with active TB, additional tests should be performed. If symptoms compatible with TB are present, the healthcare worker should be excluded from the workplace until a diagnosis of active TB is ruled out or it has been established that the healthcare worker is being treated and a determination has

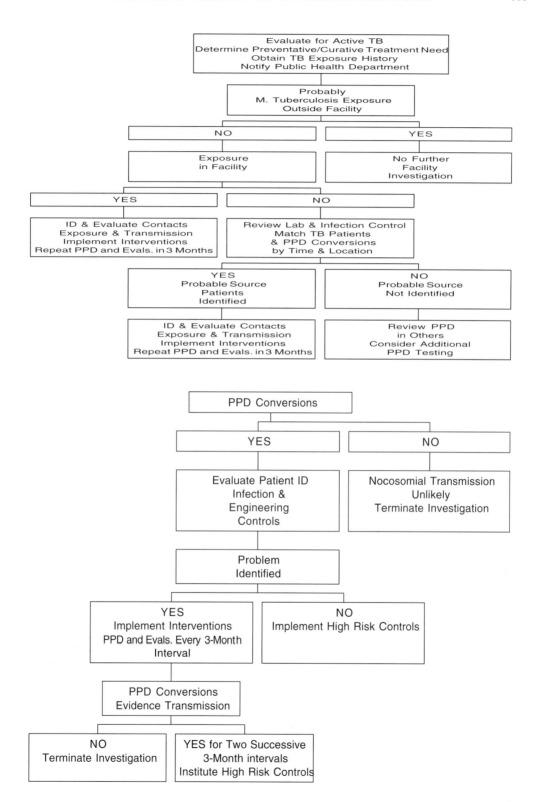

FIGURE 14.2   These charts describe the PPD conversion to intervention decision logic. (From CDC, Federal Register, October 28, 1994.)

been made that the healthcare worker is noninfectious. Healthcare workers who do not have active TB should be evaluated for preventive therapy according to published guidelines. If a healthcare worker's PPD test result converts to positive, a confirmed or suspected TB exposure history should be obtained to determine the potential source. When the exposure source is known, the drug-susceptibility pattern of the *M. tuberculosis* isolated from the source should be identified so that the correct curative or preventive therapy can be initiated. The drug-susceptibility pattern should be recorded in the healthcare worker's medical record, in case the healthcare worker subsequently develops active TB and needs therapy specific for the drug-susceptibility pattern. All healthcare workers, including those with histories of positive PPD test results, should be reminded periodically about TB symptoms and pulmonary symptoms suggestive of TB and the need for prompt evaluation. (CDC, 1994)

14.11.5 Chest Radiographs

That the pulmonary system is emphasized in both the medical history and physical examination ensures that the employee is evaluated with specific attention to the most common site of infectious TB. Although extrapulmonary tuberculosis can occur (e.g., in bone, meninges of the brain, and draining abscesses), it is not usually a source of infection for others. The phrase "with emphasis on the pulmonary system" is used to indicate that, while the history and physical examinations evaluate the health of the patient as a whole, particular emphasis should be placed on the pulmonary system. (OSHA, 1997)

Routine chest radiographs are not required for asymptomatic, PPD-negative healthcare workers. Healthcare workers with positive PPD test results should have a chest radiograph as part of the initial evaluation of their PPD test; if the chest radiograph is negative, repeat chest radiographs are not necessary unless symptoms develop that could be attributed to TB. More frequent monitoring for symptoms of TB may be considered for recent converters and other PPD-positive healthcare workers who are at increased risk for developing active TB (e.g., HIV-infected or otherwise severely immunocompromised healthcare workers). (CDC, 1994)

14.11.6 Active Tuberculosis

Active TB healthcare workers with pulmonary or laryngeal TB pose a risk to patients and other healthcare workers while they are infectious and should be excluded from the workplace until noninfectious. The same work restrictions apply to all healthcare workers regardless of their immune status. Before the healthcare worker who has TB can return to the workplace, the healthcare facility should have documentation from the healthcare worker's healthcare provider that the healthcare worker is receiving adequate therapy, the cough has resolved, and the healthcare worker has had three consecutive negative sputum smears collected on different days. After work duties are resumed and while the healthcare worker remains on anti-TB therapy, facility staff should receive periodic documentation from the healthcare worker's healthcare provider that the healthcare worker is being maintained on effective drug therapy for the recommended time period and that the sputum AFB smears continue to be negative.

Healthcare workers with active laryngeal or pulmonary TB who discontinue treatment before cure should be evaluated promptly for infectiousness. If the evaluation determines that they are still infectious, they should be excluded from the workplace until:

1. Treatment has resumed.
2. An adequate response to therapy has been documented.
3. Three more consecutive sputum AFB smears collected on different days have been negative.

Healthcare workers who have TB at sites other than the lung or larynx usually do not need to be excluded from the workplace if a diagnosis of concurrent pulmonary TB has been ruled out. (CDC, 1994)

14.11.7 Latent Tuberculosis Infection

Healthcare workers receiving preventive treatment for latent TB infection should not be restricted from their usual work activities. Healthcare workers with latent TB infection who cannot take or who do not accept or complete a full course of preventive therapy should not be excluded from the workplace; instead, these healthcare workers should be counseled about the risk for developing active TB and instructed regularly to seek prompt evaluation if signs or symptoms develop that could be due to TB. (CDC, 1994)

14.11.8 Problem Evaluation

Epidemiologic investigations may be indicated for several situations:

- PPD test conversions or active TB in healthcare workers
- Possible person-to-person *M. tuberculosis* transmission
- Patients or healthcare workers with active TB who are not promptly identified and isolated, thus exposing other persons in the facility to *M. tuberculosis*

The general objectives are to:

- Determine the likelihood that transmission of and infection with *M. tuberculosis* has occurred in the facility.
- Determine the extent to which *M. tuberculosis* has been transmitted.
- Identify those persons who have been exposed and infected, enabling them to receive appropriate clinical management.
- Identify factors that could have contributed to transmission and infection and to implement appropriate interventions.
- Evaluate the effectiveness of any interventions that are implemented.
- Ensure that exposure to and *M. tuberculosis* transmission has been terminated.

The exact circumstances of these situations are likely to vary considerably, and the associated epidemiologic investigations should be tailored to the individual circumstances. (CDC, 1994)

14.12 INVESTIGATING PPD TEST CONVERSIONS AND ACTIVE TUBERCULOSIS

Most infected people have a positive reaction to the TB skin test within 2 to 10 weeks after exposure; consequently, early detection of newly infected workers is critical as it permits early initiation of appropriate therapy and results in a decrease in morbidity and mortality. Proper administration of a TB skin test results in a reaction described as a classic example of a delayed (cellular) hypersensitivity reaction. This reaction indicates infection with *Mycobacterium* spp., most commonly *M. tuberculosis*. The reaction characteristically begins in 5 to 6 hours, is maximal at 48 to 72 hours, and subsides over a period of days. Proper administration and interpretation of the test are critical and can be complex.

In 1990, the American Thoracic Society revised the criteria for interpreting the TB skin test. Information such as the health status of the tested employee, history of BCG vaccination, recent close contact with persons with active TB, chest x-ray results, and other factors must be considered when interpreting the TB skin test results. The CDC has established criteria for a TB skin test conversion: When an employee's TB skin test results change from negative to positive, that change indicates a recent TB infection. In its 1994 *Core Curriculum on Tuberculosis*, CDC describes the complexities of interpreting the indurations resulting from TB skin testing. A number of factors can affect the size of a TB skin test induration relative to whether or not the test should be interpreted as being positive:

- Induration of 5 mm or more is classified as positive for persons with known or suspected HIV infection.
- An induration must be 10 mm to be classified as positive in persons who are foreign-born in high prevalence countries.
- An induration of 15 mm or more is classified as positive in certain other situations.

In addition, TB skin testing can result in both false positive and false negative results. (OSHA, 1997)

Investigating PPD test conversions in healthcare workers' PPD test conversions may be detected in healthcare workers as a result of a contact investigation when the probable source of exposure and transmission is already known or as a result of routine screening when the probable source of exposure and infection is not already known and may not be immediately apparent. (CDC, 1994)

14.12.1 Skin Test Conversion

If a skin test conversion in a healthcare worker is identified as part of routine screening, the following steps should be considered:

- The healthcare worker should be evaluated promptly for active TB.
- The initial evaluation should include a thorough history, physical examination, and chest radiograph.
- On the basis of the initial evaluation, other diagnostic procedures (e.g., sputum examination) may be indicated.
- If appropriate, the healthcare worker should be placed on preventive or curative therapy in accordance with current guidelines. (CDC, 1994)

14.12.2 History

A history of possible exposure to *M. tuberculosis* should be obtained from the healthcare worker to determine the most likely source of infection. When the source of infection is known, the drug-susceptibility pattern of the *M. tuberculosis* isolate from the source patient should be identified to determine appropriate preventive or curative therapy regimens. If the history suggests that the healthcare worker was exposed to and infected with *M. tuberculosis* outside the facility, no further epidemiologic investigation to identify a source in the facility is necessary. If the history does not suggest that the healthcare worker was exposed and infected outside the facility but does identify a probable source of exposure in the facility, then contacts of the suspected source patient should be identified and evaluated and possible reasons for the exposure and transmission should be evaluated. Interventions should be implemented to correct these causes.

Testing of PPD-negative healthcare workers should be performed immediately and repeated after 3 months. If no additional PPD test conversions are detected on follow-up testing, the investigation can be terminated. If additional PPD test conversions are detected on follow-up testing, then possible reasons for exposure and transmission should be reassessed and the appropriateness and degree of adherence to the interventions implemented should be evaluated. PPD testing of PPD-negative healthcare workers should be repeated after another 3 months. If no additional PPD test conversions are detected on the second round of follow-up testing, the investigation can be terminated.

If additional PPD conversions are detected on the second round of follow-up testing, then a high-risk protocol should be implemented in the affected area or occupational group and the public health department or other persons with expertise in TB infection control should be consulted. If the history does not suggest that the healthcare worker was exposed to and infected with *M. tuberculosis* outside the facility and does not identify a probable source of exposure in the facility, then further investigation to identify the probable source patient in the facility is warranted.

The interval during which the healthcare worker could have been infected should be estimated. Generally, this would be the interval from 10 weeks before the most recent negative PPD test through 2 weeks before the first positive PPD test (i.e., the conversion). Laboratory and infection-control records should be reviewed to identify all patients or healthcare workers who have suspected or confirmed infectious TB or who could have transmitted *M. tuberculosis* to the healthcare worker. If this process does identify a likely source patient, then contacts of the suspected source patient should be identified and evaluated. Possible reasons for the exposure and transmission should be evaluated. Interventions should be implemented to correct these causes, and PPD testing of PPD-negative healthcare workers should be repeated after 3 months. If this process does not identify a likely source case, then PPD screening results of other healthcare workers in the same area or occupational group should be reviewed for additional evidence of *M. tuberculosis* transmission.

If sufficient additional PPD screening results are not available, appropriate personnel should consider conducting additional PPD screening of other healthcare workers in the same area or occupational group. If this review and/or screening does not identify additional PPD conversions, then nosocomial transmission is less likely and the contact investigation can probably be terminated. (CDC, 1994)

14.13 EXPOSURE INCIDENTS

Under paragraph (j) of 29CFR 1910.1035, an exposure incident is an event in which an employee has been exposed to:

- An individual with confirmed infectious TB
- Air containing aerosolized *M. tuberculosis* without the benefit of all applicable exposure control measures required by this section

Investigation procedures must be in place before an incident occurs to ensure that investigations are done promptly, consistently, and thoroughly from case to case. Procedures required for reporting exposure incidents include identification of the person to whom the incident is to be reported and procedures the employer will use for evaluating the circumstances surrounding exposure incidents as required by paragraph (g)(4)(iv). In the event that unprotected employees are exposed to aerosolized *M. tuberculosis*, this exposure incident must be reported to the employer as soon as feasible in order to promptly initiate:

- Proper medical management and follow-up of the exposed employee
- Investigate the circumstances surrounding such incidents while pertinent conditions remain relatively unchanged and are fresh in the employee's memory

Paragraph (g)(4)(iv) directs the employer to investigate and document the circumstances surrounding the exposure incident to determine if changes can be instituted to prevent similar occurrences. (OSHA, 1997)

14.13.1 Potential Nosocomial Transmission

When it is uncertain whether or not the healthcare worker's PPD test conversion resulted from occupational exposure and infection, the absence of other data implicating nosocomial transmission suggests that the conversion could have resulted from:

- Unrecognized exposure to *M. tuberculosis* outside the facility
- Cross-reactivity with another antigen (e.g., nontuberculous mycobacteria)
- Errors in applying, reading, or interpreting the test
- False positivity caused by the normal variability of the test
- False positivity caused by a defective PPD preparation

If this review and/or screening identifies additional PPD test conversions, nosocomial transmission is more likely. To identify problems that could have led to exposure and transmission, evaluate:

- Patient identification (i.e., triage) process
- TB infection-control policies and practices
- Engineering controls

If no such problems are identified:

- High-risk protocol should be implemented in the affected area or occupational group.
- The public health department or other persons with expertise in TB infection control should be consulted.

If such problems are identified:

- Appropriate interventions should be implemented to correct the problems.
- PPD skin testing of PPD-negative healthcare workers should be repeated after 3 months.

If no additional PPD conversions are detected on follow-up testing, the investigation can be terminated. If additional PPD conversions are detected on follow-up testing,

- Possible reasons for exposure and transmission should be reassessed.
- Appropriateness of and adherence to the interventions implemented should be evaluated.
- PPD skin testing of PPD-negative healthcare workers should be repeated after another 3 months.

If no additional PPD test conversions are detected on this second round of follow-up testing, the investigation can be terminated. If additional PPD test conversions are detected on the second round of follow-up testing, a high-risk protocol should be implemented in the affected area or occupational group, and the public health department or other persons with expertise in TB infection control should be consulted. (CDC, 1994)

14.13.2 Investigating Active Tuberculosis Cases

If a healthcare worker develops active TB, the case should be evaluated epidemiologically in a manner similar to PPD test conversions in healthcare workers to determine potential occupational transmission probability, to identify possible causes, and to implement appropriate interventions. Contacts of the healthcare worker should be identified and evaluated for TB infection and disease. The public health department should be notified immediately for consultation to allow for investigation of community contacts who were not exposed in the healthcare facility. The public health department should notify facilities when physicians report healthcare workers with TB so that an investigation of contacts can be conducted in the facility. The information provided should be in accordance with state or local laws to protect the confidentiality of the healthcare worker. (CDC, 1994)

14.13.3 Investigating *M. tuberculosis* Patient-to-Patient Transmission

Surveillance of active TB cases in patients should be conducted. Patient-to-patient transmittal should be suspected if:

- A high proportion of TB patients had prior admissions during the year preceding onset of their TB.
- The number of patients with drug-resistant TB has increased suddenly.
- Isolates obtained from multiple patients had identical and characteristic drug-susceptibility, or DNA fingerprint patterns.

If this surveillance suggests the possibility of patient-to-patient *M. tuberculosis* transmission, the following steps should be taken:

- Review the healthcare worker PPD test results and patient surveillance data for the suspected areas to detect additional patients or healthcare workers with PPD test conversions and active disease.
- Look for possible exposures that patients with newly diagnosed TB could have had to other TB patients during previous admissions — were the patients admitted to the same room or area, or did they receive the same procedure or go to the same treatment area on the same day?

If the evaluation thus far suggests transmission has occurred, then:

- Evaluate possible causes of the transmission, such as a problem with patient detection, institutional barriers to implementing appropriate isolation practices, or inadequate engineering controls.
- Ascertain whether other patients or healthcare workers could have been exposed; if so, evaluate these persons for TB infection and disease.
- Notify the public health department to begin a community contact investigation if necessary. (CDC, 1994)

14.13.4 Investigating Contacts and Infectious Tuberculosis

Investigations should proceed if:

- A patient who has active TB is examined in a healthcare facility.
- The illness is not diagnosed correctly, resulting in a failure to apply appropriate precautions, leading to inappropriate isolation.
- A healthcare worker develops active TB and exposes other persons in the facility.

The following steps should be taken when the illness is later diagnosed correctly:

- Identify other patients and healthcare workers who were exposed to the source patient before isolation procedures were begun, interview the source patient and all applicable personnel and review that patient's medical record.
- Determine areas of the facility in which the source patient was hospitalized, visited, or worked before being placed in isolation.
- Identify healthcare workers who may have been exposed during that time.

Note that the contact investigation should first determine if *M. tuberculosis* transmission has occurred from the source patient to those persons with whom the source patient had the most intense contact. (CDC, 1994)

14.13.5 PPD Tests

Administer PPD tests to the most intensely exposed healthcare workers and patients as soon as possible after the exposure has occurred. If transmission did occur to the most intensely exposed persons, then those persons with whom the patient had less contact should be evaluated. If the initial PPD test result is negative, a second test should be administered 12 weeks after the exposure was terminated. Those persons who were exposed to *M. tuberculosis* and who have either a PPD test conversion or symptoms suggestive of TB should receive prompt clinical evaluation and, if indicated, chest radiographs and bacteriologic studies. Those persons who have evidence of newly acquired infection or active disease should be evaluated for preventive or curative therapy. Persons who have previously had positive PPD test results and who have been exposed to an infectious TB patient do not require a repeat PPD test or a chest radiograph unless they have symptoms suggestive of TB.

14.13.6 Initial Diagnosis Evaluation

Determine why the diagnosis of TB was delayed. If the correct diagnosis was made but the patient was not isolated promptly, the reasons for the isolation delay need to be defined and corrective actions must be taken. (CDC, 1994)

14.14 PUBLIC HEALTH DEPARTMENT

As soon as a patient or healthcare worker is known or suspected to have active TB, the patient or healthcare worker should be reported to the public health department so that appropriate follow-up can be arranged and a community contact investigation can be performed. The health department should be notified well before patient discharge to facilitate follow-up and continuation of therapy. A discharge plan coordinated with the patient or healthcare worker, the health department, and the inpatient facility should be implemented. The public health department should protect the confidentiality of the patient or healthcare worker in accordance with state and local laws. Healthcare facilities and health departments should coordinate their efforts to perform appropriate contact investigations on patients and healthcare workers who have active TB.

In accordance with state and local laws and regulations, results of all AFB-positive sputum smears, cultures positive for *M. tuberculosis*, and drug-susceptibility results on *M. tuberculosis* isolates should be reported to the public health department as soon as these results are available. The public health department may be able to assist facilities with planning and implementing various aspects of a TB infection-control program. The state health department may be able to provide names of experts to assist with the engineering aspects of TB infection control. (CDC, 1994)

14.15 OPERATING ROOMS

Elective operative procedures on patients who have TB should be delayed until the patient is no longer infectious. If operative procedures must be performed, operating rooms should have anterooms. For operating rooms without anterooms, doors to the operating room should be closed. Traffic into and out of the rooms should be minimal to reduce the frequency of opening and closing the doors. Attempts should be made to perform procedures at times when other patients are not present in the operating suites and when a minimum number of personnel are present. Placing bacterial filters on patient endotracheal tubes (or at the expiratory side of the breathing circuit of ventilators or anesthesia machines, if these are used) may be required. During postoperative recovery, patients should be monitored and placed in private rooms that meet recommended standards for ventilating TB isolation rooms. When operative procedures (or other procedures requiring a sterile field) are performed on patients who may have infectious TB, respiratory protection worn by the healthcare worker must protect the operative field from the respiratory secretions of the healthcare worker and the healthcare worker from the infectious droplet nuclei generated by the patient. (CDC, 1994)

14.16 AUTOPSY ROOMS

Because infectious aerosols are likely to be present in autopsy rooms, such areas should be at negative pressure with respect to adjacent areas. The room air should be exhausted directly to the outside of the building. The American Society of Heating, Refrigeration, and Air Conditioning Engineers (ASHRAE) recommends that autopsy rooms have ventilation that provides airflow of 12 air changes per hour (ACH). Where possible, this level should be increased by means of ventilation system design or by auxiliary methods, such as recirculation of air within the room through HEPA filters. Recirculation of HEPA-filtered air within the room or ultraviolet germicidal irradiation (UVGI) may be used. Respiratory protection should be worn by personnel while performing autopsies on deceased persons who may have had TB at the time of death. (CDC, 1994)

14.17 LABORATORIES

14.17.1   CDC and NIH

Prior to 1984, no single code of practice, standards, or guidelines or other publications provided detailed descriptions of techniques or equipment for laboratory activities involving pathogens. In that year, the CDC and the National Institutes of Health (NIH) published guidelines entitled *Biosafety in Microbiological and Biomedical Laboratories*. These biosafety guidelines are based on combinations of standard and laboratory practices and techniques, safety equipment, and laboratory facilities appropriate for the operations performed and the hazard posed and are applicable to work with any infectious agent. The basic format for the biosafety guidelines categorizes infectious agents and laboratory activities into four classes or levels denoted as Biosafety Levels 1 through 4. The requirements in paragraph (e) of 29CFR 1910.1035, including those regarding biosafety cabinets, are derived primarily from the CDC–NIH recommendations found in *Biosafety in Microbiological and Biomedical Laboratories*.

Only those provisions that relate to the health and safety of employees are required by the standard. The provisions in paragraph (e) are a minimal program, and OSHA anticipates that employers affected by this paragraph will continue to follow any other appropriate portions of the above recommendations in addition to the requirements of this standard. In addition, the employer is responsible for following this entire standard (e.g., training employees, medical surveillance). (OSHA, 1997)

14.17.2 Tuberculosis Hazards

Microbiology laboratories are special, often unique, work environments that may pose special infectious disease risks to persons in or near them. Personnel have contracted infections in the laboratory throughout the history of microbiology. Aerosols present the greatest hazard in laboratories. Tubercle bacilli may be present in sputum, gastric lavage fluids, cerebrospinal fluid, urine, and lesions from a variety of tissues. Bacilli grown in culture may amplify beyond what would normally be found in the sample. Such culturing is used for identification and susceptibility testing. Artificial concentrations may also occur during laboratory preparation. These concentrations increase the transmission risk per sample volume if the bacteria are not contained. The bacilli may survive in heat-fixed smears and may be aerosolized in the preparation of frozen sections and during manipulation of liquid cultures. Under paragraph (e) of 29CFR 1910.1035, Clinical and Research Laboratories, the proposed standard requires a number of provisions to eliminate or minimize exposure in clinical and research laboratory settings. (OSHA, 1997)

14.17.3 Laboratory Director's Assessment

For facilities with clinical or research laboratories, paragraph (c)(2)(iv) of 29CFR 1910.1035 requires that the Exposure Control Plan contain a determination from the director of the laboratory as to whether the laboratory facility should operate at Biosafety Level 2 or 3 containment according to CDC–NIH recommendations. Occasions will arise when the laboratory director should select a biosafety level higher than that recommended. A higher biosafety level may be indicated by

- Special containment requirements for experimentally generated aerosols for inhalation studies
- Proximity of the laboratory to areas of special concern
- Absence of certain recommended safeguards

OSHA states that the laboratory director may make adaptations to the CDC–NIH recommendations. Regardless of adaptations, OSHA requires the laboratory director to determine and document the need for controlled access, anterooms, sealed windows, directional airflow, preventing recirculation of laboratory exhaust air, filtration of exhaust air before discharge outside, and thimble exhaust connections for biological safety cabinets. The Exposure Control Plan must include determinations, along with any adaptations to the CDC–NIH biosafety level, and information to the laboratory employees as to adaptations to and changes in recommended biosafety levels. (OSHA, 1997)

14.17.4 Clinical and Research Laboratories

Paragraph (e) of 29CFR 1910.1035, Clinical and Research Laboratories, addresses requirements that must be met by clinical and research laboratories engaged in the culture, production, concentration, experimentation, and manipulation of *M. tuberculosis*. These requirements apply in addition to the other requirements of the standard. Paragraph (e) applies to two types of facilities that OSHA has designated as *clinical laboratories* and *research laboratories*. For the purpose of this standard, a clinical laboratory is a laboratory or area of a facility that conducts routine and repetitive operations for the diagnosis of TB, such as preparing acid-fast smears and culturing sputa or other clinical specimens for identification, typing, or susceptibility testing. A research laboratory is a laboratory that propagates and manipulates cultures of *M. tuberculosis* in large volumes or high concentrations that exceed those used for the identification and typing activities common to clinical laboratories.

The proposed standard requires, in paragraphs (e)(2)(i)(A) through (D), that both clinical and research laboratories follow several standard microbiological practices. All procedures are to be performed in a manner that minimizes the creation of aerosols. In view of the mode of *M. tuberculosis* transmission (that is, through inhalation of airborne organisms), this provision is extremely important in eliminating or minimizing employee exposure. The employer must evaluate laboratory tasks and institute the measures necessary to minimize the creation of aerosols. (OSHA, 1997)

14.17.5 Mouth Pipetting

OSHA proposes to adopt the good laboratory and infection control practice of prohibiting pipetting or suctioning by mouth. The use of cotton plugs or other barriers does little to reduce the hazards of mouth pipetting. Even a technician who is skilled in mouth pipetting may inadvertently suck fluids containing *M. tuberculosis* into the mouth. In addition to containing *M. tuberculosis* when the fluid is expelled, these fluids may also contain bloodborne pathogens that would have contacted the employee's mucous membranes (i.e., the mouth) as well as any blisters, cuts, or other lesions in the mouth or on the lips. (OSHA, 1997)

14.17.6 Access

Limiting access to these laboratories assures that unauthorized individuals are not placed at risk and do not distract or otherwise interfere with the activity of the authorized employees. This provision works in concert with the requirement for signs in paragraph (h)(2)(iv) and ensures that only employees who meet the special requirements set forth by the laboratory director, which will include training, personal protective equipment, and other requirements, could enter the area. Research laboratories working with *M. tuberculosis* are held to several additional requirements. Paragraph (e)(3)(i)(A) requires that research facilities keep laboratory doors closed when working with *M. tuberculosis*. Paragraph (e)(3)(i)(B) requires that access to the work area be limited to persons who comply with specified entry and exit requirements. These provisions are adopted from the CDC–NIH recommendations, *Biosafety in Microbiological and Biomedical Laboratories*. (OSHA, 1997)

14.17.7 Biosafety Manual

The requirement for a biosafety manual helps ensure that any additional procedures are developed to address situations that are unique to a particular facility and provide appropriate protection to exposed employees. The manual must be reviewed as necessary and at least annually and should be updated as necessary to reflect changes in the work setting. The phrase "as necessary" has been used to indicate that updating of the manual to reflect work setting changes is to be done as soon as possible and is not to be postponed until the annual review. Employees are required to read the biosafety manual's sections on potential hazards and practices and procedures. (OSHA, 1997)

14.17.8 Respiratory Protection

Paragraph (e)(3)(i)(C) of 29CFR 1910.1035 requires that respiratory protection be worn in research laboratories when aerosols cannot be safely contained (e.g., when aerosols are generated outside a biological safety cabinet). Research laboratories are working with larger volumes and higher concentrations of *M. tuberculosis* than clinical laboratories. The risk to employees from aerosolized bacilli is increased, thus these employees should be protected from lapses in containment. An example of when aerosols would be generated would be a flask containing *M. tuberculosis* being dropped and breaking outside of the biosafety cabinet. Another example would be centrifugation of cultures containing *M. tuberculosis* cultures in an open centrifuge without aerosol-proof centrifuge safety containers or utilizing such containers but then opening them outside of the biosafety cabinet. (OSHA, 1997)

14.17.9 Containerization

The requirement that contaminated material removed from the work area be placed in a container that prevents leakage during collection, handling, processing, storage, transport, or shipping is intended to ensure that no accidental spills or other contamination place other employees at risk. Although the proposed standard requires proper containerization of laboratory wastes, it includes no such requirement for wastes originating from the provision of care or services to individuals with suspected or confirmed infectious TB (e.g., facial tissues that the individual has used). The reason for this is that items such as facial tissues capture and contain the liquids generated by the individual. Once captured, the liquid is not readily aerosolized. In their guidelines, the CDC states: "Disposable items contaminated with respiratory secretions are not associated with *M. tuberculosis* transmission." In the laboratory, however, the liquids containing *M. tuberculosis* are generally not captured or contained on an item but exist as an individual specimen or culture. Also, in some instances, the bacilli have been concentrated. The possibility, therefore, for formation of droplet nuclei from these wastes is increased. Proper containerization and labeling of laboratory wastes is required to assist in preventing droplet nuclei formation and possible infection. Proper containerization and labeling of wastes to be decontaminated outside a laboratory not only help prevent employee exposure but also warn employees who come in contact with this waste of the hazard within the container. (OSHA, 1997)

14.17.10   Work Surfaces and Spills

Work surfaces and laboratory equipment must be decontaminated at the end of each shift and after any spill of viable material. Paragraph (e)(2)(ii)(D) of 29CFR 1910.1035 requires that spills be cleaned up immediately by employees trained and equipped to work with potentially concentrated *M. tuberculosis*. Because *M. tuberculosis* can become aerosolized during cleanup procedures, the task cannot be done by someone who is not skilled and properly equipped. Exposure incidents must be reported so that the postexposure management and follow-up required by paragraph (g) can be initiated and the circumstances surrounding the exposure incidents can be investigated.

14.17.11   Signage

Paragraph (e)(2)(ii)(E) of 29CFR 1910.1035 requires that, when materials or animals infected with *M. tuberculosis* are present in the laboratory, a hazard warning sign, in accordance with paragraph (h)(2)(iv), Communication of Hazards and Training, incorporating the universal biohazard symbol shall be posted on all laboratory and animal room access doors.

14.17.12   Biosafety Cabinets (BSCs)

The CDC–NIH manual *Biosafety in Microbiological and Biomedical Laboratories* recommends Biosafety Level 2 or 3 for such laboratories depending on the procedures being performed (Ex. 7–72). The requirements of paragraph (e)(2)(iii)(A) of 29CFR 1910.1035 stipulate that a certified Class 2 BSC must be used whenever activities with the potential for generating aerosols of *M. tuberculosis* are conducted and when high concentrations or volumes of *M. tuberculosis* are used. Such materials may be centrifuged in the open laboratory (i.e., outside of a biosafety cabinet) if sealed rotor heads or centrifuge safety cups are used. These requirements protect employees from exposure during the performance of procedures by ensuring that aerosolized *M. tuberculosis* will be contained and kept away from a worker's breathing zone.

Paragraph (e)(2)(iii)(B) requires that BSCs be certified when installed, annually after initial installation certification, whenever moved, or whenever filters are changed. Biological safety cabinets must be certified to ensure that proper protection is provided. The National Sanitation Foundation (NSF) Standard 49 describes design, construction, and performance criteria for biosafety cabinets. Moreover, this NSF standard is subject to periodic review by the NSF in order to keep the requirements consistent with new technology. OSHA has incorporated the current NSF Standard 49 performance criteria into the OSHA standard. For example, Standard 49 requires that "each cabinet be tested and performance evaluated on site, assuring that all physical containment criteria are met at the time of installation, prior to use, and periodically thereafter (Ex. 7–135)."

Paragraph (e)(3)(iii)(D) requires that the HEPA-filtered exhaust from Class 2 or 3 BSCs be discharged to the outside of the building or through the building exhaust system. If it is discharged through the building exhaust system, it must be connected to this system in a manner that avoids any interference with the air balance of the cabinets or the building exhaust system. This is required to ensure that BSCs and the building exhaust system continue to function as intended. (OSHA, 1997)

14.17.13   Wastes

Standard 29CFR 1910.1035 requires that all cultures, stocks, and other wastes contaminated with *M. tuberculosis* be decontaminated before disposal by a decontamination method, such as autoclaving, known to effectively destroy *M. tuberculosis*. Materials to be decontaminated outside of the immediate laboratory are to be placed in a durable leakproof container that is closed to prevent leakage for transport from the laboratory and labeled or color-coded in accordance with paragraph (h)(1)(ii). Decontamination before disposal helps ensure that other employees are not inadvertently exposed to the bacterium.

Paragraph (e)(2)(iv) requires that a method for decontamination of wastes contaminated with *M. tuberculosis* (e.g., autoclave, chemical disinfection, incinerator, or other approved decontamination system known to effectively destroy *M. tuberculosis*) must be available within or as near as feasible to the work area. The availability of such methods of decontamination is required for inactivating or destroying *M. tuberculosis* in or on a variety of media, including culture fluids, plastic ware, and equipment. These materials must be decontaminated to prevent potential aerosolization of *M. tuberculosis* and inadvertent exposure of employees outside of the laboratory.

14.17.14   Containment Equipment

Paragraph (e)(3)(ii) of 29CFR 1910.1035 requires that employers ensure that employees manipulating cultures and clinical or environmental materials that may generate *M. tuberculosis*-containing aerosols, challenging animals with *M. tuberculosis* aerosols, harvesting tissues or fluids from infected animals, or performing necropsies on infected animals use the appropriate containment equipment and/or devices when performing these activities. Such equipment and devices include Class 2 or 3 BSCs, or appropriate combinations of personal protective equipment and physical containment devices (such as respirators, centrifuge safety cups, sealed centrifuge rotors, and containment caging for animals). This requirement, like the others in this paragraph, is intended to ensure that employees are protected during the performance of these potentially high-hazard procedures.

14.17.15   Construction

Research laboratories are also held to additional requirements with regard to facility construction. Paragraph (e)(3)(iii)(A) of 29CFR 1910.1035 requires that the laboratory be separated from areas that are open to unrestricted traffic flow within the building. Passage through two sets of self-closing doors is the requirement for entry into the work area from access corridors or other contiguous areas. This type of entrance reduces the likelihood of untrained employees accidentally entering the work area, as such entry necessitates deliberate action on the part of the individual. Paragraph (e)(3)(iii)(B) requires that windows in the laboratory be closed and sealed. This helps ensure containment of any aerosols and helps maintain proper operation of BSCs through minimization of cross drafts.

14.17.16   Ventilation System

Paragraph (e)(3)(iii)(C) of 29CFR 1910.1035 requires that a ducted exhaust air ventilation system be provided that creates directional airflow that draws air from clean areas into the laboratory toward contaminated areas. The proper direction of the airflow (i.e., into the work area) should be verified by the employer at least every 6 months. The requirement that research laboratories verify directional airflow into the work area is to ensure that air is drawn into the laboratory toward contaminated areas to assist in maintaining containment of aerosols within the laboratory. The exhaust air should not be recirculated to any other area of the building, should be discharged to the outside, and should be dispersed away from occupied areas and air intakes.

14.17.17   Continuous Flow Centrifuges

Paragraph (e)(3)(iii)(E) of 29CFR 1910.1035 requires that continuous flow centrifuges or other equipment that may produce aerosols must be contained in devices that exhaust air through a HEPA filter before discharge into the laboratory. This ensures that any aerosols that may contain *M. tuberculosis* are effectively filtered from the exhaust air before discharge into the laboratory, thereby protecting employees against inadvertent exposure.

14.18 CONTRACTORS

Paragraph (d)(6) of 29CFR 1910.1035 requires that the employer must inform outside contractors who provide temporary or contract employees who may incur occupational exposure of the hazard so that such contractors can institute precautions to protect their employees. OSHA is concerned that contractors be aware of the existence of TB hazards so that appropriate actions can be undertaken to prevent the contractors' employees from being unwittingly exposed. By conveying such information to contractors, accountability for these employees is established. If the contractors are aware of the hazards, then it is the responsibility of the contractors to institute procedures to protect their employees from occupational exposure to *M. tuberculosis*.

14.18.1   Contractor Employees

Standard 29CFR 1910.1035 applies in situations when an employer has part-time employees or where employees of other employers are working in a covered facility. These employees are covered by the standard in the same manner as other employees who have occupational exposure to tuberculosis. For example, they would be provided with the same protections as full-time on-site employees, such as being included in the exposure determination, being trained, being provided with medical surveillance, and being issued respiratory protection if necessary.

With regard to employers who provide employees to other employers (e.g., personnel providers, temporary help agencies, nurse registries), a shared responsibility for worker protection exists between the provider and the client or host employer. The safety and health rights of temporary or leased or contracted employees are the same as the rights of those who are employed directly by the host employer. The host employer is generally responsible for safety and health measures taken to address hazards that are an integral part of the workplace the host employer controls. Where other employers are involved, contractors or other providers, a joint employer–employee relationship may exist in which both (or more) employers share responsibility for the safety and health of the employees. OSHA's concern is ensuring that workers receive full protection under this standard, and determining who provides which protections to the various employees may be specified as a matter of contract or employment agreement existing between the client/host and the contractor/provider. In a typical arrangement, for example, the provider employer might provide the generic training required by the standard and assure that proper follow-up medical evaluation occurs after an exposure incident. Host employers would typically control potential exposure conditions and fulfill other requirements of the standard, such as site-specific training and respiratory protection.

This note also clarifies that repair, replacement, or maintenance personnel, working in any of the work settings covered under paragraphs (a)(1) through (a)(8) of 29CFR 1910.1035, who service air systems or equipment or who renovate, repair, or maintain areas of buildings that may reasonably be anticipated to contain aerosolized *M. tuberculosis*, are also covered under the scope of the standard. The standard requires the use of engineering controls, such as isolation rooms, to reduce the concentration of droplet nuclei and therefore reduce the likelihood of TB infection and subsequent illness. The ventilation systems that exhaust air from isolation rooms may reasonably be anticipated to contain aerosolized *M. tuberculosis*. Maintenance and other workers who are responsible for the servicing and repair of ventilation systems that handle air that may contain aerosolized *M. tuberculosis* are at risk for occupational exposure when, as the result of performing their duties, they are exposed to TB-contaminated air moving through the ventilation system. Also at risk are employees who are responsible for renovating, repairing, or maintaining areas of buildings where exposure to aerosolized *M. tuberculosis* may occur other than those associated with the ventilation systems. Maintenance staff who need to repair fixtures in an isolation room or contractor personnel hired to provide housekeeping in isolation rooms or areas are examples of such employees who would also be covered under the standard. OSHA expects that such exposures would occur only rarely. In many circumstances, minor nonemergency maintenance activities could be performed by healthcare personnel required to enter the isolation rooms or areas for other reasons, such as to care for a patient. However, there may be activities that necessitate the expertise of certain maintenance employees, which could place those employees at risk of occupational exposure.

14.18.2 Temporary Employees

Employers who provide temporary employees to any of the other employers covered under the scope of 29CFR 1910.1035 (e.g., temporary nurses hired to work at a hospital, temporary lab technicians working in a clinical laboratory) must comply with the OSHA requirements. Employees in these situations are covered by the standard in the same manner as other employees who have occupational exposure to tuberculosis. A shared responsibility for worker protection exists between the personnel service employer and the client (or host) employer. These matters may be specified as a matter of contract or employment agreement existing between the personnel service employer and the host employer. OSHA has assumed that a typical contract or employment agreement exists between the two employers, with the personnel provider accepting responsibility for the general requirements and the host employer being responsible for site-specific measures. Therefore, the personnel service provider is complying with non-site-specific provisions such as exposure determination, medical surveillance, and non-site-specific employee training. The host employer would comply with more site-specific provisions such as procedures for early identification, engineering controls, and site-specific employee training. In addition, OSHA assumes that the personnel service provider has accepted the responsibility for respiratory protection. OSHA requires that workers in these situations receive full protection under the standard.

14.19 VOLUNTEERS

Several of the sectors covered by the proposed standard may be utilizing volunteers for assistance in the workplace. Under the Occupational Safety and Health Act, OSHA is mandated to protect employees against workplace hazards. Consequently, volunteers are not covered by OSHA standards because they are not employees. However, employers should be aware that simply labeling a person as a volunteer does not determine whether an employer/employee relationship exists, if the person is compensated for his or her services. Some states or localities may decide to extend the protections of OSHA standards to volunteers; however, such action is the independent decision of these jurisdictions and is not a requirement of the Occupational Safety and Health Act.

14.20 WORK LOCALITY DEFINITIONS

Although 29CFR 1910.1035 permits the exposure determination to list job classifications, grouping job classifications according to location would not be sufficient to meet the requirement for identifying job classifications with occupational exposure. For example, identifying job classifications by using the descriptor "Emergency Department" would not fulfill the requirement because the specific employee job classifications that have occupational exposure are not identified. An employer who has determined that employees in the Emergency Department warrant coverage under the standard would have to list the job classifications that involve occupational exposure and identify the tasks and procedures that result in occupational exposure. OSHA believes that merely grouping employees by location (e.g., designating all employees who work in the Emergency Department) may exclude employees who have occupational exposure, as such a grouping could overlook employees who may occasionally enter the Emergency Department but are not routinely assigned there. OSHA seeks comment about the protectiveness of permitting exposure determinations to be made by location within a work setting in certain specific instances where the employer believes such a delineation is useful and will not misclassify employees and specifically requests examples of regulatory language that could achieve these objectives.

14.21 OTHER SETTINGS

Tuberculosis precautions may be appropriate in a number of other types of settings.

14.21.1 Homeless Shelters

Residents of shelters for the homeless comprise a population that is also at increased risk for tuberculosis. Members of this population are more likely to have risk factors that are associated with TB than the general population, although the exact prevalence of TB in this population is unknown. The data quoted in the 1992 CDC document, *Prevention and Control of Tuberculosis Among Homeless Persons*, indicated a prevalence of clinically active tuberculosis among homeless adults ranging from 1.6 to 6.8%. The prevalence of latent tuberculosis ranged from 18 to 51%, with a point prevalence of active TB of 968 cases/100,000 homeless adults. Similar to the population in correctional facilities, residents of homeless shelters have a high prevalence of HIV infection and intravenous drug use, factors that increase the likelihood that their infections will progress to active TB. In addition, environmental factors such as overcrowding and poor ventilation promote the transmission of disease.

14.21.2 Drug Treatment Facilities

Based on tuberculin skin testing reported in 1993, 13.3% of the clients of drug treatment facilities had evidence of TB infection. Many of these persons have a history of intravenous drug use and have or are at risk for HIV infection. These persons are at increased risk for developing active TB and transmitting the disease to others. Many of these individuals may discontinue treatment prematurely even if they are diagnosed and started on effective drug treatment. In addition, the CDC reported that studies in some areas have shown that over 20% of selected inner city intravenous drug user populations have tuberculous infection.

The specific precautions that are applied will vary depending on the setting. At a minimum:

- Risk assessment should be performed yearly.
- Written TB infection-control plans should be developed, evaluated, and revised on a regular basis.
- Protocols for identifying and managing patients who may have active TB should be in place.
- Healthcare workers should receive appropriate training and education.
- Screening; protocols for problem evaluation should be in place.
- Coordination with the public health department should be arranged when necessary. (CDC, 1994)

14.21.3 Emergency Medical Services

Emergency medical service (EMS) employees may provide emergency treatment and transportation for individuals with suspected or confirmed tuberculosis. Emergency medical services are often used to transport individuals who have been identified as having either suspected or confirmed infectious tuberculosis from a facility with inadequate isolation capabilities to another facility better equipped to isolate these individuals. Paragraph (c)(2)(ii) of 29CFR 1910.1035 applies to employers who transfer individuals with suspected or confirmed infectious TB to a facility with AFB isolation capabilities. Provisions:

- Apply to employers who operate a facility from which an individual with suspected or confirmed infectious TB is transferred
- Do not apply to employers whose employees provide certain services such as social welfare services to individuals who have been isolated and in settings where home healthcare and home hospice care are provided

Proximity to the patient and time spent within an ambulance or other emergency vehicle affects the likelihood of occupational exposure as the result of breathing droplet nuclei generated when the patient coughs or speaks.

When EMS personnel or others must transport patients who have confirmed or suspected active TB, a surgical mask should be placed, if possible, over the patient's mouth and nose. Because administrative and engineering controls during emergency transport situations cannot be ensured, EMS personnel should wear respiratory protection when transporting such patients. If feasible, the windows of the vehicle should be kept open. The heating and air-conditioning system should be set on a nonrecirculating cycle. EMS personnel should be included in a comprehensive PPD screening program, receive a baseline PPD test and follow-up testing as indicated by the risk assessment, and be included in the follow-up of contacts of an infectious TB patient. (CDC, 1994)

14.21.4 Hospices and Long-Term-Care Facilities

For employers who provide home healthcare or home-based hospice care, paragraph (c)(2)(v) of 29CFR 1910.1035 specifies the elements that are to be included in the Exposure Control Plan. In recognition of the uniqueness of home-based work settings, OSHA has limited the elements of the Exposure Control Plan for an employer who provides home healthcare and home-based hospice care. For home healthcare and home-based hospice care, individuals are in their private homes receiving healthcare and other services, thus the employer has limited control over the work site in which those services are provided. Employers providing such home-based care will not be transferring individuals identified as having suspected or confirmed infectious TB from their homes to facilities with isolation capabilities, nor will the employer be initiating isolation precautions in the home. The employer must include procedures for prompt identification of individuals with suspected or confirmed infectious TB and for minimizing employee exposure to such individuals.

The home healthcare employer may already know that the individual has been identified as having suspected or confirmed infectious TB and has been confined to their home. An individual may be suffering from other immunocompromised conditions and may develop active TB. Because employees in home healthcare and home-based hospice care may be providing services to individuals at risk of developing active TB, procedures must be in place for identifying those individuals. The Exposure Control Plan must include procedures for minimizing employee exposure, minimizing the time spent in the home by combining tasks to limit the number of entries, and minimizing the number of employees who must enter the home along with the time they spend there.

Paragraph (c)(2)(v) also requires that the Exposure Control Plan include a list of high-hazard procedures, if any, performed in the workplace and procedures for delaying elective high-hazard procedures until the individual is noninfectious. Listing the high-hazard procedures will serve to identify those procedures that may require special considerations. In the home setting, this would not include the use of AFB isolation precautions. To the extent possible the employer should also include procedures for when these types of procedures can be delayed. This will decrease the exposure of employees to aerosolized *M. tuberculosis* that might be generated performing these procedures.

Patients who have confirmed or suspected TB should be managed as are patients in hospitals. General-use and specialized areas (e.g., treatment or TB isolation rooms) should be ventilated in the same manner as described for similar hospital areas. (CDC, 1994)

14.21.5 Dental Offices

In general, the symptoms for which patients seek treatment in a dental-care setting are not likely to be caused by infectious TB. During dental procedures, patients and dental workers share the same air for varying periods of time. Coughing may be stimulated occasionally by oral manipulations, although no specific dental procedures have been classified as cough inducing. Because the potential exists for *M. tuberculosis* transmission in dental settings, the following recommendations should be followed:

- A risk assessment should be done periodically.
- TB infection-control policies for each dental setting should be based on the risk assessment; policies should include provisions for detection and referral of patients who may have undiagnosed active TB, urgent dental care and active TB patient management, and employer-sponsored healthcare worker education, counseling, and screening.
- While taking patients' initial medical histories and at periodic updates, dental healthcare workers should routinely ask all patients to provide any history of TB and symptoms suggestive of TB.
- Patients with a medical history or symptoms suggestive of undiagnosed active TB should be referred promptly for medical evaluation of possible infectiousness; such patients should not remain in the dental-care facility any longer than required to arrange a referral, should wear surgical masks, and should be instructed to cover their mouths and noses when coughing or sneezing.
- Elective dental treatment should be deferred until a physician confirms that the patient does not have infectious TB.
- If the patient is diagnosed as having active TB, elective dental treatment should be deferred until the patient is no longer infectious.
- If urgent dental care must be provided for a patient who has, or is strongly suspected of having, infectious TB, such care should be provided in facilities that can provide TB isolation.
- Dental healthcare workers should use respiratory protection while performing procedures on such patients.

Any dental healthcare worker who has a persistent cough, especially in the presence of other signs or symptoms compatible with active TB, should be evaluated promptly for TB and should not return to the workplace until a diagnosis of TB has been excluded or a determination has been made that the healthcare worker is noninfectious, either through successful therapy or other means. Dental-care facilities that provide care to populations at high risk for active TB should use appropriate engineering controls. (CDC, 1994)

14.21.6 Home Healthcare

The OSHA requirements for home healthcare are included with those for hospice care (see previous discussion on hospices and long-term care). Healthcare workers (HCWs) who provide medical services in the homes of patients who have suspected or confirmed infectious TB should:

- Instruct such patients to cover their mouths and noses with a tissue when coughing or sneezing.
- Wear respiratory protection when entering these patients' homes.

Precautions in the home may be discontinued when the patient is no longer infectious. Healthcare workers who provide healthcare services in their patients' homes can assist in preventing transmission of *M. tuberculosis* by educating their patients regarding the importance of taking medications as prescribed and administering DOT.

Cough-inducing procedures performed on patients who have infectious TB should not be done in the patients' homes unless absolutely necessary. When medically necessary cough-inducing procedures (e.g., AFB sputum collection for evaluation of therapy) must be performed on patients who may have infectious TB, the procedures should be performed in a healthcare facility in a room or booth that has the recommended ventilation for such procedures or in a well-ventilated area away from other household members. If feasible, the healthcare worker should consider either opening a window to improve ventilation or collecting the specimen while outside the dwelling. The healthcare worker collecting these specimens should wear respiratory protection during the procedure. Healthcare workers who provide medical services in their patients' homes should be included in comprehensive employer-sponsored TB training, education, counseling, and screening programs. These programs must include provisions for identifying healthcare workers who have active TB, such as:

- Baseline PPD skin testing
- Follow-up PPD testing at intervals appropriate to the degree of risk

Patients who are at risk for developing active TB and the healthcare workers who provide medical services in the homes of such patients should be reminded periodically of the importance of having pulmonary symptoms evaluated promptly to permit early detection of and treatment for TB. (CDC, 1994)

14.21.7   Medical Offices

In general, the symptoms of active TB are symptoms for which patients are likely to seek treatment in a medical office. Furthermore, the populations served by some medical offices, or the healthcare workers in the office, may be at relatively high risk for TB. Because of the potential for *M. tuberculosis* transmission, the following recommendations should be considered:

- Conduct a periodic risk assessment.
- Develop TB infection-control policies based on results of the risk assessment; the policies should include provisions for identifying and managing patients who may have undiagnosed active TB, managing active TB patients, and educating, training, counseling, and screening healthcare workers.

While taking patients' initial medical histories and at periodic updates, healthcare workers who work in medical offices should routinely ask all patients to provide any history of TB disease or symptoms suggestive of TB. Patients with a medical history and symptoms suggestive of active TB should receive an appropriate diagnostic evaluation for TB and should be evaluated promptly for possible infectiousness in a facility that has TB isolation capability during the evaluation. At a minimum, the patient should be provided with and asked to wear a surgical mask and should be instructed to cover the mouth and nose with a tissue when coughing or sneezing. They should be separated as much as possible from other patients.

Medical offices that provide evaluation or treatment services for TB patients should follow the recommendations for managing patients in ambulatory-care settings. If cough-inducing procedures are to be administered, appropriate precautions should be followed. Any healthcare worker who has a persistent cough, especially in the presence of other signs or symptoms, should be evaluated promptly for TB. Such workers should not return to the workplace until a diagnosis of TB has been excluded or a determination has been made that the healthcare worker is noninfectious, either through successful therapy or other means. Healthcare workers who work in medical offices in which there is a likelihood of exposure to patients who have infectious TB should be included in employer-sponsored education, training, and counseling and in PPD testing programs appropriate to the level of risk in the office. In medical offices that provide care to populations at relatively high risk for active TB, use of engineering controls as described in this document for general-use areas (e.g., waiting rooms) may be appropriate. (CDC, 1994)

14.21.8 Correctional Facilities

Prison medical facilities should also follow the recommendations outlined in 29CFR 1910.1035. Facilities such as prisons, jails, and detainment centers operated by the Immigration and Natural-ization Service (INS) would be included in the scope of the OSHA standard. The CDC considers TB to be a major problem in correctional institutions, with cases occurring at a frequency three times that of the general population. In addition to a number of outbreaks that have occurred, the overall incidence of tuberculosis in the prison population is increasing. This can be attributed to over-representation of populations at high risk for TB in prisons and jails and environmental factors that promote the transmission of TB. Compared to the general population, inmates have a higher prevalence of TB infection. The population of correctional facilities is also characterized as having a high prevalence of individuals with HIV infection and intravenous drug users, factors that place these inmates at a higher risk of developing active TB. In addition, many prisons and jails are old and overcrowded and have inadequate ventilation.

Inmates may be moved frequently within a facility and between facilities, increasing the number of persons, both inmates and employees, exposed to an infected individual and making contact tracing difficult. Detention facilities, such as those operated by the INS, may house persons who are entering this country from countries with a prevalence of TB many times that of the U.S. population. In addition, a substantial number of individuals being housed in these facilities currently awaiting deportation have an additional increased risk of TB because they have been previously incarcerated in correctional institutions. In 1995, the CDC reported that approximately 36% of the total reported cases of active TB were among the foreign-born.

14.21.9 Construction

Construction operations occurring in the work settings are covered by the scope of 29CFR 1910.1035 where there is a reasonable anticipation of exposure to aerosolized *M. tuberculosis* (e.g., while rebuilding an HVAC system that would connect to an existing one that is in use). The OSHA standard is not intended to cover employees involved in other construction operations where they would not have occupational exposure to air which may reasonably be anticipated to contain aerosolized *M. tuberculosis* (e.g., a crane operator constructing a new wing of a hospital). The OSHA standard would apply only to construction employees who would incur occupational expo-sure to tuberculosis.

Such a case might arise during maintenance operations on an air system that carries air that may reasonably be anticipated to contain aerosolized *M. tuberculosis* or during renovation, repair, or alteration of areas of buildings that may reasonably be anticipated to contain aerosolized *M. tuberculosis*. The probability of exposure to *M. tuberculosis* during these activities may be high and it is necessary, therefore, for employees performing the work to wear respirators, receive medical surveillance, and be protected by the other provisions of the proposed TB standard. Employees of such contractors are subject to the same levels of TB exposure and need the same protection as other exposed employees.

Although the impact of the OSHA standard will be limited, OSHA believes that construction should not be exempted from the proposed standard. OSHA believes that a loophole would be opened in the enforcement of the OSHA standard if construction were exempted. The distinction between maintenance and construction is often an ambiguous one. If construction were excluded, contractors, such as HVAC contractors, might argue that their work is construction and that they are not covered by the OSHA standard. By covering construction, this ambiguity does not arise. This approach is consistent with that taken in other standards (e.g., Ethylene Oxide, 29 CFR 1910.1047; Benzene, 29 CFR 1910.1028).

14.22 DETERMINING INFECTIOUSNESS

The infectiousness of patients with TB correlates with the number of organisms expelled into the air, which, in turn, correlates with the following factors:

- Disease in the lungs, airways, or larynx
- Presence of cough or other forceful expiratory measures
- Presence of AFB in the sputum
- Failure of the patient to cover the mouth and nose when coughing
- Presence of cavitation on chest radiograph
- Inappropriate or short duration of chemotherapy
- Administration of procedures that can induce coughing or cause aerosolization of *M. tuberculosis* (e.g., sputum induction)

The most infectious persons are most likely those who have:

- Not been treated for TB
- Either pulmonary or laryngeal TB and a cough or are undergoing cough-inducing procedures
- A positive AFB sputum smear
- Cavitation on chest radiograph

Persons with extrapulmonary TB usually are not infectious unless they have:

- Concomitant pulmonary disease
- Nonpulmonary disease located in the respiratory tract or oral cavity
- Extrapulmonary disease that includes an open abscess or lesion in which the concentration of organisms is high, especially if drainage from the abscess or lesion is extensive (Hutton et al., 1990; Lundgren et al., 1987)
- Coinfection with HIV does not appear to affect the infectiousness of TB patients (Manoff et al., 1988; Cauthen et al., 1991; Klausner et al., 1993)

In general, children who have TB may be less likely than adults to be infectious; however, transmission from children can occur. Children with TB should be evaluated for infectiousness using the same parameters as for adults (i.e., pulmonary or laryngeal TB, presence of cough, positive sputum AFB smear, cavitation on chest radiograph, or adequacy and duration of therapy). Pediatric patients who may be infectious include those who are not on therapy, have just been started on therapy, or are on inadequate therapy and who have:

- Laryngeal or extensive pulmonary involvement
- Pronounced cough or are undergoing cough-inducing procedures
- Positive sputum AFB smears
- Cavitary TB

Children who have typical primary tuberculous lesions and do not have any of the indicators of infectiousness listed here usually do not need to be placed in isolation. Because the source case for pediatric TB patients often occurs in a member of the infected child's family (Wallgren, 1937), parents and other visitors of all pediatric TB patients should be evaluated for TB as soon as possible.

Infection is most likely to result from exposure to persons who have unsuspected pulmonary TB and are not receiving anti-TB therapy or who have been diagnosed with TB and are not receiving adequate therapy.

Administration of effective anti-TB therapy has been associated with decreased infectiousness among persons who have active TB (Riley et al., 1962). Effective therapy reduces coughing, the amount of sputum produced, and the number of organisms in the sputum. The period of time a patient must undergo effective therapy before becoming noninfectious varies among patients (Noble, 1981). For example, some TB patients are never infectious, whereas those with unrecognized or inadequately treated drug-resistant TB may remain infectious for weeks or months (Beck-Sague et al., 1992). Thus, decisions about infectiousness should be made on an individual basis.

In general, patients who have suspected or confirmed active TB should be considered infectious if they:

- Are coughing
- Are undergoing cough-inducing procedures
- Have positive AFB sputum smears
- Are not on chemotherapy or have just started chemotherapy
- Have a poor clinical or bacteriologic response to chemotherapy

A drug-susceptible TB patient is probably no longer infectious after having:

- Adequate chemotherapy
- A significant clinical and bacteriologic response to therapy (i.e., reduction in cough, resolution of fever, and progressively decreasing quantity of bacilli on smear)

However, because drug-susceptibility results are not usually known when the decision to discontinue isolation is made, all TB patients should remain in isolation while hospitalized until three consecutive negative sputum smears have been collected on three different days and clinical improvement has been demonstrated. (CDC, 1994)

14.23 DIAGNOSIS AND TREATMENT OF LATENT AND ACTIVE TUBERCULOSIS

A diagnosis of TB may be considered for any patient who has a persistent cough (i.e., a cough lasting more than 3 weeks) or other signs or symptoms compatible with TB (e.g., bloody sputum, night sweats, weight loss, anorexia, or fever). The index of suspicion for TB will vary in different geographic areas, depend on the prevalence of TB and other characteristics of the population served by the facility, and be very high in areas or among groups of patients in which TB prevalence is high. Persons for whom a diagnosis of TB is being considered should receive appropriate diagnostic tests, which may include PPD skin testing, chest radiography, and bacteriologic studies (e.g., sputum microscopy and culture). (CDC, 1994)

14.23.1 PPD Skin Test

The PPD skin test is the only method available for demonstrating infection with *M. tuberculosis*. Currently available PPD tests are less than 100% sensitive and specific for detection of infection with *M. tuberculosis*, but no better diagnostic methods have yet been devised. Interpretation of PPD test results requires knowledge of the antigen used, the immunologic basis for the reaction to this antigen, the technique used to administer and read the test, and the results of epidemiologic and clinical experience with the test. (ATS/CDC, 1990)

The PPD test, like all medical tests, is subject to variability, but many of the variations in administering and reading PPD tests can be avoided by proper training and careful attention to details. The intracutaneous (Mantoux) administration of a measured amount of PPD-tuberculin is currently the preferred method for doing the test. One-tenth milliliter of PPD (5 TU) is injected just beneath the surface of the skin on either the volar or dorsal surface of the forearm. A discrete, pale elevation of the skin (i.e., a wheal) that is 6 to 10 mm in diameter should be produced. PPD test results should be read by designated, trained personnel between 48 and 72 hours after injection. Self-reading of PPD test results by patients or healthcare workers should not be accepted (Howard and Solomon, 1988). PPD tests results are based on the presence or absence of an induration at the injection site. Redness or erythema should not be measured. The induration transverse diameter should be recorded in millimeters.

The interpretation of a PPD reaction should be influenced by the purpose for which the test was given (e.g., epidemiologic vs. diagnostic purposes), the TB infection prevalence in the population being tested, and the consequences of false classification. Establishing an appropriate definition of a positive reaction can minimize errors in classification.

The positive predictive value of PPD tests (i.e., the probability that a person with a positive PPD test is actually infected with *M. tuberculosis*) is dependent on the prevalence of TB infection in the population being tested and the specificity of the test (Snider, 1982; Huebner, 1993). In populations with a low prevalence of TB infection, the probability that a positive PPD test represents true infection with *M. tuberculosis* is very low if the cut-point is set too low (i.e., the test is not adequately specific). In populations with a high prevalence of TB infection, the probability that a positive PPD test using the same cut-point represents true infection with *M. tuberculosis* is much higher. Different cut-points are used to separate positive reactions from negative reactions for different populations, depending on the risk for TB infection in that population to ensure that few persons infected with tubercle bacilli will be misclassified as having negative reactions and few persons not infected with tubercle bacilli will be misclassified as having positive reactions. A lower cut-point (e.g., 5 mm) is used for persons in the highest risk groups, which include:

- HIV-infected persons
- Recent close contacts of persons with TB (e.g., in the household or in an unprotected occupational exposure similar in intensity and duration to household contact)
- Persons who have abnormal chest radiographs with fibrotic changes consistent with inactive TB

A higher cut-point (e.g., 10 mm) is used for:

- Persons who are not in the highest risk group but who have other risk factors
- Injecting-drug users known to be HIV seronegative
- Persons with certain medical conditions that increase the risk for progression from latent TB infection to active TB
- Medically under-served, low-income populations
- Persons born in foreign countries having a high TB prevalence
- Residents of correctional institutions and nursing homes

An even higher cut-point (e.g., 15 mm) is used for all other persons who have none of the above risk factors. Recent PPD converters are considered members of a high-risk group.

A ≥10-mm increase in the size of the induration within a 2-year period is classified as a conversion from a negative to a positive test result for persons less than 35 years of age. A ≥15-mm increase in the size of induration within a 2-year period is classified as a conversion for persons 35 years of age or older.

Prevalence of TB in the facility should be considered when choosing the appropriate cut-point for defining a positive PPD reaction. In facilities with essentially no risk for exposure to TB patients, an induration ≥15 mm may be an appropriate cut-point for healthcare workers who have no other risk factors. Where TB patients receive care, the appropriate cut-point for healthcare workers who have no other risk factors may be ≥10 mm. A recent PPD test conversion in a healthcare worker should be defined generally as an increase of ≥10 mm in the size of induration within a 2-year period. For healthcare workers in facilities where exposure to TB is very unlikely, an increase of ≥15 mm within a 2-year period may be more appropriate for defining a recent conversion because of the lower positive predictive value of the test in such groups. (CDC, 1994)

14.23.2 Anergy Testing

Persons infected with HIV may have suppressed reactions to PPD skin tests because of anergy, particularly if their CD4+ T-lymphocyte counts decline (Canessa et al., 1989). Persons with anergy will have a negative PPD test regardless of infection with *M. tuberculosis*. HIV-infected persons should be evaluated for anergy in conjunction with PPD testing (CDC, 1991). Two companion antigens (e.g., *Candida* antigen and tetanus toxoid) should be administered in addition to PPD. Persons with ≥3 mm of induration to any of the skin tests (including tuberculin) are not considered to be anergic. Reactions to PPD of ≥5 mm are considered to be evidence of TB infection in HIV-infected persons regardless of the reactions to the companion antigens. If there is no reaction (i.e., < 3-mm induration) to any of the antigens, the person being tested is considered anergic. Determination of whether such persons are likely to be infected with *M. tuberculosis* must be based on other epidemiologic factors, such as the proportion of other persons with the same level of exposure who have positive PPD test results or the intensity or exposure duration to infectious TB patients that the anergic person experienced. (CDC, 1994)

14.23.3 Pregnancy and PPD Skin Testing

Although thousands (perhaps millions) of pregnant women have been PPD skin tested since the test was devised, thus far no documented episodes of fetal harm have resulted from use of the tuberculin test (Snider, 1985). Pregnancy should not exclude a female healthcare worker from being skin tested as part of a contact investigation or as part of a regular skin-testing program. (CDC, 1994)

14.23.4 BCG Vaccination and PPD Skin Testing

Vaccination with BCG may produce a PPD reaction that cannot be distinguished reliably from a reaction caused by infection with *M. tuberculosis*. For a person who was vaccinated with BCG, the probability that a PPD test reaction results from infection with *M. tuberculosis* increases:

- As the size of the reaction increases
- When the person is a contact of a person with TB
- When the person's country of origin has a high TB prevalence
- As the length of time between vaccination and PPD testing increases

For example, a PPD test reaction of ≥10 mm can probably be attributed to *M. tuberculosis* infection in an adult who was vaccinated with BCG as a child and who is from a country with a high prevalence of TB (Snider, 1985; CDC, 1988). (CDC, 1994)

14.23.5 Booster Phenomenon

The ability of persons who have TB infection to react to PPD may gradually wane. For example, if tested with PPD, adults who were infected during their childhood may have a negative reaction. PPD could boost the hypersensitivity, and the size of the reaction could be larger on a subsequent test. This boosted reaction may be misinterpreted as a PPD test conversion from a newly acquired infection. Misinterpretation of a boosted reaction as a new infection could result in unnecessary investigations of laboratory and patient records in an attempt to identify the source case and in unnecessary prescription of preventive therapy for healthcare workers. Although boosting can occur among persons in any age group, the likelihood of the reaction increases with the age of the person being tested (American Thoracic Society/CDC, 1990; Thompson et al., 1979).

When PPD testing of adults is to be repeated periodically (as in healthcare worker skin-testing programs), two-step testing can be used to reduce the likelihood that a boosted reaction is misinterpreted as a new infection. Two-step testing should be performed on all newly employed healthcare workers who have initial negative PPD test results at the time of employment or who have not had a documented negative PPD test result during the 12 months preceding the initial test. A second test should be performed 1 to 3 weeks after the first test. If the second test result is positive, this is most likely a boosted reaction and the healthcare worker should be classified as previously infected. If the remainder of the tests are negative, the healthcare worker is classified as uninfected, and a positive reaction to a subsequent test is likely to represent a new infection with *M. tuberculosis*. (CDC, 1994)

14.23.6 Chest Radiography

Patients who have positive skin-test results or symptoms suggestive of TB should be evaluated with a chest radiograph regardless of PPD test results. Radiographic abnormalities that strongly suggest active TB include upper-lobe infiltration, particularly if cavitation is seen (Des Prez and Heim, 1990), and patchy or nodular infiltrates in the apical or subapical posterior upper lobes or the superior segment of the lower lobe. If abnormalities are noted, or if the patient has symptoms suggestive of extrapulmonary TB, additional diagnostic tests should be conducted. The radiographic presentation of pulmonary TB in HIV-infected patients may be unusual (Pitchenik and Rubinson, 1985). Typical apical cavitary disease is less common among such patients. Infiltrates may be present in any lung zone, a finding that is often associated with mediastinal and/or hilar adenopathy, or a normal chest radiograph may be noted, although this latter finding occurs rarely. (CDC, 1994)

14.23.7 Bacteriology Smear

Smear and culture examinations of at least three sputum specimens collected on different days are the main diagnostic procedure for pulmonary TB. Sputum smears that fail to demonstrate AFB do not exclude the diagnosis of TB. In the United States, approximately 60% of patients with positive sputum cultures have positive AFB sputum smears. HIV-infected patients who have pulmonary TB may be less likely than immunocompetent patients to have AFB present on sputum smears, which is consistent with the lower frequency of cavitary pulmonary disease observed among HIV-infected persons. Specimens for smear and culture should contain an adequate amount of expectorated sputum but not much saliva. If a diagnosis of TB cannot be established from sputum, a bronchoscopy may be necessary. In young children who cannot produce an adequate amount of sputum, gastric aspirates may provide an adequate specimen for diagnosis. A culture of sputum or other clinical specimen that contains *M. tuberculosis* provides a definitive diagnosis of TB. Conventional laboratory methods may require 4 to 8 weeks for species identification; use of radiometric culture techniques and nucleic acid probes facilitates more rapid detection and identification of mycobacteria. Mixed mycobacterial infection, either simultaneous or sequential, can obscure the identification of *M. tuberculosis* during the clinical evaluation and the laboratory analysis (Burnens and Vurma-Rapp, 1989). The use of nucleic acid probes for both *M. avium* complex and *M. tuberculosis* may be useful for identifying mixed mycobacterial infections in clinical specimens. (CDC, 1994)

14.24 PREVENTIVE THERAPY FOR LATENT TUBERCULOSIS INFECTION

Determining whether a person with a positive PPD test reaction or conversion is a candidate for preventive therapy must be based on:

- Likelihood that the reaction represents true infection with *M. tuberculosis* (as determined by the cut-points)
- Estimated risk for progression from latent infection to active TB
- Risk for hepatitis associated with taking isoniazide (INH) preventive therapy (as determined by age and other factors)

Healthcare workers with positive PPD test results should be evaluated for preventive therapy regardless of their ages if they:

- Are recent converters
- Are close contacts of persons who have active TB
- Have a medical condition that increases the risk for TB
- Have HIV infection
- Use injected drugs

Healthcare workers with positive PPD test results who do not have these risk factors should be evaluated for preventive therapy if they are less than 35 years of age. Preventive therapy should be considered for anergic persons who are known contacts of infectious TB patients and persons from populations in which the TB infection prevalence is very high (e.g., a prevalence ≥ 10%). (CDC, 1994)

14.24.1   Isoniazide (INH)

No evidence suggests that INH poses a carcinogenic risk to humans (Glassroth, White, and Snider, 1977; Glassroth, Snider, and Comstock, 1977; Costello and Snider, 1980). The usual preventive therapy regimen is 300 mg oral INH daily for adults and 10 mg/kg/day for children (CDC, 1990). The recommended duration of therapy is 12 months for persons with HIV infection and 9 months for children. Other persons should receive INH therapy for 6 to 12 months. Because the risk for INH-associated hepatitis may be increased during the peripartum period, the decision to use preventive therapy during pregnancy should be made on an individual basis and should depend on the patient's estimated risk for progression to active disease. Note that, in general, preventive therapy can be delayed until after delivery; however, for pregnant women who were probably infected recently or who have high-risk medical conditions, especially HIV infection, INH preventive therapy should begin when the infection is documented (Moulding et al., 1989; Snider et al., 1980; Snider, 1984; Hamadeh and Glassroth, 1992).

For persons who have silicosis or a chest radiograph demonstrating inactive fibrotic lesions and who have no evidence of active TB, acceptable regimens include:

- 4 months of INH plus rifampin
- 12 months of INH, providing that infection with INH-resistant organisms is unlikely (American Thoracic Society/CDC, 1994)

For persons likely to be infected with MDR-TB, alternative multidrug preventive therapy regimens should be considered (CDC, 1992).

All persons placed on preventive therapy should be:

- Educated regarding the possible adverse reactions associated with INH use
- Questioned carefully at monthly intervals by qualified personnel for signs or symptoms consistent with liver damage or other adverse effects

Because INH-associated hepatitis occurs more frequently among persons older than 35 years of age, a transaminase measurement should include persons in this age group before initiation of INH therapy and should be conducted monthly until treatment has been completed.

Other factors associated with an increased risk for hepatitis include daily alcohol use, chronic liver disease, and use of injected drugs. In addition, postpubertal black and Hispanic women may be at greater risk for hepatitis or drug interactions (Snider and Caras, 1992). More careful clinical monitoring of persons with these risk factors and possibly more frequent laboratory monitoring should be considered. If any of these tests exceeds three to five times the upper limit of normal, discontinuation of INH should be strongly considered. Liver function tests are not a substitute for monthly clinical evaluations or prompt assessment of signs or symptoms of adverse reactions that could occur between the regularly scheduled evaluations. Persons who have latent TB infection should be advised that they could be reinfected with another strain of *M. tuberculosis* (Small et al., 1993). (CDC, 1994)

14.25 TREATMENT OF ACTIVE TUBERCULOSIS PATIENTS

Drug-susceptibility testing should be performed on all initial isolates from patients with TB; however, test results may not be available for several weeks, making selection of an initial regimen difficult, especially in areas where drug-resistant TB has been documented. Current recommendations for therapy and dosage schedules for the treatment of drug-susceptible TB should be followed. Streptomycin is contraindicated in the treatment of pregnant women because of the risk of toxicity to the fetus. In geographic areas or facilities in which drug-resistant TB is highly prevalent, the initial treatment regimen used while results of drug-susceptibility tests are pending may have to be expanded. This decision should be based on analysis of surveillance data.

When results from drug-susceptibility tests become available, the regimen should be adjusted appropriately (Iseman and Madsen, 1989; Gobel, 1986; Gobel et al., 1993; Simone and Iseman, 1992). If drug resistance is present, clinicians unfamiliar with the management of patients with drug-resistant TB should seek expert consultation. For any regimen to be effective, adherence to the regimen must be ensured. The most effective method of ensuring adherence is the use of DOT after the patient has been discharged from the hospital (CDC, 1993; American Thoracic Society/CDC, 1992). This practice should be coordinated with the public health department. (CDC, 1994)

14.26 ENGINEERING CONTROLS

According to paragraph (d) of 29CFR 1910.1035, Work Practices and Engineering Controls, employee protection is most effectively attained by elimination or minimization of the hazard at its source, which engineering controls and work practices are both designed to do. Industrial hygiene principles also teach that control methods that depend upon the vagaries of human behavior are inherently less reliable than well-maintained mechanical methods. For these reasons, OSHA has preferred engineering and work practice controls and has required, under paragraph (d)(1), their use to eliminate or minimize employee exposure to *M. tuberculosis*. Nevertheless, OSHA recognizes that situations may exist in which neither of these control methods is feasible and that, in these circumstances, employee protection must be achieved through the use of personal protective equipment, primarily respirators. In other situations, personal protective equipment may have to be utilized in conjunction with engineering controls and/or work practices to obtain a further reduction in employee exposure.

Engineering controls serve to reduce employee exposure in the workplace by either removing the hazard or isolating the worker from exposure. These controls include:

- Process or equipment redesign
- Process or equipment enclosure (e.g., biosafety cabinets)
- Employee isolation

In general, engineering controls act on the source of the hazard and eliminate or reduce employee exposure without reliance on the employee to take self-protective action.

In comparison, work practice controls reduce the likelihood of exposure through alteration of the manner in which a task is performed (e.g., closing the door of an AFB isolation room immediately upon entering or exiting). Although work practice controls also act on the source of the hazard, the protection provided is based upon employer and employee behavior rather than installation of a physical device.

In many instances these two control methodologies work in tandem, as in using work practice controls to assure effective operation of engineering controls. The employer is required to develop a number of work practices relative to controlling occupational exposure to TB. In paragraph (d)(2), these work practices are required to be implemented in the work setting.

Because the source of the hazard is frequently a living person, typical methods of reducing or eliminating the hazard at the source may not always be feasible. For example, in an industrial operation, a process may be entirely enclosed and operated or monitored by an employee at a remote location, a situation that would rarely, if ever, occur in the work settings covered by this standard. OSHA believes, therefore, that prevention of exposures to *M. tuberculosis* will often require use of a combination of control methods to achieve adequate protection of employees. Paragraph (d)(1) requires work practices and engineering controls to be used to eliminate or minimize employee exposures.

14.26.1 Recommended Ventilation

To prevent the spread and reduce the concentration of infectious droplet nuclei:

1. Utilize direct source control through local exhaust ventilation.
2. Control airflow direction and patterns to prevent air contamination in areas adjacent to the infectious source.
3. Dilute and remove contaminated air via general ventilation.
4. Clean the air via HEPA filtration or ultraviolet germicidal irradiation (UVGI).

The first two levels of the hierarchy minimize areas in the healthcare facility where exposure to infectious TB may occur and reduce, but do not eliminate, risk in those few areas where exposure to *M. tuberculosis* can still occur. Because persons entering such rooms may be exposed to *M. tuberculosis*, personal respiratory protective equipment may also be needed.

14.26.2 Recordkeeping

In paragraph (i)(4) of 29CFR 1910.1035, OSHA proposes requiring engineering control maintenance and monitoring records to be kept that include:

- Date
- Equipment identification
- Task performed
- Sign-off

The performance monitoring records must include:

- Date and time
- Location
- Parameter measured
- Results of monitoring
- Sign-off

Schedules can be used to by maintenance employees as an engineering controls checklist or timetable. In general, OSHA has left the time frame up to the employer, as the employer is familiar with the characteristics of the workplace that could affect the performance of these controls (e.g., dusty conditions, high heat and humidity, seasonal variations).

Proposed paragraph (i)(4)(iii) requires engineering control maintenance and monitoring records to be maintained for 3 years. The 3-year period is a reasonable period of time and will enable the employer to develop and sustain a proper maintenance program and to track the effectiveness of the controls. Moreover, the records will aid the OSHA compliance officer in enforcing the standard's requirements for engineering controls.

For employers who have work settings where TB cases are isolated, paragraph (c)(2)(iii)(D) requires the employer to develop a schedule for the inspection, maintenance, and performance monitoring of engineering controls. Engineering controls required by the proposed standard play an essential role in reducing employee exposures to *M. tuberculosis*. Thus, it is necessary that these controls be appropriately maintained, inspected, and monitored in order to ensure that they are functioning properly.

Because engineering controls are mechanical systems, they are prone to occasional lapses in performance caused by occurrences such as clogged filters, slipping or broken drive belts, burned-out motors, obstructed ducts, and so forth. Because these situations cannot be predicted, it is necessary to regularly inspect engineering controls for proper functioning.

Results of engineering maintenance measures should be reviewed at regular intervals. Data from the most recent evaluation and from maintenance procedures and logs should be reviewed carefully as part of the risk assessment. Treatment and procedure rooms in which patients who have infectious TB or who have an undiagnosed pulmonary disease and are at high risk for active TB receive care should meet the ventilation recommendations for isolation rooms.

14.26.3 Maintenance and Inspection Schedules

Paragraph (d)(5)(iii) of 29CFR 1910.1035 stipulates that engineering controls must be maintained and inspected and performance monitored for filter loading and leakage every 6 months, whenever filters are changed, or more often if necessary to maintain effectiveness. The primary intent of this provision is to ensure that engineering controls are maintained in such a manner that they continue to function effectively. A number of factors can affect the functioning of engineering controls, such as frozen bearings, broken belts, and burned out motors.

The employer is responsible for the maintenance of engineering controls in proper working condition. Repairs cannot be delayed until the next inspection. This provision does, however, stipulate a maximum time period of 6 months between inspections and performance monitoring of engineering controls and HEPA filters in air systems carrying air that contains aerosolized *M. tuberculosis*. The employer's maintenance schedule may specify more frequent inspection, maintenance, and performance monitoring based upon conditions found in that particular work site. When filters are changed, performance monitoring must be conducted to ensure that the filter has been correctly installed and is functioning properly.

In view of the importance of these systems in reducing the concentration of droplet nuclei and thereby the risk of TB transmission, OSHA believes that 6 months is the longest period that these systems should be allowed to operate without inspection and performance monitoring. This maximum 6-month period of time between consecutive inspections and performance monitoring of HEPA filters is supported by the CDC.

14.26.4 Ventilation

New construction or renovation of existing healthcare facilities should have TB isolation and treatment rooms that have airflow:

- $\geq$ 12 air changes per hour (ACH)
- Exhausted to the outside
- Separate from the general ventilation

Ventilation systems for healthcare facilities should be designed and modified when necessary by ventilation engineers in collaboration with infection-control and occupational health staff. ASHRAE (ASHRAE, 1991), the American Institute of Architects (AIA) (AIA, 1987), and the American Conference of Governmental Industrial Hygienists (ACGIH) (ACGIH, 1992) have published recommendations for designing and operating ventilation systems.

As part of the TB infection-control plan, healthcare facility personnel should determine the numbers needed of the following:

- TB isolation rooms
- Treatment rooms
- Local exhaust devices (e.g., for cough-inducing or aerosol-generating procedures)

The locations of these rooms and devices will depend on where in the facility the ventilation conditions recommended can be achieved. Grouping isolation rooms together in one area of the facility may facilitate care of TB patients and installation and maintenance of optimal engineering controls (particularly ventilation).

Periodic evaluation reviews include:

- Number of TB isolation rooms, treatment rooms, and local exhaust devices needed
- Regular maintenance and monitoring of the local and general exhaust systems (including HEPA filtration systems, if they are used)

Engineering control methods must be tailored to each facility based on:

- Needs
- Feasibility of using the ventilation and air-cleaning concepts (CDC, 1994)

14.26.5 Local Source Control

Local source control ventilation:

- Captures airborne contaminants at or near their source (Figure 14.3)
- Removes some contaminants without exposing persons in the area to infectious agents
- Can prevent or reduce the spread of infectious droplet nuclei into the general air circulation by entrapping infectious droplet nuclei emitted by the patient (i.e., the source)
- Is especially important when performing procedures likely to generate aerosols containing infectious particles and when infectious TB patients are coughing or sneezing
- Should be used, if feasible, wherever aerosol-generating procedures are performed

Two basic types of local exhaust devices using hoods are:

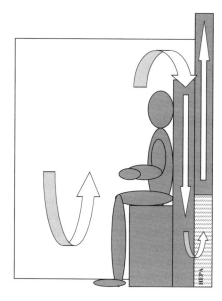

FIGURE 14.3 Local source control in this case is control of droplet nuclei emitted by the patient.

- Enclosing type, in which the hood either partially or fully encloses the infectious source
- Exterior type, in which the infectious source is near but outside the hood

Fully enclosed hoods, booths, or tents are always preferable to exterior types because of their superior ability to prevent contaminants from escaping into the healthcare worker's breathing zone. (CDC, 1994)

14.26.6 Enclosing Devices

The enclosing type of local exhaust ventilation device includes:

- Laboratory hoods used for processing specimens that could contain viable infectious organisms
- Booths used for sputum induction or administration of aerosolized medications, such as aerosolized pentamidine (the most complex device is an enclosure that has a sophisticated self-contained airflow and recirculation system)
- Tents or hoods made of vinyl or other materials used to enclose and isolate a patient (the tent placed over the patient has an exhaust connection to the room discharge exhaust system)

These devices are available in various configurations. Both tents and booths should have sufficient airflow to remove at least 99% of airborne particles during the interval between the departure of one patient and the arrival of the next (Mutchler, 1973). The time required for removing a given percentage of airborne particles from an enclosed space depends on several factors:

- Number of air changes per hour, which is determined by the number of cubic feet of air in the room or booth and the rate at which air is entering the room or booth at the intake source
- Ventilation inlet and outlet location
- Physical configuration of the room or booth (CDC, 1994)

14.26.7 Exterior Devices

The exterior type of local exhaust ventilation device is usually a hood very near, but not enclosing, the infectious patient. The airflow produced by these devices should be sufficient to prevent the escape of droplet nuclei due to crosscurrents of air near the patient's face. Whenever possible, the patient should face directly into the hood opening so that any coughing or sneezing is directed into the hood, where the droplet nuclei are captured. The device should maintain an air velocity of ≥ 200 ft/min at the patient's breathing zone to ensure capture of droplet nuclei. (CDC, 1994)

14.26.8 Discharge Exhaust

Air from booths, tents, and hoods may be discharged into the room in which the device is located or exhausted to the outside. If the air is discharged into the room, a HEPA filter should be incorporated at the discharge duct or vent of the device. The exhaust fan should be located on the discharge side of the HEPA filter to ensure that the air pressure in the filter housing and booth is negative with respect to adjacent areas. Uncontaminated air from the room will flow into the booth through all openings, thus preventing infectious droplet nuclei in the booth from escaping into the room. Most commercially available booths, tents, and hoods are fitted with HEPA filters, in which case additional HEPA filtration is not needed. If the device does not incorporate a HEPA filter, the air from the device should be exhausted to the outside in accordance with recommendations for isolation room exhaust. (CDC, 1994)

14.26.9 General Ventilation

General ventilation is used in general-use areas of healthcare facilities (e.g., waiting-room areas and emergency departments). Healthcare facilities should either include as part of their staff an engineer or other professional with expertise in ventilation or have this expertise available from an expert in ventilation engineering who also has hospital experience. These persons should work closely with infection-control staff to assist in controlling airborne infections. Ventilation system designs in healthcare facilities should meet any applicable federal, state, and local requirements. Re-entrainment of air due to localized exterior air patterns must be avoided (Figure 14.4). The direction of airflow in healthcare facilities should be designed, constructed, and maintained so that air flows from clean areas to less-clean areas.

Healthcare facilities serving populations that have a high prevalence of TB may need to supplement the general ventilation or use additional engineering approaches (e.g., HEPA filtration or UVGI) in general-use areas where TB patients are likely to be (e.g., waiting-room areas, emergency departments, and radiology suites). A single-pass, nonrecirculating system that exhausts air to the outside, a recirculation system that passes air through HEPA filters before recirculating it to the general ventilation system, or upper-air UVGI may be used in such areas.

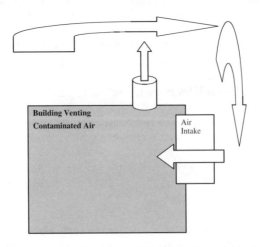

FIGURE 14.4 Designs must include recognition of potential downdraft or other exterior air distribution patterns.

General ventilation can be used for several purposes:

- Diluting and removing contaminated air
- Controlling airflow patterns within rooms
- Controlling the direction of airflow throughout a facility (CDC, 1994)

14.26.10   Dilution and Removal

General ventilation maintains air quality by two processes:

- Dilution — Uncontaminated supply air (incoming air) mixes with the contaminated room air.
- Removal of airborne contaminants — Air is removed from the room by the exhaust system.

These processes reduce the concentration of droplet nuclei in the room air. Two types of general ventilation systems can be used for dilution and removal of contaminated air:

- Single-pass system
- Recirculating system (CDC, 1994)

14.26.10.1   Single-Pass System

Supply air is either outside air that has been appropriately heated and cooled or air from a central system that supplies a number of areas. After air passes through the room (or area), 100% of that air is exhausted to the outside. The single-pass system is preferred when infectious airborne droplet nuclei are known to be in room and prevents contaminated air from being recirculated to other areas of the facility. (CDC, 1994)

14.26.10.2   Recirculating System

A small portion of exhaust air is discharged to the outside and is replaced with fresh outside air. Fresh air mixes with the portion of exhaust air that was not discharged to the outside. The resulting mixture, which can contain a large proportion of contaminated air, is then recirculated to the areas serviced by the system. This air mixture could be recirculated into the general ventilation, in which case contaminants may be carried from contaminated areas to uncontaminated areas, or within a specific room or area, in which case other areas of the facility will not be affected. (CDC, 1994)

14.26.11   Ventilation Rates

Recommended general ventilation rates for healthcare facilities are usually expressed in number of air changes per hour. This number is the ratio of the volume of air entering the room per hour to the room volume and is equal to the exhaust airflow, Q (cubic feet per minute), divided by the room volume, V (cubic feet), multiplied by 60:

$$ACH = Q/V \times 60$$

The feasibility of achieving specific ventilation rates depends on the construction and operational requirements of the ventilation system (e.g., the energy requirements to move and to heat or cool the air) and may also be different for retrofitted facilities and newly constructed facilities. The expense and effort of achieving specific higher ventilation rates for new construction may be reasonable, but achieving similar ventilation rates for an existing facility may be more difficult.

Achieving higher ventilation rates by using auxiliary methods (e.g., room-air recirculation) in addition to exhaust ventilation may be feasible in existing facilities. (CDC, 1994)

14.26.12   Airflow Patterns (Air Mixing)

Air mixing (Figure 14.5) is used to:

- Provide optimum airflow patterns
- Prevent both stagnation and short-circuiting of air

General ventilation systems should be designed to:

- Provide optimal patterns of airflow within rooms
- Prevent air stagnation or short-circuiting of air from the supply to the exhaust (i.e., passage of air directly from the air supply to the air exhaust)

To provide optimal airflow patterns, the air supply and exhaust should be located such that clean air first flows to parts of the room where healthcare workers are likely to work, then flows across the infectious source and into the exhaust. In this way, the healthcare worker is not positioned between the infectious source and the exhaust location. Ways to achieve this airflow pattern include supplying air at the side of the room opposite the patient and exhausting the air from the side where

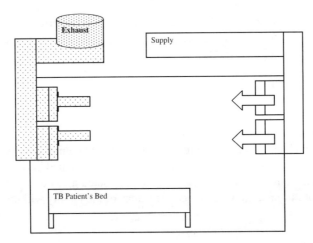

Room Air Mixing Techniques – Dilution Ventilation

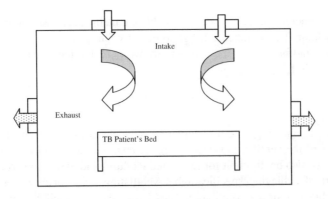

FIGURE 14.5   Air mixing may be used to facilitate correct airflow patterns.

the patient is located. Another method is to supply air near the ceiling and exhaust air near the floor; this method works best when the supply air is cooler than the room air.

Airflow patterns are affected by:

- Large air temperature differentials
- Precise location of the supply and exhausts
- Location of furniture
- Movement of healthcare workers and patients
- Physical configuration of the space

Smoke tubes can be used to visualize airflow patterns in a manner similar to that described for estimating room air mixing. Adequate air mixing, which requires that an adequate number of ACHs be provided to a room, must be ensured to prevent air stagnation within the room. However, the air will not usually be changed the calculated number of times per hour because the airflow patterns in the room may not permit complete mixing of the supply and room air in all parts of the room. This results in an effective airflow rate for which the supplied airflow may be less than required for proper ventilation. To account for this variation, a mixing factor (which ranges from 1 for perfect mixing to 10 for poor mixing) is applied as a multiplier to determine the actual supply airflow; that is, the recommended ACH multiplied by the mixing factor equals the actual required ACH (Galson and Goddard, 1968; ACGIH, 1992). The air supply and exhaust system should be designed to achieve the lowest mixing factor possible. The mixing factor is determined most accurately by experimentally testing each space configuration, but this procedure is complex and time consuming.

A reasonably good qualitative measure of mixing can be estimated by an experienced ventilation engineer who releases smoke from smoke tubes at a number of locations in the room and observes the movement of the smoke. Smoke movement in all areas of the room indicates good mixing. Stagnation of air in some room areas indicates poor mixing, and movement of the supply and exhaust openings or redirection of the supply air is necessary.

14.26.13 Airflow Direction

The intent of airflow direction control is to:

1. Contain contaminated air in localized areas in a facility
2. Prevent the spread of contaminated air to uncontaminated areas

The general ventilation system should be designed and balanced so that air flows from less contaminated to more contaminated areas (SHRAE, 1991; AIA, 1987). For example, air should flow from corridors (cleaner areas) into TB isolation rooms (less clean areas) to prevent the spread of contaminants to other areas. In some special treatment rooms in which operative and invasive procedures are performed, the direction of airflow is from the room to the hallway to provide cleaner air during these procedures. Cough-inducing or aerosol-generating procedures (e.g., bronchoscopy and irrigation of tuberculous abscesses) should not be performed on patients who may have infectious TB in rooms with this type of airflow.

Creating a lower (negative) pressure in the area into which the flow of air is desired controls the direction of airflow. For air to flow from one area to another, the air pressure in the two areas must be different. Air will flow from a higher pressure area to a lower pressure area. The lower pressure area is described as being at negative pressure relative to the higher pressure area. Negative pressure is attained by exhausting air from an area at a higher rate than air is being supplied. The level of negative pressure necessary to achieve the desired airflow will depend on the physical configuration of the ventilation system and area airflow path and flow openings and should be determined on an individual basis by an experienced ventilation engineer. (CDC, 1994)

14.26.14   Negative Pressure

In conjunction with this provision, paragraph (d)(5)(i) of 29CFR 1910.1035 requires that negative pressure be maintained in AFB isolation rooms or areas. A pressure differential of 0.001 inch of water and an inward air velocity of 100 feet per minute (fpm) are minimum acceptable levels. The pressure difference necessary to achieve and maintain negative pressure in a room is very small and may be difficult to measure accurately. Negative pressure can be achieved by balancing the room supply and exhaust flows to set the exhaust flow to a value of 10% (but no less than 50 cubic feet per minute [cfm]) greater than the supply.

As stated previously, the negative pressure principle plays an important role in controlling the spread of *M. tuberculosis* to other areas of the facility where unprotected workers may be exposed. In isolation rooms and areas and in areas where high hazard procedures (including autopsies) are performed, engineering controls creating negative pressure will prevent the escape of droplet nuclei from the room, thus preventing dispersion of *M. tuberculosis* into the corridor and other areas of the facility where unprotected employees may be working. In addition, negative pressure fulfills the secondary purpose of general ventilation by reducing the concentration of contaminants in the air.

General ventilation maintains air quality by two processes, dilution and removal of airborne contaminants. Dilution reduces the concentration of contaminants in a room by supplying air that does not contain those contaminants. The supply air mixes with and then displaces some of the contaminated room air, which is subsequently removed from the room by the exhaust system. This process reduces the concentration of droplet nuclei in the room air and the risk of TB transmission. Paragraph (d)(5)(ii) requires that in those areas where negative pressure is required (e.g., AFB isolation rooms or areas), maintenance of negative pressure must be qualitatively demonstrated (e.g., by smoke trails) daily while in use for tuberculosis isolation. In Supplement 3 of its 1994 guidelines, the CDC states that "TB isolation rooms should be checked daily for negative pressure while being used for TB isolation."

Proper maintenance of negative pressure will prevent the contaminated air from escaping from the room or area and exposing unprotected employees. Means of qualitatively demonstrating negative pressure include smoke trail testing, flutter strips, and continuous monitoring devices (Figure 14.6). Paragraph (d)(5)(v) states that ducts carrying air that may reasonably be anticipated to contain aerosolized *M. tuberculosis* must be maintained under negative pressure for their entire length before in-duct HEPA filtration or until the ducts exit the building for discharge. Ducts maintained under negative pressure will contain exhaust air within the system. Air will not escape to the outside as it would under positive pressure even if the ducts leaked. The purpose of this provision is to prevent escape of air that may contain aerosolized *M. tuberculosis* into areas where occupational exposure is not anticipated and unprotected employees may be exposed.

Pressure differential is used to control the direction of airflow between the room and adjacent areas, thereby preventing the escape of contaminated air from the room into other areas of the facility. The actual level of negative pressure achieved will depend on the difference in the ventilation exhaust and supply flows and the physical room configuration, including the airflow path and flow openings. The minimum pressure difference necessary to achieve and maintain negative pressure that will result in airflow into the room is 0.001 inH_2O of water. Higher pressures (≥ 0.001 inH_2O) are satisfactory; however, these higher pressures may be difficult to achieve. If rooms are well sealed, negative pressures greater than the minimum of 0.001 inH_2O may be readily achieved. If rooms are not well sealed, as may be the case in many facilities (especially older facilities), achieving higher negative pressures may require exhaust/supply flow differentials beyond the capability of the ventilation system.

To establish negative pressure in a room that has a normally functioning ventilation system, room supply and exhaust airflows are first balanced to achieve an exhaust flow of whichever is greater: 10%, or 50 cfm greater than the supply. If the minimum of 0.001 inH_2O is not achieved

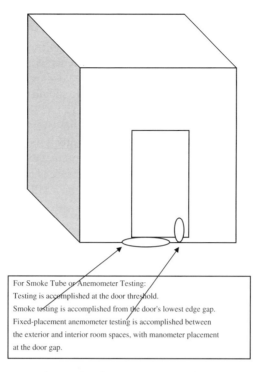

For Smoke Tube or Anemometer Testing:
Testing is accomplished at the door threshold.
Smoke testing is accomplished from the door's lowest edge gap.
Fixed-placement anemometer testing is accomplished between
the exterior and interior room spaces, with manometer placement
at the door gap.

FIGURE 14.6   Various means must be used to document proper airflow.

and cannot be achieved by increasing the flow differential (within the limits of the ventilation system), the room should be inspected for leakage (e.g., through doors, windows, plumbing, and equipment wall penetrations). Corrective action should be taken to seal the leaks. Negative pressure in a room can be altered by changing the ventilation system operation or by opening and closing the doors, corridor doors, or windows of the room. When an operating configuration has been established, all doors and windows must remain properly closed in the isolation room and other areas except when persons need to enter or leave. (CDC, 1994)

14.26.15   Anterooms

An anteroom may increase the effectiveness of the isolation room and minimize the potential escape of droplet nuclei into the corridor. The anteroom should have positive air pressure in relation to the isolation room. The pressure relationship between the isolation room, anteroom, and general corridor may vary according to ventilation design. Although an anteroom is not a substitute for negative pressure in a room, an anteroom may be used to reduce escape of droplet nuclei during opening and closing of the isolation room door. Some anterooms have their own air supply duct, but others do not. The TB isolation room should have negative pressure relative to the anteroom. If the existing ventilation system is incapable of achieving the desired negative pressure, steps should be taken to provide a means to discharge air from the room if the room lacks a separate ventilation system or if the system cannot provide the proper airflow. (CDC, 1994)

14.26.16   Fixed Air Recirculation

Systems that recirculate the air in an entire room may be designed to achieve negative pressure by discharging air outside the room. Some portable room-air recirculation units are designed to discharge air to the outside to achieve negative pressure. Air cleaners that can accomplish this must be designed specifically for this purpose. A small centrifugal blower (i.e., exhaust fan) can be used to exhaust air to the outside through a window or outside wall. This approach may be used as an interim measure to achieve negative pressure, but it provides no fresh air and suboptimal dilution. Another approach to achieving the required pressure difference is to pressurize the corridor. The general ventilation system of the corridor is balanced to create a higher air pressure in the corridor than in the isolation room. Ideally, the corridor air supply rate should be increased while the corridor exhaust rate is not increased. If this is not possible, the exhaust rate should be decreased by resetting appropriate exhaust dampers. Caution should be exercised, however, to ensure that the exhaust rate is not reduced below acceptable levels. This approach requires proper maintenance of all settings used to achieve the pressure balance, including doors. This method may not be desirable if the corridor being pressurized has rooms in which negative pressure is not desired. In many situations, this system is difficult to achieve and should be considered only after careful review by ventilation personnel. (CDC, 1994)

14.26.17   Monitoring Negative Pressure

The negative pressure in a room can be monitored by visually observing the direction of airflow (e.g., using smoke tubes) or by measuring the differential pressure between the room and its surrounding area (Figure 14.7). Smoke from a smoke tube can be used to observe airflow between areas or airflow patterns within an area. To check the negative pressure in a room by using a smoke tube, hold the smoke tube near the bottom of the door and approximately 2 inches in front of the door or at the face of a grille or other opening, if the door has such a feature. Generate a small amount of smoke by gently squeezing the bulb. The smoke tube should be held parallel to the door, and the smoke should be issued from the tube slowly to ensure that the velocity of the smoke from the tube does not overpower the air velocity. The smoke will travel in the direction of airflow. If the room is at negative pressure, the smoke will travel under the door and into the room (e.g., from higher to lower pressure). Not at negative pressure, the smoke will be blown outward or will stay stationary. This test must be performed while the door is closed; room air cleaners used in the room should be running.

Differential pressure-sensing devices also can be used to monitor negative pressure through periodic (noncontinuous) pressure measurements or continuous pressure monitoring. A visible and/or audible warning signal that air pressure is low may provide a pressure readout signal that can be recorded for later verification and used to automatically adjust the facility's ventilation control system. Pressure-measuring devices should sense the room pressure just inside the airflow path into the room (e.g., at the bottom of the door). Unusual airflow patterns within the room can cause pressure variations. If the pressure-sensing ports cannot be located directly across the airflow path, validation will be required that the negative pressure at the sensing point remains the same as the negative pressure across the flow path. Pressure-sensing devices should incorporate an audible warning with a time delay to indicate that a door is open. When the door to the room is opened, the negative pressure will decrease. The time-delayed signal should allow sufficient time for persons to enter or leave the room without activating the audible warning.

A potential problem with using pressure-sensing devices is that the pressure differentials used to achieve the low negative pressure necessitate the use of very sensitive mechanical devices, electronic devices, or pressure gauges to ensure accurate measurements. Use of devices that cannot measure these low pressures (i.e., pressures as low as 1 inH_2O) will require setting higher negative pressures that may be difficult and, in some instances, impractical to achieve. Periodic checks are required to ensure that desired negative pressure is present and that continuous monitoring devices, if used, are operating properly. If smoke tubes or other visual checks are used, TB isolation rooms and treatment rooms should be checked frequently for negative pressure. Rooms undergoing changes to the ventilation system should be checked daily. TB isolation rooms should be checked daily for negative pressure while being used for TB isolation. If these rooms are not being used

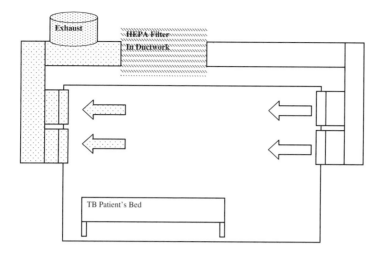

TB isolation room with HEPA filter in ductwork

Filter is placed in the ductwork.

Air is drawn into the ductwork intake grates.

Air travels through the ductwork to the HEPA filtration unit.

Air is exhausted back into the ductwork which carries the filtered air back into the room.

Air movement dynamics will be more complex than those shown here. However, the general trends will be as shown. All of the air and droplet nuclei from the patient will not be captured in the HEPA filter system. The end result will be some clearing of the air of droplet nuclei without 100% effectiveness.

Contamination of the ductwork leading to the HEPA filter should be expected and planned for in any maintenance events. Exhaust unit on this side should exhaust all air prior to the HEPA filter unit, assuming no other exhaust means is present.

FIGURE 14.7 Negative pressure measurements must be conducted with proper equipment and proper placement of probes.

for patients who have suspected or confirmed TB but potentially could be used for such patients, the negative pressure in the rooms should be checked monthly. If pressure-sensing devices are used, negative pressure should be verified at least once a month by using smoke tubes or taking pressure measurements. (CDC, 1994)

14.26.18 HEPA Filtration

HEPA-type filters clean air through the physical removal of particulates from the airstream. These filters have a minimum removal efficiency of 99.97% for particles ≥ 0.3 μm in diameter. Droplet nuclei of *M. tuberculosis* range in size from 1 to 5 μm in diameter; therefore, HEPA filtration can be expected to remove most droplet nuclei from the air.

To remove contaminants from the air, HEPA filtration can be used to supplement other recommended ventilation measures. HEPA filters are defined as air-cleaning devices that have a demonstrated and documented minimum removal efficiency of 99.97% of particles ≥ 0.3 μm in diameter. HEPA filters have been shown to be effective in reducing the concentration of *Aspergillus* spores (which range in size from 1.5 to 6 μm) to below measurable levels (Sherertz et al., 1987; Rhame et al., 1984; Opal et al., 1986). The ability of HEPA filters to remove tubercle bacilli from the air has not been studied, but *M. tuberculosis* droplet nuclei probably range from 1 to 5 μm in diameter. Should air recirculation into the general ventilation system be unavoidable, HEPA filters should be installed in the exhaust duct leading from the room to the general ventilation system and should be designed to remove infectious organisms and particulates the size of droplet nuclei from the air. (CDC, 1994)

14.26.18.1   Placement

The HEPA device should only be used in systems that recirculate air from rooms in which procedures are performed on patients who may have infectious TB if the following criteria for the units have been proven by the manufacturer: (1) the device is not completely passive and utilizes techniques such as electrostatics, and (2) failure of the electrostatic component will not permit a drop in filtration efficiency to less than 99.97%. (CDC, 1994)

14.26.18.2   Maintenance

In any application, HEPA filters should be installed carefully and maintained meticulously to ensure adequate functioning. The manufacturers of in-room air cleaning equipment should provide documentation of the HEPA filter efficiency and the efficiency of the device in lowering room air contaminant levels. (CDC, 1994)

14.26.18.3   Exhausting Air to the Outside

A HEPA filter can be used as an added safety measure to clean air from isolation rooms and local exhaust devices (e.g., booths, tents, or hoods used for cough-inducing procedures) before exhausting air directly to the outside. Such use is unnecessary if the exhaust air cannot reenter the ventilation system supply. The use of HEPA filters should be considered wherever exhaust air could possibly reenter the system. In many instances, exhaust air is not discharged directly to the outside; rather, the air is directed through heat-recovery devices (e.g., heat wheels). Heat wheels are often used to reduce the costs of operating ventilation systems. If such units are used with the system, a HEPA filter should also be used. As the wheel rotates, energy is transferred into or removed from the supply inlet air stream. The HEPA filter should be placed upstream from the heat wheel because of the potential for leakage across the seals separating the inlet and exhaust chambers and the theoretical possibility that droplet nuclei could be impacted on the wheel by the exhaust air and subsequently stripped off into the supply air. (CDC, 1994)

14.26.19   Recirculation

Air from TB isolation rooms and treatment rooms used to treat patients who have confirmed or suspected infectious TB should be exhausted to the outside in accordance with applicable federal, state, and local regulations. The air should not be recirculated into the general ventilation. In some instances, recirculation of air into the general ventilation system from such rooms is unavoidable (e.g., in existing facilities in which the ventilation system or facility configuration makes venting the exhaust to the outside impossible). In such cases, a HEPA filter should be installed in the exhaust duct leading from the room to the general ventilation system to remove infectious organisms and particulates the size of droplet nuclei from the air before the air is returned to the general ventilation system (Figure 14.8). Air from TB isolation rooms and treatment rooms in new or renovated facilities should not be recirculated into the general ventilation system. (CDC, 1994)

14.26.19.1   Recirculation within a Room

Air recirculation for individual rooms can be used in areas where no general ventilation system is present or an existing system is incapable of providing adequate airflow. An increase in ventilation is desired without affecting the fresh air supply or negative pressure system already in place.

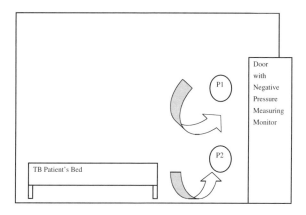

TB isolation room with negative pressure measuring monitor

Pressure as illustrated by P1 and P2 will vary within the room.

The Pressure Monitor intake must be placed to receive the air from P2 locations rather than P1 locations. Negative pressure must be maintained in droplet nuclei outfall areas.

FIGURE 14.8 HEPA filtration within ductwork can be used to maintain appropriate isolation.

Recirculation of HEPA-filtered air within a room can be achieved in several ways: (1) by exhausting air from the room into a duct, filtering the air through a HEPA filter installed in the duct, and returning the air to the room; (2) by filtering air through HEPA recirculation systems mounted on the wall or ceiling; and (3) by filtering air through portable HEPA recirculation systems. The first two approaches are fixed air recirculation systems, because the HEPA filter devices are fixed in place and are not easily movable. (CDC, 1994)

14.26.19.2 *Fixed Air Recirculation Systems*

The preferred method of recirculating HEPA-filtered air within a room is a built-in system (Figure 14.9), in which air is exhausted from the room into a duct, filtered through a HEPA filter, and returned to the room. This technique may be used to add air changes in areas where a recommended minimum ACH is difficult to meet with general ventilation alone. The air does not have to be conditioned, other than by the filtration, and this permits higher airflow rates than the general ventilation system can usually achieve. An alternative is the use of HEPA filtration units that are mounted on the wall or ceiling of the room. Fixed recirculation systems are preferred over portable freestanding) units and can be installed and maintained with a greater degree of reliability. (CDC, 1994)

14.26.19.3 *Portable Air Recirculation Units*

Portable HEPA filtration units may be considered for recirculating air within rooms in which no general ventilation system is present, where the system is incapable of providing adequate airflow, or where increased effectiveness in room airflow is desired. Effectiveness depends on circulating as much of the air in the room as possible through the HEPA filter, which may be difficult to achieve and evaluate, and can vary depending on the configuration of the room, the furniture and persons in the room, and the placement of the HEPA filtration unit and supply and exhaust grilles. The effectiveness of the portable unit may vary considerably in rooms with different configurations or in the same room if moved from one location to another in the room. If portable

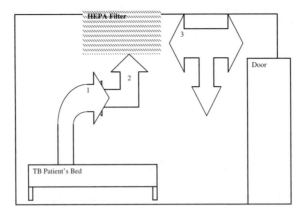

TB Isolation Room with HEPA filter.

Filter is placed 1/3 way between patient's bed and the door:

Arrow 1 illustrates the air containing droplet nuclei from the patient

Arrow 2 is this air as the HEPA filtration unit draws the air into the HEPA filter beds

Arrow 3 is the air exhausted from the HEPA filters

Air movement dynamics will be more complex thanthose shown here. However, the general trends will be as shown. All of the air and droplet nuclei from the patient will not be captured in the HEPA filter system. So the end result will be some clearing of the air of droplet nuclei without 100% effectiveness.

Opening the door will change the air patterns. The exhaust from the HEPA unit and the air nearest the door should be the intial air out the door. However, air patterns cannot be assured.

FIGURE 14.9 Air may be recirculated through a HEPA filtration unit.

units are used, caution should be exercised to ensure they can recirculate all or nearly all of the room air through the HEPA filter. Some commercially available units may not be able to meet this requirement because of design limitations or insufficient airflow capacity. In addition, units should be designed and operated to ensure that persons in the room cannot interfere with or otherwise compromise the functioning of the unit.

Portable HEPA filtration units have not been evaluated adequately to determine their role in TB infection-control programs. Portable HEPA filtration units should be designed to achieve the equivalent of ≥12 ACH to ensure adequate air mixing in all areas of the hospital room and not to interfere with the current ventilation system. Some HEPA filtration units employ UVGI for disinfecting air after HEPA filtration. However, whether exposing the HEPA-filtered air to ultraviolet irradiation further decreases the concentration of contaminants is not known. (CDC, 1994)

14.26.19.4   Evaluation

An estimate of the unit's ability to circulate the air in the room can be made by visualizing airflow patterns as was described previously for estimating room air mixing. If the air movement is good in all areas of the room, the unit should be effective. Portable HEPA units' effectiveness may not have been evaluated adequately by the manufacturer, and considerable variation in their effectiveness should be expected. (CDC, 1994)

14.26.20   Installing, Maintaining, and Monitoring HEPA Filters

Proper installation and testing and meticulous maintenance are critical if a HEPA filtration system is used, especially if the system recirculates air to other parts of the facility. Improper design, installation, or maintenance could allow infectious particles to circumvent filtration and escape into the general ventilation system. HEPA filters should be installed to prevent leakage between filter segments and the filter bed and frame. A regularly scheduled maintenance program is required to monitor the HEPA filter for possible leakage and for filter loading. A quantitative leakage and filter performance test (e.g., the dioctal phthalate [DOP] penetration test should be performed at the initial installation and every time the filter is changed or moved. The testing should be repeated every 6 months for filters in general-use areas and in areas with systems that exhaust air that is likely to be contaminated with *M. tuberculosis* (e.g., TB isolation rooms). A manometer or other pressure-sensing device should be installed in the filter system to provide an accurate and objective means of determining the need for filter replacement. The filter manufacturer should supply information as to the pressure drop characteristics of the filter. Installation of the filter should allow for maintenance that will not contaminate the delivery system or the area served.

For general infection-control purposes, special care should be taken not to jar or drop the filter element during or after removal. The scheduled maintenance program should include procedures for installation, removal, and disposal of filter elements. Only adequately trained personnel should perform HEPA filter maintenance. Appropriate respiratory protection should be worn while performing maintenance and testing procedures. Filter housings and ducts leading to the housings should be labeled clearly with the words *Contaminated Air* (or a similar warning). When a HEPA filter is used, one or more lower efficiency disposable prefilters installed upstream will extend the useful HEPA filter time interval. A disposable filter can increase the life of a HEPA filter by 25%. If the disposable filter is followed by a 90% extended surface filter, the life of the HEPA filter can be extended almost 900% (ACGIH, 1992). These prefilters should be handled and disposed of in the same manner as the HEPA filter. (CDC, 1994)

14.26.21   AFB Isolation Rooms and Treatment Rooms

Paragraph (d)(5) of 29CFR 1910.1035 requires that engineering controls be used in facilities that admit or provide medical services or AFB isolation to individuals with suspected or confirmed infectious TB except in settings where home healthcare or home-based hospice care is being provided. For example, engineering controls must be used in isolation rooms or areas, areas where high hazard procedures are performed, and autopsy rooms where *M. tuberculosis* may be aerosolized. This provision specifically excepts settings where home healthcare or home-based hospice care is being provided. In such situations, the employer is not in control of the employee's work setting because the setting is the private home of the individual being provided with care. In view of this, an employer providing home healthcare or home-based hospice care would not be required to implement engineering controls in the individuals' homes.

To prevent the escape of droplet nuclei, the TB isolation room should be maintained under negative pressure. Doors to isolation rooms should be kept closed, except when patients or personnel must enter or exit the room, so that negative pressure can be maintained. Negative pressure in the room should be monitored daily while the room is being used for TB isolation. ASHRAE, AIA, and the Health Resources and Services Administration recommend a minimum of 6 ACH for TB isolation and treatment rooms, based on comfort and odor control considerations. The effectiveness of this level of airflow in reducing the concentration of droplet nuclei in the room, thus reducing the transmission of airborne pathogens, has not been evaluated directly or adequately.

For the purposes of reducing the concentration of droplet nuclei, TB isolation and treatment rooms in existing healthcare facilities should have airflow ≥6 ACH and increased to ≥12 ACH by adjusting or modifying the ventilation system or by using auxiliary means (e.g., recirculation of air through fixed HEPA filtration systems or portable air cleaners). Paragraph (d)(5)(vi) requires that, while in use for TB isolation, doors and windows of AFB isolation rooms or areas must be kept closed except when doors are opened for the purpose of entering or exiting and when windows are part of the ventilation system being used to achieve negative pressure. For example, the window may be serving as the exit for the exhaust from an in-room HEPA filtration unit. AFB isolation rooms and areas are to be maintained under negative pressure while in use for TB isolation. Negative pressure in a room can be altered by small changes in the ventilation system operation or by the opening and closing of the isolation room doors or windows. In order to assure that the ventilation system functions as intended, once an operating configuration has been established, doors and windows be opened only when necessary.

Paragraph (d)(5)(vii) stipulates that when an AFB isolation room or area is vacated by an individual with suspected or confirmed infectious TB, the room or area must be ventilated for an appropriate period of time, according to current CDC recommendations for a removal efficiency of 99.9%, before permitting employees to enter without respiratory protection. The time required for removing airborne particles from an enclosed space depends on several factors:

- Number of air changes per hour (which is determined, in part, by the number of cubic feet of air in the room or booth)
- Rate at which air is entering the room or booth at the intake source vs. the rate at which it is being exhausted
- Location of the ventilation inlet and outlet
- Physical configuration of the room or booth

Perfect mixing of air normally does not occur because a number of factors, such as room configuration, may influence the movement of air.

Because perfect air mixing is not likely to occur, the necessary time required for a specific removal efficiency may be underestimated. In order to compensate for this shortcoming, OSHA has proposed that the most conservative (i.e., protective) removal efficiency (99.9%) be used to determine the appropriate amount of time an AFB isolation room or area must be ventilated before permitting employees to enter without respiratory protection. Using this conservative approach will help to assure that an appropriate time has passed before unprotected employees enter the area, even in situations where perfect air mixing has not occurred.

Ventilation of the room would not be necessary if the room was previously occupied by an individual with suspected infectious tuberculosis and that individual was medically determined to be noninfectious, as there would be no droplet nuclei present.

These isolation rooms are used to separate patients who are likely to have infectious TB from other persons, to provide an environment that will allow reduction of the concentration of droplet nuclei through various engineering methods, and to prevent the escape of droplet nuclei from such rooms into the corridor and other areas of the facility using directional airflow.

A hierarchy of ventilation methods used to achieve a reduction in the concentration of droplet nuclei and to achieve directional airflow using negative pressure has been developed. The method selected will depend on the configuration of the isolation room and the ventilation system in the facility, and the determination should be made in consultation with a ventilation engineer. (CDC, 1994)

14.26.21.1 Escape of Droplet Nuclei

Rooms used for TB isolation should be single-patient rooms and have negative pressure relative to the corridor or other areas connected to the room. Doors between the isolation room and other areas should remain closed except for entry into or exit from the room. Openings in the room (e.g., windows and electrical and plumbing entries) should be sealed as much as possible. However, a small gap of 1/8 to 1/2 inch should be maintained at the bottom of the door to provide a controlled airflow path. Proper use of negative pressure will prevent contaminated air from escaping the room. (CDC, 1994)

14.26.21.2   Reducing Droplet Nuclei

A minimum of 6 ACH for TB isolation rooms and treatment rooms is recommended by ASHRAE (ASHRAE, 1991), AIA (AIA, 1987), and the Health Resources and Services Administration (HRSA, 1984). This ventilation rate is based on comfort- and odor-control considerations. The effectiveness of this level of airflow in reducing the concentration of droplet nuclei in the room, thus reducing the transmission of airborne pathogens, has not been evaluated directly or adequately. Ventilation rates greater than 6 ACH are likely to produce an incrementally greater reduction in the concentration of bacteria in a room than are lower rates. To reduce the concentration of droplet nuclei, TB isolation rooms and treatment rooms in existing healthcare facilities should have airflow of ≥6 ACH. Where feasible, this airflow rate should be increased to ≥12 ACH by adjusting or modifying the ventilation system or using auxiliary means (e.g., air recirculation through fixed HEPA filtration units or portable air cleaners). New construction or renovation of existing healthcare facilities should be designed so that TB isolation rooms achieve an airflow of ≥12 ACH. (CDC, 1994)

14.26.21.3   Exhaust

Paragraph (d)(5)(iv) requires that air from AFB isolation rooms or areas be exhausted directly outside, away from intake vents and employees. If the air from these areas cannot be exhausted in such a manner or must be recirculated, the air must pass through HEPA filters before discharge or recirculation. In order for the air to be safely discharged, exhaust ducts must not be located near areas that may be populated (e.g., sidewalks or windows that may be opened). Ventilation system exhaust discharges must be designed to prevent reentry of exhaust air. Wind blowing over a building creates a highly turbulent recirculation zone, which can cause reentry of the exhaust into the building. Exhaust flow needs to be discharged above such a zone. The employer should be aware that exhausting of this air may also fall under federal, state, and local regulations concerning environmental discharges. This provision also states that, if a portion of this air is recirculated, it must pass through a properly designed, installed, and maintained HEPA filter before being discharged back into general facility ventilation. Whenever feasible, exhaust air from the AFB isolation rooms or areas must be exhausted to the outside. In its 1994 guidelines, the CDC states:

> Air from TB isolation rooms and treatment rooms used to treat patients who have confirmed or suspected infectious TB should be exhausted to the outside in accordance with applicable federal, state, and local regulations. The air should not be recirculated into the general ventilation. In some instances, recirculation of air into the general ventilation system from such rooms is unavoidable (i.e., in existing facilities in which the ventilation system or facility configuration makes venting the exhaust to the outside impossible). In such cases, HEPA filters should be installed on the exhaust duct leading from the room to the general ventilation system to remove infectious organisms and particulates the size of droplet nuclei from the air before it is returned to the general ventilation system (Section II.F; Suppl. 3). Air from TB isolation rooms and treatment rooms in new or renovated facilities should not be recirculated into the general ventilation system.

OSHA agrees with the CDC that exhaust air should be vented to the outside. However, OSHA recognizes that instances may exist where outside discharge may not be feasible and, therefore, permits recirculation with HEPA filtration of the recirculated air, in such instances.

Air from TB isolation rooms and treatment rooms in which patients with infectious TB may be examined should be exhausted directly to the outside of the building away from air-intake vents, persons, and animals in accordance with federal, state, and local regulations concerning environmental discharges. Exhaust ducts should not be located near areas that may be populated (e.g., near sidewalks or windows that could be opened). Ventilation system exhaust discharges and inlets should be designed to prevent exhausted air reentry. Wind blowing over a building creates a highly turbulent recirculation zone, which can cause exhausted air to reenter the building. Exhaust flow should be discharged above such a zone. Design guidelines for proper placement of exhaust ducts can be found in the 1989 ASHRAE *Fundamentals Handbook* (ASHRAE, 1989). If recirculation of air from such rooms into the general ventilation system is unavoidable, the air should be passed through HEPA filter before recirculation. (CDC, 1994)

14.26.22   Alternatives to Tuberculosis Isolation Rooms

Isolation can also be achieved by use of negative-pressure enclosures (e.g., tents or booths) to provide patient isolation in areas such as emergency rooms and medical testing and treatment areas and to supplement isolation in designated isolation rooms. (CDC, 1994)

14.27 ULTRAVIOLET GERMICIDAL IRRADIATION (UVGI)

OSHA is not proposing to allow the use of UVGI in place of ventilation for controlling aerosolized *M. tuberculosis*. Although the germicidal properties of certain wavelengths of ultraviolet light (UV-C) are generally recognized, the Agency has not included UVGI as a primary engineering control in the proposed standard. With regard to the use of UVGI, the CDC states:

Because the clinical effectiveness of UV systems varies, and because of the risk for *M. tuberculosis* transmission if a system malfunctions or is maintained improperly, UVGI is not recommended for the following specific applications: 1. Duct systems using UVGI are not recommended as a substitute for HEPA filters if air from isolation rooms must be recirculated to other areas of a facility. 2. UVGI alone is not recommended as a substitute for HEPA filtration or local exhaust of air to the outside from booths, tents, or hoods used for cough-inducing procedures. 3. UVGI is not a substitute for negative pressure.

The CDC goes on to discuss a number of factors that affect the effectiveness of UVGI and UV lamps in killing airborne tubercle bacilli. These factors include the:

- Intensity of UVGI
- Duration of irradiation of the organism
- Relative humidity of the environment
- Age of the UV lamp
- Amount of dust on the surface of the lamp

In light of this information, OSHA does not believe that UVGI can reliably and uniformly control airborne tubercle bacilli; consequently, UVGI is not acceptable as a primary engineering control. However, some employers may choose to use UVGI as a supplement to ventilation or HEPA filtration.

Upper-room air UVGI may be used as an adjunct to general ventilation in the isolation room. Air in the isolation room may be recirculated within the room through HEPA filters or UVGI devices to increase the effective ACH thermal efficiency. Research has demonstrated that UVGI is effective in killing or inactivating tubercle bacilli under experimental conditions (Riley et al., 1962, 1957; Riley and Nardell, 1989; Riley, 1988; Stead,1989) and in reducing transmission of other infections in hospitals (McLean, 1961), military housing (Willmon et al., 1948) and classrooms (Wells et al., 1942; Wells and Holla, 1950; Perkins et al., 1947). UVGI has been recommended as a supplement to other TB infection-control measures in settings where the need for killing or inactivating tubercle bacilli is important (CDC, 1990, 1982; ATS/CDC, 1983; NTRDA, 1969; CDC, 1971; Schieffelbein and Snider, 1988; CDC 1989).

Ultraviolet radiation is defined as that portion of the electromagnetic spectrum described by wavelengths from 100 to 400 nm. The UV spectrum has been separated into three different wavelength bands: UV-A (long wavelengths; range, 320–400 nm), UV-B (midrange wavelengths; range, 290–320 nm); and UV-C (short wavelengths; range, 100–290 nm) (ICI, 1987). Commercially available UV lamps used for germicidal purposes are low-pressure mercury vapor lamps (Nagy, 1964) that emit radiant energy in the UV-C range, predominantly at a wavelength of 253.7 nm (IES, 1966). (CDC, 1994)

14.27.1   Placement

For general-use areas in which the risk for *M. tuberculosis* transmission is relatively high, UVGI lamps may be used as an adjunct to ventilation for reducing the concentration of infectious droplet nuclei. Ultraviolet (UV) units can be installed in:

- A room or corridor to irradiate the air in the upper portion of the room (e.g., upper-room air irradiation)
- Ducts to irradiate air passing through the ducts
- Ducts that recirculate air back into the same room

Ultraviolet units installed in ducts should not be substituted for HEPA filters in ducts that discharge air from TB isolation rooms into the general ventilation system. (CDC, 1994)

14.27.2   Maintenance

Lamps should be installed properly and maintained adequately, which includes the monitoring of irradiance levels. Tubes should be changed according to the manufacturer's instructions or when meter readings indicate tube failure. An employee trained in the use and handling of UV lamps should be responsible for these measures and for keeping maintenance records. Applicable safety guidelines should be followed. Caution should be exercised to protect healthcare workers, patients, visitors, and others from excessive exposure to UV radiation. (CDC, 1994)

14.27.3   Duct Irradiation

To inactivate tubercle bacilli without exposing persons to UVGI in duct irradiation systems may be used. UV lamps are placed inside ducts that remove air from rooms to disinfect the air before the air is recirculated. When UVGI duct systems are properly designed, installed, and maintained, high levels of UV radiation may be produced in the ductwork. The only potential for human exposure to this radiation occurs during maintenance operations. Duct irradiation may be used in a TB isolation room or treatment room to recirculate air:

1. From the room
2. Through a duct containing UV lamps
3. Back into the room

This recirculation method can increase the overall room airflow but does not increase the supply of fresh outside air to the room.

In other patients' rooms and in waiting rooms, emergency rooms, and other general-use areas of a facility, where patients with undiagnosed TB could potentially contaminate the air, this method may recirculate partially contaminated air back into the general ventilation. Duct-irradiation systems are dependent on airflow patterns within a room that ensure that all or nearly all of the room air circulates through the duct. (CDC, 1994)

14.27.4 Upper Air Irradiation

Upper air irradiation is used to inactivate tubercle bacilli in the upper part of the room, while minimizing radiation exposure to persons in the lower part of the room. UVGI lamps are suspended from the ceiling or mounted on the wall. The bottom of the lamp is shielded to direct the radiation upward but not downward. The system depends on air mixing to take irradiated air from the upper to the lower part of the room and nonirradiated air from the lower to the upper part. The irradiated air space is much larger than that in a duct system. UVGI has been effective in killing bacteria under conditions where air mixing was accomplished mainly by convection. For example, BCG was atomized in a room that did not have supplemental ventilation (Riley et al., 1976), and in another study a surrogate bacteria, *Serratia marcesens*, was aerosolized in a room with a ventilation rate of 6 ACH (Kethley and Branch, 1972). These reports estimated the effect of UVGI to be equivalent to 10 and 39 ACH, respectively, for the organisms tested, which are less resistant to UVGI than *M. tuberculosis* (Riley et al., 1976).

The addition of fans or some heating/air conditioning arrangements may double the effectiveness of UVGI lamps (Riley, Permutt, and Kaufman 1971a; Riley and Permutt, 1971; Riley, Permutt, and Kaufman, 1971b). Greater rates of ventilation, however, may decrease the length of time the air is irradiated, thus decreasing the killing of bacteria (Collins, 1971; Kethley and Branch, 1972). The optimal relationship between ventilation and UVGI is not known. Air irradiation lamps used in corridors have been effective in killing atomized *S. marcesens* (Riley and Kaufman, 1971). Use of UVGI lamps in an outpatient room has reduced culturable airborne bacteria by 14 to 19%; however, the irradiation did not reduce the concentration of Gram-positive, rod-shaped bacteria. Although fast-growing mycobacteria were cultured, *M. tuberculosis* could not be recovered from the room's air samples because of fungal overgrowth of media plates (Macher et al., 1992). UVGI irradiation of upper room air may be used in isolation or treatment rooms as a supplemental method of air cleaning and in other patients' rooms and in waiting rooms, emergency rooms, corridors, and other central areas of a facility where patients with undiagnosed TB could potentially contaminate the air. Determinants of UVGI effectiveness include:

- Room configuration
- UV lamp placement
- Adequacy of airflow patterns in bringing contaminated air into contact with the irradiated upper room space

Air mixing may be facilitated by supplying cool air near the ceiling in rooms where warmer air (or a heating device) is present below. The ceiling should be high enough for a large volume of upper-room air to be irradiated without healthcare workers and patients being overexposed to UV radiation. Because the clinical effectiveness of UV systems varies and because of the risk for *M. tuberculosis* transmission if a system malfunctions or is maintained improperly, UVGI is not recommended for a number of specific applications. Duct systems using UVGI are not recommended as a substitute for HEPA filters if air from isolation rooms must be recirculated to other areas of a facility. UVGI alone is not recommended as a substitute for HEPA filtration or local exhaust of air to the outside from booths, tents, or hoods used for cough-inducing procedures. UVGI is not a substitute for negative pressure.

The use of UV lamps and HEPA filtration in a single unit would not be expected to have any infection-control benefits not provided by use of the HEPA filter alone. The effectiveness of UVGI in killing airborne tubercle bacilli depends on the intensity of the UVGI, duration of contact the organism has with the irradiation, and relative humidity (Riley et al., 1962; Riley and Nardell, 1989; McLean, 1961). Humidity can have an adverse effect on UVGI effectiveness at levels greater than 70% relative humidity for *Serratia marcescens* (Riley and Kaufman, 1972). The interaction of these factors has not been fully defined, however, making precise recommendations for individual

UVGI installations difficult to develop. Old lamps or dust-covered UV lamps are less effective; therefore, regular maintenance of UVGI systems is crucial. Short-term overexposure to UV radiation can cause erythema and keratoconjunctivitis (NIOSH, 1972; Everett et al., 1969). Broad-spectrum UV radiation has been associated with increased risk for squamous and basal cell carcinomas of the skin (IARC, 1992). UV-C was recently classified by the International Agency for Research on Cancer (IARC) as "probably carcinogenic to humans" (Group 2A) (IARC, 1992). This classification is based on studies suggesting that UV-C radiation can induce skin cancers in animals; DNA damage, chromosomal aberrations and sister chromatid exchange, and transformation in human cells *in vitro*; and DNA damage in mammalian skin cells *in vivo*. In the animal studies, a contribution of UV-B to the tumor effects could not be excluded, but the effects were greater than expected for UV-B alone (IARC, 1992).

Although some recent studies have demonstrated that UV radiation can activate HIV gene promoters (i.e., the genes in HIV that prompt replication of the virus) in laboratory samples of human cells (Valerie et al., 1988; Valerie and Rosenberg, 1990; Zmudzka and Beer, 1990; Wallace and Lasker, 1992; Stein et al., 1989; Clerici and Shearer, 1992), the implications of these *in vitro* findings for humans are unknown. In 1972, the National Institute for Occupational Safety and Health (NIOSH) published a recommended exposure limit (REL) for occupational exposure to UV radiation (NIOSH, 1972). The REL is intended to protect workers from the acute effects of UV exposure (e.g., erythema and photokeratoconjunctivitis). Photosensitive persons and those exposed concomitantly to photoactive chemicals may not be protected by the recommended standard. If proper procedures are not followed, healthcare workers performing maintenance on such fixtures are at risk for exposure to UV radiation. Because UV fixtures used for upper air irradiation are present in rooms, rather than hidden in ducts, safety may be much more difficult to achieve and maintain. Fixtures must be designed and installed to ensure that UV exposure to persons in the room (including healthcare workers and inpatients) are below current safe exposure levels.

Recent health hazard evaluations conducted by the CDC have noted problems with overexposure of healthcare workers to UVGI and with inadequate maintenance, training, labeling, and use of personal protective equipment (NIOSH, 1992a,b,c). The current number of persons who are properly trained in UVGI system design and installation is limited. The CDC strongly recommends that a competent UVGI system designer be consulted to address safety considerations before such a system is procured and installed. Experts who might be consulted include industrial hygienists, engineers, and health physicists. Principles for the safe installation of UV lamp fixtures have been developed and can be used as guidelines (Macher, 1989; Riley, 1991). If UV lamps are being used in a facility, the general TB education of healthcare workers should include the basic principles of UVGI systems (e.g., how they work and what their limitations are), the potential hazardous effects of UVGI if overexposure occurs, the potential for photosensitivity associated with certain medical conditions or use of some medications, and the importance of general maintenance procedures for UVGI fixtures.

Exposure to UV intensities above the REL should be avoided. Lightweight clothing made of tightly woven fabric and UV-absorbing sunscreens with solar protection factors (SPFs) of at least 15 may help protect photosensitive persons. Healthcare workers should be advised that occupational health staff should examine any eye or skin irritation that develops after UV exposure. (CDC, 1994)

14.27.5 Exposure Criteria for UV Radiation

The NIOSH recommended exposure limit for UV radiation is wavelength dependent because different wavelengths of UV radiation have different adverse effects on the skin and eyes (NIOSH, 1972). The relative spectral effectiveness (S_1) is used to compare various UV sources with a source producing UV radiation at 270 nm, the wavelength of maximum ocular sensitivity. For example, the S_1 at 254 nm is 0.5; therefore, twice as much energy is required at 254 nm to produce an identical biologic effect at 270 nm (NIOSH, 1972). At 254 nm, the REL is 0.006 Joules per square centimeter (J/cm^2); at 270 nm, the REL is 0.003 J/cm^2. For germicidal lamps that emit radiant energy predominantly at a wavelength of 254 nm, proper use of the REL requires that the measured irradiance level (E) in microwatts per square centimeter (W/cm^2) be multiplied by the relative spectral effectiveness at 254 nm (0.5) to obtain the effective irradiance (E_{eff}). The maximum permissible exposure time (t) can then be calculated (in seconds) by dividing 0.003 J/cm^2 (the REL at 270 nm) by E_{eff} in W/cm^2 (NIOSH, 1972; ACGIH, 1991). To protect healthcare workers who are exposed to germicidal UV radiation for 8 hours per workday, the measured irradiance (E) should be $\geq$ 0.2 $\mu W/cm^2$. This is calculated by using the following from 29CFR 1910.1035 E_{eff} (0.1 $\mu W/cm^2$) to obtain and then divide this value by S_1 (0.5). (CDC, 1994)

Maximum Permissible Exposure Times for Selected Values of Effective Irradiance

Permissible exposure times* per day	Effective irradiance (E_{eff}) + ($\mu W/cm2$)
8 hrs	0.1
4 hrs	0.2
2 hrs	0.4
1 hr	0.8
30 min	1.7
15 min	3.3
10 min	5.0
5 min	10.0
1 min	50.0
30 sec	100.0

* Permissible exposure times are designed to prevent acute effects of irradiation to skin and eyes (136). These recommended limits are wavelength dependent because different wavelengths of ultraviolet (UV) radiation have different adverse effects on these organs.

+ Relative spectral effectiveness (S sub 1) is used to compare various UV sources with a source producing UV radiation at 270 nm, the wavelength of maximum ocular sensitivity. For example, the relative spectral effectiveness at 254 nm is 0.5; therefore, twice as much energy is required at 254 nm to produce an identical biologic effect at 270 nm. At 254 nm, the NIOSH REL is 0.006 joules per square centimeter ($J/cm2$); and at 270 nm, it is 0.003 $J/cm2$. For germicidal lamps that emit radiant energy predominantly at a wavelength of 254 nm, proper use of the REL requires that the measured irradiance level (E) in microwatts per square centimeter ($\mu W/cm2$) be multiplied by the relative spectral effectiveness at 254 nm (0.5) to obtain E_{eff}. The maximum permissible exposure time can be calculated (in seconds) by dividing 0.003 $J/cm2$ (the NIOSH REL at 270 nm) by E_{eff} in $\mu W/cm2$ (136,150). To protect health-care workers who are exposed to germicidal UV radiation for 8 hours per work day, the measured irradiance (E) should be $\leq$ 0.2 $\mu W/cm2$, which is calculated by obtaining E_{eff} (0.1 $\mu W/cm2$), then dividing this value by S sub 1 (0.5).

14.27.6 Labeling and Posting

Warning signs should be posted on UV lamps and wherever high-intensity (i.e., UV exposure greater than the REL) germicidal UV irradiation is present (e.g., upper air space or accesses to ducts, if duct irradiation is used) to alert maintenance staff or other healthcare workers of the hazard. Some examples are shown below:

CAUTION

CAUTION: ULTRAVIOLET ENERGY

ULTRAVIOLET ENERGY

TURN OFF LAMPS AND PROTECT EYES & SKIN BEFORE ENTERING UPPER ROOM

14.27.7 Maintenance

Because the intensity of UV lamps fluctuates as they age, a schedule for replacing the lamps should be developed. The schedule can be determined from either a time/use log or a system based on cumulative time. The tube should be checked periodically for dust build-up, which lessens the output of UVGI. If the tube is dirty, it should be allowed to cool and then cleaned with a damp cloth. It should be replaced if glowing stops or flicker occurs to an objectionable extent. Maintenance personnel must turn off all UV tubes before entering the upper part of the room or accessing ducts for any purpose. Only a few seconds of direct exposure to the intense UV radiation in the upper air space or in ducts can cause burns. (CDC, 1994)

14.27.8 Protective Equipment

Gloves and goggles (and/or face shields) should be worn if exposure greater than the recommended standard is anticipated. Banks of UVGI tubes can be installed in ventilating ducts. Safety devices should be used on access doors to eliminate hazard to maintenance personnel. For duct irradiation systems, the access door for servicing the lamps should have:

- An inspection window through which the lamps are checked periodically for dust build-up and malfunctioning
- A warning sign written in languages appropriate for maintenance personnel to alert them to the health hazard of looking directly at bare tubes
- A lock with an automatic electric switch or other device that turns off the lamps when the door is opened

Two types of fixtures are used in upper air irradiation:

- Wall-mounted fixtures that have louvers to block downward radiation
- Ceiling-mounted fixtures that have baffles to block radiation below the horizontal plane of the UV tube

The actual UV tube in either type of fixture must not be visible from any normal position in the room. Light switches that can be locked should be used, if possible, to prevent injury to personnel who might unintentionally turn the lamps on during maintenance procedures. In most applications, properly shielding the UV lamps to provide protection from most, if not all, of the direct UV radiation is difficult. Radiation reflected from glass, polished metal, and high-gloss ceramic paints can be harmful to persons in the room, particularly if more than one UV lamp is in use. Surfaces in irradiated rooms that can reflect UVGI into occupied areas of the room should be covered with non-UV-reflecting material. (CDC, 1994)

14.27.9 Monitoring

A regularly scheduled evaluation of the UV intensity to which healthcare workers, patients, and others are exposed should be conducted. UV measurements should be made in various locations within a room using a detector designed to be most sensitive at 254 nm. Equipment used to measure germicidal UV radiation should be maintained and calibrated on a regular schedule. An industrial hygienist or other person knowledgeable in making UV measurements must carefully check a new UV installation for hot spots (i.e., areas of the room where the REL is exceeded). UV radiation levels should not exceed those in the recommended guidelines. (CDC, 1994)

14.28 RESPIRATORY PROTECTION

Paragraph (f) Respiratory Protection of 29 CFR 1910.1035 describes respirators as follows:

Respirators serve as supplemental protection to reduce employee exposures when engineering and work practice controls are not sufficient to provide adequate protection against airborne contaminants.

During hearings for OSHA's General Industry Respiratory Standard, 29 CFR 1910.134, OSHA stated that "all aspects of respirator use for protection against TB would be addressed in the rulemaking for Occupational Exposure to Tuberculosis." Past OSHA standards have referred to the Respirator Standard 29 CFR 1910.134 for the general requirements for respirator use. OSHA has included provisions relative to all aspects of respirator use for TB to provide interested parties with review and comment opportunities. OSHA is, however, considering including in the final TB rule cross-referencing to the general requirements 29 CFR 1910.134. Thus, the final TB rule will include only provisions specific to respirator use for TB. The following section provides a discussion of the current OSHA proposed respiratory provisions that are currently in 29 CFR 1910.1035. (OSHA, 1997)

14.28.1 Employer Responsibility

Paragraph (f)(1)(iii) of 29 CFR 1910.1035 states that the employer shall provide needed respirators to employees at no cost to the employee and will assure correct use. Paragraph (f)(1)(iv) stipulates further that the employer must ensure that the employee dons a respirator before entering the work setting or performing the tasks set forth in paragraphs (f)(1)(i) and (f)(1)(ii). The employer must also ensure that the employee uses the respirator until leaving the work setting or completing the task, regardless of other control measures in place. The employee is protected for the entire period of occupational exposure. (OSHA, 1997)

14.28.2 CDC Risk Assessment and Respirators

The CDC/NIH document *Biosafety in Microbiological and Biomedical Laboratories* recommends that respiratory protection be worn whenever aerosols of organisms such as *M. tuberculosis* cannot be safely contained. Employers must determine which occupationally exposed employees need to wear a respirator. Those settings where respirators are required should, according to CDC, be identified on the basis of the facility's risk assessment. The precise level of effectiveness in protecting HCWs from *M. tuberculosis* transmission in healthcare settings has not been determined. Information concerning *M. tuberculosis* transmission is incomplete and the following have not been defined:

- Smallest infectious dose of *M. tuberculosis.*
- Highest level of exposure to *M. tuberculosis* at which transmission will not occur.
- Size distribution of droplet nuclei and the number of particles containing viable *M. tuberculosis* that are expelled by infectious TB patients.
- Accurate concentration of infectious droplet nuclei measurement methods in a room have not been developed.

Nevertheless, in certain settings, administrative and engineering controls may not adequately protect HCWs from airborne droplet nuclei. Respiratory protective devices used in these settings should have characteristics that are suitable for the organism and the settings. (CDC, 1994)

OSHA agrees that the quantity of *M. tuberculosis* that, when inhaled, will result in infection (i.e., infectious dose) has not been determined conclusively. The number of droplet nuclei expelled into a room by an infectious individual or aerosol-producing procedure and the concentration of droplet nuclei in a room or area are unknown. OSHA agrees with the CDC that engineering and administrative controls cannot be assumed to protect employees against exposure to airborne TB droplet nuclei. Therefore, the use of respiratory protection is necessary. (OSHA, 1997)

14.28.3 Employee Use

Paragraph (f)(1)(i) of 29 CFR 1910.1035 states that "each employer must provide a respirator to each employee who:

- Enters an AFB isolation room or area in use for TB isolation. An *AFB isolation room or area* is defined in paragraph (j), Definitions. This definition clarifies that the requirement refers to entering any area where high-hazard procedures are being performed and entering an autopsy room where *M. tuberculosis* may be aerosolized.
- Is present during performance of procedures or services for an unmasked individual with suspected or confirmed infectious TB. Paragraph (f)(1)(i)(B) is intended to cover those situations outside of an AFB isolation room or area. For example, a facility may not have a portable x-ray and may, therefore, perform this procedure in a standard x-ray room. If the individual is not masked in such a situation, all employees present (i.e., the x-ray technician and any other employees in the room) must utilize respiratory protection.
- Transports an unmasked individual with suspected or confirmed infectious TB in an enclosed vehicle or who transports an unmasked individual with suspected or confirmed infectious TB within the facility. Paragraph (f)(1)(i)(C) is needed for those special circumstances in which the individual may not be masked (e.g., individual is combative and will not wear a mask). The employee transporting the individual would most likely spend an extended period of time in close proximity to the individual. Employees rendering emergency medical services may spend an extended time period within an enclosed vehicle in very close proximity to individuals with suspected or confirmed infectious TB. Droplet nuclei that escape capture in the patient's mask are contained within the vehicle and increase the risk that these employees will breathe the droplet nuclei generated.
- Repairs, replaces, or maintains air systems or equipment that may reasonably be anticipated to contain aerosolized *M. Tuberculosis*. Paragraph (f)(1)(i)(D) addresses this protective requirement given the potential of exposure to air that could contain aerosolized bacilli.
- Is working in an area where an unmasked individual with suspected or confirmed infectious TB has been segregated or otherwise confined (e.g., while awaiting transfer). Paragraph (f)(1)(i)(E) requirements and the underlying rationale are the same as that for Paragraph (f)(1)(i)(C) requirements.
- Is working in a residence where an individual with suspected or confirmed infectious TB is present. Paragraph (f)(1)(i)(F) requirements acknowledge that because the individual has been releasing droplet nuclei into the residence airspace, whether or not that individual is masked is not a factor. Other employees in addition to medical personnel who are entering these residences fall under this provision. (OSHA, 1997). The CDC discusses this provision as a component of home health-care and states:

Healthcare workers who provide medical services in the homes of patients who have suspected or confirmed infectious TB should instruct such patients to cover their mouths and noses with a tissue when coughing or sneezing. Until such patients are no longer infectious, HCWs should wear respiratory protection when entering these patients' homes. (CDC, 1994)

Paragraph (f)(1)(ii) also requires that each employer who operates a research laboratory must provide a respirator to each employee who is present when *M. tuberculosis* aerosols cannot be safely contained. This requirement is consistent with CDC/NIH recommendations regarding respirator use in research laboratories. (OSHA, 1997; CDC, 1994)

CDC states that personal respiratory protection should be used by persons:

- Entering rooms where patients with known or suspected infectious TB are being isolated
- Present during cough-inducing or aerosol-generating procedures performed on such patients
- Where administrative and engineering controls are not likely to protect them from inhaling infectious airborne droplet nuclei. These other settings include
- Transporting patients who may have infectious TB in emergency transport vehicles
- Providing urgent surgical or dental care to patients who may have infectious TB before a determination has been made that the patient is noninfectious
- Who are performing maintenance and testing procedures on HEPA filtration systems

14.28.4 Criteria Standards

A respirator shall be, at a minimum, either a HEPA respirator selected from among those jointly approved as acceptable by NIOSH under the provisions of 30 CFR part 11 or an N95 respirator certified by NIOSH under the provisions of 42 CFR part 84. NIOSH in accordance with the current Respiratory Protection Standard 29 CFR 1910.134 is the respiratory equipment-approving federal agency.

Until recently, HEPA respirators were the only NIOSH-certified negative pressure respirators that met the CDC's filter efficiency criteria. However, on July 10, 1995, NIOSH's original respirator certification procedures for air-purifying particulate respirators, 30 CFR part 11, were replaced by revised procedures, 42 CFR part 84. Under these new procedures, all nonpowered air-purifying particulate respirators are challenged with a 0.3-μ particle (the most penetrating size) at a flow rate of 85 L per min. The underlying reasoning for the acceptability of type N95 respirators is that their filter efficiency of >95% for a 0.3-μ particle will exceed the 95% filtering efficiency for a particle three times as large (i.e., 1 μ).

OSHA permits the employer to select either a HEPA respirator certified under 30 CFR part 11 or a respirator certified under 42 CFR part 84, since particulate respirators certified under both of these regulations are currently on the market. HEPA respirators are the only nonpowered particulate respirators certified under 30 CFR part 11 that meet the CDC guideline filtration criteria. However, applications for certification of nonpowered particulate respirators under 30 CFR part 11 are no longer being accepted by NIOSH. Therefore, dwindling stocks of HEPA respirators certified under that regulation will eventually lead to their unavailability, and employers will of necessity be selecting respirators from those approved under 42 CFR part 84. (OSHA, 1997)

14.28.5 Respirator Characteristics and Criteria

Paragraph (f)(3) sets out the respirator characteristics. The criteria are presented in performance-oriented language to provide employees with a respirator that will protect them against aerosolized *M. tuberculosis*. These criteria have been drawn from CDC recommendations based on currently available information that includes: data on the effectiveness of respiratory protection against non-infectious hazardous material in workplaces other than healthcare settings with an interpretation the data's applicability to respiratory protection against *M. tuberculosis*; the efficiency of respirator filters in filtering biological aerosols; face-seal leakage; and respirator characteristics used in conjunction with administrative and engineering controls in outbreak settings where transmission to HCWs and patients was terminated.(OSHA, 1997 and CDC, 1994)

Respiratory protective devices used in healthcare settings for protection against *M. tuberculosis* should meet the following standard performance criteria:

- Filter particles 1 μm in size in the unloaded state with a filter efficiency of ≥ 95% (i.e., filter leakage of ≤ 5%), given flow rates of up to 50 L per min
- Qualitatively or quantitatively fit tested in a reliable way to obtain a face-seal leakage of ≤10% (American National Standards Institute, 1992; NIOSH, 1987)

- Fit the different facial sizes and characteristics of HCWs, which can usually be met by making the respirators available in at least three sizes
- Checked for facepiece fit in accordance with standards established by OSHA, and good industrial hygiene practice by HCWs each time the respirators are put on

Available data suggest that infectious droplet nuclei range in size from 1 to 5 µm; therefore, respirators used in healthcare settings should be able to efficiently filter the smallest particles. Fifty liters per minute is a reasonable estimate of the highest airflow rate an HCW is likely to achieve during breathing, even while performing strenuous work activities.

In some settings, HCWs may be at risk for two types of exposure: inhalation of *M. tuberculosis* and mucous membrane exposure to fluids that may contain bloodborne pathogens. In these settings, protection against both types of exposure should be used.

When operative procedures (or other procedures requiring a sterile field) are performed on patients who may have infectious TB, respiratory protection worn by the HCW should protect:

- The surgical field from the HCW respiratory secretions
- The HCW from infectious droplet nuclei that may be expelled by the patient or generated by the procedure

Note: Respirators with expiration valves and positive-pressure respirators do not protect the sterile field.

Under the NIOSH revision, filter materials will be tested at a flow rate of 85 L/min for penetration by particles with a median aerodynamic diameter of 0.3 µm and, if certified, placed in one of the following categories:

- Type A, which has $\geq$ 99.97% efficiency (similar to current HEPA filter media)
- Type B, $\geq$ 99% efficiency
- Type C, $\geq$ 95% efficiency

According to this proposed scheme, type C filter material would meet or exceed the standard performance criteria needed. (OSHA, 1997)

14.28.6 Respirator Effectiveness

Data regarding protection against *M. tuberculosis* transmission are not available. The parameters used to determine the effectiveness of a respiratory protective device are face-seal efficacy and filter efficacy.

14.28.7 Face-Seal Leakage

OSHA requires a face-seal leakage of 10% or less. Any respirator that passes a qualitative fit test meets this criterion. However, quantitative fit testing necessitates that a particular numerical value be achieved. Paragraph (f)(5)(iii) requires that when quantitative fit testing is performed, the employer shall not permit an employee to wear a tight-fitting respirator unless a minimum fit factor of 100 is obtained in the test chamber. This value corresponds to face-seal leakage of 10% or less. (OSHA, 1997)

CDC performance criteria are as follows:

- Face-seal leakage compromises the ability of particulate respirators to protect HCWs from airborne materials.
- A proper seal between the respirator's sealing surface and the face of the person wearing the respirator is essential for effective and reliable performance of any negative-pressure respirator. This seal is less critical, but still important, for positive-pressure respirators.

Face-seal leakage can result from various factors, including:

- Incorrect facepiece size or shape
- Incorrect or defective facepiece sealing lip
- Beard growth
- Perspiration or facial oils that can cause facepiece slippage
- Failure to use all the head straps
- Incorrect positioning of the facepiece on the face
- Incorrect head strap tension or position
- Improper respirator maintenance
- Respirator damage

Every time a person wearing a negative-pressure particulate respirator inhales negative pressure relative to the workplace air is created inside the facepiece and air containing contaminants can take a path of least resistance into the respirator — through leaks at the face-seal interface — thus avoiding the higher-resistance filter material.

Currently available cup-shaped, disposable particulate respirators have from 0 to 20% face-seal leakage (NIOSH, 1987; American National Standards Institute 1980). This face-seal leakage results from variability of the human face, limitations of the respirator, design, construction, and the number of sizes available

Face-seal leakage is probably higher if the respirator is not fitted properly to the HCW's face, tested for an adequate fit by a qualified person, or checked for fit by the HCW every time the respirator is put on.

Face-seal leakage may be reduced to less than 10% with improvements in design, a greater variety in available sizes, and appropriate fit testing and fit checking.

In comparison with negative-pressure respirators, positive-pressure respirators produce a positive pressure inside the facepiece under most conditions of use. For example, in a powered air purifying respirator (PAPR):

1. A blower forcibly draws ambient air through HEPA filters.
2. Delivers the filtered air to the facepiece.
3. Air is blown into the facepiece at flow rates that generally exceed the expected inhalation flow rates.
4. Positive pressure inside the facepiece reduces face-seal leakage to low levels, particularly during the relatively low inhalation rates expected in healthcare settings.

PAPRs with a tight-fitting facepiece have < 2% face-seal leakage under routine conditions (NIOSH, 1987). PAPRs with loose-fitting facepieces, hoods, or helmets have < 4% face-seal leakage under routine conditions (NIOSH, 1987) and may offer lower levels of face-seal leakage than nonpowered, half-mask respirators. Full facepiece, non-powered respirators have the same leakage (i.e., less than 2%) as PAPRs.

Another factor contributing to face-seal leakage of cup-shaped, disposable respirators is that some of these respirators are available in only one size. A single size may produce higher leakage for persons who have a smaller or difficult-to-fit face (Lowry et al., 1977). The facepieces used for some reusable (including HEPA and replaceable filter, negative-pressure) and all positive-pressure particulate air-purifying respirators are available in as many as three different sizes. (CDC, 1994)

14.28.8 Filter Leakage

Aerosol leakage through respirator filters depends on independent variables:

- Filtration characteristics for each type of filter
- Size distribution of the droplets in the aerosol
- Linear velocity through the filtering material
- Filter loading (i.e., the amount of contaminant deposited on the filter)
- Any electrostatic charges on the filter and on the droplets in the aerosol

For HEPA-filter respirators, virtually all inward leakage of droplet nuclei occurs at the respirator's faceseal. (CDC, 1994)

14.28.9 Fit Testing and Fit Check

Fit testing is part of the respiratory protection program required by OSHA for all respiratory protective devices used in the workplace. A fit test determines whether a respiratory protective device adequately fits a particular HCW. The HCW may need to be fit tested with several devices to determine which device offers the best fit. Fit tests can detect only the leakage that occurs at the time of the fit testing, and the tests cannot distinguish face-seal leakage from filter leakage.

Determination of facepiece fit can involve qualitative or quantitative tests (NIOSH, 1987). A qualitative test relies on the subjective response of the HCW being fit tested. A quantitative test uses detectors to measure inward leakage. Disposable, negative-pressure particulate respirators can be qualitatively fit tested with aerosolized substances that can be tasted, although the results of this testing are limited because the tests depend on the subjective response of the HCW being tested. Quantitative fit testing of disposable negative-pressure particulate respirators can best be performed if the manufacturer provides a test respirator with a probe for this purpose. Replaceable filter, negative-pressure particulate respirators and all positive-pressure particulate respirators can be fit tested reliably, both qualitatively and quantitatively, when fitted with HEPA filters. (CDC, 1994)

The respirators must:

- Be qualitatively or quantitatively fit tested in a reliable way to obtain a face-seal leakage of ≤ 10%
- Fit the different facial sizes and characteristics of HCWs, which can usually be met by making the respirators available in at least three sizes.
- Be checked for facepiece fit in accordance with OSHA standards and good industrial hygiene practice by HCWs each time they put on their respirators

A fit check is a maneuver that an HCW performs before each use of the respiratory protective device. It can be performed according to the manufacturer's facepiece fitting instructions by using the applicable negative-pressure or positive-pressure test. Some currently available cup-shaped, disposable negative-pressure particulate respirators cannot be fit checked reliably by persons wearing the devices because occluding the entire surface of the filter is difficult.

HCWs should undergo fit testing to identify a respirator that adequately fits each individual and should receive fitting instructions that include demonstrations and practice in how the respirator should be worn and adjusted, and how to determine if it fits properly. They need to be taught to check the facepiece fit before each use.

14.28.10 Reuse

Conscientious respirator maintenance should be an integral part of an overall respirator program. This maintenance applies to respirators with replaceable filters and respirators that are classified as disposable. Manufacturers' instructions for inspecting, cleaning, and maintaining respirators should be followed to ensure that the respirator continues to function properly (NIOSH, 1987).

In healthcare settings where respirators are used for protection against biological aerosols, the concentration of infectious particles in the air is probably low; thus, the filter material in a respirator is very unlikely to become occluded with airborne material. Particles impacting on the filter material in a respirator may not be re-aerosolized easily. Respirators with replaceable filters are reusable. Infection-control personnel need to develop standard operating procedures for storing, reusing, and disposing of respirators that have been designated as disposable and for disposing of replaceable filter elements. (CDC, 1994)

14.28.11 Respiratory Protection Program

Paragraph (f)(2)(i) requires that the employer establishes and implements a written respiratory protection program that assures respirators are properly selected, fitted, used, and maintained. The program must include the following elements:

- Procedures for selecting respirators for use in the work setting
- A determination of each employee's ability to wear a respirator, as required under paragraph (g)(3)(ii), Medical Surveillance, for each employee required to wear one
- Procedures for the proper use of respirators
- Fit testing procedures for tight-fitting respirators
- Procedures and schedules for cleaning, disinfecting, storing, inspecting, repairing, or otherwise maintaining respirators
- Training of employees to assure the proper use and maintenance of the respirators as required under paragraph (h), Communication of Hazards and Training
- Procedures for periodically evaluating the effectiveness of the program

This provision ensures that the employer establishes standardized procedures for selecting, using, and maintaining respirators in the workplace. Generic guidance developed by an outside party (e.g., a respirator manufacturer) on the general use of a particular respirator is not an appropriate substitute for a respiratory protection program. (OSHA, 1997)

CDC states that all HCWs who need to use respirators for protection against infection with *M. tuberculosis* should be included in the respiratory protection program.

Visitors to TB patients should be given respirators to wear while in isolation rooms, and they should be given general instructions on how to use their respirators.

The number of HCWs included in the respiratory protection program in each facility will vary depending on the number of potentially infectious TB patients, how many rooms or areas to which patients with suspected or confirmed infectious TB are admitted, and the number of HCWs needed in these rooms or areas.

Respiratory protection programs should include enough HCWs to provide adequate care for a patient with known or suspected TB should such a patient be admitted to the facility.

Administrative measures should be used to limit the number of HCWs who need to enter these rooms or areas, thus limiting the number of HCWs who need to be included in the respiratory protection program. (CDC, 1994)

14.28.12 Responsibility

Paragraph (f)(2)(ii) requires the employer to designate a person qualified by appropriate training or experience to be responsible for the respiratory protection program and periodic evaluations of its effectiveness. OSHA is proposing that a qualified person be designated as responsible for the administration of the program. The person chosen needs to have sufficient knowledge of respiratory protection and the workplace to properly supervise the program. (OSHA, 1997)

According to CDC, supervisory responsibility for the respiratory protection program should be assigned to designated persons who have expertise in issues relevant to the program, including infectious diseases and occupational health. (CDC, 1994)

Employers are required, in paragraph (f)(2)(iii), to review and update the written program to reflect current workplace conditions and respirator use. As the workplace situation or respirator use changes, the program is to be revised. Paragraph (f)(2)(iv) requires that employers, upon request, make the written respiratory protection program available to affected employees, their designated representatives, the Assistant Secretary, and the Director. (OSHA, 1997)

According to the CDC, the program should be evaluated completely at least once a year, and both the written operating procedures and program administration should be revised as necessary based on the results of the evaluation. Elements of the program that should be evaluated include work practices and employee acceptance of respirator use. (CDC, 1994)

14.28.13 Standard Operating Procedures

Written standard operating procedures should contain information concerning all aspects of the respiratory protection program. (CDC, 1994)

14.28.14 Medical Screening

HCWs should not be assigned a task requiring use of respirators unless they are physically able to perform the task while wearing the respirator. HCWs should be screened for pertinent medical conditions at the time they are hired, then rescreened periodically. Screening begins with a general screening questionnaire for pertinent medical conditions. Results should then be used to identify HCWs who need further evaluation. (CDC, 1994)

14.28.15 Training

HCWs who wear respirators and the persons who supervise them should be informed about the necessity for wearing respirators and the potential risks associated with not doing so. This training should also include at a minimum:

- The nature, extent, and specific hazards of *M. tuberculosis* transmission in their respective health-care facility.
- A description of specific risks for TB infection among persons exposed to *M. tuberculosis* and of any subsequent treatment with INH or other chemoprophylactic agents
- The possibility of active TB disease
- A description of engineering controls and work practices and the reasons why they do not eliminate the need for personal respiratory protection.
- An explanation for selecting a particular respirator type, how the respirator is properly maintained and stored, and the operation, capabilities, and limitations of the respirator provided.
- Instruction in how the HCW wearing the respirator should inspect, don, fit check, and correctly wear the provided respirator (i.e., achieve and maintain proper face-seal fit on the HCW's face).
- An opportunity to wear the respirator and check the important parts.
- Instruction in how to recognize an inadequately functioning respirator. (CDC, 1994)

14.28.16 Respirator Inspection, Cleaning, Maintenance, and Storage

Conscientious respirator maintenance should be an integral part of an overall respirator program. Manufacturers' instructions for inspecting, cleaning, and maintaining respirators should be followed to ensure that the respirator continues to function properly. (CDC, 1994).

RESOURCES AND REFERENCES

29 CFR 1910.1035, Occupational exposure to tuberculosis: proposed rule, Federal Register, October 17, 1997.

American Conference of Governmental Industrial Hygienists, *Industrial Ventilation: A Manual of Recommended Practice*, American Conference of Governmental Hygienists, Inc., Cincinnati, 1992.

American Conference of Governmental Industrial Hygienists, Threshold limit values and biological exposure indices for 1991–1992, American Conference of Governmental Industrial Hygienist, Inc. Cincinnati, 1991.

American Institute of Architects, Committee on Architecture for Health, General hospital, in *Guidelines for Construction and Equipment of Hospital and Medical Facilities*, The American Institute of Architects Press, Washington, D.C., 1987.

American National Standards Institute, ANSI Z88.2-1980: American National Standard Practices for Respiratory Protection, American National Standards Institute, New York, 1980.

American National Standards Institute, ANSI Z88.2-1980: American National Standard Practices for Respiratory Protection, American National Standards Institute, New York, 1992.

American Society of Heating, Refrigerating and Air-Conditioning Engineers, Air flow around buildings, in *1989 Fundamentals Handbook*, American Society of Heating, Refrigerating and Air-Conditioning Engineers, Atlanta, 1989.

American Society of Heating, Refrigerating and Air-Conditioning Engineers, Health facilities, in *1991 Application Handbook*, American Society of Heating, Refrigerating and Air-Conditioning Engineers, Atlanta, 1991.

American Thoracic Society/CDC, Control of tuberculosis, *Am. Rev. Respir. Dis.*, 128, 336-342, 1983.

American Thoracic Society/CDC, Treatment of tuberculosis and tuberculosis infection in adults and children, *Am. Rev. Respir. Dis.*, 134, 355–363, 1986.

American Thoracic Society/CDC, Diagnostic standards and classification of tuberculosis, *Am. Rev. Respir. Dis.*, 142, 725–735, 1990.

American Thoracic Society/CDC, Control of tuberculosis in the United States, *Am. Rev. Respir. Dis.*, 146, 1624–1635, 1992.

American Thoracic Society/CDC, Treatment of tuberculosis and tuberculosis infection in adults and children, *Am. J. Respir. Crit. Care Med.*, 149, 1359–1374, 1994.

Beck-Sague, C., Dooley, S.W., Hutton, M.D. et al., Outbreak of multidrug-resistant mycobacterium tuberculosis infections in a hospital: Transmission to patients with HIV infection and staff, *JAMA*, 268, 1280–1286, 1992.

Burnens, A.P. and Vurma-Rapp, U. Mixed mycobacterial cultures — Occurrence in the clinical laboratory, *Int. J., Med. Microbiol.*, 27, 85–90, 1989.

Canessa, P.A., Fasano, L., Lavecchia, M.A., Torraca, A., and Schiattone, M.L., Tuberculin skin test in asymptomatic HIV seropositive carriers {Letter}, *Chest*, 96,1215–1216, 1989.

Cauthen, G.M., Dooley, S.W., Bigler, W., Burr, J., and Ihle, W., Tuberculosis (TB) transmission by HIV-associated TB cases {Abstract no. M.C.3326}, Vol. 1., VII International Conference on AIDS, Florence, Italy, June 16–21, 1991.

CDC, Guidelines for preventing the transmission of tuberculosis in health-care settings, with special focus on HIV-related issues, *MMWR*, 39 (No. RR-17), 1990.

CDC Guidelines for Preventing the Transmission of *Mycobacterium tuberculosis* in Healthcare Facilities, Federal Register, October 28, 1994.

CDC, Guidelines for Prevention of TB Transmission in Hospitals, Department of Health and Human Services, Public Health Service, DHHS publication no. (CDC) 82-8371, Atlanta, GA, 1982.

CDC, Initial therapy for tuberculosis in the era of multidrug resistance: Recommendations of the Advisory Council for the Elimination of Tuberculosis, *MMWR*, 42 (No. RR-7), 1993.

CDC, Management of persons exposed to multidrug-resistant tuberculosis, *MMWR*, 41(No. RR-11), 59–71, 1992.

CDC, Notes on air hygiene: summary of Conference on Air Disinfection, *Arch. Environ. Health*, 22, 473–474, 1971.

CDC, Prevention and control of tuberculosis in correctional institutions: Recommendations of the Advisory Committee for the Elimination of Tuberculosis, *MMWR*, 38, 313–320, 1989.

CDC, Purified protein derivative (PPD)-tuberculin anergy and HIV infection: guidelines for anergy testing and management of anergic persons at risk of tuberculosis, *MMWR,* 40 (No. RR-5), 1991.

CDC, The use of preventive therapy for tuberculous infection in the United States: recommendations of the Advisory Committee for Elimination of Tuberculosis, *MMWR,* 39 (No. RR-8), 9–12, 1990.

CDC, Use of BCG vaccines in the control of TB: A joint statement by the ACIP and the Advisory Committee for the elimination of tuberculosis, *MMWR,* 37, 663–664, 669–675, 1988.

Clerici, M. and Shearer, G.M., UV light exposure and HIV replication, *Science,* 258, 1070–1071, 1992.

Collins, F.M., Relative susceptibility of acid-fast and non-acid-fast bacteria to ultraviolet light, *Appl. Microbiol.,* 21, 411–413, 1971.

Costello, H.D. and Snider, D.E., Jr., The incidence of cancer among participants in a controlled, randomized isoniazid preventive therapy trial, *Am. J. Epidemiol.,* 111, 67–74, 1980.

Des Prez, R.M. and Heim, C.R., Mycobacterium tuberculosis, in Mandell, G.L., Douglas, R.G., Jr., and Bennett, J.E., Eds., *Principles and Practice of Infectious Diseases,* 3rd ed., Churchill Livingstone, New York, 1877–1906, 1990.

Everett, M.A., Sayre, R.M., and Olson, R.L., Physiologic response of human skin to ultraviolet light, in *The Biologic Effects of Ultraviolet Radiation,* Urbach, F., Ed., Pergamon Press, Oxford, England, 1969.

Galson, E. and Goddard, K.R., Hospital air conditioning and sepsis control, *ASHRAE,* July, 33–41, 1968.

Glassroth, J.L., White, M.C., and Snider, D.E., Jr., An assessment of the possible association of isoniazid with human cancer deaths, *Am. Rev. Respir. Dis.,* 116, 1065–1074, 1977.

Glassroth, J.L., Snider, D.E., Jr., and Comstock, G.W., Urinary tract cancer and isoniazid, *Am. Rev. Respir. Dis.,* 116, 331–333, 1977.

Gobel, M., Drug-resistant tuberculosis, *Semin. Respir. Infect.,* 1, 220–229, 1986.

Gobel, M., Iseman, M.D., Madsen, L.A., Waite, D., Ackerson, L., and Horsburgh, C.R., Jr., Treatment of 171 patients with pulmonary tuberculosis resistant to isoniazid and rifampin, *N. Engl. J. Med.,* 328, 527–532, 1993.

Hamadeh, M.A. and Glassroth, J., Tuberculosis and pregnancy, *Chest,* 101, 1114–1120, 1992.

Health Resources and Services Administration (HRSA), Guidelines for Construction and Equipment of Hospital and Medical Facilities, U.S. Department of Health and Human Services, PHS Publication No. (HRSA)84-14500, Public Health Service, Rockville, MD, 1984.

Howard, T.P. and Solomon, D.A., Reading the tuberculin skin test: Who, when, and how? *Arch. Intern. Med.,* 148, 2457–2459, 1988.

Huebner, R.E., Schein, M.F., and Bass, J.B., Jr., The tuberculin skin test, *Clin. Infect. Dis.,* 17, 968–975, 1993.

Hutton, M.D., Stead, W.W., Cauthen, G.M. et al. Nosocomial transmission of tuberculosis associated with a draining tuberculous abscess, *J. Infect. Dis.,* 161, 286–295, 1990.

Illuminating Engineering Society (IES), *IES Lighting Handbook,* 4th ed., Illuminating Engineering Society, New York, 25–27, 1966.

International Agency for Research on Cancer, IARC Monographs on the Evaluation of Carcinogenic Risks to Humans: Solar and Ultraviolet Radiation, Vol. 55, World Health Organization, International Agency for Research on Cancer, Lyon, France, 1992.

International Commission on Illumination (ICI), International Lighting Vocabulary {French}, 4th ed., CIE publication No. 17.4, Bureau Central de la Commission Electrotechnique Internationale, Geneva, Switzerland, 1987.

Iseman, M.D. and Madsen, L.A., Drug-resistant tuberculosis, *Clin. Chest Med.,* 10, 341–353, 1989.

Kethley, T.W. and Branch, K., Ultraviolet lamps for room air disinfection: Effect of sampling location and particle size of bacterial aerosol, *Arch. Environ. Health,* 25, 205–214, 1972.

Klausner, J.D., Ryder, R.W., Baende, E. et al. Mycobacterium tuberculosis in household contacts of human immunodeficiency virus type 1-seropositive patients with active pulmonary tuberculosis in Kinshasa, Zaire, *J. Infect. Dis.,* 168,106–111, 1993.

Lowry, P.L., Hesch, P.R., and Revoir, W.H., Performance of single-use respirators, *Am. Ind. Hyg. Assoc. J.,* 38, 462–467, 1977.

Lundgren, R., Norrman, E., and Asberg, I., Tuberculous infection transmitted at autopsy, *Tubercle,* 168, 147–50, 1987.

Macher, J.M., *Ultraviolet Radiation and Ventilation to Help Control Tuberculosis Transmission,* Guidelines prepared for California Indoor Air Quality Program, Air and Industrial Hygiene Laboratory, Berkeley, CA, 1989.

Macher, J.M., Alevantis, L.E., Chang, Y.-L., and Liu, K.-S., Effect of ultraviolet germicidal lamps on airborne microorganisms in an outpatient waiting room, *Appl. Occup. Environ. Hyg.,* 7, 505–513, 1992.

Manoff, S.B., Cauthen, G.M., Stoneburner, R.L., Bloch, A.B., Schultz, S., and Snider, D.E., Jr., TB patients with AIDS: Are they more likely to spread TB? {Abstract no. 4621}, Book 2, IV International Conference on AIDS, Stockholm, Sweden, June 12–16, 216, 1988.

McLean, R.L., General discussion: The mechanism of spread of Asian influenza, *Am, Rev, Respir, Dis.,* 83, 36–38, 1961.

Moulding, T.S., Redeker, A.G., and Kanel, G.C., Twenty isoniazid-associated deaths in one state, *Am. Rev. Respir. Dis.,* 140, 700–705, 1989.

Mutchler, J.E., Principles of ventilation, in The Industrial Environment — Its Evaluation and Control, U.S. Department of Health, Education, and Welfare, Public Health Service, NIOSH, Washington, D.C., 1973.

National Tuberculosis and Respiratory Disease Association, Guidelines for the general hospital in the admission and care of tuberculous patients, *Am. Rev. Respir. Dis.,* 99, 631–633,1969.

Nagy, R., Application and measurement of ultraviolet radiation, *Am. Ind. Hyg. Assoc. J.,* 25, 274–281,1964.

NIOSH, Criteria for a Recommended Standard...Occupational Exposure to Ultraviolet Radiation, U.S. Department of Health, Education, and Welfare, Public Health Service, Publication No. (HSM)73-110009, Washington, D.C., 1972.

NIOSH, Guide to Industrial Respiratory Protection: Morgantown, WV, U.S. Department of Health and Human Services, Public Health Service, CDC, DHHS publication no. (NIOSH) 87–116, 1987.

NIOSH, Hazard Evaluation and Technical Assistance Report: Onondaga County Medical Examiner's Office, Syracuse, New York, U.S. Department of Health and Human Services, Public Health Service, CDC, Cincinnati, NIOSH Report No. HETA 92-171-2255, 1992a.

NIOSH, Hazard Evaluation and Technical Assistance Report: John C. Murphy Family Health Center, Berkeley, Missouri. U.S. Department of Health and Human Services, Public Health Service, CDC, Cincinnati, NIOSH Report No. HETA 91-148-2236, 1992b.

NIOSH, Hazard Evaluation and Technical Assistance Report: San Francisco General Hospital and Medical Center, San Francisco, California, U.S. Department of Health and Human Services, Public Health Service, CDC, Cincinnati: NIOSH Report No. HETA 90-122-L2073, 1992c.

Noble, R.C., Infectiousness of pulmonary tuberculosis after starting chemotherapy: Review of the available data on an unresolved question, *Am. J. Infect. Control,* 9, 6–10, 1981.

NTRDA, National Tuberculosis and Respiratory Disease Assoc., Guidelines for the general hospital in the admission and care of tuberculous patients, *Am. Rev. Respir. Dis.,* 99, 631–633, 1969.

Opal, S.M., Asp, A.A., Cannady, P.B., Morse, P.L., Burton, L.J., and Hammer, P.G., Efficacy of infection control measures during a nosocomial outbreak of disseminated *Aspergillus* associated with hospital construction, *J. Infect. Dis.,* 153, 63–67, 1986.

OSHA Section 6(b) of the Occupational Safety and Health Act of 1970, 29 USC 655.

Perkins, J.E., Bahlke, A.M., and Silverman, H.F., Effect of ultraviolet irradiation of classrooms on spread of measles in large rural central schools, *Am. J. Public Health Nations Health,* 137, 529–537, 1947.

Pitchenik, A.E. and Rubinson, H.A., The radiographic appearance of tuberculosis in patients with the acquired immune deficiency syndrome (AIDS) and pre-AIDS, *Am. Rev. Respir. Dis.,* 131, 393–396, 1985.

Rhame, F.S., Streifel, A.J., Kersey, J.H., and McGlave, P.B., Extrinsic risk factors for pneumonia in the patient at high risk of infection, *Am. J. Med.,* 76, 42–52, 1984.

Riley, R.L,. Airborne infection, *Am. J. Med.,* 57, 466–475, 1974.

Riley, R.L., Ultraviolet air disinfection for control of respiratory contagion, in Kundsin, R.B., Ed., *Architectural Design and Indoor Microbial Pollution,* Oxford University Press, New York, 1988, 175–197.

Riley, R.L., *Principles of UV Air Disinfection,* Johns Hopkins University, School of Hygiene and Public Health, Baltimore, MD, 1991.

Riley, R.L. and Kaufman, J.E., Air disinfection in corridors by upper air irradiation with ultraviolet, *Arch. Environ. Health,* 22, 551–553, 1971.

Riley, R.L. and Kaufman, J.E., Effect of relative humidity on the inactivation of airborne *Serratia marcescens* by ultraviolet radiation, *Appl. Microbiol.,* 23,1113–1120, 1972.

Riley, R.L., Knight, M., and Middlebrook, G., Ultraviolet susceptibility of BCG and virulent tubercle bacilli, *Am. Rev. Respir. Dis.,* 113, 413–418, 1976.

Riley, R.L., Mills, C.C., O'Grady, F., Sultan, L.U., Wittstadt, F., and Shivpuri, D.N., Infectiousness of air from a tuberculosis ward, *Am. Rev. Respir. Dis.*, 85, 511–525, 1962.

Riley, R.L. and Nardell, E.A., Clearing the air: The theory and application of UV air disinfection, *Am. Rev. Respir. Dis.*, 139, 1286–1294, 1989.

Riley, R.L. and Permutt, S., Room air disinfection by ultraviolet irradiation of upper air, *Arch. Environ. Health*, 22, 208–219, 1971.

Riley, R.L., Permutt, S., and Kaufman, J.E., Convection, air mixing, and ultraviolet air disinfection in rooms, *Arch. Environ. Health*, 22, 200–207, 1971a.

Riley, R.L., Permutt, S., and Kaufman, J.E., Room air disinfection by ultraviolet irradiation of upper air: Further analysis of convective air exchange, *Arch. Environ. Health*, 23, 35–39, 1971b.

Riley, R.L., Wells, W.F., Mills, C.C., Nyka, W., and McLean, R.L., Air hygiene in tuberculosis: Quantitative studies of infectivity and control in a pilot ward, *Am. Rev. Tubercul.*, 75, 420–431, 1957.

Schieffelbein, C.W., Jr. and Snider, D.E. Jr., Tuberculosis control among homeless populations, *Arch. Intern. Med.*, 148,1843–1846, 1988.

Sherertz, R.J., Belani, A., Kramer, B.S. et al., Impact of air filtration on nosocomial *Aspergillus* infections. *Am. J. Med.*, 83, 709–718, 1987.

Simone, P.M. and Iseman, M.D., Drug-resistant tuberculosis: A deadly — and growing — danger, *J. Respir. Dis.*, 13, 960–971, 1992.

Small, P.M., Shafer, R.W., Hopewell, P.C. et al., Exogenous infection with multi-drug-resistant mycobacterium tuberculosis in patients with advanced HIV infection, *N. Engl. J. Med.*, 328,1137–1144, 1993.

Snider, D., Pregnancy and tuberculosis, *Chest*, 86(suppl.),10S–13S, 1984.

Snider, D.E., Jr., The tuberculin skin test, *Am. Rev. Respir. Dis.*, 125, 108–118, 1982.

Snider, D.E. and Farer, L.S., Package inserts for antituberculosis drugs and tuberculins, *Am. Rev. Respir. Dis.*, 131, 809–810, 1985.

Snider, D.E., Jr., Layde, P.M., Johnson, M.W., and Lyle, M.A., Treatment of tuberculosis during pregnancy, *Am. Rev. Respir. Dis.*, 122, 65–79, 1980.

Snider, D.E. Jr. and Caras, G.J., Isoniazid-associated hepatitis deaths: A review of available information, *Am. Rev. Respir. Dis.*, 145, 494–497, 1992.

Snider, D.E., Jr., Bacille Calmette-Guerin vaccinations and tuberculin skin test, *JAMA*, 253, 3438–3439, 1985.

Stein, B., Rahmsdorf, H.J., Steffen, A., Litfin, M., and Herrlich, P., UV-induced DNA damage is an intermediate step in UV-induced expression of human immuno-deficiency virus type 1, collagenase, C-Fos, and metallathionein, *Mol. Cell. Biol.*, 9, 5169–5181, 1989.

Thompson, N.J., Glassroth, J.L., Snider, D.E. Jr., and Farer, L.S., The booster phenomenon in serial tuberculin testing, *Am. Rev. Respir. Dis.*, 119, 587–597, 1979.

Valerie, K. and Rosenberg, M., Chromatin structure implicated in activation of HIV-1 gene expression by ultraviolet light, *New Biol.*, 2, 712–718, 1990.

Valerie, K., Delers, A., Bruck, C. et al., Activation of human immunodeficiency virus type 1 by DNA damage in human cells, *Nature*, 333, 78–81, 1988.

Wallace, B.M. and Lasker, J.S., Awakenings...UV light and HIV gene activation, *Science*, 257, 1211–1212, 1992.

Wallgren, A., On contagiousness of childhood tuberculosis, *Acta Pediatr. Scand.*, 22, 229–234, 1937.

Wells, W.F. and Holla, W.A., Ventilation in the flow of measles and chickenpox through a community: Progress report, January 1, 1946 to June 15, 1949 Airborne Infection Study, Westchester County Department of Health, *JAMA*, 142, 1337–1344, 1950.

Wells, W.F., Wells, M.W., and Wilder, T.S., The environmental control of epidemic contagion, I. An epidemiologic study of radiant disinfection of air in day schools, *Am. J. Hyg.*, 35, 97–121, 1942.

Willmon, T.L., Hollaender, A., and Langmuir, A.D., Studies of the control of acute respiratory diseases among naval recruits, I. A review of a four-year experience with ultraviolet irradiation and dust suppressive measures, 1943 to 1947, *Am. J. Hyg.*, 48, 227–232, 1948.

Zmudzka, B.Z. and Beer, J.Z., Activation of human immunodeficiency virus by ultraviolet radiation (yearly review), *Photochem. Photobiol.*, 52, 1153–1162, 1990.

CHAPTER 15

Security

Martha J. Boss and Dennis W. Day

CONTENTS

1-56670-606-8/03/$0.00+$1.50
© 2003 by CRC Press LLC

Our discussion about security focuses on the efforts made in the United States to ensure biosecurity. These efforts, while not perfect, form the basis for improvements to be made in security. Presentation of current governmental structures and assumed response actions provide a backbone to assist in the decisions needed to integrate these responses. Other countries may have similar methodologies; however, given that the United States responds to worldwide situations, coordination must occur with standing United States agencies and conceptual paradigms.

15.1 SECURITY AND BIOTERRORISM

The report Bioterrorism: Federal Research and Preparedness Activities (U.S. General Accounting Office (GAO)-01–915, Sept. 28, 2001) report mandated by the Public Health Improvement Act of 2000 (P.L. 106-505, sec. 102) was presented October 5, 2001. This testimony was presented before the Subcommittee on Government Efficiency, Financial Management, and Intergovernmental Relations, Committee on Government Reform, House of Representatives. The testimony described the following:

- Research and preparedness activities being undertaken by federal departments and agencies to manage the bioterrorist attack consequences
- Coordination of these activities
- Findings on the preparedness of state and local jurisdictions to respond to a bioterrorist attack

Bioterrorism is the threat or intentional release of biological agents (viruses, bacteria, or their toxins) for the purposes of influencing the conduct of government or intimidating or coercing a civilian population. Federal departments and agencies are participating in a variety of research and preparedness activities, including:

- Improving the detection of biological agents
- Developing a national pharmaceuticals stockpile to treat disaster victims

Coordination among federal departments and agencies is fragmented. Concerns are emerging about the preparedness of state and local jurisdictions, including:

- Insufficient state and local planning for response to terrorist events
- Inadequacies in the public health infrastructure

- Lack of hospital participation in training on terrorism and emergency response planning
- Insufficient capabilities for treating mass casualties
- Lack of timely availability of medical teams and resources in an emergency

15.1.1 Federal Response Plan

In an emergency that required federal disaster assistance, federal departments and agencies would respond according to responsibilities outlined in the Federal Response Plan. The Federal Response Plan, originally drafted in 1992 and updated in 1999, is authorized under the Robert T. Stafford Disaster Relief and Emergency Assistance Act (Stafford Act; P.L. 93–288, as amended). The plan outlines the planning assumptions, policies, operation concepts, organizational structures, and specific assignment of responsibilities to lead departments and agencies in providing federal assistance once the President has declared an emergency requiring federal assistance.

15.1.2 Potential Attacks

Two types of attacks were discussed in the testimony:

- Weapons of mass destruction
- Biological agents — bioterrorism

A biological attack was considered unique in that detection time from the attack to initiation of symptoms that alert to an attack may be several days. During the delay interval, infection of others not immediately associated with the initial attack event may occur. The initial and subsequent infections may also be misdiagnosed, leading to further spread and lack of effective governmental emergency response. Consequently, in order to successfully respond, the following coordinated activities are required:

- Infectious disease surveillance
- Epidemiological investigation
- Laboratory identification of biological agents
- Distribution of antibiotics to large population segments to prevent the spread of an infectious disease
- Providing emergency medical services
- Continuing healthcare services delivery
- Managing mass fatalities

15.2 CRISIS AND CONSEQUENCE MANAGEMENT

Federal programs to prepare for and respond to chemical and biological terrorist attacks operate under an umbrella of various policies and contingency plans. Federal policies on combating terrorism are laid out in a series of presidential directives and implementing guidance. Federal response to terrorist attacks may overlap and run concurrently during the emergency response and are dependent upon the nature of the incident.

15.2.1 Crisis Management

Efforts to stop a terrorist attack, arrest terrorists, and gather evidence for criminal prosecution are led by the Department of Justice, through the Federal Bureau of Investigation. All federal agencies and departments, as needed, would support the Department of Justice and the Federal Bureau of Investigation on-scene commander.

15.2.2 Consequence Management

Efforts to provide medical treatment and emergency services, evacuate people from dangerous areas, and restore government services are led by the Federal Emergency Management Agency in support of state and local authorities. (Note: The federal government does not have primary responsibility for consequence management; state and local authorities do.)

15.3 CONTINGENCY PLANS

In a chemical or biological terrorist incident, the federal government would operate under one or more contingency plans. The U.S. Government Interagency Domestic Terrorism Concept of Operations Plan establishes conceptual guidelines for:

- Assessing and monitoring a developing threat
- Notifying appropriate agencies concerning the nature of the threat
- Deploying necessary advisory and technical resources to assist the lead federal agency in facilitating interdepartmental coordination of crisis and consequence management activities

In the event that the President declares a national emergency, the Federal Emergency Management Agency would coordinate the federal response using a generic disaster contingency plan called the Federal Response Plan. The Federal Response Plan is authorized by the Robert T. Stafford Disaster Relief and Emergency Assistance Act (P.L. 93–288, as amended). It provides a broad framework for coordinating the delivery of federal disaster assistance to state and local governments when an emergency overwhelms their ability to respond effectively and designates primary and supporting federal agencies for a variety of emergency support operations. The Plan includes:

- Federal agency roles in consequence management during terrorist attacks
- Planning assumptions
- Policies
- Operation concepts
- Organizational structures
- Specific responsibility assignment of responsibilities to lead departments and agencies in providing federal assistance
- Categories of assistance types of specific emergency support functions
- Mass care
- Health and medical services

Several individual agencies have their own contingency plans or guidance specific to their activities.

15.4 COORDINATION

In May 1998, the President established a National Coordinator within the National Security Council to better lead and coordinate these federal programs. The position's functions were never detailed in either an executive order or legislation. Many of the overall leadership and coordination functions that the GAO has identified as critical were not given to the National Coordinator. Several agencies performed interagency functions that the GAO believed would have been performed more appropriately above the level of individual agencies. The interagency roles of these various agencies were not always clear and sometimes overlapped, which led to a fragmented approach. The Department of Justice, the National Security Council, the Federal Bureau of Investigation, and the

Federal Emergency Management Agency all had been developing or planning to develop potentially duplicative national strategies to combat terrorism.

The President announced the creation of an Office of Homeland Security on September 20, 2001, and specified its functions in Executive Order 13228 on October 8, 2001. These actions represent potentially significant steps toward improved coordination of federal activities and are generally consistent with recent recommendations. Some questions that remain to be addressed include:

- How will this new office be structured?
- What authority will the director have?
- How can this effort be institutionalized and sustained over time?

15.5 EVENT PROBABILITY

The Federal Bureau of Investigation had identified the largest domestic threat to be the lone-wolf terrorist, an individual who operates alone. U.S. intelligence agencies have reported an increased possibility that terrorists would use chemical or biological weapons in the next decade; however, terrorists would have to overcome significant technical and operational challenges to successfully produce and release chemical or biological agents of sufficient quality and quantity to kill or injure large numbers of people without substantial assistance from a foreign government sponsor.

In most cases, specialized knowledge is required in the manufacturing process and in improvising an effective delivery device for most chemical and nearly all biological agents that could be used in terrorist attacks. Some of the required components of chemical agents and highly infective strains of biological agents are difficult to obtain. Terrorists may have to overcome other obstacles to successfully launch an attack that would result in mass casualties, such as unfavorable meteorological conditions and personal safety risks. The term *weapon of mass destruction* (WMD) generally refers to chemical, biological, radiological, or nuclear agents or weapons. As clearly shown on September 11, a terrorist attack would not have to fit that definition to result in:

- Mass casualties
- Critical infrastructures destruction
- Economic losses
- Disruption of daily life nationwide

The attack increased the uncertainties regarding the threat, given that the attacks:

- Were conducted by a large group of conspirators rather than one individual
- Constituted long-planned coordinated efforts, showing a level of sophistication that may not have been anticipated by the Federal Bureau of Investigation
- Were implemented by individuals willing to commit suicide in the attacks, showing no concern for their own personal safety, which was considered one of the barriers to using chemical or biological agents

15.6 FEDERAL RESPONSE OF THE UNITED STATES

The preparedness efforts of federal departments and agencies have included:

- Increasing federal, state, and local response capabilities
- Developing response teams of medical professionals

- Increasing availability of medical treatments
- Participating in and sponsoring terrorism response exercises
- Planning to aid victims and providing support during special events such as presidential inaugurations, major political party conventions, and the Superbowl.

(*Note:* Presidential Decision Directive 62, issued May 22, 1998, created a category of special events called National Security Special Events, which are events of such significance that they warrant greater federal planning and protection than other special events.)

15.6.1 State and Local Outreach

Several federal departments and agencies, such as the Federal Emergency Management Agency (FEMA) and the Centers for Disease Control and Prevention (CDC), have programs to increase the ability of state and local authorities to successfully respond to an emergency, including a bioterrorist attack. These departments and agencies contribute to state and local jurisdictions by:

- Paying for equipment
- Developing emergency response plan elements
- Providing technical assistance
- Increasing communications capabilities
- Conducting training courses

15.6.2 Federal Response Teams and Exercises

Some federal departments and agencies have developed teams to directly respond to terrorist events and other emergencies. Federally initiated bioterrorism response exercises have been conducted across the country.

15.6.3 Special Events

Special events include presidential inaugurations, major political party conventions, and the Superbowl. Federal departments and agencies also provide support at special events to improve response in case of an emergency. Besides improving emergency response at the events, participation by departments and agencies gives them valuable experience working together to develop and practice plans to combat terrorism.

15.6.4 Bioterrorism Identification and Medical Response

Federal departments and agencies have also been increasing their own capacity to identify and deal with a bioterrorist incident. For example, the CDC, U.S. Department of Agriculture (USDA), and Food and Drug Administration (FDA) are:

- Improving surveillance methods for detecting disease outbreaks in humans and animals
- Establishing laboratory response networks to maintain state-of-the-art capabilities for biological agent identification and characterization of human clinical samples

Several agencies are involved in increasing the availability of medical supplies that could be used in an emergency, including a bioterrorist attack. The CDC's National Pharmaceutical Stockpile contains pharmaceuticals, antidotes, and medical supplies that can be delivered anywhere in the United States within 12 hours of the decision to deploy. The stockpile was deployed for the first time on September 11, 2001, in response to the terrorist attacks on New York City and Washington, D.C.

The GAO identified over 20 departments and agencies as having a role in preparing for or responding to the public health and medical consequences of a bioterrorist attack. The potential redundancy of these federal efforts highlights the need for scrutiny. In the GAO report on combating terrorism issued on September 20, 2001, the GAO recommended that the President, working closely with the Congress, consolidate some of the activities of DOJ's Office of Justice Programs (OJP) under the Federal Emergency Management Agency (FEMA).

15.7 U.S. DEPARTMENT OF AGRICULTURE

15.7.1 Agricultural Research Service

The Agricultural Research Service (ARS) is

- The principal in-house research agency of the USDA
- One of the four component agencies of the Research, Education, and Economics (REE) mission area

Congress first authorized federally supported agricultural research in the Organic Act of 1862, which established what is now the USDA. That statute directed the Commissioner of Agriculture to acquire and preserve in his department all information he could obtain by means of books and correspondence and by practical and scientific experiments. The USDA's agricultural research programs scope has been expanded and extended many times since the department was first created. ARS has about 1200 research projects working at over 100 locations across the country and at 4 overseas laboratories. The National Agricultural Library and the National Arboretum are also part of Animal and Plant Health Inspection Service (APHIS), the mission of which is to protect America's animal and plant resources by:

- Safeguarding resources from exotic invasive pests and diseases
- Monitoring and managing agricultural pests and diseases existing in the United States
- Resolving and managing trade issues related to animal or plant health
- Ensuring the humane care and treatment of animals

The APHIS mission is an integral part of the USDA's efforts to provide the nation with safe and affordable food. Without APHIS protecting America's animal and plant resources from agricultural pests and diseases, threats to our food supply would be quite significant.

Congress has passed several laws that give APHIS the authority to implement its protection mission. APHIS employees are organized into five main operational divisions: Animal Care, International Services, Plant Protection and Quarantine, Veterinary Services, and Wildlife Services. The principal legislative authorities of APHIS include the Organic Act of 1944, the Plant Protection Act (as contained in the Agricultural Risk Protection Act of 2000), Sections 12–14 of the Federal Meat Inspection Act, the Bureau of Animal Industry Act of 1884, the Tariff Act of 1930, the Animal Damage Control Act of 1931, the Animal Welfare Act of 1966, the Horse Protection Act of 1970, and the Virus–Serum–Toxin Act of 1913. Several laws authorize the collection of user fees for agricultural quarantine inspection and other APHIS services.

15.7.2 Food Safety Inspection Service

The Food Safety Inspection Service (FSIS) inspects meat, poultry, and egg products and conducts strategic planning.

15.7.3 Office of Crisis Planning and Management

The Office of Crisis Planning and Management (OCPM), under the Assistant Secretary for Administration, Office of the Secretary, serves as USDA's focal point for coordinating national security, natural disaster, other emergencies, and agriculture-related international civil emergency planning and related activities, and acts as the primary USDA representative for antiterrorism activities. Duties include:

- Coordinating with agencies and offices within the USDA to identify USDA intelligence requirements and convey them to the intelligence community
- Organizing and distributing specialized intelligence reports to individual agencies and offices within the USDA
- Providing staff support for the USDA Counter-Terrorism Policy Council and the Director of OCPM who chairs the USDA Biosecurity Committee
- Serving as primary contact with the Federal Emergency Management Agency (FEMA) and all other federal departments and agencies having emergency responsibilities
- Establishing, maintaining, and managing emergency management policies and programs for the department to ensure that an emergency structure is in place to respond swiftly to a disaster or other crisis situation (the emergency structure is required to assess the impact of the disaster on food production, processing, and food distribution and to ensure that assistance programs are operating in the affected area)
- Coordinating USDA participation in disaster-related exercises and conducting training sessions for USDA State Emergency Boards
- Facilitating coordination of USDA agencies within the department and with other federal departments and organizations on matters concerning crisis planning and management
- Managing the USDA's critical situations emergency response effort, including its Continuity of Operations and Continuity of Government plans under Presidential Decision Directives (PDDs) as well as the USDA Emergency Coordination Center management

15.8 DEPARTMENT OF COMMERCE

15.8.1 National Institute of Standards and Technology

The National Institute of Standards and Technology (NIST) is a nonregulatory federal agency within the U.S. Commerce Department Technology Administration. The mission of NIST is to develop and promote measurements, standards, and technology to enhance productivity, facilitate trade, and improve the quality of life. Duties include:

- Conducting projects that support law enforcement, military operations, emergency services, airport and building security, cyber security, and efforts to develop new types of security technologies
- Developing cutting-edge science and technology infrastructure necessary to strengthen and safeguard America's economic foundations and security capabilities
- Conducting research that advances the nation's technology infrastructure and is needed by U.S. industry to continually improve products and services

The Advanced Technology Program accelerates the development of innovative technologies for broad national benefit by co-funding research and development partnerships with the private sector.

15.8.2 National Oceanic and Atmospheric Administration

The National Oceanic and Atmospheric Administration (NOAA) improves technology and provides available backups in the event of power outages or security for technology operation centers.

15.8.3 National Telecommunications and Information Administration

The National Telecommunications and Information Administration (NTIA) provides spectrum management and telecommunications research capabilities and tests new technology applications for radiofrequency spectrum use.

15.8.4 U.S. Patent and Trademark Office

The U.S. Patent and Trademark Office (PTO) improves patent and trademark quality and reduces dependency.

15.8.5 Bureau of Export Administration

The Bureau of Export Administration (BXA) coordinates the Department of Commerce's overall critical infrastructure protection and homeland security efforts.

15.8.6 Critical Infrastructure Assurance Office

The Critical Infrastructure Assurance Office (CIAO) promotes federal initiatives and public/private partnerships across industry sectors to protect the nation's critical infrastructures and creates within CIAO the Homeland Security Information Technology and Evaluation Program to promote the coordinated information technology for homeland security purposes.

15.9 DEPARTMENT OF ENERGY

The Department of Energy (DOE) develops technologies for detecting and responding to a bioterrorist attack and models of the spread of and exposure to a biological agent after release.

15.10 DEPARTMENT OF DEFENSE

According to the GAO, neither the Department of Defense (DOD) nor the military services has systematically examined the current medical personnel distribution across specialties with respect to adequacy for chemical and biological defense. In general, the DOD has not successfully adapted its conventional medical planning to chemical and biological warfare.

In medical planning, the DOD has used software, evaluations, and review processes that address conventional threats but have not fully incorporated chemical and biological threats. Medical planners have lacked the information on casualty rates or qualified medical personnel required to address the appropriateness of the current medical personnel distribution across specialties. Joint protocols for treating chemical and biological casualties have recently been completed. However, agreement has not been reached as to the appropriate medical personnel to provide treatment for different casualties caused by chemical or biological agents. DOD officials attribute the lack of systematic efforts to:

- Failure to establish chemical and biological readiness as a medical priority in defense planning guidance (particularly for biological warfare)
- Complex assumptions required to predict casualties
- Poor data availability on affects of particular agents
- Disagreements among the military services about how quickly troops could actually be evacuated
- Pessimism that medical personnel could effectively treat substantial numbers of chemical and biological casualties

Training, as well as testing and exercises, for chemical and biological casualties medical management remains limited. Military services officials for medical planning maintain that specialized training in the military is the appropriate way to address any need for additional medical skills. Courses are essentially voluntary. (*Note:* The majority of uniformed medical personnel have not completed any specialized military medical training for chemical and biological casualties Only the Army includes an introduction to chemical and biological casualty management in training required of medical personnel.) Medical personnel who have been trained may not be readily identified in the event of an emergency as tracking systems do not exist or are not currently functioning.

Except for the Army's Medic 2000 study (which found that the lowest proficiency scores among medics were for nuclear, biological, and chemical skills), the military services have not defined standards for treatment of chemical and biological casualties or tested the medical personnel proficiency, and no realistic field exercise of chemical or biological defense has been conducted. The surgeons general from the military services have begun integrating chemical and a few biological scenarios into their medical exercises. Only two joint military exercises planned since 1993 have included both medical support activities and chemical or biological warfare. Key evaluations used to advise the President on readiness to implement the national security strategy have never set a scenario for the unified commanders requiring medical personnel to respond to the effects of weapons of mass destruction (WMD). Exercises involving medical support for chemical and biological casualties have been rare because of conflicting priorities encountered by both war fighters and medical personnel and because of difficulty and expense.

1. The military services and joint staff should reach an agreement about which medical personnel are qualified to provide specific treatments. These medical personnel should be validated by proficiency testing of the identified personnel to help further refine requirements for training and distribution of medical personnel across specialties.
2. The military services should develop medical training requirements for chemical and biological contingencies, assess the training effectiveness with rigorous proficiency standards and tests, and track individual training and proficiency.
3. The joint staff, commanders-in-chief, and the military services should increase chemical and biological exercises involving medical personnel to an extent commensurate with current chemical and biological threat assessments. Given the threat of mass casualties, exercises should explore the extent of medical capabilities and the full consequences of scenarios that overwhelm them.

15.10.1 Defense Advanced Research Projects Agency (DARPA)

The Defense Advanced Research Projects Agency (DARPA) develops imaginative, innovative, and often high-risk research ideas offering a significant technological impact that will go well beyond the normal evolutionary developmental approaches. It pursues these ideas from the demonstration of technical feasibility through the development of prototype systems.

15.10.2 Joint Task Force for Civil Support

The Joint Task Force for Civil Support plans and, when directed, commands and controls the DOD's WMD and high-yield explosive consequence management capabilities in support of FEMA.

15.10.3 National Guard

The National Guard manages response teams that would enter a contaminated area to gather samples for on-site evaluation.

15.10.4   U.S. Army

The U.S. Army maintains a repository of information about chemical and biological weapons and agents, detectors, and protection and decontamination equipment.

15.11 DEPARTMENT OF HEALTH AND HUMAN SERVICES

The Department of Health and Human Services (HHS) coordinates federal assistance in response to public health and medical care needs in an emergency. The HHS could receive support from other agencies and organizations, such as DOD, USDA, and FEMA, to assist state and local jurisdictions.

15.11.1   Agency for Healthcare Research and Quality

The Agency for Healthcare Research and Quality examines clinical training and the ability of frontline medical staff to detect and respond to a bioterrorist threat. It studies the use of information systems and decision support systems to enhance preparedness for medical care in the event of a bioterrorist event.

15.11.2   Centers for Disease Control and Prevention

The Centers for Disease Control and Prevention (CDC), under the Federal Response Plan, is the lead HHS agency providing assistance to state and local governments for these functions:

1. *Health surveillance* — Assist in establishing surveillance systems to monitor the general population and special high-risk population segments; carry out field studies and investigations; monitor injury and disease patterns and potential disease outbreaks; and provide technical assistance and consultations on disease and injury prevention and precautions.
2. *Worker health and safety* — Assist in monitoring health and well-being of emergency workers; perform field investigations and studies; and provide technical assistance and consultation on worker health and safety measures and precautions.
3. *Radiological, chemical, and biological hazard consultation* — Assess health and medical effects of radiological, chemical, and biological exposures on the general population and on high-risk population groups; conduct field investigations, including collection and analysis of relevant samples; advise on protective actions related to direct human and animal exposure and on indirect exposure through radiologically, chemically, or biologically contaminated food, drugs, water supply, and other media; and provide technical assistance and consultation on medical treatment and decontamination of radiologically, chemically, or biologically injured or contaminated victims.
4. *Public health and disease and injury prevention information* — Transmit information to members of the general public who are located in or near areas affected by a major disaster or emergency; assess the threat of vector-borne diseases following a major disaster or emergency; conduct field investigations, including the collection and laboratory analysis of relevant samples; provide vector control equipment and supplies; provide technical assistance and consultation on protective actions regarding vector-borne diseases; and provide technical assistance and consultation on medical treatment of victims of vector-borne diseases.

In its FY2002–FY2006 Plan for Combating Bioterrorism, the HHS notes that potential sources for data on morbidity trends include:

- 911 emergency calls
- Reasons for emergency department visits
- Hospital bed usage
- Purchase of specific products at pharmacies

The HHS is currently leading an effort to work with governmental and nongovernmental partners to upgrade the nation's public health infrastructure and capacities to respond to bioterrorism. As part of this effort, several CDC centers, institutes, and offices work together in the agency's Bioterrorism Preparedness and Response Program. The principal priority of the CDC's program is to upgrade infrastructure and capacity to respond to a large-scale epidemic, regardless of whether it is the result of a bioterrorist attack or a naturally occurring infectious disease outbreak. The program was started in fiscal year 1999 and was tasked with building and enhancing national, state, and local capacity; developing a national pharmaceutical stockpile; and conducting several independent studies on bioterrorism.

The CDC's counter-bioterrorism activities are focused on building and expanding the public health infrastructure at the federal, state, and local levels. In addition to preparing for a bioterrorist attack, these activities also prepare the agency to respond to other challenges, such as identifying and containing a naturally occurring emerging infectious disease.

The CDC provides grants, technical support, and performance standards to support bioterrorism preparedness and response planning at the state and local levels. It has worked with the Department of Justice to complete a public health assessment tool, which is being used to determine the ability of state and local public health agencies to respond to release of biological and chemical agents, as well as other public health emergencies. Ten states (Florida, Hawaii, Maine, Michigan, Minnesota, Pennsylvania, Rhode Island, South Carolina, Utah, and Wisconsin) have completed the assessment and others are currently completing it.

The CDC provides state and local grants, technical support, and performance standards to support bioterrorism preparedness through:

- Increasing staff, thus enhancing the capacity to detect the release of a biological agent or an emerging infectious disease
- Improving the communications infrastructure
- Preparing bioterrorism response plans

Research activities focus on detection, treatment, vaccination, and emergency response equipment. Rapid identification and confirmatory diagnosis of biological agents are critical to ensuring that prevention and treatment measures can be implemented quickly. A Laboratory Response Network of federal, state, and local laboratories maintains state-of-the-art capabilities for biological agent identification and characterization of human clinical samples such as blood. Technical assistance and training in identification techniques are provided to state and local public health laboratories. The CDC is upgrading its epidemiological and disease surveillance system to provide increased surveillance and epidemiological capacities before, during, and after special events. Besides improving emergency response at these special events, the agency gains valuable experience in developing and practicing plans to combat terrorism.

The CDC monitors unusual clusters of illnesses, such as influenza in June; although unusual clusters are not always a cause for concern, they can indicate a potential problem. The CDC also provides increased surveillance of disease outbreaks in animals and improved surveillance methods for detecting disease outbreaks. Communication capabilities are increased in order to improve the gathering and exchanging of information related to bioterrorist incidents. According to the HHS, the epidemiological capacity at CDC needs to be improved. A standard system of disease reporting would better enable CDC to:

- Monitor disease
- Track trends
- Intervene at the earliest sign of unusual or unexplained illness

15.11.3 Rapid Response and Advance Technology Laboratory

The CDC operates a Rapid Response and Advance Technology Laboratory, which screens samples for the presence of suspicious biological agents and evaluates new technology and protocols for the detection of biological agents. These technology assessments and protocols, as well as reagents and reference samples, are being shared with state and local public health laboratories. Among the duties are:

- Developing and validating new diagnostic tests
- Creating agent-specific detection protocols
- Developing equipment performance standards
- Conducting research on smallpox and anthrax viruses and therapeutics

15.11.4 Laboratory Response Network for Bioterrorism

Research can provide testing of biological samples for detection and confirmation of biological agents. Hospital and commercial laboratories that have state-of-the-art equipment and well-trained staff need to be added to this network to provide additional surge capacity. Currently, 104 laboratories are in the network. The Laboratory Response Network (LRN) is building a multilevel network of local, state and federal partners to prepare and respond to an act of terrorism. Following are the designations for these partners:

- Level A — Community hospitals with diagnostic facilities equipped with reagents and assays screen samples collected by the FBI or local law enforcement agencies.
- Level B — State health departments confirm sample credibility; a credible sample shifts to a Level C center.
- Level C — This level is designed for specific agents, such as plague, anthrax, or smallpox; from here, an identified sample moves to Level D labs, but if the FBI detects a credible threat (e.g., smallpox) the sample bypasses the first three tiers and shoots directly to Level D.
- Level D — Much like forensic labs, this level of labs documents and files records in the event of a court subpoena. In one such facility, the SAS Air Sampler will be used to ensure that no aerosolized cross-contamination exists between laboratory areas. The sampling plan design intent is to ensure safety procedure work for quality control monitoring: (1) Begin with a clean laboratory with disinfected countertops, floors, and hoods; and (2) sample the air and repeat the disinfecting cycle until the site is clean and ready for the empirical test. This protocol is designed to affirm whether the safety procedures work.

The CDC is developing a crisis communications/media response curriculum for bioterrorism, as well as core capabilities guidelines to assist states and localities in their efforts to build comprehensive anti-bioterrorism programs.

15.11.5 Health Alert Network

The CDC is currently developing the Health Alert Network, which will support the key information exchange over the Internet and provide a means to conduct distance training that could potentially reach a large segment of the public health community. Currently, 13 states are connected to all of their local jurisdictions. The CDC is directly connected to groups such as the American Medical Association to reach healthcare providers.

The CDC has described the Health Alert Network as a highway on which programs such as the National Electronic Disease Surveillance System (NEDSS) and the Epidemic Information Exchange (Epi-X) will run. NEDSS is designed to facilitate the development of an integrated, coherent national system for public health surveillance. Ultimately, it is meant to support the automated collection, transmission, and monitoring of disease data from multiple sources (for example, clinician's offices and laboratories) from local to state health departments to the CDC. Epi-X is a secure, Web-based exchange for public health officials to rapidly report and discuss disease outbreaks and other health events potentially related to bioterrorism as they are identified and investigated.

In collaboration with the Association of Public Health Laboratories and the Department of Defense, the CDC has started a secure Web-based network that allows state, local, and other public health laboratories access to guidelines for analyzing biological agents. Authenticated users can order critical reagents necessary for performing laboratory sample analysis. The network provides emergency contact information for state and local officials in the event of possible bioterrorism incidents and lists critical biological and chemical agents. It provides summaries of state and local bioterrorism projects and contains general information about the CDC's bioterrorism initiative. The network also provides links to documents on bioterrorism preparedness and response. Note that one aspect of this work is developing, testing, and implementing standards that will permit surveillance data from different systems to be easily shared. During the West Nile virus outbreak, while a secure electronic communication network was in place at the time of the initial outbreak, not all involved agencies and officials were capable of using this system at the same time. Because the CDC's laboratory was not linked to the New York State network, the New York State Department of Health had to act as an intermediary in sharing the CDC's laboratory test results with local health departments. The CDC and New York State Department of Health laboratory databases were not linked to the database in New York City, and laboratory results consequently had to be manually entered there. These problems slowed the investigation of the outbreak.

15.11.6 National Bioterrorism Response Training Plan

The CDC is implementing the National Bioterrorism Response Training Plan. This plan focuses on preparing CDC officials to respond to bioterrorism and includes the development of exercises to assess progress in achieving bioterrorism preparedness at the federal, state, and local levels.

15.11.7 Food and Drug Administration

Duties of the Food and Drug Administration (FDA) include:

- Improving capabilities to identify and characterize foodborne pathogens
- Identifying biological agents using animal studies and microbiological surveillance
- Licensing of vaccines for anthrax and smallpox
- Determining procedures for allowing use of not-yet-approved drugs and specifying data needed for approval and labeling

15.11.8 National Institutes of Health

The National Institutes of Health (NIH) develops new therapies for smallpox virus and smallpox and bacterial antigen detection systems.

15.11.9 National Institute for Occupational Safety and Health

The National Institute for Occupational Safety and Health (NIOSH) develops standards for respiratory protection equipment used against biological agents by firefighters, laboratory technicians, and other potentially affected workers.

15.11.10   Office of Emergency Preparedness

Duties of the Office of Emergency Preparedness (OEP) include:

- Developing and managing response teams that can provide support at the disaster site
- Overseeing a study on response systems
- Providing contracts to increase local emergency response capabilities, which involves entering into contracts to enhance medical response capability; the program includes a focus on response to bioterrorism, including early recognition, mass postexposure treatment and mass casualty care, and mass fatality management for local jurisdictions (fire, police, and emergency medical services; hospitals; public health agencies; and other services)
- Creating disaster medical assistance teams to provide medical treatment and assistance in the event of an emergency; four of these teams (known as National Medical Response Teams) are specially trained and equipped to provide medical care to victims of WMD events, such as bioterrorist attacks

15.11.11   National Center for Environmental Health

The National Center for Environmental Health (NCEH) helps local, state, federal, and international agencies plan their responses to emergency situations. It responds to requests for emergency and recovery assistance after technologic disasters and established and now maintains the national pharmaceutical stockpile, which is designed to ensure the rapid deployment of life-saving pharmaceuticals for treating victims of terrorist attacks. It also provides technical support for public health activities during international emergencies, including civil strife, disasters, and famine.

The NCEH provides a number of environmental health services that help other agencies, environmental health programs, and professionals better anticipate, identify, and respond to environmental problems and their consequences on human health. NCEH's services include helping to protect the public's health within U.S. national parks and on international cruise vessels that enter U.S. ports, ensuring the health of the public and workers during disposal of chemical weapons and providing information and consultation on a wide range of environmental health issues.

15.12 DEPARTMENT OF JUSTICE (DOJ)

15.12.1   Federal Bureau of Investigation

The Federal Bureau of Investigation (FBI) conducts work on detection and characterization of biological materials.

15.12.2   Office of Justice Programs

The Office of Justice Programs (OJP) helps prepare state and local emergency responders by:

- Providing training, exercises, technical assistance, and equipment programs
- Assisting states in developing strategic plans, including funding for training, equipment acquisition, technical assistance, exercise planning, and execution to enhance state and local capabilities (for fire, law enforcement, emergency medical, and hazardous materials response services; hospitals; public health departments; and other services) to respond to terrorist incidents
- Developing a data collection tool to assist states in conducting their threat, risk, and needs assessments and in developing their preparedness strategy for terrorism, including bioterrorism
- Developing a biological agent detector

15.13 DEPARTMENT OF TRANSPORTATION

The Department of Transportation (DOT) ensures that various modes of transportation operate safely on an individual basis and together as an interlinked transportation system.

15.13.1 Civil Aviation Security

Civil Aviation Security protects the commercial air transportation users against terrorist and other criminal acts.

15.13.2 Federal Aviation Administration Office of System Safety

The Federal Aviation Administration (FAA) develops and implements improved tools and processes; facilitates more effective use of safety data, both inside and outside the agency; and helps improve aviation safety.

15.13.3 Federal Transit Administration Safety and Security Office

The Federal Transit Administration Safety and Security Office is concerned with matters relating to the safety and security of our nation's mass transit systems.

15.13.4 Hazardous Materials Safety

Hazardous Materials Safety coordinates a national safety program for the transportation of hazardous materials by air, rail, highway, and water.

15.13.5 National Response Center

The National Response Center is the sole federal point of contact for reporting oil and chemical spills.

15.13.6 Research and Special Programs Administration (RSPA) Office of Pipeline Safety

The RSPA Office of Pipeline Safety administers the DOT national regulatory programs to ensure the safe transportation of natural gas, petroleum, and other hazardous materials by pipeline.

15.13.7 U.S. Coast Guard Marine Safety Center

The U.S. Coast Guard Marine Safety Center works directly with the marine industry, the Commandant, and Coast Guard field units in the evaluation and approval of commercial vessel and systems designs, development of safety standards and policies, response to maritime casualties, and oversight of delegated third parties in support of the Coast Guard's marine safety and environmental protection program. The Coast Guard's homeland security role includes:

- Protecting ports, the flow of commerce, and the marine transportation system from terrorism
- Maintaining maritime border security against illegal drugs, illegal aliens, firearms, and WMD
- Ensuring rapid deployment and resupply of our military assets, both by keeping Coast Guard units at a high state of readiness and by keeping marine transportation open for the transit of assets and personnel from other branches of the armed forces
- Protecting against illegal fishing and indiscriminate destruction of living marine resources

- Prevention of and response to oil and hazardous material spills, both accidental and intentional
- Coordinating efforts and intelligence with federal, state, and local agencies

15.14 DEPARTMENT OF THE TREASURY

15.14.1 U.S. Secret Service

The U.S. Secret Service (USSS) develops biological agent detectors.

15.15 ENVIRONMENTAL PROTECTION AGENCY

Duties of the Environmental Protection Agency (EPA) include:

- Improving detection of biological agents
- Providing technical assistance in identifying and decontaminating biological agents
- Conducting assessments of water supply vulnerability to terrorism, including contamination with biological agents

15.16 FEDERAL EMERGENCY MANAGEMENT AGENCY

The Federal Emergency Management Agency (FEMA) provides grant assistance and guidance to support state and local consequence management planning, training, and exercises for all types of terrorism, including bioterrorism. FEMA works with state emergency management agencies and maintains databases of safety precautions for biological, chemical, and nuclear agents.

15.16.1 Office of National Preparedness: Consequence Management of WMD Attack

In May 2001, the President asked the director of FEMA to establish an office to coordinate listed agency activities that address consequence management resulting from the use of WMD.

15.17 OFFICE OF MANAGEMENT AND BUDGET OVERSIGHT OF TERRORISM FUNDING

The Office of Management and Budget (OMB) established a reporting system on the budgeting and expenditure of funds to combat terrorism, with goals to reduce overlap and improve coordination as part of the annual budget cycle.

15.18 DEPARTMENT OF VETERANS AFFAIRS

The 1982 Veterans Affairs (VA)/Department of Defense (DOD) Health Resources Sharing and Emergency Operations Act (P.L. 97–174) authorized the VA to:

- Ensure hospital backup to DOD in war or other emergencies
- Support communities following domestic terrorist incidents and other major disasters

Concern about the lack of a medical response plan for civilians led to a 1984 administrative establishment of a national medical system that would back up DOD and handle domestic disasters.

The role of the VA as part of the federal government's response for disasters has grown with the reduction of medical capacity in the Public Health Service and military medical facilities. The VA established an Emergency Management Strategic Healthcare Group with responsibility for the following six emergency response functions:

1. *Ensuring the continuity of VA medical facility operations* — Prior to emergency conditions, VA emergency management staff are responsible for minimizing disruption in the treatment of veterans by (a) developing, managing, and reviewing plans for disasters and evacuations; and (b) coordinating mutual aid agreements for patient transfers among VA facilities. During emergency conditions, these staff are responsible for ensuring that these plans are carried out as intended.

2. *Backing up DOD's medical resources following an outbreak of war or other emergencies involving military personnel* — In 2001, the VA had plans in place for the allocation of up to 5500 of its staffed operating beds for DOD casualties within 72 hours of notification. In total, 66 VA medical centers are designated as primary receiving centers for treating DOD patients. In turn, these centers must execute plans for early release or movement of VA patients to 65 other VA medical centers designated as secondary support centers.

3. *Jointly administering the National Disaster Medical System (NDMS)* — In 1984, the VA, DOD, FEMA, and HHS created a federal partnership to administer and oversee NDMS, which is a joint effort between the federal and private sectors to provide backup to civilian health care if disaster events produce mass casualties. The system divides the country into 72 areas selected for their concentration of hospitals and proximity to airports. Nationwide, more than 2000 civilian and federal hospitals participate in the system. One of the VA's roles in NDMS is to help coordinate VA hospital capacity with the nonfederal hospitals participating in the system.

4. *Carrying out Federal Response Plan efforts to assist state and local governments in coping with disasters* — Under FEMA's leadership, the VA and other agencies are responsible for carrying out the Federal Response Plan. The VA is one of several federal agencies sharing responsibility for providing public works and engineering services, mass care and sheltering, resource support, and health and medical services. The VA is involved with other agencies in positioning medical resources at high-visibility public events requiring enhanced security, such as national political conventions. The VA maintains a database of deployable VA medical personnel that is intended to help the agency to quickly locate medical personnel (such as nurses, physicians, and pharmacists) for deployment to a disaster site.

5. *Carrying out Federal Radiological Emergency Response Plan efforts to respond to nuclear hazards* — Depending on the type of emergency involved, the VA is responsible for supporting the designated lead federal agency in responding to accidents at nuclear power stations or to terrorist acts intended to spread radioactivity in the environment. The VA has its own medical emergency radiological response team of physicians and other health specialists. When requested by the lead agency, the VA's response team is expected to be ready to deploy to an incident site within 12 to 24 hours to provide technical advice, radiological monitoring, decontamination expertise, and medical care as a supplement to local authorities' efforts.

6. *Supporting efforts to ensure the continuity of government during national emergencies* — The VA maintains the agency's relocation site and necessary communication facilities to continue functioning during a major national emergency.

15.18.1 Medical Supplies

Under a memorandum of agreement between the VA and the HHS Office of Emergency Preparedness, the VA maintains at designated locations medical stockpiles containing antidotes, antibiotics, and medical supplies and smaller stockpiles containing antidotes, which can be loaned to local governments or predeployed for special events, such as the Olympic Games. The VA would play a key support role in the nation's stockpiling of pharmaceuticals and medical supplies in the event of large-scale disaster events caused by WMD. Under contract with the CDC, the VA purchases drugs and medical supplies for the National Pharmaceutical Stockpile because of VA's purchasing power and ability to negotiate large discounts. It also manages a spectrum of contracts for the storage, rotation, security, and transportation of stockpiled items.

The VA maintains stockpiles of pharmaceuticals for another HHS agency, the CDC, and maintains an inventory of pharmaceutical and medical supplies (called 12-hour push packages) that can be delivered to any location in the nation within 12 hours of a federal decision to deploy them. The VA maintains a larger stock of antibiotics, antidotes, other drugs, medical equipment, and supplies known as vendor-managed inventory that can be deployed within 24 to 36 hours of notification. The GAO has recommended additional steps that the VA, in concert with the OEP and CDC, should take to further tighten the security of the nation's stockpiles: (1) finalize and implement approved operating plans, and (2) ensure compliance with these plans through periodic quality reviews.

The VA has a substantial medical infrastructure of 163 hospitals and 800 outpatient clinics strategically located throughout the United States and has the largest pharmaceutical and medical supply procurement systems in the world. The VA maintains a nationwide register of skilled VA medical personnel and a network of 140 treatment programs for posttraumatic stress disorder. It has well-established relationships with the nation's medical schools and has expanded physician-training slots in disciplines associated with WMD preparedness. The VA will augment the resources of state and local responders as VA hospital emergency plans are included in local community emergency response plans.

It should be noted that the VA does not have the capability to process and treat mass casualties resulting from WMD incidents. VA hospitals and most private sector medical facilities are better prepared for treating injuries resulting from chemical exposure than those resulting from biological agents or radiological material. VA hospitals, like community hospitals, lack decontamination equipment, routine training to treat mass casualties, and adequate on-hand medical supplies.

15.19 FEDERAL WORKING GROUPS

To coordinate their activities, federal departments and agencies are using:

- Interagency plans
- Interagency work groups, which are used to minimize duplication of funding and effort in federal activities to combat terrorism
- Formal agreements between departments and agencies to share resources and knowledge

15.19.1 Collaborative Funding of Smallpox Research

The following agencies conduct research on vaccines for smallpox: CDC, FDA through the Center for Biologics Evaluation, and NIH through the National Institute of Allergy and Infectious Diseases (NIAID).

15.19.2 Cooperative Work on Rapid Detection of Biological Agents in Animals, Plants, and Food

Work is underway to develop a system to improve on-site rapid detection of biological agents in animals, plants, and food.

15.19.3 Food Safety Surveillance Systems

FoodNet and PulseNet are surveillance systems for identifying and characterizing contaminated food.

15.19.4 Force Packages Response Team

Designated groups of military units respond to incidents.

15.19.5 Informal Working Group — Equipment Request Review

This working group reviews state and local jurisdiction equipment requests to ensure that duplicate funding is not being given for the same activities.

15.19.6 Interagency Board for Equipment Standardization and Interoperability

This working group develops and maintains a standardized equipment list of essential items for responding to a terrorist WMD attack.

15.19.7 National Medical Response Team Caches

These caches form a stockpile of drugs for OEP's National Medical Response Teams.

15.19.8 National Disaster Medical System

The National Disaster Medical System is a partnership between federal agencies, state and local governments, and the private sector that ensures that resources are available to provide medical services following a disaster that overwhelms the local health care resources.

15.19.9 National Pharmaceutical Stockpile Program

The CDC's National Pharmaceutical Stockpile Program (NPSP) is intended to ensure the availability of lifesaving pharmaceuticals, antibiotics, and chemical interventions, as well as medical, surgical, and patient support supplies and equipment for prompt delivery to the site of a disaster, including a possible biological or chemical terrorist event anywhere in the United States. The program is available to supplement the initial response to an incident of biological or chemical terrorism. That response will come from local and state emergency, medical and public health personnel.

The primary purpose is to provide critical drugs and medical material that would otherwise be unavailable to local communities. The program has a cache of vaccines available to address smallpox threats. In addition to medications and supplies for intravenous administration, the NPSP includes medical equipment that would be essential for treatment, including airway supplies, bandages and dressings, and other emergency medications. These are items that local clinicians may find in short supply during a terrorism incident event. The CDC has established relationships with various national security agencies to facilitate continuous updates and analyses of threat agents and to ensure that the NPSP reflects current needs.

The program maintains repositories of lifesaving pharmaceuticals, antidotes, and medical supplies that can be delivered to the site of a biological (or other) attack. A decision to deploy the stockpile is based on the best epidemiologic, laboratory, and public health information regarding the nature of the threat. The NPSP has two basic components: push packages and vendor-managed inventory.

15.19.9.1 Push Packages

Twelve-hour push packages are intended for immediate response. They are preassembled sets of supplies, pharmaceuticals, and medical equipment for immediate deployment to reach any affected area within 12 hours of the federal decision to release the assets. They are ready for quick delivery to the field and are stored in fully stocked, environmentally controlled, secured warehouses. Each package consists of 50 tons of material intended to address a mass casualty incident. The packages would permit emergency medical staff to treat a variety of different agents, as the actual threat may not have been identified at the time of the stockpile.

15.19.9.2 *Vendor-Managed Inventory*

If the incident requires a larger or multiphased response, follow-on vendor-managed inventory (VMI) packages will be shipped to arrive within 24 to 36 hours. They are composed of pharmaceuticals and supplies that can be tailored to provide pharmaceuticals, supplies, and/or products specific for the suspected or confirmed agent or combination of agents.

15.19.10 National Response Teams

National Response Teams constitute a national planning, policy, and coordinating body to provide guidance before and assistance during an incident.

15.19.11   Technical Support Working Group

This group coordinates interagency research and development requirements across the federal government in order to prevent duplication of effort between agencies. It also helps to identify research needs and funds a project to detect biological agents in food that can be used by both DOD and USDA.

15.19.12   Standing Agreements, Plans, and Programs

The following agreements, plans, and programs are in place for the Federal Working Groups.

15.19.12.1 *Agreement on Tracking Diseases in Animals That Can Be Transmitted to Humans*

This group is negotiating an agreement to share information and expertise on tracking diseases that can be transmitted from animals to people and could be used in a bioterrorist attack.

15.19.12.2 *Federal Response Plan — Health and Medical Services Annex*

This annex in the Federal Response Plan states that the HHS is the primary agency for coordinating federal assistance to supplement state and local resources in response to public health and medical care needs in an emergency, including a bioterrorist attack.

15.19.12.3 *Domestic Preparedness Program*

This program was formed in response to the National Defense Authorization Act of Fiscal Year 1997 (P.L. 104–201) and required the DOD to enhance the capability of federal, state, and local emergency responders regarding terrorist incidents involving WMD and high-yield explosives. As of October 1, 2000, the DOD and DOJ share responsibilities under this program.

15.20 SOUND THREAT AND RISK ASSESSMENTS

The GAO has recommended that the federal government conduct multidisciplinary and analytically sound threat and risk assessments to define and prioritize requirements and properly focus programs and investments in combating terrorism. Such assessments would be useful in addressing the fragmentation that is evident in the different threat lists of biological agents developed by federal departments and agencies. Understanding which biological agents are considered most likely to be used in an act of domestic terrorism is necessary to focus the investment in new technologies, equipment, training, and planning.

15.21 FRAGMENTATION

Overall coordination of federal programs to combat terrorism is fragmented. Several agencies have coordination functions, including the DOJ, FBI, FEMA, and OMB. Several different agencies are responsible for various coordination functions, which limits accountability and hinders unity of effort. Several key agencies have not been included in bioterrorism-related policy and response planning. The programs that agencies have developed to provide assistance to state and local governments are similar and potentially duplicative. Officials from a number of the agencies that combat terrorism report that the coordination roles of these various agencies are not always clear and sometimes overlap, leading to a fragmented approach. The GAO has found that the overall coordination of federal research and development efforts to combat terrorism is still limited by a number of factors, including the compartmentalization or security classification of some research efforts.

Fragmentation has also hindered unity of effort. Officials at the DOT report that the department has been overlooked in bioterrorism-related planning and policy. DOT officials noted that even though the nation's transportation centers account for a significant percentage of the nation's potential terrorist targets, the DOT was not part of the founding group of agencies that worked on bioterrorism issues and has not been included in bioterrorism response plans. DOT officials also told us that the department is supposed to deliver supplies for FEMA under the Federal Response Plan, but DOT was not brought into the planning early enough to understand the extent of its responsibilities in the transportation process. The department learned what its responsibilities would be during TOPOFF 2000.

15.22 BIOLOGICAL AGENT THREAT LISTS

Several different agencies have developed or are in the process of developing biological agent threat lists, which differ based on the focus of each agency. The CDC collaborated with law enforcement, intelligence, and defense agencies to develop a critical agent list that focuses on the biological agents that would have the greatest impact on public health. The FBI, the National Institute of Justice, and the Technical Support Working Group are completing a report that lists biological agents that may be more likely to be used by a terrorist group working in the United States that is not sponsored by a foreign government. The USDA's Animal and Plant Health Inspection Service uses two lists of agents of concern for a potential bioterrorist attack developed through an international process (although only some of these agents are capable of making both animals and humans sick). According to agency officials, separate threat lists are appropriate because the agency's charters differ. In the GAO's view, the existence of competing lists makes the assignment of priorities difficult for state and local officials.

15.23 LOCAL PREPAREDNESS

Nonprofit research organizations, congressionally chartered advisory panels, government documents, and articles in peer-reviewed literature have identified concerns about the preparedness of states and local areas to respond to a bioterrorist attack. These concerns include:

- Insufficient state and local planning for response to terrorist events
- Inadequacies in the public health infrastructure
- Lack of hospital participation in training on terrorism and emergency response planning
- Insufficient capacity for treating mass casualties from a terrorist act
- Questions regarding the timely availability of medical teams and resources in an emergency

Questions exist regarding how effectively federal programs have prepared state and local governments to respond to terrorism. All 50 states and approximately 255 local jurisdictions have received or are scheduled to receive at least some federal assistance, including training and equipment grants, to help them prepare for a terrorist WMD incident.

In 1997, FEMA identified planning and equipment for response to nuclear, biological, and chemical incidents as an area in need of significant improvement at the state level. However, an October 2000 report concluded that even those cities receiving federal aid are still not adequately prepared to respond to a bioterrorist attack.

15.23.1 Infectious Disease Surveillance System

The nation's infectious disease surveillance system components are not well prepared to detect or respond to a bioterrorist attack. Reductions in public health laboratory staffing and training have affected the ability of state and local authorities to identify biological agents. Even the initial West Nile virus outbreak in 1999, which was relatively small and occurred in an area with one of the nation's largest local public health agencies, taxed the federal, state, and local laboratory resources. Both the New York State and CDC laboratories were inundated with requests for tests, and the CDC laboratory handled the bulk of the testing due to the limited capacity at the New York State laboratories. Officials indicated that the CDC laboratory would have been unable to respond to another outbreak, had one occurred at the same time. In fiscal year 2000, the CDC awarded approximately $11 million to 48 states and four major urban health departments to improve and upgrade their surveillance and epidemiological capabilities.

15.23.2 Hospitals

Inadequate training and planning for bioterrorism response by hospitals is a major problem. The Gilmore Panel concluded that the expertise level, or lack thereof, for recognizing and dealing with a terrorist attack involving a biological or chemical agent may be a problem in many hospitals. The Gilmore Panel is an advisory panel that assesses domestic response capabilities for terrorism involving WMD and the capabilities at the federal, state, and local levels to respond to a domestic terrorist incident involving a WMD (that is, a chemical, biological, radiological, or nuclear agent or weapon). A recent research report concluded that hospitals need to improve their preparedness for mass casualty incidents. Local officials told the GAO that hospitals have been reluctant to participate in local training, planning, and exercises to improve their preparedness.

Several federal and local officials reported to the GAO that little excess capacity exists in the healthcare system for treating mass casualty patients. Studies have reported that emergency rooms in some areas are routinely filled and unable to accept patients in need of urgent care. According to one local official, the healthcare system might not be able to handle the aftermath of a disaster because of the problems caused by overcrowding and the lack of excess capacity. Local officials are concerned about whether the federal government could quickly deliver enough medical teams and resources to help after a disaster. Agency officials say that federal response teams, such as Disaster Medical Assistance Teams, could be on site within 12 to 24 hours. However, local officials who have deployed with such teams say that the federal assistance probably would not arrive for 24 to 72 hours. Of concern is the time and resources required to prepare and distribute drugs from the National Pharmaceutical Stockpile during an emergency. Partially in response to these concerns, the CDC has developed training for state and local officials on using the stockpile and will deploy a small staff with the supplies to assist the local jurisdiction with distribution.

In summary, the GAO concerns address the preparedness of state and local jurisdictions, including the level of state and local planning for response to terrorist events; inadequacies in the public health infrastructure; lack of hospital participation in training on terrorism and emergency response planning; hospital capabilities for treating mass casualties; and the timely availability of medical teams and resources in an emergency.

15.24 HOMELAND SECURITY

The President recently took steps to improve oversight and coordination, including the creation of the Office of Homeland Security. Over 40 federal departments and agencies have some role in combating terrorism, and coordinating their activities is a significant challenge. In May 2001, the President asked the Vice President to oversee the development of a coordinated national effort dealing with WMDs. At the same time, the President asked the director of FEMA to establish an Office of National Preparedness to implement the results of the Vice President's effort that relate to programs within federal agencies that address consequence management resulting from the use of WMDs. This effort is intended to better focus policies and to ensure that programs and activities are fully coordinated in support of building the necessary preparedness and response capabilities. In addition, on September 20, 2001, the President announced the creation of the Office of Homeland Security to lead, oversee, and coordinate a comprehensive national strategy to protect the country from terrorism and respond to any attacks that may occur. These actions represent potentially significant steps toward improved coordination of federal activities. In a recent report, the GAO listed a number of important characteristics and responsibilities necessary for a single focal point, such as the proposed Office of Homeland Security, to improve coordination and accountability.

15.25 IDENTIFICATION OF BIOLOGICAL AGENTS

Research is currently being done to:

- Enable the rapid identification of biological agents in a variety of settings
- Develop new or improved vaccines, antibiotics, and antivirals to improve treatment and vaccination for infectious diseases caused by biological agents
- Develop and test emergency response equipment such as respiratory and other personal protective equipment

The USDA, DOD, DOE, HHS, Department of Justice (DOJ), Department of the Treasury, and EPA have all sponsored or conducted projects to improve the detection and characterization of biological agents in a variety of different settings, from water to clinical samples (such as blood). The EPA is sponsoring research to improve its ability to detect biological agents in the water supply. Some of these projects, such as those conducted or sponsored by the DOD and DOJ, are not primarily for the public health and medical consequences of a bioterrorist attack against the civilian population, but could eventually benefit from research for those purposes.

Departments and agencies are also conducting or sponsoring studies to improve treatment and vaccination for diseases caused by biological agents. HHS projects include basic research sponsored by the National Institutes of Health (NIH) to develop drugs and diagnostics and applied research sponsored by the Agency for Healthcare Research and Quality to improve healthcare delivery systems by studying the use of information systems and decision support systems to enhance preparedness for the delivery of medical care in an emergency.

Several agencies, including the Department of Commerce's NIST and the DOJ's National Institute of Justice are conducting research that focuses on developing performance standards and methods for testing the performance of emergency response equipment, such as respirators and personal protective equipment. Federal departments and agencies have also been increasing their own capacity to identify and deal with a bioterrorist incident. For example, the CDC, USDA, and FDA are improving surveillance methods for detecting disease outbreaks in humans and animals. They have also established laboratory response networks to maintain state-of-the-art capabilities for biological agent identification and the characterization of human clinical samples.

15.26 CHEMICAL AND BIOLOGICAL WEAPONS RESPONSE CHALLENGES

A terrorist attack using chemical or biological weapons presents an array of complex issues to state and local first responders. Expert panels convened by CDC prioritized the following biological agents:

- Smallpox
- Anthrax
- Pneumonic plague
- Tularemia,
- Botulinum toxin
- Viral hemorrhagic fevers

Because anthrax, plague, and tularemia can be effectively treated with antibiotics that are immediately available, purchasing these products for the NPSP formulary was given first priority. Hospitals must quickly recognize cases of anthrax or smallpox or other agents — all of which may in the early stages resemble the flu — and call federal health officials for help. Quickly discovering bioterrorism remains the weakest link, as the diseases incubate for days to weeks before someone exposed gets sick. Responders would include police, firefighters, emergency medical services, and hazardous material technicians.

To ensure backfilling capability, eight caches, each with enough drugs to treat 10,000 to 35,000 people immediately, are stored around the country, ready to ship in 12 hours. Within the next 24 hours, the government could begin shipping more tablets from other stockpiles and manufacturers' inventories, and emergency production of yet more would begin. Agents used must be identified for rapid decontamination of victims and to apply appropriate medical treatments. Overwhelming incidents may require state and local response capabilities to call on federal agencies to provide assistance. The federal agencies can offer the following:

- Special teams that can respond to terrorist incidents involving chemical or biological agents or weapons — These teams perform a wide variety of functions, such as hands-on response; providing technical advice to state, local, or federal authorities; or coordinating the other federal teams' response efforts.
- Laboratories that may support response teams by performing tests to analyze and test samples of chemical and biological agents — In some incidents, these laboratories may perform functions that enable federal response teams to perform their role; when a diagnosis is confirmed at a laboratory, response teams can begin to treat victims appropriately.
- Programs to train and equip state and local authorities to respond to chemical and biological terrorism — The programs have included exercises to allow first responders to interact with themselves and federal responders.
- Research and development projects to combat terrorism
- Products to detect and identify chemical and biological weapons
- Chemical monitoring devices and new or improved vaccines
- Antibiotics, and antivirals — If the government ever discovers a bioterrorist attack, it can immediately ship 50-ton packages of medical supplies (antibiotics, intravenous fluids, and other equipment) to local hospitals struggling to contain the deadly outbreak. Current stockpile items include enough antibiotics to treat 2 million cases of anthrax; plans call for stockpiling enough for 10 million people. Also stockpiled are streptomycin and gentamicin, which are stored in huge bottles so that doses for each person would have to be counted out by hand.
- Vaccines — Experts still do not agree on how smallpox vaccine would best be distributed. Only people who have had contact with the sick may be vaccinated initially, while entire cities where smallpox was found would need inoculations. A government spokesman has said: "No one will be vaccinated in advance out of fear of bioterrorism because smallpox is a live vaccine and thus is very risky to anyone with a weak immune system."

The variety of chemical agents that could potentially be used by terrorists could be:

- Dispersed as a gas, vapor, liquid, or aerosol
- Disseminated by explosive or mechanical delivery

Some chemicals disperse rapidly while others remain toxic for days or weeks and require decontamination and clean up. Rapid exposure to a highly concentrated agent would increase the number of casualties. Federal, state, and local officials generally agree that a chemical terrorist incident would look like a major hazardous material emergency. According to the International Association of Fire Chiefs, over 600 local and state hazardous material teams will be the first to respond to a chemical incident. If local responders are unable to manage the situation or are overwhelmed, the incident commander has access to state and federal assets.

Terrorists can potentially use a variety of biological agents that must be disseminated by some means that infects enough individuals to initiate a disease epidemic. The most effective way to disseminate a biological agent is by aerosol. This method allows the simultaneous respiratory infection of a large number of people. Some biological agents (e.g., plague and smallpox) are communicable and can be spread beyond those directly affected by the weapon or dissemination device. Release may not be known for several days until victims present themselves to medical personnel, when the symptoms might easily be confused with less virulent illnesses. The critical detection of the biological agent begins with the public health infrastructure that detects illness outbreaks, identifies the sources and modes of transmission, and performs rapid agent laboratory identification. Once diagnosis of a biological agent is confirmed, treating victims may require the use of federal consequence management teams and items from the National Pharmaceutical Stockpile.

15.26.1 Generic Problems and Solutions

Generic problems and solutions would apply to any type of terrorist incident, major accident, or natural disaster. They would apply not only to chemical and biological terrorism but also to all hazards including emergencies unrelated to terrorism, such as major accidents or natural disasters. Such solutions might include:

- *Command and control* — The roles, responsibilities, and legal authority to plan and carry out a response to a weapon of mass destruction terrorist incident are not always clear, which could result in a delayed and inadequate response.
- *Planning and operations* — State and local emergency operations plans do not always conform to federal plans. The operational procedures for requesting federal assistance are not always compatible with state and local procedures.
- *Resource management and logistics* — State and local governments can be overwhelmed with the resource management and logistical requirements of managing a large incident, particularly after the arrival of additional state and federal assets. State and local officials could have difficulty providing support to numerous military units that might be needed.
- *Communication* — Interoperability difficulties exist at the interagency and intergovernmental levels. The public health community lacks robust communication systems, protocols, equipment, and facilities.
- *Exercises* — Many exercises focus primarily on crisis management, which often ends in a successful tactical resolution of the incident and do not include more likely scenarios where terrorist attacks are successful, requiring a consequence management exercise component.
- *Mass casualties* — Overall planning and integration among agencies are needed for mass casualty management, including conventional terrorist incidents. Medical surge capacity for any type of weapon of mass destruction event may be limited. Disposition of bodies would also be an issue.

15.26.2 Problems and Solutions for Chemical and Biological Terrorist Events

Problems and solutions for chemical and biological terrorist events are not relevant in a conventional, radiological, or nuclear terrorist incident but would be relevant in other chemical or biological events not related to terrorism, such as an accidental chemical release or a natural disease outbreak. Such approaches vary in their level of applicability, with some only being applicable to specific chemical or biological agents. They might include:

- *Public health surveillance* — A basic capacity for public health surveillance is lacking. Improved public-health-coordinated surveillance for biological terrorism and emerging infectious diseases is an urgent preparedness requirement at the local level.
- *Detection and risk assessment* — The capability of first responders and specialized response teams to rapidly and accurately detect, recognize, and identify chemical or biological agents and assess the associated health risks can be slow. Following the release of a chemical or biological agent, emergency hazardous material teams do not always conduct a downwind analysis of the toxic cloud, which could delay a decision to evacuate potentially affected populations.
- *Protective equipment and training* — First responders often lack special personal protective equipment (EPA Level A protective clothing and masks) to safeguard them from chemical or biological agents. Without adequate PPE and decontamination, the responders could become contaminated themselves, thus becoming additional vectors for agent transmittal.

Training curricula deal with the technical response level, such as treatment protocols, but do not describe operational guidelines and strategies for responding to large-scale public health emergencies. Physicians sometimes lack adequate training to recognize chemical and biological agents.

Concerning chemical and biological-specific planning, emergency operations plans and all-hazard plans do not adequately address the response to a large-scale chemical or biological terrorism event. Plans often do not address chemical or biological incidents.

For hospital notification and decontamination, delays could occur in the notification of local hospitals that a biological incident has occurred. By the time the hospitals are notified, they could already have become contaminated by self-referred patients, putting them in the position of having to close and not treat other victims. First responders could become victims themselves and contaminate emergency rooms.

With regard to pharmaceutical distribution, state and local health officials have found it difficult to break down and distribute tons of medical supplies contained in push packages from the National Pharmaceutical Stockpile. Some pharmaceuticals, such as antibiotics, are generic and can be used to treat several different biological agents, whereas others, such as vaccines, are agent specific.

Even a small outbreak of an emerging disease would strain the resources of laboratories. Needs include:

- Broadening laboratory capabilities
- Ensuring adequate staffing and expertise
- Improving the ability to deal with surges in testing needs

Problems also exist in communication between public health officials and veterinary officials. The local and state veterinary disaster response plan may not adequately address the biological incident impact on the animal population, which could have dramatic health, economic, and public relations implications.

Quarantine could be resource intensive and would require a well-planned strategy to implement and sustain. Questions that have to be addressed include:

- Implementation authority
- Enforcement
- Logistics

- Financial support
- Psychological ramifications of quarantine

A risk-management approach is needed to help focus resource investments. Efforts to better prepare for chemical and biological attacks include solutions that have broad applicability across a variety of contingencies and solutions that are applicable to only a specific type of attack. Efforts to improve public health surveillance would be useful in any disease outbreak, whereas efforts to provide vaccines for smallpox would be useful only if terrorists used smallpox in a biological attack.

15.27 NIOSH AND CDC PERSONAL PROTECTIVE EQUIPMENT RECOMMENDATIONS

The use of antibiotics or other medications should be decided in consultation with local public health authorities. For a biological agent, the air concentration of the infectious particles will depend upon the method used to release the agent. Some devices used for intentional biological terrorism may have the capacity to disseminate large quantities of biological materials in aerosols. Biological weapons may expose people to bacteria, viruses, or toxins as fine airborne particles. Biological agents are infectious through one or more of the following mechanisms of exposure, depending upon the particular agent type:

- Inhalation, with infection through respiratory mucosa or lung tissues
- Ingestion, through contact with the mucous membranes of the eyes or nasal tissues
- Skin penetration, through open cuts (even very small cuts and abrasions)

Self-contained breathing apparatus (SCBA), which first responders currently use for entry into potentially hazardous atmospheres, can provide responders with respiratory protection against biological exposures associated with a suspected biological terrorism act. The SCBA respirators with a full facepiece would be operated in the most protective, positive pressure (pressure demand) mode during emergency responses and would provide the highest level of protection against airborne hazards when properly fitted to the user's face and properly used. Such equipment could reduce the user's exposure to the hazard by a factor of at least 10,000 according to NIOSH. This reduction is true whether the hazard is from airborne particles, a chemical vapor, or a gas.

Protective clothing, including gloves and booties, may also be required for the response to a suspected biological terrorism act. Protective clothing may be needed to prevent skin exposures and/or contamination of other clothing. The type of protective clothing needed will depend upon the:

- Agent type
- Concentration
- Exposure route

15.27.1   Level A

Level A protection requires a continuous-flow, positive-pressure, pressure-demand SCBA in conjunction with a totally enclosing (covers the SCBA and the individual wearing the SCBA) impermeable protective suit in responding to a suspected biological incident if:

- Type of airborne agent is unknown.
- Dissemination method is via aerosolization.
- Duration and exposure concentration warrants such use.
- Event is uncontrolled.

15.27.2   Level B

Level B protection requires a continuous-flow, positive-pressure, pressure-demand SCBA in conjunction with an enclosing (covers the individual wearing the SCBA) impermeable protective suit if:

- The suspected biological aerosol is no longer being generated.
- Other conditions may present a splash hazard.

15.27.3   Level C

Level C protection requires a full facepiece respirator with a P100 filter or powered air-purifying respirator (PAPR) with a HEPA filter if:

- An aerosol-generating device was not used to create high airborne concentration.
- Dissemination was by a letter or package that can be easily bagged.

Care should be taken when bagging letters and packages to minimize creating a puff of air that could spread pathogens. Avoid large bags and work very slowly and carefully when placing objects in bags. This type of respirator reduces the user's exposure by a factor of 50 if the user has been properly fit tested. Disposable hooded coveralls, gloves, and foot coverings also should be used. NIOSH recommends against wearing standard firefighter turnout gear into potentially contaminated areas when responding to reports involving biological agents.

15.27.4 Decontamination

Decontamination of protective equipment and clothing is an important precaution to make sure that any particles that might have settled on the outside of protective equipment are removed before taking off gear. Decontamination sequences currently used for hazardous material emergencies should be used as appropriate for the level of protection employed. Equipment can be decontaminated using soap and water; a 0.5% hypochlorite solution (1 part household bleach to 10 parts water) can be used, as appropriate, or if gear has any visible contamination. (*Note:* Bleach may damage some types of firefighter turnout gear, which is one reason why it should not be used for biological agent response actions.) After taking off gear, response workers should shower using copious quantities of soap and water.

15.28 SECURITY AND EMERGENCY RESPONSE BY MICROBIOLOGICAL AND BIOMEDICAL LABORATORIES

Traditional laboratory biosafety guidelines have emphasized the use of good work practices, appropriate containment equipment, well-designed facilities, and administrative controls to minimize risks of accidental infection or injury for laboratory workers and prevent contamination of the environment outside the laboratory.

15.28.1   Applicability and Relevance

The traditional guidelines as discussed in the CDC's "Biosafety in Microbiological and Biomedical Laboratories" (1999) are:

- Address laboratory security issues including those that prevent unauthorized entry to laboratory areas or accidental or intentional removal of dangerous biological agents from laboratory confines
- Are for laboratories using biological agents or toxins capable of causing serious or fatal illness to humans or animals; most of these laboratories would be working under the BSL-3 or -4 conditions
- Should be followed by research, clinical, and production laboratories working with newly identified human pathogens, high-level animal pathogens, or toxins not covered by BSL-3 or -4 recommendations

15.28.2 Guidelines

The guidelines provided here recognize that laboratory security is related to but different from laboratory safety. They involve both safety and security experts in the evaluation and development of facility or laboratory protocols, and they review safety policies and procedures regularly. Management should review policies to ensure adequacy given current conditions and consistency given other facility-wide policies and procedures. Laboratory supervisors should ensure that all laboratory workers and visitors:

- Understand security requirements
- Are trained and equipped to follow established procedures
- Review safety policies and procedures whenever an incident occurs or a new threat is identified
- Adhere to controlled access to areas where biologic agents or toxins are used and stored

Laboratories and animal care areas should be separate from the public areas of the buildings, and locked at all times. Card-keys or similar devices should be used to permit entry to laboratory and animal care areas. All entries (including entries by visitors, maintenance workers, repairmen and others needing one-time or occasional entry) should be recorded, either by the card-key device (preferable) or by signature in a logbook. Only workers required to perform a job should be allowed in laboratory areas, and workers should be allowed only in areas and at hours required to perform their particular jobs. Access for students and visiting scientists should be limited to hours when regular employees are present. Access for routine cleaning, maintenance, and repairs should also be limited to hours when regular employees are present.

Freezers, refrigerators, cabinets, and other containers where stocks of biological agents, hazardous chemicals, or radioactive materials are stored should be locked when they are not in direct view of workers (e.g., when located in unattended storage areas).

Know who is in the laboratory area; all workers should be known to facility administrators and laboratory directors. Depending on the biological agents involved and the type of work being done, a background check and/or security clearance may be appropriate before new employees are assigned to the laboratory area. All workers (including students, visiting scientists, and other short-term workers) should wear visible identification badges that should include, at a minimum, a photograph, the wearer's name, and an expiration date. Colored markers or other easily recognizable design symbols may be used on the identification badges to indicate clearance to enter restricted areas. Guests should be issued identification badges and escorted or cleared for entry using the same procedures as for regular workers.

Know what materials are being brought into the laboratory area. All packages should be screened (visually or by x-ray) before being brought into the laboratory area. Packages containing specimens, bacterial or virus isolates, or toxins should be opened in a safety cabinet or other appropriate containment device. Know what materials are being removed from the laboratory area. Biological materials/toxins for shipment to other laboratories should be packaged and labeled in conformance with all applicable local, federal, and international shipping regulations. Required permits (e.g., PHS, DOT, DOC, USDA) should be in hand before materials are prepared for shipment. The recipient (preferably) or receiving facility should be known to the sender, and the sender should make an effort to ensure that materials are shipped to a facility equipped to handle those materials safely. Hand carrying of microbiological materials and toxins to other facilities is rarely appropriate. If biological materials or toxins are to be hand carried on common carriers, all applicable regulations must be followed. Contaminated or possibly contaminated materials should be decontaminated before they leave the laboratory area. Chemicals and radioactive materials should be disposed of in accordance with local, state, and federal regulations.

Have an emergency plan. Control of access to laboratory areas can make an emergency response more difficult. This must be considered when emergency plans are developed. An evaluation of the laboratory area by appropriate facility personnel, with outside experts if necessary, to identify both safety and security concerns should be conducted before an emergency plan is developed. Facility administrators, laboratory directors, principal investigators, laboratory workers, the facility safety office, and facility security officials should be involved in emergency planning. Police, fire, and other emergency responders should be informed as to the types of biological materials in use in the laboratory areas and assisted in planning their responses to emergencies in the laboratory areas. Plans should include provision for immediate notification of (and response by) laboratory directors, laboratory workers, safety office personnel, or other knowledgeable individuals when an emergency occurs, so they can deal with biosafety issues if they occur.

Laboratory emergency planning should be coordinated with facility-wide plans. Such factors as bomb threats, severe weather (hurricanes, floods), earthquakes, power outages, and other natural (or unnatural) disasters should be considered when developing laboratory emergency plans.

Have a protocol for reporting incidents. Laboratory directors, in cooperation with facility safety and security officials, should have policies and procedures in place for reporting and investigation of incidents or possible incidents (e.g., undocumented visitors, missing chemicals, unusual or threatening phone calls).

RESOURCES

Atlas, R.M., Biological weapons pose challenge for microbiology community, *ASM News*, 64, 383–389, 1998.

Bioterrorism Coordination and Preparedness Report, Bioterrorism: Federal Research and Preparedness Activities, U.S. General Accounting Office (GAO)-01–915, Sept. 28, 2001, report mandated by the Public Health Improvement Act of 2000 (P.L. 106-505, sec. 102), presented October 5, 2001.

National Institutes of Health (NIH) Issuance of director's Decision: The NIH Incident, Federal Register, 62(185), 50018–50033, 1997.

Ruys, T., Laboratory design principles, in *Handbook of Facilities Planning*, Ruys, T., Ed., Van Nostrand Reinhold, New York, 1990, pp. 257–264.

U.S. Department of Health and Human Services, Biosafety in Microbiological and Biomedical Laboratories, 4th ed., HHS Publication No. (CDC) (99-xxxx), May, 1999, HHS Public Health Service Centers for Disease Control and Prevention and National Institutes of Health, U.S. Government Printing Office, Washington, D.C., 1999.

U.S. Public Health Service, Final Rule: Additional Requirements for Facilities Transferring or Receiving Select Agents, Federal Register, Oct. 24, 1996 (61 FR 29327).

Index

To: Aunt Leona

4-5-2003

We hope you will
enjoy reading this book.

From

Philip & Tracey
Rusch